MARINE NAVIGATION

PROCEEDINGS OF THE INTERNATIONAL CONFERENCE ON MARINE NAVIGATION AND SAFETY OF SEA TRANSPORTATION (TRANSNAV 2017), GDYNIA, POLAND, 21–23 JUNE 2017

Marine Navigation

Marine Navigation and Safety of Sea Transportation

Editor

Adam Weintrit

Gdynia Maritime University, Gdynia, Poland

CRC Press
Taylor & Francis Group
Boca Raton London New York

CRC Press is an imprint of the
Taylor & Francis Group, an **informa** business

A BALKEMA BOOK

Published by:
CRC Press/Balkema
P.O. Box 447, 2300 AK Leiden, The Netherlands
e-mail: Pub.NL@taylorandfrancis.com
www.crcpress.com – www.taylorandfrancis.com

First issued in paperback 2020

Typeset by V Publishing Solutions Pvt Ltd., Chennai, India

ISBN 13: 978-0-367-73615-6 (pbk)
ISBN 13: 978-1-138-29762-3 (hbk)

**Visit the Taylor & Francis Web site at
http://www.taylorandfrancis.com**

**and the CRC Press Web site at
http://www.crcpress.com**

Contents

General Chair of TransNav Conference
Prof. Dr. Adam **Weitntrit**, FRIN, FNI, Master Mariner

Executive Chair of TransNav Conference
Dr. Tomasz **Neumann**

Honorary Committee Members
President of the Nautical Institute: Captain David (Duke) **Snider** FNI
Rector of Gdynia Maritime University: Prof. Dr. Janusz **Zarębski**

Host of TransNav Conference
Prof. Dr. Leszek **Smolarek,** Dean of Faculty of Navigation of Gdynia Maritime University

List of Scientific Program Committee Members of the 12th International Conference on Marine Navigation and Safety of Sea Transportation – TransNav 2017
21-23 June 2017, Gdynia, Poland

*Prof. Agnar **Aamodt**, Norwegian University of Science and Technology, Trondheim, Norway*
*Prof. Ajith **Abraham**, Scientific Network for Innovation and Research Excellence, Auburn, Washington, USA*
*Prof. Teresa **Abramowicz-Gerigk**, Gdynia Maritime University, Gdynia, Poland*
*Prof. Michele **Acciaro**, Kühne Logistics University, Hamburg, Germany*
*Prof. Sauli **Ahvenjarvi**, Satakunta University of Applied Sciences, Rauma, Finland*
*Prof. Paolo **Alfredini**, University of São Paulo, Polytechnic School, São Paulo, Brazil*
*Prof. Daniel N. **Aloi**, Oakland University, Rochester, Michigan, The United States*
*Prof. Anatoli **Alop**, Estonian Maritime Academy, Tallin, Estonia*
*Prof. Karin **Andersson**, Chalmers University of Technology, Gothenburg, Sweden*
*Prof. Yasuo **Arai**, Marine Technical College, Ashiya, Hyogo, Japan*
*Prof. Terje **Aven**, University of Stavanger (UiS), Norway*
*Prof. Michael **Baldauf**, Word Maritime University, Malmö, Sweden*
*Prof. Andrzej **Banachowicz**, West Pomeranian University of Technology, Szczecin, Poland*
*Prof. Marek **Banaszkiewicz**, Space Research Center, Polish Academy of Sciences, Warsaw, Poland*
*Prof. Marcin **Barlik**, Warsaw University of Technology, Poland*
*Prof. Eugen **Barsan**, Constanta Maritime University, Romania*
*Prof. Milan **Batista**, University of Ljubljana, Ljubljana, Slovenia*
*Prof. Ghiorghe **Batrinca**, Constantza Maritime University, Romania*
*Prof. Raphael **Baumler** , World Maritime University, Malmö, Sweden*
*Prof. Angelica **Baylon**, Maritime Academy of Asia & the Pacific, Philippines*
*Prof. Knud **Benedict**, University of Wismar, University of Technology, Business and Design, Germany*
*Prof. Christophe **Berenguer**, Grenoble Institute of Technology, Saint Martin d'Hères, France*
*Prof. Heinz Peter **Berg**, Bundesamt für Strahlenschutz, Salzgitter, Germany*
*Prof. Tor Einar **Berg**, Norwegian Marine Technology Research Institute, Trondheim, Norway*
*Prof. Carmine Giuseppe **Biancardi**, The University of Naples „Parthenope", Naples, Italy*
*Prof. Vitaly **Bondarev**, Baltic Fishing Fleet State Academy, Kaliningrad, Russia*
*Prof. Neil **Bose**, Australian Maritime College, University of Tasmania, Launceston, Australia*
*Prof. Jarosław **Bosy**, Wroclaw University of Environmental and Life Sciences, Wroclaw, Poland*
*Prof. Alexey **Boykov**, Moscow State Academy of Water Transport, Moscow, Russia*
*Prof. Andrzej **Bujak**, WSB Schools of Banking, Wrocław, Poland*
*Prof. Zbigniew **Burciu**, Gdynia Maritime University, Gdynia, Poland*
*Sr. Jesus **Carbajosa Menendez**, President of Spanish Institute of Navigation, Spain*
*Prof. Doina **Carp**, Constanta Maritime University, Romania*
*Prof. Ayşe Güldem **Cerit**, Dokuz Eylül University, Izmir, Turkey*
*Prof. Shyy Woei **Chang**, National Kaohsiung Marine University, Taiwan*
*Prof. Adam **Charchalis**, Gdynia Maritime University, Gdynia, Poland*
*Prof. Wu **Chen**, Hong Kong Polytechnic University, Hong Kong*
*Prof. Andrzej **Chudzikiewicz**, Warsaw University of Technology, Poland*
*Prof. Frank **Coolen**, Durham University, The United Kingdom*
*Prof. Thomas **Cottier**, University of Bern, Switzerland*
*Prof. Kevin **Cullinane**, University of Newcastle upon Tyne, The United Kingdom*
*Prof. Jerzy **Czajkowski**, Gdynia Maritime University, Gdynia, Poland*
*Prof. Krzysztof **Czaplewski**, Gdynia Maritime University, Gdynia, Poland*
*Prof. Ireneusz **Czarnowski**, Gdynia Maritime University, Gdynia, Poland*
*Prof. Mirosław **Czechowski**, Gdynia Maritime University, Gdynia, Poland*
*Prof. German **de Melo Rodriguez**, Polytechnic University of Catalonia, Barcelona, Spain*

Prof. Robert **De Souza**, National University of Singapore NUS, Singapore
Prof. Decio Crisol **Donha**, Polytechnic School, University of São Paulo, Brazil
Prof. Patrick **Donner**, World Maritime University, Malmö, Sweden
Prof. Eamonn **Doyle**, Irish Institute of Master Mariners, Cork, Ireland
Prof. Branislav **Dragović**, University of Montenegro, Kotor, Montenegro
Prof. Daniel **Duda**, Polish Naval Academy, Polish Nautological Society, Poland
Prof. Czesław **Dyrcz**, Polish Naval Academy, Gdynia, Poland
Prof. Marek **Dzida**, Gdańsk University of Technology, Gdańsk, Poland
Prof. Milan **Džunda**, Technical University of Košice, Slovakia
Prof. Billy **Edge**, North Carolina State University, US
Prof. Bernd **Eissfeller**, Institute of Space Technology and Space Applications, Universitaet der Bundeswehr Munich, Germany
Prof. Ahmed **El-Rabbany**, University of New Brunswick; Ryerson University in Toronto, Ontario, Canada
Prof. Naser **El-Sheimy**, The University of Calgary, Canada
Prof. Akram **Elentably**, King Abdulaziz University (KAU), Jeddah, Saudi Arabia
Prof. Tarek A. **Elsayed**, Arab Academy for Science, Technology & Maritime Transport, Alexandria, Egypt
Prof. William **Emery**, Colorado University, Boulder, The United States
Prof. Sophia **Everett**, Victoria University, Melbourne, Australia
Prof. Odd M. **Faltinsen**, Norwegian University of Science and Technology, Trondheim, Norway
Prof. Jeffrey **Falzarano**, Texas A&M University, Corpus Christi, US
Prof. Alfonso **Farina**, SELEX-Sistemi Integrati, Rome, Italy
Prof. Andrzej **Fellner**, Silesian University of Technology, Katowice, Poland
Prof. Andrzej **Felski**, Polish Naval Academy, Gdynia, Poland
Prof. Yanming **Feng**, Queensland University of Technology, Brisbane, Queensland, Australia
Prof. Włodzimierz **Filipowicz**, Gdynia Maritime University, Gdynia, Poland
Prof. Renato **Filjar**, University College of Applied Sciences, Bjelovar, Croatia
Prof. Börje **Forssell**, Norwegian University of Science and Technology, Trondheim, Norway
Prof. Alberto **Francescutto**, University of Trieste, Trieste, Italy
Prof. Erik **Franckx**, Vrije Universiteit Brussel, Brussels, Belgium
Prof. Jens **Froese**, Jacobs University Bremen, Germany
Prof. Masao **Furusho**, Kobe University, Japan
Prof. Wiesław **Galor**, Maritime University of Szczecin, Poland
Prof. Yang **Gao**, University of Calgary, Canada
Prof. Aleksandrs **Gasparjans**, Latvian Maritime Academy, Riga, Latvia
Prof. Jerzy **Gaździcki**, President of the Polish Association for Spatial Information; Warsaw, Poland
Prof. Avtandil **Gegenava**, Georgian Maritime Transport Agency, Maritime Rescue Coordination Center, Georgia
Prof. Mirosław **Gerigk**, Gdańsk University of Technology, Gdańsk, Poland
Prof. Hassan **Ghassemi**, Amirkabir University of Technology (AUT), Teheran, Iran
Prof. Witold **Gierusz**, Gdynia Maritime University, Gdynia, Poland
Prof. Dariusz **Gotlib**, Warsaw University of Technology, Warsaw, Poland
Prof. Martha R. **Grabowski**, Le Moyne College; Rensselaer Polytechnic Institute, US
Prof. Dorota **Grejner-Brzezinska**, The Ohio State University, United States of America
Prof. Norbert **Gruenwald**, University of Applied Sciences Technology, Business and Design, Wismar, Germany
Prof. Marek **Grzegorzewski**, Polish Air Force Academy, Deblin, Poland
Prof. Andrzej **Grzelakowski**, Gdynia Maritime University, Gdynia, Poland
Prof. Marek **Grzybowski**, Gdynia Maritime University, Gdynia, Poland
Prof. Lucjan **Gucma**, Maritime University of Szczecin, Poland
Prof. Stanisław **Gucma**, Maritime University of Szczecin, Poland
Prof. Carlos **Guedes Soares**, Instituto Superior Técnico, Universidade Técnica de Lisboa, Portugal
Prof. Seung-Gi **Gug**, Korea Maritime and Ocean University, Pusan, Korea
Prof. Hans-Dietrich **Haasis**, University of Bremen, Germany
Prof. Jerzy **Hajduk**, Maritime University of Szczecin, Poland
Prof. Esa **Hämäläinen**, University of Turku, Finland
Prof. Jong-Khil **Han**, Sungkyul University, Anyang, Gyeonggi-do, South Korea
Prof. Kazuhiko **Hasegawa**, Osaka University, Osaka, Japan
Prof. Peter J. **Hayes**, California Maritime Academy, California State University, Vallejo, California, The United States
Prof. Bernhard **Hofmann-Wellenhof**, Graz University of Technology, Graz, Austria
Prof. Serge Paul **Hoogendoorn**, Delft University of Technology, Delft, The Netherlands
Prof. Mohammed **Hossam-E-Haider**, Military Institute of Science and Technology, Dhaka, Bangladesh
Prof. Qinyou **Hu**, Shanghai Maritime University, Shanghai, China
Prof Carl **Hult**, Kalmar Maritime Academy, Linnaeus University, Kalmar, Sweden
Prof. Marek **Idzior**, Poznań University of Technology, Poznań, Poland
Prof. Stojce Dimov **Ilcev**, Durban University of Technology, South Africa
Prof. Akio **Imai**, Kobe University, Japan
Prof. Toshio **Iseki**, Tokyo University of Marine Science and Technology, Tokyo, Japan,
Prof. Marianna **Jacyna**, Warsaw University of Technology, Poland
Prof. Jacek **Jania**, University of Silesia in Katowice, Poland
Prof. Ales **Janota**, University of Žilina, Slovakia
Prof. Maurice **Jansen**, Erasmus University Rotterdam, The Netherlands
Prof. Jacek **Januszewski**, Gdynia Maritime University, Gdynia, Poland
Prof. Piotr **Jędrzejowicz**, Gdynia Maritime University, Gdynia, Poland

Prof. Jung Sik **Jeong**, Mokpo National Maritime University, South Korea
Prof. Tae-Gweon **Jeong**, Korean Maritime and Ocean University, Pusan, Korea
Prof. Jean-Pierre **Jessel**, Institut de Recherche en Informatique de Toulouse, France
Prof. Shuanggen **Jin**, Shanghai Astronomical Observatory, Chinese Academy of Sciences, Shanghai, China
Prof. Yongxing **Jin**, Shanghai Maritime University, China
Prof. Zofia **Jóźwiak**, Maritime University of Szczecin, Poland
Prof. Mirosław **Jurdziński**, Gdynia Maritime University, Gdynia, Poland
Prof. Pawel **Kabacik**, Wroclaw University of Technology, Wrocław, Poland
Prof. Tadeusz **Kaczorek**, Warsaw University of Technology, Poland
Prof. Izzet **Kale**, University of Westminster, London, the United Kingdom
Prof. Kalin **Kalinov**, Nikola Y. Vaptsarov Naval Academy, Varna, Bulgaria
Prof. Eiichi **Kobayashi**, Kobe University, Japan
Prof. Hiroaki **Kobayashi**, Tokyo University of Marine Science and Technology, Japan
Prof. Lech **Kobyliński**, Polish Academy of Sciences, Ship Handling Research and Training Centre, Ilawa, Poland
Prof. Krzysztof **Kolowrocki**, Gdynia Maritime University, Gdynia, Poland
Prof. Zdzisław **Kopacz**, Polish Naval Academy, Gdynia, Poland
Prof. Serdjo **Kos**, University of Rijeka, Croatia
Prof. Eugeniusz **Kozaczka**, Gdańsk University of Technology, Gdańsk, Poland
Prof. Andrzej **Królikowski**, Gdynia Maritime University, Gdynia, Poland
Prof. Jan **Kryński**, Institute of Geodesy and Cartography, Warsaw, Poland
Prof. Nobuaki **Kubo**, Tokyo University of Marine Science and Technology, Tokyo, Japan
Prof. Pentti **Kujala**, Aalto University, Helsinki, Finland
Prof. Jan **Kulczyk**, Wroclaw University of Technology, Wroclaw, Poland
Prof. Krzysztof **Kulpa**, Warsaw University of Technology, Warsaw, Poland
Prof. Sashi **Kumar**, U.S. Maritime Administration (MARAD)
Prof. Uday **Kumar**, Luleå University of Technology, Luleå, Sweden
Prof. Alexander **Kuznetsov**, Admiral Makarov State University of Maritime and Inland Shipping, St. Petersburg, Russia
Prof. Bogumił **Łączyński**, Gdynia Maritime University, Gdynia, Poland
Prof. Siu Lee (Jasmine) **Lam**, Nanyang Technological University, Singapore
Prof. David **Last**, The Royal Institute of Navigation, London, the United Kingdom
Prof. Bogusław **Łazarz**, Silesian University of Technology, Katowice, Poland
Prof. Joong-Woo **Lee**, Korea Maritime and Ocean University, Pusan, South Korea
Prof. Andrzej **Lenart**, Gdynia Maritime University, Gdynia, Poland
Prof. Nadav **Levanon**, Tel Aviv University, Tel Aviv, Israel
Prof. Barrie **Lewarn**, Australian Maritime College, University of Tasmania, Launceston, Tasmania, Australia
Prof. Andrzej **Lewiński**, University of Technology and Humanities in Radom, Poland
Prof. Józef **Lisowski**, Gdynia Maritime University, Gdynia, Poland
Prof. Kezhong **Liu**, Wuhan University of Technology, Wuhan, Hubei, China
Prof. Zhengjiang **Liu**, Dalian Maritime University, Dalian, China
Prof. Zhizhao **Liu**, Hong Kong Polytechnic University, Hong Kong
Prof. Vladimir **Loginovsky**, Admiral Makarov State University of Maritime and Inland Shipping, St. Petersburg, Russia
Prof. Pierfrancesco **Lombardo**, University of Rome La Sapienza, Rome, Italy
Prof. Dieter **Lompe**, Hochschule Bremerhaven, Germany
Prof. Chin-Shan **Lu**, Hong Kong Polytechnic University, Hong Kong
Prof. Cezary **Łuczywek**, Gdynia Maritime University, Gdynia, Poland
Prof. Mirosław **Luft**, University of Technology and Humanities in Radom, Poland
Prof. Zbigniew **Łukasik**, University of Technology and Humanities in Radom, Poland
Prof. Tihomir **Luković**, University of Dubrovnik, Croatia
Prof. Evgeniy **Lushnikov**, Maritime University of Szczecin, Poland
Prof. Margareta **Lützhöft**, Australian Maritime College, University of Tasmania, Launceston, Australia
Prof. Scott **MacKinnon**, Chalmers University of Technology, Gothenburg, Sweden
Prof. Melchor M. **Magramo**, John B. Lacson Foundation Maritime University, Iloilo City, Philippines
Prof. Prabhat K. **Mahanti**, University of New Brunswick, Saint John, Canada
Prof. Artur **Makar**, Polish Naval Academy, Gdynia, Poland
Prof. Jerzy **Manerowski**, Warsaw University of Technology, Warsaw, Poland
Prof. Michael Ekow **Manuel**, World Maritime University, Malmoe, Sweden
Prof. Aleksey **Marchenko**, The University Centre in Svalbard, Longyearbyen, Norway
Prof. Eduardo **Marone**, Universidade Federal do Parana, Curitiba, Parana, Brazil
Prof. Francesc Xavier **Martinez de Oses**, Polytechnical University of Catalonia, Barcelona, Spain
Prof. Mustafa **Massad**, Jordan Academy for Maritime Studies (JAMS), Amman, Jordan
Prof. Jerzy **Matusiak**, Helsinki University of Technology, Helsinki, Finland
Prof. Boyan **Mednikarov**, Nikola Y. Vaptsarov Naval Academy,Varna, Bulgaria
Prof. Max **Mejia**, World Maritime University, Malmö, Sweden
Prof. Jerzy **Merkisz**, Poznań University of Technology, Poznań, Poland
Prof. Jerzy **Mikulski**, University of Economics in Katowice, Poland
Prof. Waldemar **Mironiuk**, Polish Naval Academy, Gdynia, Poland
Prof. Mykhaylo V. **Miyusov**, Odessa National Maritime Academy, Odessa, Ukraine
Prof. Sergey **Moiseenko**, Kaliningrad State Technical University, Kaliningrad, Russian Federation
Prof. Jakub **Montewka**, Gdynia Maritime University, Gdynia, Poland
Prof. Daniel Seong-Hyeok **Moon**, World Maritime University, Malmoe, Sweden

List of reviewers

*Prof. Teresa **Abramowicz-Gerigk**, Gdynia Maritime University, Gdynia, Poland*
*Prof. Michele **Acciaro**, Kühne Logistics University, Hamburg, Germany*
*Prof. Daniel N. **Aloi**, Oakland University, Rochester, Michigan, The United States*
*Prof. Yasuo **Arai**, Marine Technical College, Ashiya, Hyogo, Japan*
*Prof. Michael **Baldauf**, Word Maritime University, Malmö, Sweden*
*Prof. Andrzej **Banachowicz**, West Pomeranian University of Technology, Szczecin, Poland*
*Prof. Raphael **Baumler**, World Maritime University, Malmö, Sweden*
*Prof. Knud **Benedict**, University of Wismar, University of Technology, Business and Design, Germany*
*Prof. Christophe **Berenguer**, Grenoble Institute of Technology, Saint Martin d'Hères, France*
*Prof. Heinz Peter **Berg**, Bundesamt für Strahlenschutz, Salzgitter, Germany*
*Prof. Jarosław **Bosy**, Wroclaw University of Environmental and Life Sciences, Wroclaw, Poland*
*Prof. Alexey **Boykov**, Moscow State Academy of Water Transport, Moscow, Russia*
*Prof. Zbigniew **Burciu**, Gdynia Maritime University, Gdynia, Poland*
*Prof. Wu **Chen**, Hong Kong Polytechnic University, Hong Kong*
*Prof. Krzysztof **Czaplewski**, Gdynia Maritime University, Gdynia, Poland*
*Prof. Ireneusz **Czarnowski**, Gdynia Maritime University, Gdynia, Poland*
*Prof. Decio Crisol **Donha**, Polytechnic School, University of São Paulo, Brazil*
*Prof. Czesław **Dyrcz**, Polish Naval Academy, Gdynia, Poland*
*Prof. Milan **Džunda**, Technical University of Košice, Slovakia*
*Prof. Ahmed **El-Rabbany**, University of New Brunswick; Ryerson University in Toronto, Ontario, Canada*
*Prof. Akram **Elentably**, King Abdulaziz University (KAU), Jeddah, Saudi Arabia*
*Prof. William **Emery**, Colorado University, Boulder, The United States*
*Prof. Andrzej **Felski**, Polish Naval Academy, Gdynia, Poland*
*Prof. Włodzimierz **Filipowicz**, Gdynia Maritime University, Gdynia, Poland*
*Prof. Börje **Forssell**, Norwegian University of Science and Technology, Trondheim, Norway*
*Prof. Alberto **Francescutto**, University of Trieste, Trieste, Italy*
*Prof. Masao **Furusho**, Kobe University, Japan*
*Prof. Wiesław **Galor**, Maritime University of Szczecin, Poland*
*Prof. Witold **Gierusz**, Gdynia Maritime University, Gdynia, Poland*
*Prof. Dariusz **Gotlib**, Warsaw University of Technology, Warsaw, Poland*
*Prof. Lucjan **Gucma**, Maritime University of Szczecin, Poland*
*Prof. Jerzy **Hajduk**, Maritime University of Szczecin, Poland*
*Prof. Peter J. **Hayes**, California Maritime Academy, California State University, Vallejo, California, The United States*
*Prof. Toshio **Iseki**, Tokyo University of Marine Science and Technology, Tokyo, Japan,*
*Prof. Jacek **Jania**, University of Silesia in Katowice, Poland*
*Prof. Jacek **Januszewski**, Gdynia Maritime University, Gdynia, Poland*
*Prof. Piotr **Jędrzejowicz**, Gdynia Maritime University, Gdynia, Poland*
*Prof. Tae-Gweon **Jeong**, Korean Maritime and Ocean University, Pusan, Korea*
*Prof. Jean-Pierre **Jessel**, Institut de Recherche en Informatique de Toulouse, France*
*Prof. Mirosław **Jurdziński**, Gdynia Maritime University, Gdynia, Poland*
*Prof. Izzet **Kale**, University of Westminster, London, the United Kingdom*
*Prof. Eiichi **Kobayashi**, Kobe University, Japan*
*Prof. Lech **Kobyliński**, Polish Academy of Sciences, Ship Handling Research and Training Centre, Iława, Poland*
*Prof. Jan **Kryński**, Institute of Geodesy and Cartography, Warsaw, Poland*
*Prof. Pentti **Kujala**, Aalto University, Helsinki, Finland*
*Prof. Bogumił **Łączyński**, Gdynia Maritime University, Gdynia, Poland*
*Prof. David **Last**, The Royal Institute of Navigation, London, the United Kingdom*
*Prof. Andrzej **Lenart**, Gdynia Maritime University, Gdynia, Poland*
*Prof. Nadav **Levanon**, Tel Aviv University, Tel Aviv, Israel*
*Prof. Józef **Lisowski**, Gdynia Maritime University, Gdynia, Poland*
*Prof. Zhizhao (George) **Liu**, Hong Kong Polytechnic University, Hong Kong*
*Prof. Vladimir **Loginovsky**, Admiral Makarov State University of Maritime and Inland Shipping, St. Petersburg, Russia*
*Prof. Prabhat K. **Mahanti**, University of New Brunswick, Saint John, Canada*
*Prof. Artur **Makar**, Polish Naval Academy, Gdynia, Poland*
*Prof. Aleksey **Marchenko**, The University Centre in Svalbard, Longyearbyen, Norway*
*Prof. Eduardo **Marone**, Universidade Federal do Parana, Curitiba, Parana, Brazil*
*Prof. Sergey **Moiseenko**, Kaliningrad State Technical University, Kaliningrad, Russian Federation*
*Prof. Jakub **Montewka**, Gdynia Maritime University, Gdynia, Poland*
*Prof. Wacław **Morgaś**, Polish Naval Academy, Gdynia, Poland*
*Prof. Janusz **Narkiewicz**, Warsaw University of Technology, Poland*
*Prof. Nikitas **Nikitakos**, University of the Aegean, Chios, Greece*
*Prof. Andy **Norris**, The Royal Institute of Navigation, University of Nottingham, the United Kingdom*
*Prof. Tomasz **Nowakowski**, Wroclaw University of Technology, Poland*
*Prof. Dimos **Pantazis**, Technological Educational Institute of Athens, Greece*
*Prof. Gyei-Kark **Park**, Mokpo National Maritime University, Mokpo, Korea*
*Prof. Jin-Soo **Park**, Korea Maritime and Ocean University, Pusan, Korea*
*Prof. Francisco **Piniella**, University of Cadiz, Spain*

Marine Navigation
Introduction

A. Weintrit
Gdynia Maritime University, Gdynia, Poland
Chairman of the Poland Branch of the Nautical Institute
Chairman of TransNav Conference

I am pleased to present another volume of the proceedings related to the TransNav International Conference on Marine Navigation and Safety of Sea Transportation, published by CRC Press/Balkema, Taylor and Francis Group.

The 12th International Conference TransNav 2017 was held from 21 to 23 June 2017, organized jointly by the Faculty of Navigation of the Gdynia Maritime University and The Nautical Institute (http://transnav.am.gdynia.pl).

The Conference Proceedings are addressed to scientists and professionals in order to share their expert knowledge, experience, and research results concerning all aspects of navigation, safety of navigation and sea transportation. The book contains original papers contributing to the science of broadly defined navigation: from the highly technical to the descriptive and historical, over land through the sea, air and space, to promote young researchers, new field of maritime sciences and technologies.

The Twelfth Edition of the most innovative World conference on maritime transport research is designed to find solutions to challenges in waterborne transport, navigation and shipping, mobility of people and goods with respect to energy, infrastructure, environment, safety and security as well as to the economic issues.

The focus of TransNav Conference is high-quality, scholarly research that addresses development, application and implications, in the field of maritime education, maritime safety management, maritime policy sciences, maritime industries, marine environment and energy technology. Subjects of papers include nautical science, electronics, automation, robotics, geodesy, astronomy, mathematics, cartography, hydrography, communication and computer sciences, command and control engineering, psychology, operational research, risk analysis, theoretical physics, ecology, operation in hostile environments, instrumentation, administration, ergonomics, economics, financial planning and law. Also of interest are logistics, transport, mobility and ocean technology. The TransNav Conference provides a forum for transportation researchers, scientists, engineers, navigators, ergonomists, and policy-makers with an interest in maritime researches.From contemporary issues to the scientific, technological, political, economic, cultural and social aspects of maritime shipping, transportation and navigation, the TransNav publishes innovative, interdisciplinary and multidisciplinary research on marine navigation subjects and is set to become the leading international scholarly journal specialising in debate and discussion on maritime subjects. The TransNav is especially concerned to set maritime studies in a broad international and comparative context.

The content of the **TransNav 2017** Conference Proceedings, **Part 1**, with subtitle *Marine Navigation*, is partitioned into nineteen separate chapters (sections) titled: *Electronic Navigation* (covering three papers), *Route Planning* (covering two papers), *Mathematical Models, Methods and Algorithms* (covering two papers), *Ships Manoeuvring - Practical Aspects* (covering two papers), *Navigational Risk* (covering two papers), *Global Navigation Satellite Systems* (covering four papers), *Automatic Identification System* (covering two papers), *Marine Radar* (covering three papers), *Anti-Collision* (covering four papers), *Dynamic Positioning* (covering two papers), *Visualization of Data* (covering two papers), *Hydrometeorological Aspects and Weather Routing* (covering five papers), *Safety at Sea* (covering six papers), *Inland Navigation* (covering three papers), *Autonomous Water Transport* (covering five papers), *Communications and Global Maritime Distress and Safety System* (covering three papers), *Ships Manoeuvring - Theoretical Study* (covering two papers), *Port and Routes Optimum Location* (covering two papers), and *Magnetic Compasses* (covering one paper).

In each of them readers can find a few papers. Papers collected in the first chapter, titled *Electronic*

Navigation, concern the assessment of electronic navigation equipment's effect to mental workload by utilising revised NASA task load index, observations on ECDIS education and training, and the use of eLoran system for transmission of the national time signal.

In the second chapter there are described problems related to route panning: inductive mining in modeling of the ship's route, and route optimization in the restricted area taking into account ship safety zones.

The third chapter deals with mathematical models, methods and algorithms. In this section there is described positional game passing a greater number of ships with varying degree of cooperation and mathematical principles for vessel's movement prediction.

In the fourth chapter there are described practical aspects of ships manoeuvring: ship course planning and course keeping in close proximity to banks based on optimal control theory, and fuzzy self-tuning PID controller for a ship autopilot.

The fifth chapter deals with navigational risk. The content of the fifth chapter concerns estimated risks of navigation of LNG vessels through the Ob River Bay and Kara Sea, and some research on concept of ship safety domain.

The sixth chapter, titled *Global Navigation Satellite Systems (GNSS)*, covers the following issues: evaluation of the influence of atmospheric conditions on the quality of satellite signal, reliable vessel navigation system based on multi-GNSS, satellite multi-constellation identification techniques for liable enhanced applications (the SMILE Project), and EGNOS Poland market analysis in SHERPA Project.

In the seventh chapter there are described problems related to the use of Automatic Identification System: AIS as a tool to study maritime traffic (the case of the Baltic Sea), and evaluation method of collision risk based on actual ship behaviours extracted from AIS data.

In the eighth chapter there are described problems related to the use of marine radars. There is presented the joint waveform and filter design for marine radar tasks, improved compound multiphase waveforms with additional amplitude modulation for marine radars, and radar radiation pattern linear antennas array with controlling value of directivity coefficient.

The ninth chapter deals with anti-collision and collision avoidance. The content of the ninth chapter covers: a framework of a ship domain-based collision alert system, interaction between manned and unmanned ships (with main question: are the Colregs enough?), model research of navigational support system cooperation in collision scenario, and an analysis of vessel traffic flow before and after the grounding of the m/v Rena.

In the tenth chapter there are described problems related to Dynamic Positioning (DP): innovation methodology for safety of dynamic positioning under man-machine system control, and the visual system in a DP simulator at Maritime University of Szczecin.

The eleventh chapter deals with visualization of data. The content of the eleventh chapter concerns the system of the supervision and the visualization of multimedia data for BG (the Border Guard), and geoinformation structures in navigation according to ISO series 19100 standards.

In the twelfth chapter there are described hydro-meteorological aspects of navigation and problems related to weather routing. There are presented the following issues: optimal weather routing considering seakeeping performance based on the model test, avoidance of the tropical cyclone in ocean navigation, prediction method and calculation procedure of resistance and propulsion performance for the weather routing system, a mariners guide to numerical weather prediction, and investigation of ocean currents in navigational straits of Spitsbergen.

The thirteenth chapter deals with safety at sea. In this section there is described navigation safety and risk assessment challenges in the High North, safety management on the bridge: safety cultural factors for crew member's safety behaviour, naval artificial intelligence, software updating regime for ships necessity for cyber security and safe navigation, efficient and extremely fast transport including search and rescue units using ground effect, and belief assignments in nautical science.

In the fourteenth chapter there are described inland shipping problems: the analysis of the possibility of navigation the sea-river ships on the Odra River, risk assessment in inland navigation, and the technology of container transportation on the Odra Waterway.

The fifteenth chapter deals with autonomous water transport. The content of the fifteenth chapter concerns the following issues: safely navigating the oceans with unmanned ships, the concept of autonomous coastal transport, platform for development of the autonomous ship technology, safety qualification process for an autonomous ship prototype (a goal-based safety case approach), and optical target recognition for drone ships.

The sixteenth chapter, titled *Communications and Global Maritime Distress and Safety System (GMDSS)*, covers the following issues: voice subtitle transmission in the marine VHF radiotelephony, VHF/DSC – ECDIS/AIS communication on the base of lightweight Ethernet, and performance evaluation for maritime data communication - LF band radio wave.

In the seventeenth chapter there are described theoretical aspects of ships manoeuvring: optimal path planning of an unmanned surface vehicle in a real-time marine environment using a Dijkstra algorithm, and performance of the second-order linear Nomoto model in terms of zigzag curve parameters.

In the eighteenth chapter there are described problems related to the port and routes optimum location. There are presented models and methods for locating LNG distributing routes in the Baltic Sea area, and mathematical approaches for finding a dry port optimum location on the level of intermodal transport networks.

The last chapter deals with the problems related to magnetic compasses. The content of the nineteenth chapter covers the contemporary considerations of change regulations regarding the use of magnetic compasses in the aspect of the technical progress.

Each paper was reviewed at least by three independent reviewers. As Editor I would like to express my gratitude to distinguished authors and reviewers of submitted papers for their great contribution for expected success of the publication. Let me congratulate the authors and reviewers for their excellent work.

Electronic Navigation

Proceedings of 12th International Conference on Marine Navigation and Safety of Sea Transportation, TransNav 2017
21-23 June 2017, Gdynia, Poland

Assessment of Electronic Navigation Equipment's Effect to Mental Workload by Utilising Revised NASA Task Load Index

C. Kartoğlu & S. Kum
Istanbul Technical University, Istanbul, Turkey

ABSTRACT: Maritime navigation had huge advances in technology in past decade and is still under development. As a result of this, ship masters/officers must perform increasing number of tasks in shorter periods by using electronic equipment on the bridge. Assessing the mental workload imposed on operator is particularly critical in high technology systems. NASA Task Load Index (NASA-TLX) is one of the most common methods for measuring mental workload. The original NASA-TLX has six factors which do not contain that are related to the target evaluation system. For this reason Revised version of NASA-TLX (RNASA-TLX) is designed. In the study, RNASA-TLX and the questionnaire for electronic navigation equipment were carried out with ship masters/officers working on different types of ships in different areas whilst navigating, berthing/unberthing and anchoring. This study explains the fundamental information for understanding the electronic navigation equipment's effect to mental workload on the bridge.

1 INTRODUCTION

There have been huge technology advances in last decade and these are also in maritime navigation. New generation of ship officer has been trained to use technological advances very well and to make these advances in assisting safe and precise navigation (Tetley & Calcutt 2007). Technological advances bring increases in system complexity and increases in the number of human computer interaction challenges (Evans & Fendley 2017).

Seafaring is always a joint undertaking between human and technology. Maritime transportation is still under development with respect to technological advances and considerable number of navigation functionalities (Perera & Soares 2015). As a result of these changes/developments, ship masters/officers must perform increasing number of tasks in shorter time periods by using electronic equipment on the bridge (Grabowski & Wallace 1993) and workload is getting higher (van de Merwe et al. 2016). The more demanding and complex the task, the more navigators must work to accomplish the task (Vidulich & Tsang 2015).

An operator needs to have been heavy demands on the information processing capability to use of sophisticated control and display technologies. Such technologies often require the rapid sampling and integration of large volumes of information. The resulting demands can approach or exceed the limited information processing capacities of the operator. Consequently, the need to assess the load imposed on operator processing capacities is particularly critical in high technology systems. Mental workload refers to the degree of processing capacity that is expended during task performance (Eggemeier 1988) when inadequate (either too low or too high) may lead to imperfect perception, insufficient attention, and inadequate information processing (Brookhuis & de Waard 2010).

The tools used to measure workload can be divided into three main categories: subjective, performance, and psychophysiological measures (Evans & Fendley 2017). This study clarifies electronic navigation equipment's effect to navigators' mental workload by utilising subjective mental workload measurement tool as Revised NASA-TLX. In addition, the questionnaire for electronic navigation equipment was carried out with ship masters/officers working on different types of ships in different areas. In the analysing part, relationship among profile variables (age, rank, sea experience, etc.), and navigation equipment's effect to workload are evaluated. Comparison of usage difficulties of RADAR, ARPA, ECDIS, AIS and Auto-Pilot is also provided with the result of effecting to navigators' mental workload.

2 MENTAL WORKLOAD – MWL

Workload is affected by both operator context and external factors. Operator contexts are individual capabilities (both physical and mental). External factors are task quantity, task difficulty, time availability, and environmental factors such as temperature and lighting (Rusnock & Borghetti 2016).

Mental workload (MWL) is one of the most important issues in work systems due to the development of mechanization and automation (Nachreiner 1995) and has long been recognized as a crucial factor on safety and effectiveness of human performance in complex interactive systems (Fernandes & Braarud 2015). Effectively applying an adaptive MWL to operator is important. The ability of a mental workload measure to evaluate operator's effort correctly and continuously is valuable in achieving convenience and satisfaction in the workplace, in strengthening the industrial safety, in improving the usability of human machine interfaces, and in designing appropriate adaptive automation strategies.

Commonly used subjective methods to measure mental workload include the Subjective Workload Assessment Technique (SWAT), the Cooper-Harper Rating Scale, the Bedford Scale, Overall Workload, Workload Profile and the Integrated Workload Scale. NASA Task Load Index is one of the most validated widely used subjective method (Rusnock & Borghetti 2016, Evans & Fendley 2017).

2.1 NASA Task Load Index – NASA-TLX

The NASA-TLX was developed over a three years development cycle included more than forty laboratory simulations by The Human Performance Group at NASA's Ames Research Centre. The NASA-TLX consists of six component scales (multidimensional rating procedure) which are Mental Demand (MD), Physical Demand (PD), Temporal Demand (TD), Performance (P), Effort (E), and Frustration Level (FL). An average of these six scales, weighted to reflect the contribution of each factor to the workload of a specific activity from the perspective of operator's self-report assessments, is proposed as an integrated measure of perceived overall workload (Yoshimura et al. 2007, Ogreten & Reinerman-Jones 2015).

The NASA-TLX is a two part evaluation procedure consisting of both weights and ratings. The first requirement is for each subject to evaluate the contribution of each factor (its weight) to the workload of a specific task. There are 15 pair-wise comparisons of the six scales. Subjects mark the one of each pair that contributed more to workload. The number of times each factor is selected is tailed. None of them will be greater than 5 and total

number must be 15 (National Aeronautics and Space Administration n.d).

The second procedure is to obtain numerical ratings for each scale that reflect the magnitude of that factor. Subject gives a score between 0 and 100.

The overall workload score for each subject is computed by multiplying each rating by the weight given to this factor. The sum of the weighted ratings for task is divided by 15 (National Aeronautics and Space Administration n.d).

2.2 Revised NASA-TLX

In NASA-TLX, the six scales do not contain any concept that is related to the target evaluation system, such as operation related and information receiving organs (Cha & Park 1997a). The main difficulty for this type of investigation is good understanding of the meanings of suggesting scales (Cha & Park 1997b). Therefore, well defined and clear meanings of words to express investigated concept are required to measure mental workload.

In NASA-TLX, the PD is defined as: How much physical activity was required (e.g. pushing, pulling, turning, controlling, activating, etc.). It appears that this question would not be relevant when considering the use of electronic navigation equipment by navigator. Also, the subjective evaluation of the P scale can lead discrepancies with the measure of mental workload (Cha & Park 1997a).

Navigators are heavily exposed to multitasking and high mental workload (Murai et al. 2004). When they are on the bridge, they perform concurrent tasks. The act of navigating is mainly processed in the visual–spatial–manual pathway, and it is also affected by engaging in the auditory–verbal–vocal task. Different electronic navigation equipment places different kinds of demand on navigator (Rouzikhah et al. 2013). Many alarms included various visual or audio warnings from this equipment often occur at the same time, making it difficult for the navigator to select a correct response efficiently. During usage of these equipment, the navigator continually acquires and processes various information from his/her senses, especially much from eyes (visual information) and ears (auditory information) (Murai et al. 2004). Much visual/auditory information imposes a heavy burden that is experienced as stressful, threatening or adaptive responses on navigator in a time-critical situation (Sauer et al. 2013).

To make more adaptive of NASA-TLX applying to the usage of electronic navigation equipment, Revised NASA-TLX (RNASA-TLX) is designed. Visual Demand (VD) and Auditory Demand (AD) superseded PD and P dimensions. RNASA-TLX's six dimensions are MD, VD, AD, TD, E, and FL.

The questions of RNASA-TLX's six scales are shown in Table 1.

Table 1. Description and questions of RNASA-TLX's six scales

Title	Description
Mental Demand	How much mental and perceptual activity was required? (thinking, deciding, calculating, remembering, looking, searching, etc.)
Visual Demand	How much visual activity was required? (display of information and explanations on the screen, screen brilliance, lightening, visual warnings, etc.)
Auditory Demand	How much auditory activity was required? (auditory warnings, etc.)
Temporal Demand	How much time pressure did you feel? (in operating and menu selecting, etc.)
Effort	How hard did you have to work (mentally or physically)? (too many operations, etc.)
Frustration Level	How insecure, discouraged, irritated, stressed and annoyed versus secure, gratified, content, relaxed and complacent did you feel?

In pair-wise comparison of RNASA-TLX, the subjects were asked to select one of the two dimensions of workload that they felt, was more critical during usage of electronic navigation equipment on the bridge (Kum et al. 2008).

Weights are obtained from the pair-wise comparisons. Then subjects' workload is determined by summing the rating for each dimension, multiplied by its weighting and divided by the sum of all weights. For the mental workload of all subjects, the average value of subjects' mental workload is calculated by summing their overall workload and divided to the number of subjects (Kum et al. 2008).

3 PAIR-WISE COMPARISON OF ELECTRONIC NAVIGATION EQUIPMENT

After applying RNASA-TLX questionnaire, subjects marked the pair-wise comparison of electronic navigation equipment consisting of RADAR, ARPA, ECDIS, AIS and Auto-Pilot. In this section they are asked *"Which one is the most effective electronic equipment on the bridge that affecting your mental workload?"* as shown in Figure 1.

	Extreme	Very strong	Strong	Moderate	Equal	Moderate	Strong	Very strong	Extreme	
Radar	9 8	7 6	5 4	3 2	1	2 3	4 5	6 7	8 9	ARPA
Radar	9 8	7 6	5 4	3 2	1	2 3	4 5	6 7	8 9	ECDIS
Radar	9 8	7 6	5 4	3 2	1	2 3	4 5	6 7	8 9	AIS
Radar	9 8	7 6	5 4	3 2	1	2 3	4 5	6 7	8 9	Auto-Pilot
ARPA	9 8	7 6	5 4	3 2	1	2 3	4 5	6 7	8 9	ECDIS
ARPA	9 8	7 6	5 4	3 2	1	2 3	4 5	6 7	8 9	AIS
ARPA	9 8	7 6	5 4	3 2	1	2 3	4 5	6 7	8 9	Auto-Pilot
ECDIS	9 8	7 6	5 4	3 2	1	2 3	4 5	6 7	8 9	AIS
ECDIS	9 8	7 6	5 4	3 2	1	2 3	4 5	6 7	8 9	Auto-Pilot
AIS	9 8	7 6	5 4	3 2	1	2 3	4 5	6 7	8 9	Auto-Pilot

Figure 1. Pair-wise comparison of electronic navigation equipment.

4 RESULTS AND CONSIDERATION

In the study, totally 92 questionnaires were carried out with ship masters/officers working on different types of ships in different areas whilst navigating, berthing/unberthing and anchoring.

65.2% of the participants were between 31 and 40 years old. This was followed by the below 31 and over 40 age ranges with same percentages of 17.4%.

The participants were mostly Ship Master (62%) and 22.8% of them were Chief Officer. Experience is one of the most essential data due to understanding of the capability of the ratings on board. The more experienced officers are, the more productive they get. So the participants were also categorized by their sea experience. 53.3% of them were experienced 11-20 years on board whilst 31.5% of them were 6-10 years. 10.9% was worked on board for less than 6 years and last of all 4.3% of them were worked more than 20 years.

Most of the participants worked on High Speed Craft (HSC) and Tanker, former rate was 32.6% and latter was 28.3%. 23.9% of them worked on Dry Bulk Carrier and 6.5% on Ro-ro ships. The rest was divided into two different ships; LPG carrier and Container with same percentages of 4.3%.

21.7% of the participants were at anchorage operation, 16.3% was navigating in open sea whilst 15.2% was at port. 14.1% of them were navigating in narrow waters. Last operations were berthing, unberthing and navigation in Traffic Separation Scheme (TSS) with each rate of 10.9%.

In this study, the participants mostly navigated in Aegean Sea with 19.6%, after than, China and Mediterranean areas (with same rate 16.3%). The follower areas were Marmara Sea, İstanbul Strait and India, Singapore and Malaysia, etc.

Ship masters/officers' overall average mental workload by RNASA-TLX score is determined as 64.8 (in the scale of 100), that means their overall average mental workload level is 64.8%. Figure 2 shows the average rate of each dimension. VD was the highest as the importance of information from eyes and also display/keypad design. E was the lowest dimension that subjects did not have to work so hard. MD was the second highest and followed by TD, FL, and AD.

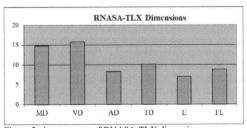

Figure 2. Average rate of RNASA-TLX dimensions.

According to distribution of the mental workload on ages, over 40 years old range had the highest mental workload. It seems that the subject feels importance of what is done highly due to maturement of profession. However the second highest ratio was belonged to ages between 31 and 40. The lowest mental workload was belonged to under 31 years old. The reason might be interpreted as younger subjects are better adapting to new technology.

In the rank distribution, Chief Officers and Masters' values were almost same whilst Chief Officers had the highest mental workload. It can be the result of taking responsibility and getting more fatigue. 2nd Officers had the lowest mental workload. 2nd Officers are responsible for navigation plan so a potential tendency of using electronic navigation equipment more than others.

3rd Officers had chosen E the most and TD the least. The rates for MD, VD and FL were also high as close to E. However AD was low as TD. On 2nd Officers' scale, AD was the highest, and TD was the lowest. In Chief Officers' scale VD and MD came in order on the top. AD and TD had the same value. There was a big difference of FL and E among others. Masters' VD was the highest, after than MD, TD and FL. E and AD were the lowest.

Sea experience was also differs per participants. The participants between 6 and 10 years' experience had more mental workload than others. The value of 11-20 years' experience was also near 6-10 years' experience. After these groups, the participants who had less than 6 years and more than 20 years of experience came. These could be interpreted that in the beginning of profession when sea experience is low level and in the future years when sea experience comes to high level, value of mental workload might be getting lower. On the other hand, in the middle years of profession when sea experience is getting higher, value of mental workload might be getting higher.

Navigators who had experience less than 6 years had E and VD as the most, and TD as the least. The participants between 11 and 20 years' experience had the highest VD, and MD was also high. The participants who had experience more than 20 years, their dimensions values were near to each other. From this study's result, the descending order was MD, FL, VD, AD, TD and E.

RNASA-TLX values according to ship type, mental workload was the highest in HSC and after Ro-ro ships. The reason for that might be high navigation speed. During high speed, it seems that the usage of electronic navigation equipment increased. It was the lowest MWL on LPG carrier. The reason for this could be interpreted as the high manoeuvre ability of LPG carriers or navigators who work on them might not feel mental workload as much as others due to carrying dangerous cargo

rather than the usage of electronic navigation equipment.

As a result of RNASA-TLX score, MWL was the highest during Navigation in TSS and Berthing. It could be concluded as how Berthing and Navigation in TSS were critical. At Port and Navigation in Open Sea had the lowest MWL. The reason of that might be rarely using of equipment.

Mental workload was the highest in Black Sea and North Sea. It seems that adverse weather conditions in these areas effected navigators' usage of equipment. Çanakkale Strait, Marmara Sea and Istanbul Strait had the similar workloads.

In the comparison of difficulties between electronic navigation equipment, as shown in Figure 3, ECDIS–Autopilot comparison had the most difference among others, after that it was ARPA–Autopilot. From this study's result, it can be interpreted as ECDIS and ARPA were used much more or were very difficult to use. Radar-ARPA comparison had the lowest difference. This also assumed that they have similar characteristics of usage. The order of highest to lowest mental workload by using electronic equipment were ECDIS, ARPA, Radar, AIS, and Autopilot. The results of this study is similar with the considerations of automated bridge and e-navigation. ECDIS has more functions and complicated settings than any other bridge electronic equipment. Officers/Masters use Autopilot all the time while they are sailing. However most of the times its functions are not used much. ARPA, Radar and AIS might not demand mental workload as much as ECDIS. In this study, Radar had higher mental workload than AIS due to International Regulations for Preventing Collisions at Sea (COLREG) advises Radar and not AIS for preventing collision.

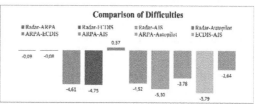

Figure 3. Comparison of difficulties between electronic navigation equipment

5 CONCLUSIONS

When workload level is high or low, human tends to make error (Kum et al. 2008, Brookhuis & de Waard 2010). The workload level should in an optimum level for protecting any tiredness due to high loading and boredom due to low loading. The main aim is to minimize any kind of human error to get "safety", especially when the automation covers all ships (Kum et al. 2008).

In case of navigators using electronic navigation equipment, when he/she makes any error, it is a high risk for a potential accident. Human is the central of bridge for protecting any undesirable situation and ship safety. That's why the important point is the mental workload required by navigators' tasks (Kum et al. 2008, Kum et al. 2007).

Electronic navigation equipment has a lot of effects on navigator' mental workload. This mental workload might have differences according to age, rank, sea experience, ship type, operation, and navigation area. In this study, it is understood that the level of navigators' mental workload is acceptable for keeping proper attention (64.8%). The results of mental workload analyse for electronic navigation equipment usage by utilising RNASA-TLX and pair-wise comparison could be summarized as follows:

1 Visual Demand had the highest value in the MWL dimensions. This might be interpreted as the importance of information from eyes and display/keypad design. Effort was the lowest, because subjects might not have to work so hard.

2 Over 40 years old navigators had the highest mental workload whilst under 31 years old had the lowest.

3 Chief Officers had the highest mental workload It can be the result of taking responsibility and getting more fatigue. 2nd Officers had the lowest mental workload. 2nd Officers are responsible for navigation plan so a potential tendency of using electronic navigation equipment more than others.

4 6-10 years' experience navigators had more mental workload than other groups. Navigators who had more than 20 years of experience had the lowest. These could be interpreted that in the future years when sea experience comes to high level, value of mental workload might be getting lower.

5 Mental workload was the highest for HSC whilst the lowest for LPG carrier.

6 During the Navigation in TSS and Berthing, navigators had the highest mental workload. At Port and Navigation in Open Sea had the lowest workload. The reason of that might be rarely using of equipment.

7 The order of highest to lowest mental workload by using electronic equipment were; ECDIS, ARPA, Radar, AIS, and Autopilot. ECDIS–Autopilot comparison had the most difference whilst Radar-ARPA had the least.

REFERENCES

Brookhuis, K. A. & de Waard, D. 2010. Monitoring drivers' mental workload in driving simulators using physiological measures. *Accident Analysis & Prevention* 42(3): 898-903.

Cha, D. W. & Park, P. 1997a. Simulator-based Mental Workload Assessment of the In-Vehicle Navigation System Driver using Revision of NASA-TLX. *IE Interfaces* 10(1): 145-154.

Cha, D. & Park, P. 1997b. User required information modality and structure of in-vehicle navigation system focused on the urban commuter. *Computers & Industrial Engineering* 33(3): 517-520.

Eggemeier, F. T. 1988. Properties of workload assessment techniques. *Advances in Psychology* 52: 41-62.

Evans, D. C. & Fendley, M. 2017. A multi-measure approach for connecting cognitive workload and automation. *Int. Journal of Human-Computer Studies* 97: 182-189.

Fernandes, A. & Braarud, P. Ø. 2015. Exploring Measures of Workload, Situation Awareness, and Task Performance in the Main Control Room. *Procedia Manufacturing* 3: 1281-1288.

Grabowski, M. & Wallace, W. A. 1993. An expert system for maritime pilots: Its design and assessment using gaming. *Management Science* 39(12): 1506-1520.

Kum, S., Furusho, M. & Fuchi, M. 2008. Assessment of VTS Operators' Mental Workload by Using NASA Task Load Index. *The Journal of JIN* 118: 307-314.

Kum, S., Furusho, M. & Iwasaki, H. 2007. Investigation on the factors of VTS operators' mental workload: case of Turkish operators. *Proceedings of 16th IMLA Conference*: 335-345.

Murai, K., Hayashi, Y., Nagata, N. & Inokuchi, S. 2004. The mental workload of a ship's navigator using heart rate variability. *Interactive Technology and Smart Education* 1(2): 127-133.

Nachreiner, F. 1995. Standards for ergonomics principles relating to the design of work systems and to mental workload. *Applied Ergonomics* 26(4): 259-263.

National Aeronautics and Space Administration. (n.d). Retrieved October 22, 2016, from https://humansystems.arc. nasa.gov/groups/tlx/downloads/TLX_pappen_manual.pdf

Ogreten, S. & Reinerman-Jones, L. 2015. Interview to Questionnaire Comparison of Workload Measurement on Nuclear Power Plant Tasks. *Procedia Manufacturing* 3: 1256-1263.

Perera, L. P., & Soares, C. G. 2015. Collision risk detection and quantification in ship navigation with integrated bridge systems. *Ocean Engineering* 109: 344-354.

Rouzikhah, H., King, M. & Rakotonirainy, A. 2013. Examining the effects of an eco-driving message on driver distraction. *Accident Analysis & Prevention* 50: 975-983.

Rusnock, C. F. & Borghetti, B. J. 2016. Workload profiles: A continuous measure of mental workload. *Int. Journal of Industrial Ergonomics*. doi:10.1016/j.ergon.2016.09.003

Sauer, J., Nickel, P. & Wastell, D. 2013. Designing automation for complex work environments under different levels of stress. *Applied Ergonomics* 44(1): 119-127.

Tetley, L. & Calcutt, D. 2007. *Electronic navigation systems. Routledge*. Retrieved from https://books.google.com.tr/books?hl=tr&lr=&id=mybjgp2gGRsC&oi=fnd&pg=PR3&ots=MiHlJ9gRTo&sig=H-k_QyLTOQWOAR5NpiCzqDngp7k&redir_esc=y#v=onepage&q&f=false

van de Merwe, F., Kähler, N. & Securius, P. 2016. Crew-centred Design of Ships–The CyClaDes Project. *Transportation Research Procedia* 14: 1611-1620.

Vidulich, M. A. & Tsang, P. S. 2015. The confluence of situation awareness and mental workload for adaptable human–machine systems. *Journal of Cognitive Engineering and Decision Making* 9(1): 95-97.

Yoshimura, K., Mitomo, N., Okazaki, T., Hikida, K. & Murai, K. 2007. Evaluating the Mariner's Workload Using the Bridge Simulator. *IFAC Proceedings* 40(16): 57-60.

Observations on ECDIS Education and Training

D. Brčić, S. Žuškin & M. Barić
University of Rijeka, Rijeka, Croatia

ABSTRACT: In Electronic Chart Display and Information System (ECDIS) handling, Officers Of Watch (OOW) are experiencing several problems which have been identified and defined so far, with varieties in their causes and nature. Issues related to education, training and (lack of) knowledge can be singled out as factors directly and indirectly associated to a number of consequent difficulties and unwanted situations. The proposed paper deals with training of seafarers as end-users of the ECDIS system, trying to present them as central subjects in the ECDIS system of stakeholders. Within the research and (using results from) the corresponding survey, opinions of OOWs are considered and discussed regarding sufficiency of stipulated hours and recommended means for the completion of the Generic ECDIS training. Respondents' profile was layered in several categories, all of which were elaborated considering survey questions. Certain features are presented regarding skills and knowledge prerequisites for which is understood that are already adopted by course attendees, which in turn opens new questions and potential issues. On the basis of the study, actually being part of larger ECDIS research, and derived inferences, possible solutions and indicative guidelines are suggested, focusing on ECDIS education and training issues and appearances. Improvement of the educational and training process is further discussed. Planned activities are presented as inevitable tasks which have been formed based on previous and proposed research.

1 INTRODUCTION

Considering different segments of ever-growing and expanding maritime industry segments, Maritime Education and Training (MET) acts as a specific type of education. Expansion of global merchant and passenger fleet and raise of new ratings are few of pronounced particularities which are taking place. The changes' reflect especially on new navigational systems' knowledge acquisition which are slowly but steadily replacing traditional navigational tools (Mindykowski *et al.* 2013, Weintrit 2009). Regarding good seamanship practice, enhancement of maritime technologies calls for proper education and skill development. The tools are changing, however navigational background persists. ECDIS acts as a trademark of recent changes, representing redesign in navigation conducting, with its implementation period slowly reaching its completion.

Recognizing opportunities in educational improvements, but also potential risks emerging from inadequate training, there appears a need of establishment of proper educational framework striving to gain an effective knowledge level. Focusing on OOWs' adaptation to the system it aims to sufficient and qualitative knowledge required for conducting safe navigation ventures, and beyond. So far, various operational, functional and educational issues were detected and addressed, comprising both 'ordinary' and 'silent' problems.

ECDIS EHO is an education-based research project introduced in times of first ECDIS implementation dates. As fulfilled so far, the respondents' profile is presented, pointing at mentioned levels of issues and considering their opinions.

The paper illustrates a cross-section of OOWs' opinion on education and training in general, as well as disagreements regarding the Generic Course. Results are showing slight differences and even opposites in this viewpoint. Mentioned statement was motivation for further analysis of respondents and their opinions regarding ECDIS education and training (EET). Overall profile of respondents was further divided in sub-categories characterized by their inherent features, always focusing on questions and comments related to EET. Special emphasis was given on the age of seafarers, comprising both sea-going and ECDIS experience. OOWs which consider that the proposed time of the course is not sufficient for generic ECDIS training were further addressed. Several possibilities in terms of EET are proposed and discussed, representing further tendencies and activities to be realized in order to improve the educational process.

2 BACKGROUND

Introduction of new technologies on-board vessels implies several issues to consider, including standardization, challenges in training for technology and the human element. In order to handle the system properly, a certain system-working knowledge should exist, however background knowledge must not be neglected. Automation, which with integration acts as one of the main features of ECDIS system, calls for new level of knowledge demands (IMO MSC/Circ. 1091 2003). This is the peak point where traditional navigational knowledge and new technology interact, with OOWs as ECDIS end-users placed in the middle.

2.1 *General considerations*

Proper usage of the system involves a range of mutually interconnected stakeholders to a greater or lesser extent, as shown on Figure 1. In order to achieve safe and effective ECDIS navigation, it is important that all of related subjects have a clear and common understanding of their roles and responsibilities (MSC.1/Circ.1503 2015). Apart from number of different regulations and requirements, OOWs are experiencing unenviable situations along their sailing journeys. The Figure 1 presents position of OOW in the ECDIS stakeholders system and their mutual relations.

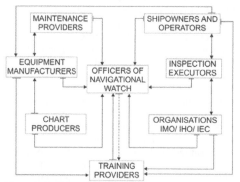

Figure 1. Position of OOWs as end-users in the system of ECDIS stakeholders.

Considering integration as most significant feature of ECDIS, it is obvious that the system will develop further in a way that even more (not necessarily navigational) devices will act as a part of it. Beside physical equipment, various computer programs and databases appear herein. As the implementation period reaches its end, several issues are appearing and efforts are made for their resolution. As from technical aspect, several standards were revised, referring to ECDIS presentation library, electronic charts protection standard, new test data sets, presentation of navigation-related information and testing standards (IHO 2014a, IHO 2014b, IHO 2014c, IHO 2015, IHO 2016, IEC 2014, IEC 2015, IEC 2016, Weintrit 2015, Wingrove 2016). These standards emerged as answers to difficulties in system handling, ECDIS experience improvement, as well as with its components. The update transition period was prolonged due to complicated and lengthy adaptation needed for ECDIS unit installation and upgrade.

New IHO and IEC standards are only one confirmation of the gravity of situation. As for training, regulations are already dictating two different, mandatory educations: Generic Course and Type Specific ECDIS course, latter of which is coming in less standardized form, and the former experiencing revisions as well (IMO MC 1.27 2012, NI 2012, Weintrit *et al.* 2012). Type Specific course, as a specific model familiarization course, is not a subject of the paper.

Generic course should ensure that OOWs understand the context of navigation with the ECDIS system, and it should deliver the knowledge and competences at least equivalent to those given in the *Model Course 1.27* (NI 2012, MC 1.27 2012).

IMO Model Courses are 'flexible in application', meaning that *'key components in the knowledge and skills transfer are training endeavors, knowledge, skills and dedication of the instructor'* (IMO MC 1.27 2012). As for ECDIS MC however, the assumption is that course attendees are experienced and have certain level of knowledge on *terrestrial navigation* as well as with *bridge-watch keeping duties*, (have at minimum some familiarization with) *visual navigation*, have a *prior completion of basic radar/ARPA course*, and that they have *considerable familiarization with personal computing* (IMO MC 1.27 2012).

When introducing new technologies on board vessels, assumed profile of seafarers will welcome novelties, while to others it would seem inaccessible. IMO, through its Model Course 1.27 addresses these OOW categories as seafarers of 20th and seafarers of 21st century (IMO MC 1.27 2012), both with their specific features. One should wonder what can experienced seafarers teach their younger colleagues, and in contrary, what new generations can teach elders?

2.2 Previous research and lessons learned

The ECDIS EHO Research (EER) started in times of first system' implementations. It consists of various educational, practical and research activities which are, during the research, branching out. The results and findings are implemented in all forms of teaching processes which take place, whether it is STCW MET or academic teaching. The EER questionnaire, placed among seafarers on international level and in different ways (during courses, on vessels, via shipping companies), acts as an additional tool for concrete feedback from OOWs as end-users. The survey comprises questions regarding Experience, Handling and Opinions related to the system (hence the *EHO* abbreviation). Every question or set of questions presents a functional topic/subject which is found and assumed to be problematic in terms of end-user aspect. For now, students as future seafarers are excluded from this part of the research, at least in the present questionnaire form.

During the research, several issues have been identified so far. Apart that it is a complex system, composed of fundamentally different elements, and ranging from internal and external equipment and databases/software, maker's different interpretations of the system are causing further problems in handling, especially to unfamiliar and not accustomed OOWs. ECDIS performance standards are in general satisfied, however many other features and functions differ significantly. Lack of knowledge and insufficient familiarization are becoming a real threat in navigation conducting. Several marine accidents were consequence of improper system usage due to poor training and ignorance (DMAIB 2015, BSU 2015, BEAmer 2015, MAIB 2015, MAC 2015, Tang & Sampson 2011).

Safety parameters' settings and primary positioning sensor back-up non-dedication were, as basic safety features, found not to be respected (Brčić, Žuškin & Kos 2015, Žuškin, Brčić & Kos 2016). ECDIS anomalies appear as independent problem, when the system behaves '*in an unexpected or unintended manner that affects the use of the equipment or navigational decisions*' (Weintrit 2015, IMO SN/Circ.312 2012). In the recent EER study (Brčić, Kos & Žuškin 2016), OOWs' statements regarding how they see ECDIS problems and difficulties, were elaborated. Apart that they were numerous, answers were diverse so that they have been divided in several categories, as shown on Figure 2.

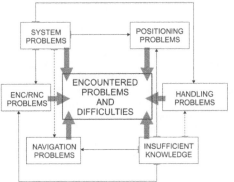

Figure 2. Identified ECDIS issues, as seen by OOWs as end-users, and their mutual linkage

Elaborating ECDIS-caused marine accidents (Brčić, Kos & Žuškin 2016), it was found that none was consequence of system failure. ECDIS anomalies are slowly but steadily resolving (Weintrit 2015, Wingrove 2016, IMO NCSR 2/22/2 2014). Model Courses and teaching frameworks are improving as well, with the last Model Course (third) edition dating in 2012. Insufficient knowledge or knowledge transfer efficiency remains as one of most relevant elements. This final output goes also for awareness of system, its main functionality, purpose and other features that have to be strongly distinguished.

3 PARTIAL STRUCTURAL ANALYSIS OF THE ECDIS EHO RESEARCH: ECDIS EET

Described issue was the motivation for the conducted research. Related questions were, where applicable, accompanied with OOWs' discussions. Answers on following question were analyzed, being the main topic of the proposed survey:

- *In accordance with IMO Model Course 1.27, the intended duration of ECDIS Generic Training course is 40 hours. Do you think that this time amount is enough for the training? You can explain the YES/NO answer if you want to.*

Several introductory/general questions were used:

- Rank/working experience/year of birth,
- ECDIS experience (categories a – i),
- Current ECDIS status on-board (AFC only, ECDIS & AFC, paperless/full ECDIS),
- Member of navigational watch/ECDIS operation included in job description (yes/no),
- ECDIS as a primary/ secondary mean of navigation on-board (primary/secondary),
- Type of vessel
- Possession of Generic Course Certificate (yes/no).

The analysis is presented in the continuation.

3.1 Overall respondents' profile

The overall profile of respondents which participated in the questionnaire comprises of 78 masters (M), 57 Chief Officers (1/o), 50 Second Officers (2/o), 9 Third Officers (3/o), 8 Apprentice Officers (A/o), 4 Staff Captains (SC), 2 Superintendents (SI), 2 Supervisors (SV), one Pilot (P), one Marine Safety Consultant (MSC), 4 Senior Dynamic Positioning Operators (SDPO) and 19 Undefined respondents (U). In total, 235 respondents fulfilled the questionnaire (In time of writing). Graphical presentation of the rank's share is shown on Figure 3.

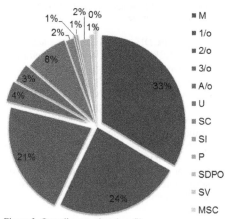

Figure 3. Overall respondents' profile

Mainly masters (33%), chief officers (24%) and second officers (21%) prevail, making the questionnaire representative regarding that they are, among other engaged ranks/professions of respondents (e.g. auditors), directly related in system handling.

3.2 Categorization of respondents

Considering general features of respondents and their experience, they were divided into four categories:

- *Profile I*: all respondents
- *Profile II*: ECDIS-experienced respondents forming part of the navigational watch
- *Profile III*: respondents who, beside conditions from Profile II, have already attended Generic ECDIS course
- *Profile IV*: respondents who have never had the opportunity to work with the system

Working experience histograms are presented in following figures. Note that presented years refer on sailing experience of respondents. ECDIS experience is considered later.

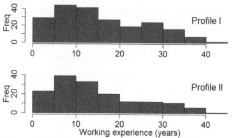

Figure 4. Working experience histogram of Profile I (top) and Profile II (bottom)

As for general overview, with minimum of 0.5 and maximum of 41 years, average sailing experience of all respondents ranges from 15.6 to 19.4 years.

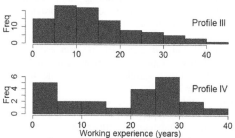

Figure 5. Working experience histogram for Profile III (top) and Profile IV (bottom)

Distribution of sailing experience is similar through all profiles except seafarers who never had the opportunity to handle the system. Nevertheless, considering the sailing experience range, all cases are covered. In Table 1, basic working experience statistics per profile is presented.

Table 1. Statistical summary of working experience per defined profile

Profile	P1	P2	P3	P4
Min (years)	0	0.5	1	0
Median (years)	15	13	13.5	24
Mean (years)	16.9	16	15.6	19.4
SD (years)	10.6	10.5	10	12.6
Max (years)	41	41	41	40
No.	235	174	111	27

3.3 Results

'We don't take sides. It is inappropriate.'

Three possible answers were marked: *YES*, 40 hours of is enough for Generic ECDIS training; *NO*, 40 hours are not enough; and *NA* (not applicable), where column remained blank, or respondents stated this way. In Table 1, answers are presented for each of defined profiles.

Table 2. Answers on sufficiency of hours required for ECDIS Generic course, per profile

Answers	P1	P2	P3	P4
Yes	206	157	100	17
No	19	15	11	3
NA	10	2	1	7
Total	235	174	111	27

Apart from *Profile 1*, it is visible that, when compared to other defined profiles, *Profile 4* makes the smallest group of respondents. At the same time, it is the profile with most dispersed answers in relation to the total respondents' number. Graphical presentation of the share of answers per profile is shown on Figure 7. In *Profile* 4, most respondents did not give any of the answer (NA). Within the same profile, the largest number of negative answers was recorded, when compared relatively to the total number of respondents per profile.

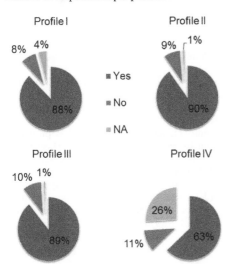

Figure 7. Share of respondents' answers per defined profile regarding sufficiency of required Generic ECDIS training hours

As for other profiles, in *Profile 3* (respondents with the possession of the Generic ECDIS Course) the negative answer (NO) was more common than in remaining profiles.

3.4 *Respondents' comments*

Additional opinion was received from 45 respondents, accompanying all answers. Those comments are summarized in the continuation. Considering the answer, hereinafter they will be called *positive* and *negative* respondents, respectively.

3.4.1 *Positive respondents*

General opinion for YES supporters is that with qualified instructor, 40 hours are enough for the system introduction and basic knowledge. They are also stating that time provided (40 hours) is not enough 'for full operation' (cited the OOW) – referring to Type Specific Course – as well as for officers who never have been on ECDIS vessel. YES respondents argue that Generic Course doesn't offer required practical skills like Type Specific Course. Same respondents are stating that it is enough in considering formal and legislative part, however that, in order to use the system correctly, in the practice it takes a lot more. Every vessel has a different system, and real learning comes with practice and real situations. For this category of respondents, it can be considered obvious.

3.4.2 *Unclassified respondents*

OOWs who stated *YES*, also stated that 40 hours is even too much, because of the fact that it has no sense that they know something if they not have a permission for entering the system. In case of any serious problem they have to wait for a serviceman, which leads to confusion. Regarding answers *YES*, the following was derived from context: Few of respondents wrote that 40 hours is enough, if the person is already familiar with and has some experience with the system (…). It comes in contradiction given that others are thinking how 40 hours are enough for the introduction and general considerations. According to this opinion, if the system was used regularly on-board 40 hours should be sufficient (should Generic course be/not be the first contact). The predicted time should be enough for the theoretical part; however a lot more hours are needed for adaptation on the system. Again, it is enough if the OOW had the opportunity to operate with the system. It depends on person.

3.4.3 *Negative respondents*

Respondents, who answered negatively on the elaborated question, accompanied their answer with argument that without previous experience there is too much of information, requiring much more time to master the system. There are lots of functions that are difficult to learn in 40 hours only. They are also proposing extension of the course to 60 or even 80 hours (!), with more practice and exercises, keeping the same quantity of theoretical part. Existing course duration is enough for commencing with the work on the system, in order to gain basis for further skills and knowledge adoption. There is more practice needed, especially for OOWs in charge of the system. For OOWs who are using the system continuously and they are *good with computers* even fewer hours is enough. However, for OOWs who are facing the system for the first time the course should be simplified and extended to 80 hours.

3.4.4 *Concluding observations*

Majority of respondents emphasize the importance of practical work and training on the system. The significance of theoretical part is recognized in very few instances.

4 FINAL CONSIDERATIONS

Figure 8 represents working experience histograms of positive and negative respondents, referring to all respondents (Profile 1).

Figure 8. Working experience histogram of NO (top) and YES (bottom) respondents

In general, relatively younger respondents consider that current course duration is not sufficient for qualitative ECDIS education and training. On average, working experience of those respondents is 11.6 years (with standard deviation of 9.1), while the average of opposite category is 16.8 years (with standard deviation of 10.5).

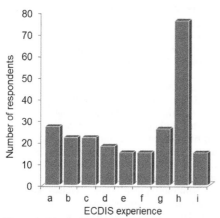

Figure 9. Distribution of respondents according to ECDIS experience: no experience (*a*), less than 6 months (*b*) between 6 months and one year (*c*), between 1 and 2 years (*d*), between 2 and 3 years (*e*), between 3 and 4 years (*f*), between 4 and 5 years (*g*), more than 5 years (*h*), other (*i* – mostly between 8 and 15 years).

Distribution of respondents according to ECDIS experience is presented on Figure 9. By analyzing

the answer per category, results in form of percentages can be shown as on Figure 10. The sample is relative; the number of respondents for each category is equivalent to 100%, while the share of possible answers (*YES/NO/NA*) is distributed within.

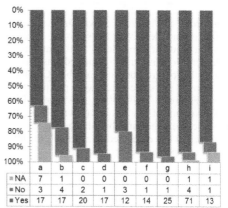

	a	b	c	d	e	f	g	h	i
NA	7	1	0	0	0	0	0	1	1
No	3	4	2	1	3	1	1	4	1
Yes	17	17	20	17	12	14	25	71	13

Figure 10. Share of answers considering ECDIS experience

As with previous results, the number of negative answers is significantly smaller than the opposite, however it is present and persistent through all categories.

On Figure 11, share of OOW rank is presented considering three of possible answers. Masters are noted for their significant share in all answers. Both *YES* and *NO* answers are expressed in all ranks, while there is no *NA* answer in case of Apprentices and Third Officers. The *NO* answer is distributed through ranks more properly than others, with positive answer always dominant.

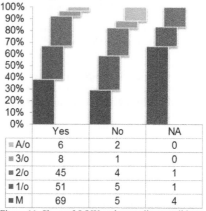

	Yes	No	NA
A/o	6	2	0
3/o	8	1	0
2/o	45	4	1
1/o	51	5	1
M	69	5	4

Figure 11. Share of OOW rank regarding possible answers

The *NO* answer extends through all ranks and through all defined categories. Total number of 19 respondents answered that time duration of 40 hours

is not sufficient for the Generic ECDIS Course. Among these seafarers:
- all officer ranks are present – from apprentice officers to masters
- all ECDIS experience categories are present – from *no experience* to 15 years
- all possible ECDIS scenarios on-board are present – paper charts only, paper charts and ECDIS systems, and paperless vessels
- different types of vessels are present – RO-RO vessels (1), container vessels (5), tugs (1), oil tankers (1), gas tankers (2), passenger vessels (1) an mega-yachts (1)
- they are both forming (16) and not forming (3) part of the navigational watch

If we filter the respondents in accordance with the last criteria, 16 of OOWs remain as members of the navigational watch. Among them:
- 10 out of 16 already possess Generic ECDIS Course, and
- For 7 out of 16 ECDIS system is primary navigational mean on-board.

ECDIS system evolves in pace with technology, however it modifies in accordance with detected deficiencies and emerging needs. There is still lack of confidence present in handling the system (Tang & Sampson 2011). New IEC and IHO standards (meaning new ECDIS models), transitions, upgrades, and other pressures do not facilitate the current situation of OOWs.

A question (one of several) that appears is as follows: if we put seafarers in the foreground, and that was the tendency in the first place: how much actually, and *what* specifically, seafarers as end-users need to know, and how much they are required to. On the other hand, how much same seafarers do not know, but it is understood that they do.

Another question is: to what extent the course is taken seriously by OOWs when attending the training. The YES answer was far more numerous. However, it can be assumed that a certain share of respondents answered in such a way because of other, less significant factors, with net obligation for 5-day attendance, to mention one. General opinion is that, with qualified ECDIS trainer, 40 hours is sufficient for the course. This is applicable to the majority of OOWs. Derived observations can be summarized as follows:
- Majority of respondents considers the recommended course duration as sufficient. Significant respondents' share states individual 'guidelines' on possible improvements
- Among negative-answer respondents, the only common feature is their age. Contrary to expectations, these OOWs are younger compared to the other group
- Several comments imply to certain insecurity present among OOWs, with traces of ignorance and even neglecting.

Generic course is standardized to certain extent, however, *in this moment* it is not applicable to all users. Practical part of generic course is essential. Handling and skill development adopt with practice and experience in handling the system. However, understanding and theoretical knowledge of the system and its components is crucial. There appears a potential risk if the theoretical background starts to be ignored, and it can be fully reflected on new generations of seafarers, the same who will master the system easier than their older colleagues. According to MC 1.27, the ECDIS Generic Course can be supplemented with eventual pre-entry course or, *alternatively*, elements of academic knowledge through the course. The course is standardized to certain extent. However, each course group consists of different members in various ways. Thus, the standardization cannot take its complete form. Individual approach is essential.

5 CONCLUSIONS AND FURTHER WORK

In the proposed paper, opinions regarding ECDIS MET were elaborated. The main topic was the sufficiency of hours stipulated for the Generic Course, as designed according to the ECDIS Model Course. Results have shown that majority of respondents consider time for the Course sufficient. This outcome was somewhat expected. The opposite answer is found to be present among all categories of seafarers, nevertheless the rank, experience or other professional trait. The only feature connecting those respondents is their relatively younger age in comparison with positive respondents. One way of interpretation of this fact could be required seamanship knowledge related not only to the system. The general interpretation should be the significance and desired output of individual approach, and high interaction during EET processes, whether it takes place as formal education, or between bridge team members.

Several activities can be proposed in order to complement ECDIS education and training:
- Introduction of new courses, adopted to particular participant,
- Applicability of ECDIS course/s on other stakeholders engaged in navigational venture, besides OOWs (where applicable),
- Development of means of stable, proper and unimpeded feedback between ECDIS stakeholders and educators, as well as between bridge teams onboard vessels.

Proposed activities should ease the ways of approach and interaction between OOWs, and complement the existing EET. ECDIS system develops further in a way that more features and elements become a part of it. It represents *navigation data collection*; however it is often misused. It is not the system that should matter solely, rather the background knowledge which should be sustained and used in conjunction with all new-coming navigational equipment.

ACKNOWLEDGEMENTS

The authors are grateful to all the officers of the navigational watch on their time and they willingness for fulfillment of the questionnaire. Their answers have an immense significance.

REFERENCES

Brčić, D., Kos, S. & Žuškin, S. (2015) Navigation with ECDIS: Choosing the proper secondary positioning source. *TransNav, the International Journal on Marine Navigation and Safety of Sea Transportation*. 9(3). pp. 317-326.

Brčić, D., Kos, S. & Žuškin, S. (2016) Partial structural analysis of the ECDIS EHO research: The handling part. *Proceedings of the 24nd International Symposium on Electronics in Transport (ISEP)*. Electrotechnical Association of Slovenia & ITS Slovenia, Ljubljana, Slovenija, 29-30. 3. 2016.

Danish Maritime Accident Investigation Board (DMAIB) (2015). Available at: http://www.dmaib.com/Sider/ Home.aspx (16.11.2015).

Federal Bureau of Maritime Casualty Investigation (BSU) (2015). Available at: http://www.bsu-bund.de/EN (01. 11.2015)

French Marine Accident Investigation Office (BEAmer) (2015). Available at: *http://www.beamer-france.org/index-en.html* (01. 11.2015).

International Electrotechnical Commission (IEC) (2014). IEC 62288, Edition 2.0. Genève: IEC.

International Electrotechnical Commission (IEC) (2015). IEC 61174, Edition 4.0. Genève: IEC.

International Electrotechnical Commission (IEC) (2017). Available at. www.iec.ch (20.10.2016)

International Hydrographic Organisation (IHO) (2014). SP S-52, Annex A, Edition 4.0. Monaco: IHO.

International Hydrographic Organisation (IHO) (2014). SP S-52, Edition 6.1. Monaco: IHO.

International Hydrographic Organisation (IHO) (2014). SP S-64, Edition 3.0. Monaco: IHO.

International Hydrographic Organisation (IHO) (2015). SP S-63, Edition 1.2. Monaco: IHO.

International Hydrographic Organisation (IHO) (2017). *Current IHO ECDIS and ENC Standards*. Available at: http://bit.ly/2e3Idpi (20.10.2016)

International Maritime Organization (IMO) (2003). *MSC/Circ. 1091*. London: IMO.

International Maritime Organization (IMO) (2012). *Model Course 1.27*. London: IMO.

International Maritime Organization (IMO) (2012). *SN.1/Circ. 312*. London: IMO.

International Maritime Organization (IMO) (2014). *NCSR 2/22/2*. London: IMO.

International Maritime Organization (IMO) (2015). *MSC.1/Circ. 1503*. London: IMO.

Marine Accident Investigation Branch (MAIB) (2015). Available at: *http://www.maib.gov.uk/* (3.11.2015).

Maritime Accident Casebook (MAC) (2015). Available at: *http://maritimeaccident.org* (04.11.2015).

Mindykowski, J., Charchalis, A., Przybylowski, P. & Weintrit, A. (2013) Maritime Edication and Research to Face the XXI-st Century Challenges in Gdynia Maritime University's Experience: Part II – Gdynia Maritime University of Experience the 21st Century Challenges. *TransNav, the International Journal on Marine Navigation and Safety of Sea Transportation*. 7(4). pp. 581-586.

The Nautical Institute (NI) (2012). *Industry recognises revised IMO Model Course for ECDIS*. Press release. London: The Nautical Institute.

Tang, L. & Sampson, H. (2011) Training and Technology: Findings from the Questionnaire Study. *Seafarers International Research Centre 2011 Symposium Proceedings*. pp. 1-20. Cardiff: SIRC.

Weintrit, A. (2009) *The Electronic Chart Display and Information System (ECDIS): An Operational Handbook*. CRC Press, Taylor & Francis Group.

Weintrit, A. (2015) ECDIS issues related to the implementation of the carriage requirements in SOLAS Convention. Archives of Transport System Telematics. 8 (1). pp. 35-40.

Weintrit, A., Kopacz, P., Bak, A., Uriasz, J. & Naus, K. (2012) Polish approach to the IMO Model Course 1.27 on operational use of ECDIS. *Annual of Navigation*. 19(2). pp. 155-170.

Wingrove, M. (Ed.) (2016) The complete guide to ECDIS. Enfield: Riviera Maritime Media Ltd.

Žuškin, S., Brčić, D. & Kos, S. (2016) Partial structural analysis of the ECDIS EHO research: The safety contour. *7th International Conference on Maritime Transport*. Universitat Politecnica de Catalunya, Barcelona. Barcelona, 27-29. 6. 2016.

The Use of eLoran System for Transmission of the National Time Signal

C. Curry
Chronos Technology Ltd., The United Kingdom; and Taviga Ltd (UK)

K. Czaplewski
Gdynia Maritime University, Gdynia, Poland

C. Schue
UrsaNav, Inc., The United States; and Taviga LLP (USA)

A. Weintrit
Gdynia Maritime University, Gdynia, Poland

ABSTRACT: A Positioning, Navigation and Timing (PNT) service using enhanced Loran (eLoran) has been transmitted experimentally in the United Kingdom for more than 5 years. The eLoran transmitter employed, at Anthorn in North-West England, is operated by a commercial company on behalf of the General Lighthouse Authorities of the United Kingdom and Ireland. It is funded in part by the Department for Transport and other UK government agencies. Chronos Technology has used these and other eLoran transmissions to conduct research into the viability of employing eLoran as a means of distributing time traceable to UTC, including for indoor applications [1]. There is growing interest internationally concerning the vulnerability of GPS and other global navigation satellite systems (GNSS) to natural and man-made interference, plus the jamming and spoofing of their transmissions. These vulnerabilities have led to a demand for sources of resilient PNT, including a robust means of distributing precise time nationally and internationally. This paper explores the ability of eLoran to disseminate UTC-traceable time to applications in GNSS-denied environments. It proposes the creation of a National Timescale with UTC distributed via eLoran signals. Practical results from a test programme are very encouraging: UTC-traceable time signals with an accuracy of better than 100 ns and with a quality comparable to that provided by GPS are received even indoors. This novel source of precise time meets the latest ITU standards for primary reference timing clocks in Internet Protocol networks.

1 INTRODUCTION

1.1 *National time standard*

Each country that contributes to Universal Coordinated Time (UTC) operates a national time standard that is independent of GNSS. Its technology will generally be based on a Hydrogen Maser. This will be adjusted using monthly corrections supplied by the Bureau International des Poids et Mesures (BIPM) in France [2].

Low-frequency eLoran is now emerging as the preferred advanced source of positioning, navigation and timing (PNT) signals alternative or complementary to global navigation satellite systems (GNSS). eLoran is globally-standardized and does not share the vulnerability of GNSS to incidental or deliberate jamming, intentional spoofing, radio-frequency interference or space weather events.

A number of countries are actively reconsidering their dependence on GNSS across multiple critical infrastructure applications and some are planning the implementation of eLoran transmitter networks. Within this context falls the question of how to deliver their national time service to those clients who have come to recognize their own vulnerability to the disruption of GNSS.

This paper discusses the concept of delivering such national time services by means of eLoran signals.

1.2 *TAI, UTC and leap seconds*

The global reference for time is International Atomic Time (TAI), a time scale calculated at the BIPM, using data from some 400 atomic clocks in over 70 national laboratories. The BIPM organizes clock comparisons for the determination of TAI through an international network of time links. Corrections to local national timing laboratory clocks are generally applied monthly or weekly and typically will be a few nanoseconds (ns).

TAI long-term stability is set by weighting participating clocks. The scale unit of TAI is kept as close as possible to the SI second by using data from those national laboratories which maintain the best primary standards. These will generally be Hydrogen Masers or high performance Caesium standards.

UTC is identical to TAI except that from time to time a leap second is added to ensure that, when averaged over a year, the Sun crosses the Greenwich meridian at noon UTC to within 0.9 s. The dates of application of the leap second are decided by the International Earth Rotation Service (IERS) [3].

2 E-LORAN AS A TIME STANDARD

2.1 *Enhanced Loran*

Enhanced Loran is an internationally-standardized positioning, navigation, and timing (PNT) service for use by many modes of transport and in other applications. It is the latest in the long-standing and proven series of low-frequency, LOng-RAnge Navigation (LORAN) systems, one that takes full advantage of 21st century technology [5].

eLoran meets the accuracy, availability, integrity, stability and continuity performance requirements for aviation non-precision instrument approaches, maritime port and harbour entrance and approach manoeuvres, land-mobile vehicle navigation, and location-based services, and is a precise source of time and frequency for applications such as telecommunications. It is an independent, dissimilar, complement to Global Navigation Satellite Systems (GNSS). It allows GNSS users to retain the safety, security, and economic benefits of GNSS, even when their satellite services are disrupted.

What is important, eLoran meets a set of worldwide standards and operates wholly independently of GPS, GLONASS, Galileo, BeiDou or any future GNSS. Each user's eLoran receiver will be operable in all regions where an eLoran service is provided. eLoran receivers work automatically, with minimal user input.

eLoran transmissions are synchronized to an identifiable, publicly-certified, source of Co-ordinated Universal Time (UTC) by a method wholly independent of GNSS. This allows the eLoran Service Provider to operate on a time scale that is synchronized with, but operates independently of, GNSS time scales. Synchronizing to a common time source will also allow receivers to employ a mixture of eLoran and satellite signals.

The principal difference between eLoran and traditional Loran-C is the addition of a data channel on the transmitted signal [5]. This conveys application-specific corrections, warnings, and signal integrity information to the user's receiver. It is this data channel that allows eLoran to meet the very demanding requirements of landing aircraft using non-precision instrument approaches and

bringing ships safely into harbour in low-visibility conditions. eLoran is also capable of providing the exceedingly precise time and frequency references needed by the telecommunications systems that carry voice and internet communications.

2.2 *Loran C vs. eLoran*

eLoran is the modern, digital-technology, version of the legacy Loran C system. It re-uses the transmitter stations of its now-obsolete forebear to deliver position fixes of much higher accuracy, integrity, availability and continuity. These transmitters radiate precisely-timed pulses, at a power level of hundreds of kilowatts, on a frequency of 100 kHz. To deliver highly-precise navigation, the pulses must be timed with an accuracy of nanoseconds. Because of this, they can fulfil the additional function of distributing precise time over long distances.

The timing of each transmitter station is derived from a local ensemble of three Caesium standard clocks that are themselves synchronized at intervals to UTC by comparison with a master standard. In this way the transmissions are locked to UTC and so provide a source of UTC-traceable timing that is totally independent of GNSS: so-called "*sky free UTC*". These low frequency transmissions propagate into and through buildings. They can be received indoors by using a magnetic-field antenna – a so-called "*H field antenna*". This capability has been extensively assessed in the course of two UK research projects, GAARDIAN and SENTINEL [5], both led by Chronos Technology.

The time and timing performance of an eLoran signal can be separated into two components: long term timing stability and phase synchronization to UTC. The long term stability of an eLoran signal has been shown to be comparable to that received of commercially-available GPS timing receivers; this will be discussed later. Phase synchronization to UTC is achieved via a "*UTC Sync*" message which is broadcast over an "*Loran Data Channel*", as will now be explained.

2.2.1 *The Loran data channel*

One of the most important differences between legacy Loran-C and the new eLoran is the addition of a Loran Data Channel (LDC) to the transmissions. The LDC offers a highly-robust, though low bit-rate, long range channel that carries digital data messages. The original purpose of these messages was: to carry differential GPS corrections, similar to those in other DGPS systems; to confirm to users the correct and safe operation of the transmission, so ensuring high navigation integrity; and to carry corrections for the small temporal variations of the timing of signals received in certain harbours where the very highest location accuracy is required.

Despite the low data rate, the LDC has sufficient capacity for authorized third parties to use it in order to broadcast high-priority data to their users.

The properties of the LDC are standardized internationally and defined in a document entitled "*Eurofix Message Format*"; the current version of which, ver.2.15, is dated March 2014 [6]. The Eurofix messages are specified by the Radio Technical Committee for Maritime Services (RTCM) Special Committee-104 (Eurofix working group [7]) and in International Telecommunication Union (ITU) Recommendation M.589-3 [8].

One message type is the "*UTC Sync*" message. This provides the information a receiver requires to derive Universal Coordinated Time of Day, Date and Leap Seconds from the eLoran transmission. The message is repeated at intervals of a few minutes. When a timing receiver is being commissioned upon installation, this message allows it to align its 1 pps output pulses to within a few microseconds of UTC. The remaining time offset is then removed in a further calibration stage, as explained below.

The Loran Data Channel employed in the UK uses the Eurofix standard described above. Other data standards have been proposed, including some with much higher data rates; future timing receivers will no doubt switch automatically to the data standard of the transmissions they receive. Eurofix is implemented by means of Pulse Position Modulation (PPM) of the pulses that constitute the eLoran transmissions. Groups of eLoran stations transmit their pulses at different rates, each defined by a "*Group Repetition Interval (GRI)*" and certain stations transmit at two such rates simultaneously. In consequence of these variants, the maximum raw bit rate of the Eurofix data channel can be as low as 50 bits per second (bps) or as high as 150 bps.

The LDC embodies strong Forward Error Correction (FEC). This makes the performance of the data channel very robust and is an important factor in allowing it to be used over substantial ranges. Radio signals at the eLoran frequency of 100 kHz propagate strongly as ground-waves; that is, as surface-waves over the Earth. In consequence, their rate of attenuation with distance depends on the electrical conductivity of the Earth's surface over which they flow, being least over sea-water and greatest over the low-conductivity terrain found in mountains and deserts. The data channel is typically usable at its full data rate out to a range of 1600 km (1000 miles) over sea-water and 800 km (500 miles) over mixed paths of land and sea. For example, the LDC from the UK Anthorn station, located on the coast of Cumbria, serves the whole of the United Kingdom, with a message data rate of approximately 35 bps and, where required, a message update interval of 2s.

Some 16 LDC message types have been defined. Of these, 8 have been assigned to existing services: they include messages concerning UTC time, differential eLoran corrections, and DGPS corrections and integrity. Additional messages are carried on behalf of third-party clients in government. The LDC is an asynchronous transmission system in which the message type is identified by each message header, allowing messages of one type to be interleaved with messages of other types. This permits flexibility, with messages of high importance (such as those that concern the health of the transmissions or the integrity of navigation fixes) to be prioritised over messages of lower urgency.

This paper proposes that one of the currently unassigned message types be used for "*regional ASF timing correction messages*".

2.2.2 *Ground wave*

It is not only the rate of attenuation of eLoran ground-wave signals that depends on the electrical conductivity of the surface over which they travel but also, to a slight but important degree, their speed of propagation. This is greatest over sea-water, slightly less over farmland and least over mountainous or desert terrain. The propagation speed of the signal through the atmosphere and over all-seawater paths is known with very high accuracy; so the time delay due to propagation of such a signal with respect to its timing at the transmitter is very precisely known. But, if a signal travels wholly or partly over land, it will experience an additional delay with respect to its propagation over sea-water. This delay is known as the Additional Secondary Factor (ASF). Its magnitude may be several microseconds. Since it directly affects the timing of arrival of signals carrying a UTC service, it must be taken into account, particularly as we are proposing time accuracies better than 100 ns.

To the extent that the electrical conductivity of the Earth in the service area of an eLoran station has been mapped, the spatial variations of ASF can be predicted using a computer model. But achieving this with sufficient accuracy for use by timing receivers is not always practicable. Happily, the ASF value at any point in the service area is very stable in time and can easily be measured with great accuracy.

If the very highest timing accuracy is required, the small temporal variations in ASFs can be measured by a reference station and corrections generated, as is already done in differential eLoran navigation systems.

Figure 1. Map of signal-to-noise ratio (SNR) values in dBs of the signals received from Anthorn (Image: Chris Hargreaves (GLA) [1])

In addition to the ground-wave component of the low-frequency eLoran signal discussed so far, a "*sky-wave*" component will be radiated. This is reflected by the ionosphere and so returned to earth, especially at night. This unwanted component has the potential to interfere with the ground-wave signal being used for timing or navigation. Fortunately, the sky-wave signal invariably reaches the receiver via a longer path than the ground-wave and so arrives later. The eLoran receiver is designed to make its timing measurements on the ground-wave pulses before any sky-wave interference arrives. This has the major benefit of rendering the eLoran signals employed for timing independent of the ionosphere. So, they are insensitive to the variations of strength and delay of the sky-wave signals which, like GNSS signals, are affected by diurnal variations of the ionosphere and the influence on it of space weather events (see Section 2.3.2).

Figure 1 maps the signal-to-noise (SNR) ratio values of the signals received from the Anthorn transmitter. There is good coverage of the whole of the UK and Ireland. The signal is also usable for timing purposes in Northern France, Belgium, Holland, Denmark and part of Germany. Similar maps are available for other European transmitter stations.

2.2.3 Accuracy, availability, integrity, continuity

eLoran's enhanced accuracy, availability, integrity and continuity meet the requirements for aviation non-precision instrument approaches, maritime harbour entrance and approach

manoeuvres, land-mobile vehicle navigation, and location-based services. It also allows absolute UTC time to be recovered with an accuracy of 50 nanoseconds as well as meeting the Stratum 1 frequency standard needed by telecommunications users [4].

Table 1. eLoran's accuracy, availability, integrity and continuity [4]

Accuracy	Availability	Integrity	Continuity
0.004 – 0.01 nautical mile (8 – 20 meters)	0.999 – 0.9999	0.999999 (1 x 10⁻⁷)	0.999 – 0.9999 over 150 seconds

Notes:
1. Accuracy to meet maritime harbour entrance and approach
2. Availability, integrity and continuity to meet aviation non-precision approach in the U.S.

2.3 Factors which could affect timing accuracy and stability

A number of factors that can affect the accuracy and stability of the eLoran timing signal are identified in this section. In each case ways of mitigating or minimising the resulting errors are discussed

2.3.1 Additional Secondary Factor (ASF)

The ASF at the receiving site (as explained in Section 4.1.2 above) can be accurately measured once and for all when an eLoran timing receiver is installed and commissioned. One technique is to employ a portable source of UTC time such as a Chronos TimePort™ unit [9]. Alternatively, the approximate UTC value from the eLoran receiver can be measured after the first LDC UTC Sync message has been received. It is then compared with the value from another traceable source of UTC, which can be GNSS if that is healthy. A third possibility is to measure the local ASF value using an eLoran Differential Timing Receiver (EDTR) (see Section 3.1 below). Whichever method is employed, the eLoran timing unit being installed is then adjusted to take the measured ASF into account, so synchronising it once-and-for all to UTC. Using these techniques, synchronisation can be achieved to within a few tens of nanoseconds.

The Chronos CTL8200 is a combination eLoran and GPS Timing Receiver which enables the user to undertake continuous simultaneous real-time evaluation and analysis of UTC as derived from both the GPS constellation of satellites and selected eLoran stations. On commissioning the unit, the GPS reading is used to establish the eLoran ASF and so synchronise the eLoran receiver. Thereafter, both sources of UTC will normally be available and when one of them is lost, the other will provide a back-up.

The CTL8200 unit employs an H field receiving antenna; that is, one in which the magnetic component of the eLoran signal is picked up by a small multi-turn loop. The magnetic field of a low-frequency radio transmission penetrates deep into buildings and below the surface of the Earth, with much lower rates of attenuation than the electric field; hence, the superiority of an H field antenna over the more conventional E field antenna for indoor use. In the SENTINEL research programme, multiple such units with data links to a control and monitoring centre [9], not only made accurate eLoran timing measurements but also provided proof-of-concept evidence of the validity of this approach (see Section 3).

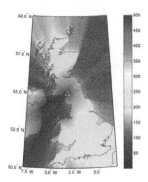

Figure 2. Estimated ASF temporal variations of the Anthorn signal (in nanoseconds) (Image: Chris Hargreaves (GLA) [1])

Temporal variations of ASF: The ASF calibration value determined when an eLoran timing receiver is first commissioned is an instantaneous value. The ASF will thereafter be subject to small variations with the season or due to changing atmospheric conditions, especially the passage of weather fronts. The greatest such temporal variations are found in those geographical regions in which the ground freezes and thaws annually; here these very slow ASF changes may well need to be considered. Elsewhere, for many timing applications, they can be ignored. Further research is needed to determine the magnitude of these temporal ASF variations.

ASF Calibration over the LDC: An elegant way to establish initial ASF calibration values and to accommodate subsequent temporal variations of ASF is to apply corrections made by an eLoran Differential Timing Receiver (EDTR) in the vicinity. Corrections for the variations measured by the EDTR are conveyed to the receiver via the Loran Data Channel. The optimal spacing of adjacent EDTRs will depend in part on the timing accuracy required, and also on the distance, proportion of land or elevated terrain, and nature of the path from the transmitter. Paths lying wholly, or substantially, over sea-water require no such adjustments. The optimal deployment of EDTRs will be a subject of the next phase of the research.

Figure 2 shows the approximate magnitude of the temporal ASF variations of the signal transmitted from Anthorn throughout its service area. Similar maps are available for other European transmitter stations.

It is envisaged that a network of regional EDTRs would supply continuous, fully-automatic adjustments for temporal ASF variations. Using these adjustments, it may well be possible for eLoran timing receivers to maintain their accuracy to within 100 ns of UTC at all times, in all weather conditions and at all seasons, and whether they operate indoors or out.

2.3.2 *Space weather*

GNSS signals are affected by extreme space weather in ways identified in the Royal Academy of Engineering publication "*Extreme space weather: impacts on engineered systems and infrastructure*" [10]. In common with all radio transmissions, eLoran signals can be affected by space weather. However, the space weather events which influence eLoran's low-frequency signals are generally not the same ones as affect the microwave signals of GNSS. In addition, it is the amplitude of the eLoran pulses, rather than their timing which are affected. As explained in Section 2.2.2 above, the eLoran receiver ignores the component of the signal received via the ionosphere, and which is vulnerable to space weather effects, and instead employs the less vulnerable ground wave component.

2.3.3 *Local electrical interference*

Broad-band electrical interference at frequencies at and around 100 kHz can be generated by devices such as switched-mode power supplies and low-voltage lighting systems. It can reduce the SNR of received eLoran signals, especially indoors. Such sources of interference are usually very localised, but may be intense. They should be identified, quantified, and minimised when a timing receiver is installed.

In many cases their effects can be greatly reduced by a combination of filtering and choice of antenna position and orientation. It may be appropriate to fit a filter to an electrically-noisy piece of machinery. Alternatively, the eLoran receiving antenna should be installed some distance from the noise source. Fortunately, an H field antenna has a horizontal polar diagram in the form of a "figure-of-eight", with two nulls, allowing it to be oriented so as to minimise interference pickup. A timing receiver displays to the installer the SNR it is experiencing, so allowing the location and orientation of its antenna to be optimised.

If the interference is of narrow bandwidth, for example a carrier-wave interferer (CWI) or a communications signal, the receiver itself will minimise its effect by deploying a notch filter at the frequency of the interferer. Notch filters are tuned and adjusted in this way automatically by the receiver, which can deploy multiple notches within the 90 ÷ 110 kHz eLoran band and the spectrum adjacent to it.

2.3.4 *Transmitter and antenna maintenance*

Experience of using the Anthorn transmissions for timing has disclosed two ways in which maintenance activities at the transmitter station impact the operation of eLoran receivers. First, during routine maintenance shut-downs the signal is lost. However, eLoran receivers pick up signals from multiple stations (often as many as 10) since at least three signals of good quality are required for eLoran navigation. While many of these stations are too distant to serve as timing sources, it is typically the case that at least one other station (and often many more) is providing a back-up timing signal; the maintenance of eLoran stations is scheduled to maximise the availability of such back-ups. Timing receivers, such as the eLoran unit in the Chronos CTL8200, switch automatically and seamlessly (that is, in a phase-coherent way) from the transmitter whose signal has been lost to one of the back-ups, reverting to the preferred station when its signal returns. This capability has been demonstrated during the SENTINEL programme [4].

The other vulnerability to transmitter station maintenance is an unwanted change in the phase of the transmission, sometimes of hundreds of nanoseconds, when certain antenna maintenance operations are performed. It is suggested that during such operations a *"Do Not Use"* message be sent via the LDC. This would cause the eLoran receiver to switch to a backup station for the duration of the maintenance operation.

3 MEASUREMENTS OF TIME ACCURACY FROM E-LORAN

A key metric for assessing the quality of a timing receiver, defined in ITU standards for telecoms synchronisation, is *"Maximum Time Interval Error (MTIE)"*. MTIE is derived by sliding windows of different observation intervals through a dataset of time interval error (TIE) values (Figure 3). As in Figure 4 shows, the resulting MTIE data points are plotted on a log-log graph.

In Figure 4 the upper curve with blue points is the MTIE mask, below which the plotted lines must fall if the receiver is to meet the ITU G.8272 standard for a primary reference timing clock (PRTC) [11]. The results show clearly that not only the GPS receiver but also the eLoran receiver both meet this specification. Indoor timing tests were undertaken using an H-field antenna with a Cs reference with daily drift of < 10 ns.

Extensive testing has been undertaken in a lab environment which shows that eLoran signals can deliver UTC traceable timing from transmitter stations that are relatively distant. Whilst it is always preferable to use the strongest signal, there may be times when the nearest transmitter will not be available.

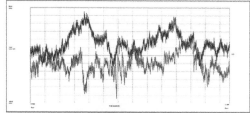

Figure 3. Time Interval Error (TIE) diagram. Red: eLoran TIE. Blue: GPS TIE. Y-Axis 10ns/div. X axis: 3 days [1]

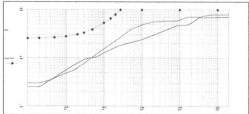

Figure 4. Maximum Time Interval Error (MTIE) diagram of Figure 3. Data collected with a CTL8200 timing receiver over 3 days. Upper curve with blue points is the G.8272 MTIE mask below which plotted lines must fall. Red: eLoran MTIE. Blue: GPS MTIE [1]

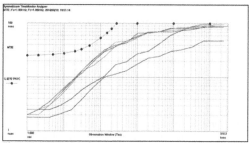

Figure 5. MTIE plots from indoor eLoran timing receiver [1]

Figure 5 shows MTIE plots for an eLoran timing receiver when using signals from stations at various ranges. The receiver was a Chronos CTL8200 operated at Chronos Technology in Gloucestershire, England. It used an H Field antenna at an indoor location unsuitable for GPS timing reception. The 5 stations monitored, in order of proximity, are presented in Table 2.

Table 2. Monitored eLoran Stations [1]

Colour	Station	Range	Location	Path
Red	Lessay	300 km	Northern France	Land and Sea
Blue	Anthorn	350 km	North-West England	Land
Cyan	Sylt	800 km	North Germany	Land and Sea
Green	Soustons	900 km	Southern France	Land and Sea
Magenta	Vaerlandet	1150 km	South-West Norway	Land and Sea

The locations of these stations relative to the UK are shown in Figure 6 below as the red markers.

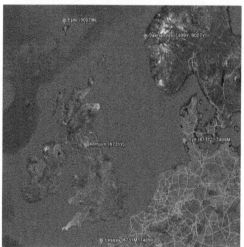

Figure 6. The locations of eLoran stations relative to the UK

The SNR values of the signals from Anthorn and Lessay were excellent; those from the other stations less than ideal. Lower SNR values like these make it harder for the receiver to acquire the signal and complete the initial synchronisation process. But that done, the MTIE plots show that all these stations, at ranges from 300 km to 1150 km, were able to supply timing readings. The errors, measured over periods ranging from 1 s to 3.5 days, were less than 100 ns. Telecom timing applications at the edges of networks and indoors currently require synchronisation with respect to UTC and stability of the order of 1 μs; these measurements show that eLoran can provide viable, reliable and resilient timing from multiple stations for such services.

3.1 CTL8200 test receiver used in trials

The Chronos CTL8200 receiver used in trials, contains an UrsaNav UN-151 eLoran timing receiver module and a GPS timing receiver. It was developed as part of the SENTINEL programme, supported by the Innovate UK, the UK's Innovation Agency. It is available for organisations that wish to evaluate eLoran timing signals aligned to UTC and can be remotely managed and configured using the SENTINEL research platform [5].

The receiver retains the valuable features of the SENTINEL GPS Jamming and interference detection Sensor including long term event analysis. Specifically, it can be networked via an Ethernet connection allowing it to be managed and monitored remotely using the SENTINEL management platform. The key feature additional to requirements of SENTINEL is its built-in ability to implement automatically the process described in Section 2.3.1 above: that is, it will measure the UTC offset of eLoran UTC due to ASF delay and synchronise the eLoran time to GPSUTC, so removing the effects of the ASF delay.

CTL8200 is also an effective EDTR, as described in Section 2.3.1 above. Operating as an eLoran Differential Timing Receiver it will continuously measure the regional eLoran UTC offset, and supply corrections to be broadcast over the LDC. These can be received by remote eLoran timing receivers that lack built-in GNSS receivers.

The CTL8200 has two 1pps outputs, one derived from eLoran and the other from GPS. These can be connected to an oscilloscope or to TIE measuring equipment for long-term analysis, either locally or remotely. This makes the unit a valuable tool for research into long-term seasonal ASF variations. Deploying multiple units will allow the spatial distribution of regional ASFs to be mapped. Results of research into seasonal and regional ASF and UTC offset variations measured in these ways will be published in due course.

The SENTINEL platform continuously monitors the relative MTIE between eLoran and GPS, thus providing valuable long term analysis of the accuracy and stability of both. If a user settable MTIE threshold is broken, alarm events are registered and email sent to designated network observers and researchers.

3.2 Conclusions of measurements of time accuracy using eLoran

The principal conclusion from these tests, and from monitoring eLoran timing signals during the GAARDIAN and SENTINEL programmes is that eLoran can provide a timing source, which we will call eLoranUTC, that is aligned to GPSUTC (and hence USNOUTC), which is available independently of GPS.

eLoran appears to be a perfectly acceptable source of precise timing for telecommunications use. It thus forms a viable means of mitigating the loss of GPS, and other GNSS, since it works in GNSS denied environments. These include in particular indoor operation and also denial of GNSS due to interference, intentional jamming or solar events.

4 NATIONAL DEPLOYMENT

Some nations are now actively considering deploying eLoran systems either by installing new transmitting stations or by upgrading legacy Loran-C systems. There is a widespread view that eLoran is the most effective complement to GNSS as a source of PNT. Among the most visionary countries is South Korea whose plans and rationale were described in a paper delivered at the 2013 European Navigation Conference by Prof. Jiwon Seo et al, titled: *"eLoran in Korea – Current Status and Future Plans"* [12].

The paper opens with this statement *"After the annual GPS jamming attacks from North Korea started from August 2010, the South Korean government realized the importance of a complementary navigation and timing system. Among various options, a high power terrestrial radio navigation system, eLoran, was considered as the most effective candidate"* [12].

This section describes how a wide-area national UTC timescale system using eLoran signals can be deployed.

4.1 *Transmissions*

Assuming a nation is a member of the TAI/UTC community linked to the BIPM, its local UTC will be derived from, and so traceable to, a national time laboratory. In Figure 7 this source of National UTC is in the lower left-hand corner. Local UTC will be transported to a master eLoran station (top left) via a resilient link and the eLoran transmissions synchronised to this source. The link could well be provided by a national telecoms carrier: such carriers are likely to have a vested interest in the provision of resilient and accurate time and timing both for their wholesale customers and for their own applications. This would be a mutually-beneficial relationship.

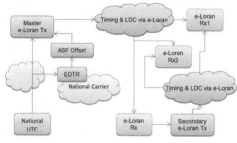

Figure 7. National UTC timescale network [1]

Local eLoran timing receivers *"eLoran Rx1 & Rx2"*, when first switched on will be synchronous with the broadcasted eLoran timing signal, but not yet synchronous. Once it has received the first UTC Sync message its time output will be close to the national UTC timescale, but will still be in error by the amount of the local ASF. To fully synchronise its 1pps to local UTC requires knowledge of that ASF.

The *"Secondary eLoran Tx"* would be set as a backup for *"eLoran Rx1 & Rx2"* and would have its own eLoran receivers monitoring other nearby transmitters and of course multiple backups can be set in this way.

As explained in Section 2.3.1 above, there are several ways of obtaining the ASF. The most attractive, in that it requires few skills and keeps the solution entirely *"sky-free"*, is to derive the ASF from a relatively local EDTR, received via the LDC.

Local EDTR units could be collocated with major synchronisation sites of the national telecoms carrier. A carrier which is already transporting national UTC will employ it as the reference time for its EDTRs, giving a continuous source of local ASF values to send over the LDC. In this way, not only will the local timescale be directly traceable to the national UTC timescale, but eLoran users who require the highest timing accuracy will remain very accurately locked to it through changing seasons and weather. Those who require lower accuracy will simply ignore the EDTR corrections.

4.2 *Station master clock*

The highly reliable master clocks of eLoran stations are usually an ensemble of triplicated Caesium atomic frequency standards with multiple redundancy throughout their architecture. Their 1 pps output will be synchronised to the national UTC timescale. This arrangement will ensure that the eLoran stations have an inherently resilient architecture. Operational characteristics of station master clocks will be the subject of future publications.

5 BENEFITS OF E-LORAN OVER OTHER PNT SYSTEMS

Let us examine the benefits of eLoran in comparison with alternative PNT systems when used for timing.

5.1 *Transmissions internationally standardised*

The eLoran system is well documented and standardised internationally. Essential characteristics are defined in the following documents:

- International Loran Association, *"Enhanced Loran (eLoran) Definition Document"*, Version 1.0, October 16, 2007 [4];
- *"Recommendation ITU-R M.589-3, Technical Characteristics of Methods of Data Transmission and Interference Protection for Radionavigation Services in the Frequency Bands between 70 and 130 kHz"*, International Telecommunication Union Recommendations [8];
- RTCM, *Minimum Performance Standards for Marine eLoran Receiving Equipment*, RTCM Special Committee SC127, Revision 2.0, March 2010 [13].

5.2 *Broadcasts national UTC over a wide area*

eLoran ground wave signals can be received indoors or outdoors over ranges of hundreds of kilometres from the transmitters. Timing receivers can maintain synchronisation to within 100 ns of UTC to the edge of coverage by receiving timing corrections broadcast over the LDC.

Transmissions from multiple eLoran stations can generally be received since these are required for the navigation use of eLoran. For timing, these multiple sources ensure that alternative signals are available even during transmitter maintenance periods, so maximising the availability of the service.

5.3 *eLoran can be received indoors*

H-Field antenna technology allows indoor reception of eLoran timing signals in areas that GPS and other GNSS signals cannot reach. This has significant benefits, chiefly resulting from the lack of a requirement for a roof antenna. It avoids the requirement for: long cable runs; specialised installation staff, materials, and processes; permission from local facilities management staff; and the risk of mutual interference between adjacent GNSS antenna installations.

The requirements and ease of deployment of H-Field antennas will be investigated further in both typical and exceptional indoor locations through collaboration with potential user community.

5.4 *Provides timing synchronous to UTC within 100 ns*

Telecom requirements: The definitive ITU document for precise time synchronisation in telecom transport networks is now "Recommendation ITU-T G.8272/ Y.1367" [11]. This specifies the requirements for primary reference time clocks (PRTCs) suitable for time and phase synchronization in packet networks under environmental conditions normal for the equipment. A typical PRTC provides the reference signal for time synchronization and/or phase synchronization for other clocks within a network or

a section of a network. In particular, a PRTC can provide the reference signal for a "Telecom Grand Master" within the network nodes where the PRTC is located. The PRTC provides a reference time signal traceable to a recognized time standard, usually UTC. The Recommendation defines the output requirements and accuracy that a PRTC must maintain.

Indoor trials with eLoran differential timing receivers have shown that they can provide precise UTC corrections in conformance to G.8272 Figure 1 (see Section 3 above). Further research needs to be undertaken to quantify seasonal ASF variations and the resulting requirement for differential timing corrections.

Financial Services requirements: After the so-called *"Flash Crash"* of 2010, when \$1T temporarily evaporated from US markets, a UK Government report *"The Future of Computer Trading in Financial Markets"* [14] recommended that all computer-based trading be synchronised to within 1 µs of UTC.

Indoor eLoran timing systems can easily meet this requirement, even without differential timing corrections.in most locations.

5.5 *Resilient against GNSS jamming and spoofing*

The high susceptibility of GNSS signals to low-power jamming is described in the SENTINEL Report and the Royal Academy of Engineering report "*Global Navigation Space Systems: reliance and vulnerabilities*" [15]. In contrast, eLoran timing signals are transmitted in a very different part of the spectrum. They are thus immune to the effects of GNSS jammers.

Like all radio signals, eLoran transmissions can be jammed. However, the power level of the signals reaching receivers (which the jammer must overcome) is many orders of magnitude greater than that of GNSS signals. Further, to transmit jamming signals at 100 kHz over all but very short ranges requires large transmitting antennas, substantial transmitter power and dangerously high voltages.

For the same reasons, eLoran is much more resilient than GNSS to spoofing attacks, of the kinds that have been studied and demonstrated recently.

5.6 *Resilient against space weather events*

The susceptibility of GNSS signals to space weather events is well described in the Royal Academy of Engineering report "*Global Navigation Space Systems: reliance and vulnerabilities*". E-Loran timing signals are transmitted in a very different part of the spectrum and employ the wholly-different, ground-wave, mode of propagation. As explained in Section 2.1.2 above, this renders them much less susceptible to space weather effects and they are not susceptible to many of the events that threaten GNSS.

5.7 *Complementary to PTP*

Precision Time Protocol (PTP) has been developed from the original IEEE 1588 specifications. ITU is developing the G.827x series of Standards. PTP is vulnerable to variable delays in Internet Protocol transmission systems. These are of greatest concern in third-party systems employing transport networks over which the carrier has no control. To transfer phase with the required accuracy, PTP employs "*on-path-support*" using embedded "*client*" clocks throughout the network.

In contrast, eLoran with its LDC is a simple and elegant stand-alone wireless technology that delivers precise UTC-traceable time indoors. If required, eLoran can work alongside PTP to mutual benefit. Trials currently underway at Chronos Technology are exploring this combination of technologies.

6 CONCLUSIONS

There are a lot of time definitions. Time is the indefinite continued progress of existence and events that occur in apparently irreversible succession from the past through the present to the future. Time is a component quantity of various measurements used to sequence events, to compare the duration of events or the intervals between them, and to quantify rates of change of quantities in material reality or in the conscious experience. Time is often referred to as the fourth dimension, along with the three spatial dimensions [3],[16],[17].

High Frequency Trading using Computer Based Trading equipment now requires UTC traceable synchronised time stamps with an accuracy of better than 1µs. GPS based Network Time Protocol (NTP) systems give only millisecond, not sub-microsecond, accuracy and are vulnerable to GPS jamming. With PTP, the delivery process is cumbersome and may require fixed delays to be calibrated out on installation. In this context, eLoran timing is an ideal solution which would work well indoors.

This paper has proposed the use of eLoran to disseminate precise time, timing and phase traceable to UTC for both indoor and outdoor applications. Continuous accuracies of better than 100 ns with respect to UTC are being achieved in current proof-of-concept and technology-readiness trials.

The paper proposes and illustrates a method of establishing national time standard services using eLoran. These would be traceable to sovereign national UTC. They would be of great benefit to a wide range of users, notably telecommunications providers and financial sector organisations for whom precise time synchronisation will be required

by future services. These organisations are also becoming concerned about their dependence on GNSS timing, given its vulnerability to jamming and interference and the complexity and expense of deploying it, especially when required indoors.

Further research, already under way, is evaluating the accuracy and optimising the delivery via the Loran Data Channel of UTC time corrections from remote differential timing receivers.

Chronos Technology is working with partners, and actively seeking additional collaborators, in both the supply and user timing community as well as Academia and Government as it widens the scope of this research [1],[18]. This study draws on research carried out over the last 8 years in the course of two projects, GAARDIAN and SENTINEL, which were supported by the Innovate UK, the UK's Innovation Agency. It demonstrates a method of employing eLoran signals to distribute a "*National Timescale*". This would be a simple and reliable way of distributing UTC traceable time for multiple applications, especially those indoors and in other GNSS denied environments that require resilient and accurate time of day, phase-synchronized and time-stabilised to UTC. The study shows how this accuracy and stability can be maintained over the long term to within 100 ns of UTC, thus meeting currently-accepted ITU standards for primary reference timing clocks in telecoms transport networks.

This paper explores the ability of eLoran to distribute UTC traceable time to applications in GNSS-denied environments, including indoors. It sets the foundation for further research into the application and dissemination of UTC using eLoran signals in geographical regions where they are available. Research into this topic has been conducted by Chronos Technology in collaboration with the General Lighthouse Authorities of the United Kingdom and Ireland (GLA), the UK National Physical Laboratory and UrsaNav, Inc [1]. This has shown that UTC-traceable time of an accuracy better than 100 ns and with a quality comparable to that provided by GPS can be received even indoors at ranges of more than 800 km (500 miles) from eLoran transmitting stations. This new time service meets the latest ITU performance standards in respect of telecommunications phase stability.

The paper proposes further research to assess spatial and temporal variations in the reception of UTC traceable time distributed in this way. It proposes this new means of disseminating national sovereign UTC for use at times and in places where GNSS is denied. It will serve critical infrastructure applications, notably telecommunications networks and financial services, in which sub-microsecond UTC-traceable time is essential to the continuity of operations. In particular it will serve these applications without the need for expensive roof mounted GNSS antenna deployments or managing complex fibre connectivity.

REFERENCES

[1] Curry C.: Delivering a National Timescale Using eLoran. Chronos Technology Ltd., Issue 1.0, 07 June 2014

[2] BIPM Annual Report on Time Activities, Volume 7, 2012

[3] Weintrit A.: The Time in the Navigation. Measures of Time: GMT - UTC - TAI - GPST (in Polish). Przegląd Telekomunikacyjny, No. 7, 2011

[4] ILA. Enhanced Loran (eLoran) Definition Document. Report Version: 1.0, International Loran Association, 16th October 2007.

[5] The SENTINEL Report, Chronos Technology Ltd, 2014

[6] Offermans G.: Eurofix Message Format, Ver.2.15, March 2014

[7] RTCM Recommended Standards for Differential GNSS (Global Navigation Satellite Systems) Service, Version 2.2, RTCM Special Committee 104, January 15, 1998

[8] Recommendation ITU-R M.589-3, Technical Characteristics of Methods of Data Transmission and Interference Protection for Radionavigation Services in the Frequency Bands between 70 and 130 kHz

[9] http://www.chronos.co.uk/index.php/en/product-groups/time-and-timing/

[10] Extreme space weather: impacts on engineered systems and infrastructure, Royal Academy of Engineering, February 2013

[11] Recommendation ITU-T G.8272/Y.1367 - Timing characteristics of primary reference time clocks, October 2013

[12] Seo J. & Kim M.: eLoran in Korea – Current Status and Future Plans, European Navigation Conference ENC 2013, Yonsei University, Korea, 2013

[13] RTCM. Minimum Performance Standards for Marine eLoran Receiving Equipment, RTCM Special Committee SC127, Revision 2.0, March 2010

[14] The Future of Computer Trading in Financial Markets, Government Office for Science, Foresight Committee 2012

[15] Global Navigation Space Systems: reliance and vulnerabilities, Royal Academy of Engineering, March 2011

[16] Davies P.C.W.: About Time: Einstein's Unfinished Revolution. Simon & Schuster Paperback, New York 2005

[17] Ridderbos K. (ed.): Time. The Darwin College Lectures, Cambridge University Press, 2002

[18] Curry C.: Long Term Time & Timing Trials with eLoran. Chronos Technology, ITSF Edinburgh, November 2011

Route Planning

Inductive Mining in Modeling of the Ship's Route

M. Dramski
Maritime University of Szczecin, Szczecin, Poland

ABSTRACT: Every ship's route can be treated as some set of sequences consisting of some activities. This set is called a process. The process can be a subject of further research leading to it's improvement and ensure e.g. the operational support. One of the techniques used in this task is inductive mining. Inductive mining is a simply and efficient approach to build a model of the process. In this paper an example of such research is described. The data was obtained by tracking the real route – the yacht going from Szczecin (Poland) to Las Palmas de Gran Canaria (Spain). The event log (the data) is very various due to the specific type of the cruise. This fact leads to observe some interesting conclusions in modeling of the process.

1 INTRODUCTION

The increasing amount of data in all the life's domains requires developing of new analysis techniques which lead to draw some conclusions from it. In every moment of our life we generate a lot of data, very often leaving the electronic traces too. It is said that during the last few years, the total amount of data generated by humanity is significantly higher then from the ancient times to the year 2000 or even later (van der Aalst, 2011).

However, a simple collection of data is not a big deal. The data can be in numeric or text form and have different meanings and forms. Some data lacks or errors can occur very often. So the analysis and the extraction of information is not so easy as it seems. First of all there is a need to precise what kind of problem is under consideration. The other approach is used to analyze images, the other to texts etc.

In this paper the data describes a process. The process is a set of cases consisting of some activities. For instance, shopping is a process consisting of few stages such order, payment, shipment etc. The exchange of a wheel in a car is a process too. The process can be simple and short but also complicated and log. Anyway, one thing must be told. Every process can be analyzed to understand it and improve if there is only such possibility. Last years of research resulted in inventing and developing the techniques appropriate for processes analysis. These techniques are described as a process mining. There is a need also to understand that the term of process mining is not equivalent with the data mining. The second term means that the main subject of research is the essential form of data such numbers, texts etc. and the aim is to extract the information. Data mining can be considered as a part of process mining. The most common use of it is the preprocessing.

In this paper the ship's route is shown as a process. This process consists of some activities ordered in cases. These activities are: in range, changed course, arrival, departure, stopped, underway, midday position and midnight position. All the data was recorder during the yacht's route from Szczecin to Las Palmas de Gran Canaria in the autumn of the year 2016.

2 THE DATA

Although the data can be given in various different forms, it is necessary to convert it into some standard formats. This stage is the first step of process mining and it means that the data has to be saved in CSV or XES format which both are standards in all the tools supporting this kind of data analysis. The data recorded during the process run is given as an event log. Because the process takes place in time, the most important field in the event log are timestamps. If there are no timestamps, it can't be said that the data has the form of an event log. The event log consists of traces and they consist of eyents. A trace is a finite sequence of events $\sigma \epsilon \varepsilon$ such that each event appears only once. The event log is a set of cases $L \subseteq \mathbb{C}$ such that each event appears at most once in the entire log. The typical event log contains: case id, event names, timestamps, resources, etc.

	Case ID	Event ID	dd-MM-yyyy:HH.mm	Activity	Resource	Costs
1	1	35654423	30-12-2010:11.02	register request	Pete	50
2	1	35654424	31-12-2010:10.06	examine thoroughly	Sue	400
3	1	35654425	05-01-2011:15.12	check ticket	Mike	100
4	1	35654426	06-01-2011:11.18	decide	Sara	200
5	1	35654427	07-01-2011:14.24	reject request	Pete	200
6	2	35654483	30-12-2010:11.32	register request	Mike	50
7	2	35654485	30-12-2010:12.12	check ticket	Mike	100
8	2	35654487	30-12-2010:14.16	examine casually	Sean	400
9	2	35654488	05-01-2011:11.22	decide	Sara	200
10	2	35654489	08-01-2011:12.05	pay compensation	Ellen	200
11	3	35654521	30-12-2010:14.32	register request	Pete	50
12	3	35654522	30-12-2010:15.06	examine casually	Mike	400
13	3	35654524	30-12-2010:16.34	check ticket	Ellen	100
14	3	35654525	06-01-2011:09.18	decide	Sara	200
15	3	35654526	06-01-2011:12.18	reinitiate request	Sara	200
16	3	35654527	06-01-2011:13.06	examine thoroughly	Sean	400
17	3	35654530	08-01-2011:11.43	check ticket	Pete	100
18	3	35654531	09-01-2011:09.55	decide	Sara	200
19	3	35654533	15-01-2011:10.45	pay compensation	Ellen	200

Figure 1. An example of the event log, source: www.processmining.org

Figure 1 illustrates a typical example of the event log. It can be easily seen that there are the characteristic data columns such case id, event id etc. Also the timestamp and other attributes are given. This fragment of example data comes of course from a spreadsheet, so it can be saved in CSV format which is supported by MS Excel.

The event log presented above is shown as a table, but also can be expressed as an equation. The example equation:

$$L = [a, b, c, a, b, d, b, b, c] \tag{1}$$

The equation (1) describes the simple event log consisting of three cases where each case has three events. These events are: a, b and c. Of course the full name of the event can be shown, but the use of shortcuts makes the equation more readable.

In this paper the original data recorded during the yacht's voyage is used. The events and the corresponding shortcuts are given in Table 1.

Table 1. The set of events based on the data obtained from Marinetraffic.com, source: (Dramski, 2016)

No.	Event name	Shortcut
1	In Range	a
2	Changed course	b
3	Arrival	c
4	Departure	d
5	Stopped	e
6	Underway	f
7	Midnight position	g
8	Midday position	h

The complete event log consists of 59 cases taken during 59 days of the voyage from Szczecin to Las

Palmas. The first 30 cases in the form of equation can be given as follows:

$$
\begin{aligned}
L = [&\langle a,f,c,h,d,e,g \rangle, \langle a,f,b,b,b,b,c,d,g \rangle, \\
&\langle g,c,e,d,c,h,d,f,c,c,d,a,g \rangle, \langle c,d,a,h,d,a,g \rangle, \\
&\langle a,h \rangle, \langle g,e,f,b,c,h,g \rangle, \langle h,d,g \rangle, \langle a,h,c \rangle, \langle a,d,h \rangle, \\
&\langle g,h,c,d,e,f,b,e,g \rangle, \langle a,f,e,g \rangle, \\
&\langle a,f,e,f,e,c,f,d,h \rangle, \ \langle g,h \rangle, \langle g,h \rangle, \langle a,h,c,g \rangle, \langle h \rangle, \\
&\langle g,h \rangle, \langle g,d,h \rangle, \langle g,e,f,c,h,g \rangle, \\
&\langle h,g \rangle, \langle d,c,e,f,h \rangle, \langle g,c,h,d \rangle, \langle g,e,f,b,b,h \rangle, \\
&\langle g,e,f,e,f,b,e,f,e,c,h \rangle, \langle g,h \rangle, \\
&\langle a,f,d,b \rangle, \langle g,h \rangle, \langle g \rangle, \langle g,a \rangle]
\end{aligned}
\tag{2}
$$

Now, when the event log is given, the model of the process can be built. In (Dramski, 2016) a Petri net using the α-algorithm is described. In this paper the inductive mining method was applied.

3 INDUCTIVE MINING

Process discovery is a quite complicated problem. Different abstractions are possible and there is a need to find a balance between overfitting and underfitting. So it leads first to create a transition system.

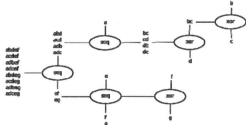

Figure 2. An example of inductive mining, source: van der Aalst, 2011

The second phase is to discover the concurrency and get the corresponding Petri net. Inductive mining is one of the alternative process discovery techniques that simply detects the main operators such: seq, par and xor.

In Figure 2 a simple example of such inductive mining is presented. There are 8 cases, each consisting of 6 events. The seq operator is easy to find because it means that there is a repeating sequence. The par operator means that two events (or more) can occur parallel and the xor operator says that only one of two given events will happen. This schema makes clear the fact that inductive mining is a kind of decomposition of the event log. Now the Petri net can be easily created.

Figure 3 presents the Petri net obtained from the example event log (Figure 2). The starting point of this net is shown by the black marker. The marker goes to transition "a" which fires and produces 2 tokens. So the event "d" occurs parallel to "c" or "b". Then the event "e" is fired and the last one is always "f" or "g". The same behavior is observable in the event log.

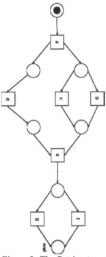

Figure 3. The Petri net created using inductive mining, source: van der Aalst, 2011

This simple process model shows the idea of inductive mining. The quality of the model strongly depends on completeness of the event log. The correctness of data recording is also very important. Let's suppose that the event "a" one time is not recorded. The model's behavior shows that this event always occurs, so the eventual lack of this event would cause huge problems and make the model completely unreliable. But if we know that the event "a" is missing we can simply add it to the model and it can be easily reconstructed.

Inductive mining is one of the tools in process discovery and is very often applied in tools supporting these techniques such ProM or Disco. The ProM tool was used in (Dramski, 2016) and showed that is very useful.

4 THE MODEL USING INDUCTIVE MINING

It is natural that the approach to process discovery depends strongly on the accessible data (van der Aalst, 2004). In this paper the use of real data is described. The short characteristics is given in the second section. There are some important facts that have to be noticed:

- The order of events is very important – it has a big influence on the operator detected;
- There are some missing values in the event log but they are not significant for the purpose of this paper;
- The real data makes the process discovery techniques more reliable;
- The aim is not only to create a process model, but also to compare this model with the reality.

To analyze the event log, the ProM 6 tool was chosen. This is one of the most popular software systems which can be used free of charge. It was developed at Technical University of Eindhoven (The Netherlands). Besides, it supports the highest number of plugins and the official data standard for process mining – the XES format.

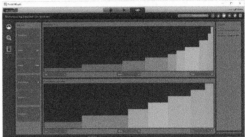

Figure 4. The analysis of the event log in ProM 6, source: own study

Figure 4 illustrates the simple analysis of the event log. ProM's interface is very easy to understand and generally there are three sections. In the first one the data can be imported to the ProM environment. Here one thing is very important – this tool supports only XES format, so if there is only CSV file available, it needs to be converted. Second section marked with the black triangle in the top of the picture says which resource will be used. In this case the event log is a resource. The third section (visible in Figure 4) marked with an eye symbol gives us the general view on data. The form of this view depends on the kind of the resource In the case of event log it simply shows all the classes of events. When other type of resource is used, the appropriate view is visible.

The statistics corresponding to our event log is given in the table below:

Table 2. The event log statistics, source: own study

Number of processes	1
Number of cases	59
Number of events	216
Number of classes of the events	8
Minimum number of events per case	1
Mean number of events per case	4
Maximum number of events per case	13
Minimum number of event classes per case	1
Mean number of event classes per case	3
Maximum number of event classes per case	7

Next the Inductive Miner plugin was chosen and applied to the event log. As a result the following Petri net was created (Figure 5).

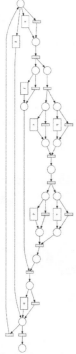

Figure 5. The Petri net using inductive mining, source: own study

Inductive mining makes possible to repeat every track from the event log. This is the natural consequence of this approach in process modeling. The grey rectangles represent silent transitions which can fire and no event occurs. It ensures that sometimes one of the events can be also omitted.

It can be seen that some events appear concurrently and some events may not happen. Let's take an event marked with the letter "a" (In range). This activity very often is observable as a first one in the given trace in the event log. But if doesn't exist – can be omitted thanks to the parallel silent transition. Similar solutions are visible in the other fragments of the model. There are also some transitions which produce two tokens. It means that every activity after this transition must occur, but their order is not set. If the transition has two inputs, it always requires to have to incoming tokens necessary to fire the transition.

Silent transitions make the model creating easier, but there is a need to think about some modifications of the event log and add other task – minimalize the total number of silent transitions. Inductive mining has a tendency to produce such transitions. It shows

that the data recording process is very important. The short analysis of the Petri net from Figure 5 is presented in the table below:

Table 3. Metrics of the Petri net, source: own study

Number of arcs	52
Number of places	18
Number of transitions	23
Number of silent transitions	15
Number of events	8

Using ProM the Petri net can be easily converted also to other models e.g. BPMN diagram like shown in Figure 6.

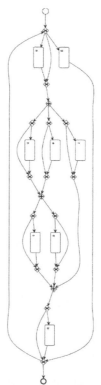

Figure 6. The BPMN model, source: own study

5 CONCLUSIONS

In this paper the use of inductive mining in creating the transport process model is described. First idea of process mining was presented. Then the definition of the event log and an example were given. The model is the Petri net obtained using a ProM tool supporting process mining techniques.

Inductive mining leads to create a very detailed model which is able to replay all the traces from the event log. Anyway, there still exists a need to do some further research on question of overfitting.

Because this methods causes that there is a high number of silent transitions, there must be some ways to reduce them and this should be the next task. In inductive mining to facts are very important:
– Find most prominent split in event log;
– Detect operators.

The model obtained is very clear and can be applied to understand the problem of the ship's route modeling. In (Dramski, 2016) the α-algorithm was described and the quality of the model was at the level of 70%. It was estimated using the footprints (van der Aalst, 2011) of the process and the model respectively.

The use of process mining techniques makes possible the operational support which can be applied in improving the whole process. If the ship's route is treated as a sequence of cases and events some observations can be made.

Anyway, there is a strong need to do further research in this domain. Petri nets, transition systems or BPMN networks and other models are excellent tools which can be applied in solving transport problems. They are very easy to understand, but also require some experience from the researchers. Very important is the question of over- and underfitting (van der Aalst, 2010). Underfitting is always easy to understand. If the model is underfitted it simply means that it's wrong. When the overfitting occurs there is a need to be very careful. Such model answers very correctly when the learn data is used. But the test data can sometimes give strange and unexpected results. In systems modeling the quality of the model is determined by two factors. First is the average error which never should have a value lower then the average error of considered system. Ending here would be a great mistake, because the second factor is even more important – the reliability. It means that the answer of the model should be similar to learn and test data.

In this paper only some activities during the ship's voyage were taken into the consideration. But there are no obstacles to consider other things such speed, directions, course etc. There is much more work to do, but it is show very clearly that process mining techniques can be satisfactorily used in modeling of transport processes.

REFERENCES

van der Aalst, W.M.P., "Process mining – discovery, conformance and enhancement of business processes", Springer-Verlag Berlin Heidelberg 2011

Dramski, M., "The alpha algorithm in the modeling of the ship's route", Transport Systems Telematics Vol. 9, Iss. 2, p.p. 8-11, 2016

van der Aalst, W.M.P. et al.,"A two-step approach to balance between underfitting and overfitting", Software and Systems Modeling, 9(1):87-111, 2010

van der Aalst, W.M.P. et al. "Discovering process models from event logs", IEEE Transactions on Knowledge and Data Engineering, 16(9):1128-1142, 2004

Route Optimization in the Restricted Area Taking into Account Ship Safety Zones

M. Wielgosz & M. Mąka
Maritime University of Szczecin, Szczecin, Poland

ABSTRACT: The article presents a method of ship's route planning and route optimization to be used for passage planning and monitoring, including plan changes resulting from unpredictable developments like newly charted dangers. A method of space discretization for seeking a safe and optimal track using the method of trapezoidal mesh is presented. The authors have implemented a model of hydrographic ship domain, crucial in the process of maintaining a safe distance from the safety contour, navigational dangers and special areas. The model takes into account basic technical parameters of the ship, such as size and speed, and is capable of incorporating other factors. The article also discusses the method used to implement algorithms. A research experiment was conducted to verify the method. The final results and an example of method application are presented along with future directions of research work.

1 INTRODUCTION

The development of e-navigation necessitates changes in and new solutions to the established shipboard practices and procedures. One such change refers to ship's voyage planning, and the ability to automatically choose tracks with user-defined basic safety parameters. These opportunities have come with the implementation and increasingly wider use of electronic chart systems, especially those based on vector format navigational charts.

Route planning principles and recommendations are precisely formulated in an IMO resolution concerning guidance for voyage planning and monitoring (IMO 1999). For the purpose of ECDIS systems they are extended in other IMO resolutions by obligatory requirements included in performance standards for ECDIS systems (IMO 1995, IMO 2006).

Automatic and optimal methods of route planning (under different, pre-defined criteria) are already known to sea navigators. They are commonly used in:
- advanced voyage planning modules of ECDIS systems;
- weather routeing systems and computer programs;
- some digital catalogues of nautical publications.

The main goal of these systems is to work out the optimal route for defined conditions and preset criteria.

A planned route (e.g. in the ECDIS system) is later automatically checked by the system (mandatory function), then corrected by those who do the planning, and finally approved by the Captain.

ECDIS carries out control within the defined width of the planned route, specified by the limits of deviation from the planned track (cross track error - XTE limits). The following items on the planned route are searched for and analysed:
- safety contour and safety depth;
- navigational dangers;
- user-selected special areas highlighted due to the occurrence of special conditions.

A properly planned, checked and approved route generally should run clear of the above mentioned hazards within the allowed trajectory. The navigator monitoring the route should take into account the above dangers if any deviation from the planned and accepted route has to be made. Special care should be taken in case of an automatic route suggestion and selection by the system.

The selected route is always a compromise between the shortest and safe trajectories. Generally, the one that is considered to be the most beneficial is chosen. Additionally, alternative routes are suggested to enable the operator to make the final decision.

In this paper we focus on the problems of route selection based on electronic charts, and the proposed method will be employed in ship navigation systems using charts encoded in the vector format.

The paper presents a novel method of route planning and selection based on the analysis of the sea area with the use of the trapezoidal mesh. As restrictions, i.e. safety parameters, these authors adopted ship safety zones required by navigators – the ship domain and above-mentioned dangers and navigational limits.

The implemented algorithms take into account standard data decoded and interpreted from vector format navigational charts. These data, a standard content in electronic navigational charts, can be enriched by additional layers of information, such as navigational warnings and hydro-meteorological information (wind direction and force, currents, storms, ice situation).

2 ROUTEING ALGORITHM

2.1 Area discretization using a trapezoidal mesh

To create a digital representation of the area we have applied a recursive discretization algorithm using a trapezoidal mesh (Mąka 2014). It allows the user to organize analysed areas and accelerate the process of checking the availability of each graph node in the algorithm determining a safe route of the ship. Such representation is particularly useful when the chart of the area is presented in the vector format. All mesh elements are trapezoidal in shape, or in special cases, when the length of one of the vertical sides is zero, they take the shape of a triangle - a degenerated trapezoid (Mąka 2014, van Kreveld 2007).

Creating the mesh starts with processing of the selected rectangular area and a single object located within. Mesh elements are formed by removing two vertical sections from each node, which is the beginning/end of the vector defining the outline of the object. Then other objects are added, and the generated mesh is locally modified to include the added objects.

The proposed solution enables local modification of the mesh depending on the position of moving objects. This is an important feature, which should also characterize algorithms developed to create graphs. This enables the execution of the whole process of ship's route choice in dynamic situations within an expected time regime.

The running mesh-creating algorithm produces the following solutions (Dramski & Mąka 2013):
1 graphical representation of the mesh (Figure 1);

2 sorted array containing the data of all mesh elements (trapezoids) including:
   – node indexes – those from which vertical sides are removed (coordinates of vertical sides);
   – line indexes limiting the trapezoid from the top and bottom;
   – coordinates of the vertices;
   – indexes of adjacent trapezoids (contacting a given trapezoid from left or right side and the vertical side created after removing the sections from mesh node);
   – trapezoid classification: prohibited or allowed;
3 adjacency matrix illustrating a possibility of transition between subsequent trapezoids and matrix graph representation.

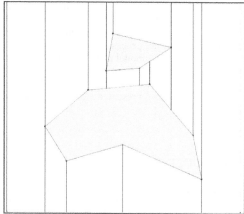

Figure 1. Graphical representation of the mesh

These data form a basis of the procedures for the deployment of nodes that further make up waypoints of the route being created and determining the available connections between them (the edge of the graph), and for the route searching algorithm.

2.2 Construction of the graph

The selection of an appropriate method of transition graph construction is one of the most important elements in the process of selecting a route in the restricted water area. One component of this process is the determination of available transitions – the edges of the graph connecting each of its vertices (nodes). A properly selected method of graph construction is of particular importance for the analysis of dynamic situations, where we have to take into account the changing positions of the own ship and target ships. Such situational changes necessitate the modification of the mesh, position of nodes and the transition graph (Figure 2). As the procedure for determining the transitions between the nodes of the graph takes more than 80% of the time of the entire area discretization and routeing algorithm operation, its optimization to shorten the computing time is an extremely important issue. It can greatly enhance the performance and shorten the duration of the whole process (Pietrzykowski & Mąka & Magaj 2014, Mąka & Dramski 2015).

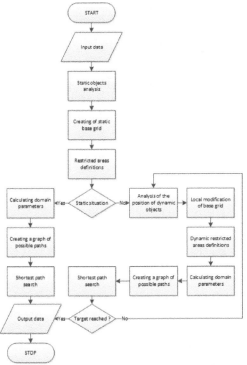

Figure 2. An algorithm of mesh generation and route selection

In the process of node deployment and specification of available connections between the nodes it was found necessary to take into account at least:

– international collision regulations (COLREGS);
– ship manoeuvrability and dynamics;
– Closest Point of Approach (CPA).

In the proposed method of the deployment of nodes defining waypoints of the route being determined (graph vertices), in a single element of the trapezoidal mesh there are up to 25 nodes generated: five nodes inside and maximum 21 at the edges. The nodes are deployed on the vertical sides of the trapezoid and on the vertical line halfway between them. There are up to five nodes formed on each section, depending on the length l (Figure 3):

– for $l > l_{gr1}$ – one node in the middle of the section;
– for $l_{gr1} \geq l > l_{gr2}$ – three nodes at distances, respectively: $\Delta\varphi = 0.1 \cdot l$, $\Delta\varphi = 0.5 \cdot l$, $\Delta\varphi = 0.9 \cdot l$ from the section ends;

– for $l \leq l_{gr2}$ – five nodes at distances, respectively: $\Delta\varphi = 0.1 \cdot l$, $\Delta\varphi = 0.3 \cdot l$, $\Delta\varphi = 0.5 \cdot l$, $\Delta\varphi = 0.7 \cdot l$, $\Delta\varphi = 0.9 \cdot l$ from the section ends.

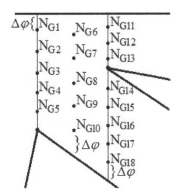

Figure 3. Distribution of the nodes in a single element of the trapezoidal mesh

Additionally, we have defined limitations resulting from the application of own ship domain (both hydrographic and anti-collision) in the process of route planning:

1 a prerequisite is that the minimum height of the vertical side of the trapezoid through which a route can run is greater than the length of the minor axis of the ellipse defining the width of the own ship domain;

2 the minimum distance $\Delta\varphi_{min}$ between the graph node and the lower and upper edges of the trapezoid has to be greater than the length of a semi-minor axis of the ellipse defining the width of the own ship domain.

Available edges of the graph were defined on the basis of the graph vertices created by the presented method. It was assumed that transitions are possible between all nodes (vertices) that can be connected by a straight line bypassing prohibited areas - lands, areas closed to navigation, located within the safety contour.

Similarly to graph vertex generation, the minimum distance $\Delta\varphi_{min}$ between the graph edge and the lower or upper edges of the trapezoid must be greater than the length of the semi-minor axis of the ellipse defining the width of the own ship domain.

Then, using Dijkstra's algorithm, the shortest track is determined between predefined points .

Figures 4 and 5 depict the results of the algorithm for node deployment and connecting edges definition. An example path between two vertices is also shown. The nodes are marked as points of the graph, while a selected path is plotted as a thick solid line in Figure 5. For better readability, the number of nodes on each side is limited to three, and only connections available inside each of the trapezoids are displayed. Connections between nodes placed in different trapezoids are not displayed.

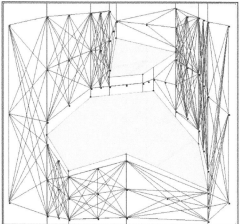

Figure 4. An example graph and available connections between nodes.

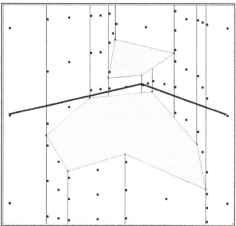

Figure 5. A selected route and graph nodes marked as black dots.

3 SHIP DOMAIN

When planning the passage of vessel through the selected area, one should consider a limit of deviation from the planned route, which will offer the vessel's navigating officers some freedom of manoeuvring by course alteration when other vessels are encountered. The planner should also foresee the need to alter course to keep clear of new navigational hazards or precautionary areas / special conditions (e.g. area temporarily restricted due to naval operations or exercises). Although the route planned in the ECDIS system takes into account the above factors, the watchkeeping navigator has to

continuously monitor navigational warnings that might introduce new objects considered dangerous to navigation during the current voyage.

3.1 *A mathematical model of the ship domain*

The problem of the ship safety zone taking account of other vessels - the ship domain - is now widely described in the literature (Goodwin 1975, Fuji 1971, Hansen et.al. 2013, Pietrzykowski & Uriasz 2009, Śmierzchalski & Weintrit 19999, Wielgosz 2015, Wielgosz 2016)

In this paper we implement the model of anti-collision domain developed by Wielgosz (Wielgosz 2015). The adopted model is an ellipse with an offset centre, described by the parametric equation (1,2,3,4,5,6) including the size and speed of the vessel.

$$x(t) = x_0 + a \cdot \cos(t) \tag{1}$$

$$y(t) = y_0 + b \cdot \sin(t) \tag{2}$$

$$a = \left(a_{1L} \cdot L^{b_{1L}} + c_{1L}\right) + a_{1v}\left(v^{b_{1v}} - 2^{b_{1v}}\right) \tag{3}$$

$$b = \left(a_{2L} \cdot L^{b_{2L}} + c_{2L}\right) + a_{2v}\left(v^{b_{2v}} - 2^{b_{2v}}\right) \tag{4}$$

$$x_0 = p_x \cdot L + q_x \cdot v + r_x \tag{5}$$

$$y_0 = p_y \cdot L + q_y \cdot v + r_y \tag{6}$$

where:
L – ship's length [m];
v – ship's speed;
a_{1L}, b_{1L}, c_{1L}, a_{2L}, b_{2L}, c_{2L} – length influence coefficients;
a_{1v}, b_{1v}, c_{1v}, a_{2v}, b_{2v} – speed influence coefficients;
t – relative bearing;
p_x, q_x, r_x – X-axis centre displacement coefficient;
p_y, q_y, r_y – Y-axis centre displacement coefficient.

3.2 *Hydrographic domain*

The concept of hydrographic domain is relatively new in the literature and is associated with the ability to interpret the navigational chart contents by an electronic chart system, if the chart is encoded in the vector format. A simplified form of hydrographic domain, e.g., safety frame, is already implemented in some ECDIS systems (Figure 6).

Although the safety frame shown in Figure 6 has a declarative character, the current research (Wielgosz 2016) shows that the actual shape and size of the effective hydrographic domain are similar to the corresponding anti-collision ellipse, with more significant displacement of its centre.

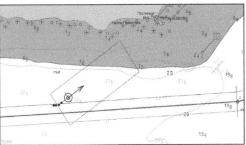

Figure 6. Safety frame in the ECDIS system (Transas ECDIS NS 4000).

A comparison of the said domains for ships of the same size and speed is presented in Figure 7.

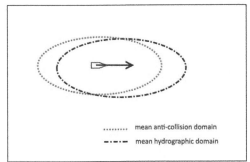

Figure 7. Examples of ship anti-collision domain and hydrographic domain

4 IMPLEMENTATION OF THE ALGORITHMS

The proposed algorithms of graph construction were implemented in several selected cases to determine the ship route in real sea areas. A trapezoidal mesh was used to create a digital model of the area. On this basis, an appropriate number of nodes (vertices) of the graph was generated.

The edges of the graph were determined on the assumption that transition is possible between nodes in both directions. Based on thus formed undirected graph, an available route is searched for between two predefined points. Dijkstra's algorithms were used for this purpose in the presented examples.

To verify the method, several experiments were carried out with the use of electronic charts of real sea and ocean areas. For the purposes of the experiments we selected sea areas featuring a large number of islands, which produces several possible routes to choose from. Three cases of algorithm implementation are described below.

4.1 Case 1

In the case illustrated in Figure 8, the vessel is planning a north-west route. Using the algorithms

under consideration, the vessel chooses a relatively narrow passage, but wide enough to accommodate the ship domain visible as an ellipse on a marked route. The selected route is the shortest one.

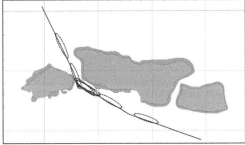

Figure 8. Case 1 – part of a chart with the selected route for 10 m safety contour

The same chart section is shown in Figure 9, with graphic presentation of discretized area and the selected areas "visible" in the algorithm as not available, i.e. restricted area.

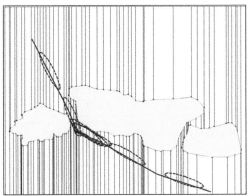

Figure 9. Case 1 – graphic representation of available and prohibited areas for 10 m safety contour

4.2 Case 2

In this case the vessel is planning a northeast route, but the graphical solution is shown for single, specific safety parameters. Figure 10 shows a chart section with a chosen route for a pre-set safety contour of 10 m. The defined route meets the shortest route requirement for the given domain size.

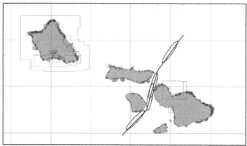

Figure 10. Case 2 – part of a chart with the selected route for 10 m safety contour

The same section of the chart with graphic presentation of allowed and prohibited sea areas in the algorithm are shown in Figure 11.

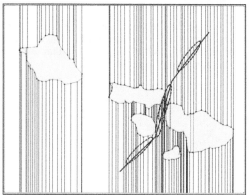

Figure 11. Case 1 – graphic representation of available (allowed) and prohibited areas for 10 m safety contour

4.3 *Case 3*

In the planned route as described in section 4.2 the basic safety parameter, safety contour, was changed to 20 m, and the experiment was repeated. Figure 12 shows a new route with visibly extended distance, a consequence of avoiding charted dangers (for the assumed parameters), that is areas which are not permitted in the algorithms.

The reason for extending the route is "virtual" elimination of passages through straits A and B, visible in Figure 12, as the result of changed safety contour. Interpretation of available and prohibited areas is shown in Figure 13, where straits A and B do not exist.

Passages through the straits "A" and "B" visible on the navigational chart (shown as available in Figures 9 and Figure10 at the preset safety contour 10 m) are not available for the algorithm, as the presented prohibited area overlaps the land as well as the adjacent area surrounded by safety contour and special area limits.

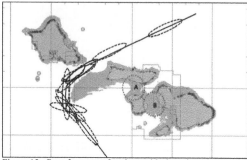

Figure 12. Case 3 – part of a chart with the selected route for 20 m safety contour

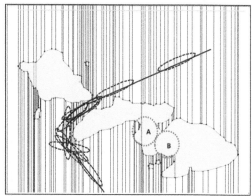

Figure 13. Case 3 – graphic representation of the available and prohibited areas for 20 m safety contour

5 CONCLUSIONS

The presented method can be used for:
- automatic search and route planning using the electronic chart system;
- approval of the planned route by ship's captain;
- verification and control of the planned route by the charterer in terms of its economic efficiency;
- improvement and/or expansion of algorithms in weather routeing optimization software;
- modification of planned and currently monitored route;
- planning of a track in anti-collision manoeuvres in restricted waters.

The proposed method can also be used en route for quick and safe changes in the current voyage plan.

It may also be a practical aid in course alteration, part of collision avoidance manoeuvres, one of the features of navigational decision support systems.

Another possible application is the implementation in electronic charts systems and navigational radars with chart functions (chart radars) fitted with the "Trial manoeuvre" function, where the proposed algorithms will indicate limitations of the planned manoeuvre and warn the user accordingly.

The described method takes into account the current safety contour set in the electronic chart system. Because of some shortcomings of currently used ENC charts encoded in the IHO S-57 standard, having relatively small amount of available safety contours to be generated, there are possible ambiguities which depth contour (isobath) is selected for the subsequent analysed charts.

Soon the problem will be solved with the implementation of navigational charts encoded in the IHO S-100 standard, containing a denser grid of available contours (isobaths). They offer to the navigator more freedom and unambiguous choice of safety contour on the entire planned route, which requires sometimes even a few dozen chart cells.

In the above cases, these authors limited the scope of the analysis by taking into account the safety contour and special condition areas.

Research is continued on the algorithms to include:
– single navigational dangers visible as charted point objects;
– extended dynamic model of the ship;
– planned and performed turns and course alterations during manoeuvres;
– in particular, prediction of trajectories of target ships in the process of calculating own ship trajectory during anti-collision manoeuvres.

It is also planned to enhance analyses of other categories of information to be entered and classified as permitted / prohibited areas - e.g. areas of poor hydro-meteorological conditions.

REFERENCES

Dramski M. & Mąka M. 2013. Algorithm of solving collision problem of two objects in restricted area. *Communications in Computer and Information Science 395 (Activities of Transport System Telematics)*, Springer-Verlag Berlin Heidelberg: 251 – 257.

Fujii, Y. & Tanaka, K. 1971. Traffic capacity. *Journal of Navigation, no. 24.* Cambridge

Goodwin, E. M. 1975. A statistical study of ship domain. *Journal of Navigation, no..28.* Cambridge

Hansen, M. & Jensen, T. & Lehn-Schiøler, T. & Melchild, K. & Rassmussen, F. & Ennemark, F. (2013) Empirical Ship Domain Based on AIS Data. *Journal of Navigation, no 66.* Cambridge.

IMO. 1995. Resolution A.817 (19) Performance standards for electronic chart display and information systems (ECDIS).

IMO. 1999. Resolution A.893(21) Guidelines for voyage planning.

IMO. 2006. Resolution 232(82) adoption of the revised performance standards for electronic chart display and information systems (ECDIS).

Mąka, M. 2014. Graphs construction method in the process of choosing the route in a restricted area. *Logistyka 3/2014*: 4250 – 4257 (in Polish).

Mąka M. & Dramski M. 2015. A proposal of the prediction algorithm of the object's position in restricted area. Archives of Transport Systems Telematics vol. 8, issue 2,2015: 13 – 16.

Pietrzykowski Z. & Mąka M. & Magaj J. 2014. Safe ship trajectory determination in the ENC environment. In J. Mikulski (Ed.): TST 2014, *Communications in Computer and Information Science 471*: 304–312, Springer-Verlag Berlin Heidelberg.

Pietrzykowski Z. & Uriasz J. 2009. The ship domain – a criterion of navigational safety assessment in an open sea area, *Journal of Navigation, no. 62.* Cambridge.

Śmierzchalski, R. & Weintrit A. 1999. Domeny obiektów nawigacyjnych jako pomoc w planowaniu trajektorii statku w sytuacji kolizyjnej na morzu (in Polish). *Proceedings of III Navigational Symposium*, Gdynia

van Kreveld, M. & de Bergh, M. & Overmars, M. & Schwarzkopf, O. 2007. Computational geometry – algorithms and applications (in Polish). *WNT*. Warsaw.

Wielgosz, M. 2015. Ship domain in navigational safety assessment. Unpublished PhD Thesis (in Polish). *Maritime University of Szczecin*.

Wielgosz, M. 2016. The ship safety zones in vessel traffic monitoring and management systems. *Scientific Journals, Maritime University of Szczecin, No.48(120)/2016: 153-158.*

Wielgosz, M., Pietrzykowski, Z. 2012. Ship domain in the restricted area – analysis of the influence of ship speed on the shape and size of the domain, *Scientific Journals, Maritime University of Szczecin, no. 30(102)/2012*

Mathematical Models, Methods and Algorithms

Positional Game Passing a Greater Number of Ships with Varying Degree of Cooperation

J. Lisowski
Gdynia Maritime University, Gdynia, Poland

ABSTRACT: Using as an example the process of safe ship control, the paper presents the problem of applying a positional non-cooperative and cooperative game of j ships for the description of the process considered as well as for the synthesis of optimal strategies. The approximated mathematical model of differential game in the form of triple linear programming problem is used for the synthesis of safe ship trajectory as a multistage process decision. The considerations have been illustrated an example of program computer simulation to determine the safe ship trajectories in situation of passing a many of the ships encountered.

1 INTRODUCTION

A large part in increasing the safety of navigation is the use of ARPA anti-collision system, which enables to track automatically at least 20 encountered j ships, determination of their movement parameters: speed V_j, course ψ_j and elements of approach: D_j – distance, N_j – bearing, $D_{j,\min} = DCPA_j$ - Distance of the Closest Point of Approach, $T_{j,\min} = TCPA_j$ - Time to the Closest Point of Approach (Fig. 1) (Bole et al. 2006).

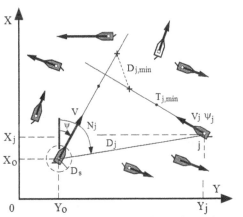

Figure 1. The own ship moving with speed V and course ψ during of passing j encountered ships.

The functional scope of a standard ARPA system ends with manoeuvre simulation to achieve the safe passing distance D_s by altering course $\pm\Delta\psi$ or speed $\pm\Delta V$ selected by the navigator (Bist 2000, Cockroft & Lameijer 2006, Cahill 2000).

The most general description of the own control object passing the j number of other encountered objects is the model of a differential game of a j number of objects (Basar & Olsder 1982, Engwerda 2005, Isaacs 1965, Mesterton-Gibbons 2001).

This model consists both of the kinematics and the dynamics of the ship's movement, the disturbances, the strategies of the own ship and encountered ships and the quality control index (Baker 2016, Clarke 2003, Fang & Luo 2005, Fossen 2011, Kula 2015, Perez 2005, Osborne 2004).

The diversity of possible models directly affects the synthesis of ship control algorithms which are afterwards affected by the ship control device, directly linked to the ARPA system and consequently determines effects of safe and optimal control (Fadali & Visioli 2009, Fletcher 1987).

2 THE POSITIONAL GAME MODEL PROCESS

The differential game described by state equation:

$$\dot{x} = f(x,u,t) \tag{1}$$

is reduced to a positional multistage game of a j number of participants (Fig. 2) (Keesman 2011).

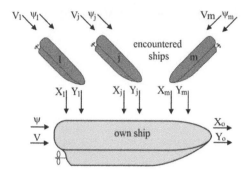

Figure 2. Block diagram of the positional game model of passing the own ship and encountered j ships.

The state x and control u variables are represented by (Lisowski & Lazarowska 2013):

$$x_{0,1} = X_o,\ x_{0,2} = Y_o,\ x_{j,1} = X_j,\ x_{j,2} = Y_j$$
$$u_{0,1} = \psi,\ u_{0,2} = V,\ u_{j,1} = \psi_j,\ u_{j,2} = V_j \qquad (2)$$
$$j = 1, 2, ..., m$$

The making of a continuous positional game discrete and reducing it to a multistage positional game is determined by own ship and depends on: the maximum relative speed of the own ship under the current navigational situation, the range of the situation and the dynamic characteristics of the own ship (Gałuszka & Świerniak 2005, Gluwer & Olsen 1998, Isil & Koditschek 2001, Landau et al. 2011).

The essence of the positional game is to make the strategies of the own ship dependent on current positions $p(t_k)$ of the ships encountered at the current step k. In this way possible course and speed alterations of the objects encountered are considered in the process model during the steering performance (Lazarowska 2012, Luus 2000, Mehrotra 1992).

The current state of the process is determined by the co-ordinates for the position of the own ship and of the ships encountered (Pantoja 1988, Zio 2009):

$$x_0 = (X_o, Y_o)$$
$$x_j = (X_j, Y_j) \qquad (3)$$
$$j = 1, 2, ..., m$$

The system generates its steering at the moment t_k on the basis of the data which are obtained from the ARPA anti-collision system concerning the current positions of the own and encountered ships:

$$p(t_k) = \begin{bmatrix} x_0(t_k) \\ x_j(t_k) \end{bmatrix}$$
$$j = 1, 2, ..., m \qquad (4)$$
$$k = 1, 2, ..., K$$

It is assumed, according to the general concept of the multistage positional game, that at each discrete

moment of the time t_k the position of the encountered ships $p(t_k)$ is known on the own ship (Gierusz & Lebkowski 2012, Lisowski 2012, Malecki 2013, Mohamed-Seghir 2016, Zak 2013).

The constraints of the state co-ordinates:

$$\{ x_0(t), x_j(t) \} \in P \qquad (5)$$

constitute the navigational constraints, while the steering constraints:

$$u_0 \in Uo$$
$$u_j \in U_j \qquad (6)$$
$$j = 1, 2, ..., m$$

take into consideration the kinematics of the ship movement, the recommendations of the COLREGS Rules and the condition to maintain the safe passing distance D_s:

$$D_{j,min} = \min D_j(t) \geq D_s \qquad (7)$$

The closed sets $U_{o,j}$ and $U_{j,o}$ defined as the sets of the acceptable strategies of the ships as players:

$$U_{o,j}[p(t)] = S_{o1,j} \cup S_{o2,j}$$
$$U_{j,o}[p(t)] = S_{j1,o} \cup S_{j2,o} \qquad (8)$$

are depended on the position $p(t)$, which means that the choice of the steering u_j by the j-th encountered ship alter the sets of the acceptable strategies of the other ships (Fig. 3).

Let refer to the set of the acceptable strategies of the own ship while passing the j-th encountered ship at a safe distance D_s (Jaworski, Kuczkowski, Smierzchalski 2012; Lebkowski 2015, Nisan et al. 2007, Straffin 2001, Tomera 2015, Lisowski 2014a).

The area, when maintaining stability in time of the course and speed of the own ship and the ship encountered is static and is comprised within the semicircle of a radius equal to the set reference speed of the own ship V within the arrangement of the co-ordinates $0X'Y'$ with the axis X' directed to the direction of the reference course (Millington & Funge 2009, Modarres 2006, Szlapczynski 2012).

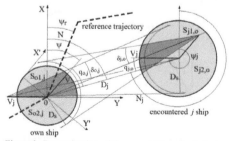

Figure 3. Determination of the acceptable safe strategies areas of the own ship $U_{o,j} = S_{o1,j} \cup S_{o2,j}$ and the encountered j ship $U_{j,o} = S_{j1,o} \cup S_{j2,o}$.

The set $U_{o,j}$ is determined with the inequalities:

$$a_{o,j}\, u_{o,x'} + b_{o,j}\, u_{o,y'} \leq c_{o,j}$$

$$u_{o,x'}^2 + u_{o,y'}^2 \leq V^2 \tag{9}$$

where:

$$\vec{V} = \vec{u}_o (u_{o,x'}, u_{o,y'})$$
$$a_{o,j} = -\lambda_{o,j}\, \cos(q_{o,j} + \lambda_{o,j}\, \delta_{o,j})$$
$$b_{o,j} = \lambda_{o,j}\, \sin(q_{o,j} + \lambda_{o,j}\, \delta_{o,j})$$
$$c_{o,j} = -\lambda_{o,j} \begin{bmatrix} V_j \sin(q_{j,o} + \lambda_{o,j}\, \delta_{o,j}) + \\ V \cos(q_{o,j} + \lambda_{o,j}\, \delta_{o,j}) \end{bmatrix} \tag{10}$$

$$\lambda_{o,j} = \begin{cases} -1 & for\ \ S_{o1,j} \quad (Port\ side) \\ 1 & for\ \ S_{o2,j} \quad (Starboard\ side) \end{cases}$$

The value $\lambda_{o,j}$ is determined by using an appropriate logical function F_j characterising any particular recommendation referring to right of way contained in COLREGS Rules (Lisowski 2014b).

The form of function F_j depends of the interpretation of the above recommendations for the purpose to use them in the control algorithm, when:

$$F_j = \begin{cases} 1 & then\ \ \lambda_{o,j} = \ \ 1 \\ 0 & then\ \ \lambda_{o,j} = -1 \end{cases} \tag{11}$$

Interpretation of the COLREGS Rules in the form of appropriate manoeuvring diagrams enables to formulate a certain logical function F_j as a semantic interpretation of legal regulations for manoeuvring.

Each particular type of the situation involving the approach of the ships is assigned the logical variable value equal to one or zero (Lisowski 2015a, 2015b):
A – encounter of the ship from bow or from any other direction,
B – approaching or moving away of the ship,
C – passing the ship astern or ahead,
D – approaching of the ship from the bow or from the stern,
E – approaching of the ship from the starboard or port side (Lisowski 2015c, 2015d) .

By minimizing logical function F_j by using a method of the Karnaugh's Tables the following is obtained:

$$F_j = A \cup \overline{A}(B\ \overline{C} \cup \overline{D}\ \overline{E}) \tag{12}$$

The resultant area of acceptable manoeuvres of the own ship in relation to the m encountered ships is:

$$U_o = \bigcap_{j=1}^{m} U_{o,j} \tag{13}$$

$$j = 1, 2, ..., m$$

is determined by an arrangement of inequalities (9).

On the other hand, however, the set of the acceptable strategies of the j-th object in relation to the own ship is determined by the following inequalities:

$$a_{j,o}\, u_{j,x'} + b_{j,o}\, u_{j,y'} \leq c_{j,o}$$

$$u_{j,x'}^2 + u_{j,y'}^2 \leq V_j^2 \tag{14}$$

where:

$$\vec{V}_j = \vec{u}_j (u_{j,x'}, u_{j,y'})$$
$$a_{j,o} = -\lambda_{j,o}\, \cos(q_{j,o} + \lambda_{j,o}\, \delta_{j,o})$$
$$b_{j,o} = \lambda_{j,o}\, \sin(q_{j,o} + \lambda_{j,o}\, \delta_{j,o})$$
$$c_{j,o} = -\lambda_{j,o} V \sin(q_{o,j} + \lambda_{j,o}\, \delta_{j,o}) \tag{15}$$

$$\lambda_{j,o} = \begin{cases} -1 & for\ \ S_{j1,o} \quad (Port\ side) \\ 1 & for\ \ S_{j2,o} \quad (Starboard\ side) \end{cases}$$

The symbol $\lambda_{j,o}$ is determined by analogy to the determination of $\lambda_{o,j}$ with the use of the logical function F_j described by the equation (11).

Consideration of the navigational constraints, as shallow waters and coastline, generate additional constraints to the set of acceptable strategies:

$$a_{n,k}\, u_{o,x'} + b_{n,k}\, u_{o,x'} \leq c_{n,k} \tag{16}$$

where:
k – is the nearest point of intersection of the straight lines approximating the coastline.

3 ALGORITHMS OF POSITIONAL GAME CONTROL

The optimal control $u_o^*(t)$ of the own ship, equivalent for the current position $p(t)$ to the optimal positional control $u_o^*(p)$, is determined in the following way:
– from the relationship (14) for the measured position $p(t_k)$, the control status at the moment t_k sets of the acceptable strategies $U_{j,o}[p(t_k)]$ are determined for the encountered ships in relation to the own ship, and from the relationship (9) the output sets $U_{o,j}[p(t_k)]$ of the acceptable strategies of the own ship in relation to each one of the encountered ships,
– a pair of vectors $u_{j,o}$ and $u_{o,j}$, are determined in relation to each j encountered ship and then the optimal positional strategy of the own ship $u_o^*(p)$ from the condition of optimum value I^* quality index control:

- when the encountered ships non-cooperate:

$$I_{nc}^* = \min_{\substack{u_o^* \in U_o = \bigcap_{j=1}^{m} U_{o,j}}} \left\{ \max_{u_{j,o} \in U_{j,o}} \min_{u_{o,j} \in U_{o,j}(u_j)} L[x_o(t_k), L_k] \right\} = L_{o,nc}^* \qquad (17)$$

$$j = 1, 2, ..., m$$

- when the encountered ships cooperate:

$$I_c^* = \min_{\substack{u_o^* \in U_o = \bigcap_{j=1}^{m} U_{o,j}}} \left\{ \min_{u_{j,o} \in U_{j,o}} \min_{u_{o,j} \in U_{o,j}(u_j)} L[x_o(t_k), L_k] \right\} = L_{o,c}^* \qquad (18)$$

$$j = 1, 2, ..., m$$

- for the non-game optimal control:

$$I_{oc}^* = \min_{\substack{u_o^* \in U_o = \bigcap_{j=1}^{m} U_{o,j}}} \left\{ L[x_o(t_k), L_k] \right\} = L_{o,oc}^* \qquad (19)$$

$$j = 1, 2, ..., m$$

where:

$$L[x_o(t_k), L_k] = \int_{t_0}^{t_K} V(t)\,dt + r_o(t_K) + d(t_K) \qquad (20)$$

refers to the goal control function of the own ship in the form of the payments – the integral payment and the final one (Lisowski 2016a).
The integral payment determines the distance of the own ship to the nearest turning point L_k on the assumed route of the voyage and the final one determines: $r_o(t_K)$ - the final risk of collision and $d(t_K)$ - final game trajectory deflection from reference trajectory.

The criteria for the selection of the optimal trajectory of the own ship is reduced to the determination of her course and speed, which ensure the smallest losses of way for the safe passing of the encountered ships at a distance not smaller than the assumed safe value D_s, having regard to the ship's dynamic in the form of the advance time t_m to the manoeuvre (Lisowski 2016b).

At the time advance maneuver t_m consists of element $t_m^{\Delta\psi}$ during course manoeuvre $\Delta\psi$ or element $t_m^{\Delta V}$ during speed manoeuvre ΔV.

The dynamic features of the ship during course alteration by an angle $\Delta\psi$ is described in a simplified manner with the use of transfer function:

$$G_\psi(s) = \frac{\Delta\psi(s)}{\alpha(s)} = \frac{k_\psi(\alpha)}{s(1 + T_\psi s)} \cong \frac{k_\psi(\alpha) \cdot e^{-T_{o\psi}s}}{s} \qquad (21)$$

where:

$T_{o\psi} \cong T_\psi$ - manoeuvre delay time which is approximately equal to the time constant of the ship as a course control object,

$k_\psi(\alpha)$ - gain coefficient the value of which results from the non-linear static characteristics of the rudder steering.

The course manoeuvre delay time is as follows:

$$t_m^{\Delta\psi} \cong T_{o\psi} + \frac{\Delta\psi}{\dot\psi} \qquad (22)$$

In practice, depending on the size and type of vessel advance time to the anti-collision manoeuvre through a change of course is: $t_m^{\Delta\psi} \cong 60 \div 720\,s$.

Differential equation of the second order describing the ship's behaviour during the change of the speed by ΔV is approximated with the use of the inertia of the first order with a time delay:

$$G_V(s) = \frac{\Delta V(s)}{\Delta n(s)} = \frac{k_v e^{-T_{ov}s}}{1 + T_v s} \qquad (23)$$

where:
T_{ov} - time of delay equal approximately to the time constant for the propulsion system: main engine-propeller shaft-screw propeller,
T_v - the time constant of the ship's hull and the mass of the accompanying water.

The speed manoeuvre delay time is as follows:

$$t_m^{\Delta V} \cong T_{ov} + 3T_v \qquad (24)$$

In practice, depending on the size and type of vessel advance time to the anti-collision manoeuvre through a change of speed is: $t_m^{\Delta V} \cong 120 \div 900\,s$.

The smallest losses of way are achieved for the maximum projection of the speed vector of the own ship on the direction of the assumed course leading to the nearest turning L_k point.

The optimal control of the own ship is calculated at each discrete stage of the ship's movement by applying triple linear programming SIMPLEX method, assuming the relationship (20) as the goal function and the constraints are obtained by including the arrangement of the inequalities (8), (14) and (16).

The above problem is then reduced to the determination the function of control goal as the maximum of the projection of the own ship speed vector on reference direction of the movement:

$$\min L = \max \left[V\left(u_{o,x'}, u_{o,y'}\right) = u_{o,x'} \right] \qquad (25)$$

70

with linear constraints approximating the joint set of the safe strategies of the own ship $U_{o,j}$:

$$a_{o,1}u_{o,x'} + b_{o,1}u_{o,y'} \leq c_{o,1}$$

$$a_{o,j}u_{o,x'} + b_{o,j}u_{o,y'} \leq c_{o,j}$$

$$a_{o,m}u_{o,x'} + b_{o,m}u_{o,y'} \leq c_{o,m} \tag{26}$$

$$a_{o,m+1}u_{o,x'} + b_{o,m+1}u_{o,y'} \leq c_{o,m+1}$$

$$a_{o,m+k+p}u_{o,x'} + b_{o,m+k+p}u_{o,y'} \leq c_{o,m+k+p}$$

where:

m – number on encountered ships,

k – number of constraints approximating coastline,

p - number of segments approximating a semi-circle with a radius equal to own ship speed.

After the interval of time t_k the current fixing of the ship position is carried out and then comes the solving of the problem using the algorithm for the positional control.

Using the function of lp – linear programming from the Optimization Toolbox contained in the Matlab/Simulink the positional multistage game manoeuvring programs: $mpgame\_nc$ for criterion (17), $mpgame\_c$ for criterion (18) and $mpngame\_oc$ for criterion (19) has been designed for determination of the safe ship trajectory in a collision situation.

4 COMPUTER SIMULATION

Computer simulation of $mpgame\_nc$, $mpgame\_c$ and $mpngame\_oc$ algorithms was carried out in Matlab/Simulink software on an example of the real navigational situation of passing $j=19$ encountered ships in the Skagerrak Strait in good visibility $D_s=0.5 \div 1.0$ nm and restricted visibility $D_s=1.5 \div 2.5$ nm (nautical miles), (Fig. 4 and Tab. 1).

Figure 4. The place of identification of navigational situations in Skagerrak and Kattegat Straits.

Table 1. Movement parameters of the own ship and encountered 19 ships.

j	D_j nm	N_j deg	V_j kn	ψ_j deg
0	-	-	20	0
1	9	320	14	90
2	15	10	16	180
3	8	10	15	200
4	12	35	17	275
5	7	270	14	50
6	8	100	8	6
7	11	315	10	90
8	13	325	7	45
9	7	45	19	10
10	15	23	6	275
11	15	23	7	270
12	4	175	4	130
13	13	40	0	0
14	7	60	16	20
15	8	120	12	30
16	9	150	10	25
17	8	310	12	135
18	10	330	10	140
19	9	340	8	150

The situation was registered on board r/v HORYZONT II, a research and training vessel of the Gdynia Maritime University, on the radar screen of the ARPA anti-collision system Raytheon (Fig. 5 and 6).

Figure 5. The research-training ship of Gdynia Maritime University r/v HORYZONT II.

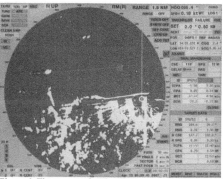

Figure 6. The screen of anti-collision system ARPA Raytheon, installed on the research-training ship of Gdynia Maritime University r/v HORYZONT II.

Examined the navigational situation, illustrated in the form of navigation velocity vectors of own ship and 19 met ships is shown in Figure 7.

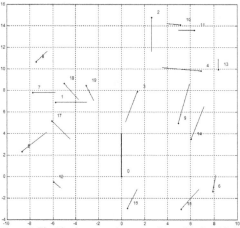

Figure 7. The 12 minute speed vectors of own ship 0 and j=19 encountered ships in navigational situation in Skagerrak Strait.

Fig. $8 \div 13$ shows the safe and optimal trajectory of the own ship in collision situation, which is determined using the algorithms of non-cooperative and cooperative positional game and non-game optimal control in good visibility at sea, for two values safe distance passing ships: D_s=0.5 nm and D_s=1.0 nm (nautical mile).

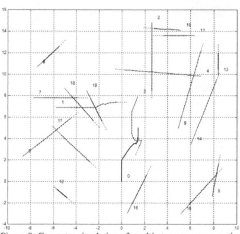

Figure 8. Computer simulation of multi-stage non-cooperative positional game algorithm *mpgame_nc* for safe own ship control in situation of passing 19 encountered ships in good visibility at sea, D_s=0.5 nm, $d(t_k)$=1.44 nm (nautical mile).

At presented below figures show that the value of a safe distance in conditions of a good visibility at sea, and the degree of cooperation between the own ship and the encountered ships have significantly affect to the value of the final deviation designated safe trajectory of its reference form.

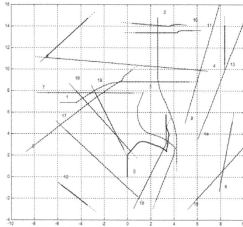

Figure 9. Computer simulation of multi-stage non-cooperative positional game algorithm *mpgame_nc* for safe own ship control in situation of passing 19 encountered ships in good visibility at sea, D_s=1.0 nm, $d(t_k)$=3.34 nm (nautical mile).

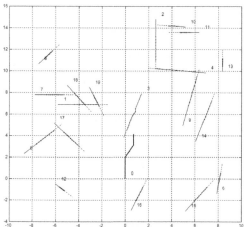

Figure 10. Computer simulation of multi-stage cooperative positional game algorithm *mpgame_c* for safe own ship control in situation of passing 19 encountered ships in good visibility at sea, D_s=0.5 nm, $d(t_k)$=0.73 nm (nautical mile).

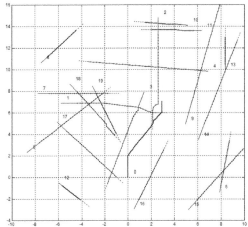

Figure 11. Computer simulation of multi-stage cooperative positional game algorithm *mpgame_c* for safe own ship control in situation of passing 19 encountered ships in good visibility at sea, D_s=1.0 nm, $d(t_k)$=2.94 nm (nautical mile).

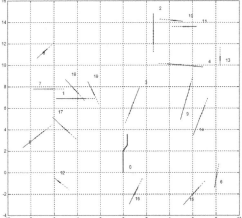

Figure 12. Computer simulation of non-game optimal control algorithm *mpngame_oc* for safe own ship control in situation of passing 19 encountered ships in good visibility at sea, D_s=0.5 nm, $d(t_k)$=0.35 nm (nautical mile).

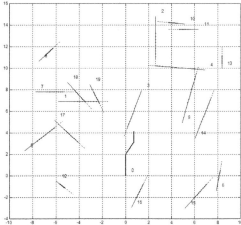

Figure 13. Computer simulation of non-game optimal control algorithm *mpngame_oc* for safe own ship control in situation of passing 19 encountered ships in good visibility at sea, D_s=1.0 nm, $d(t_k)$=0.72 nm (nautical mile).

Fig. 14÷19 shows the safe and optimal trajectory of the own ship in collision situation, which is determined using the algorithms of cooperative and non-cooperative positional game and non-game optimal control in restricted visibility at sea, for two values safe distance passing ships: D_s=1.5 nm and D_s=2.5 nm (nautical mile).

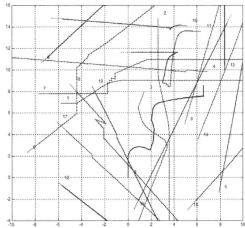

Figure 14. Computer simulation of multi-stage non-cooperative positional game algorithm *mpgame_nc* for safe own ship control in situation of passing 19 encountered ships in restricted visibility at sea, D_s=1.5 nm, $d(t_k)$=6.56 nm (nautical mile).

73

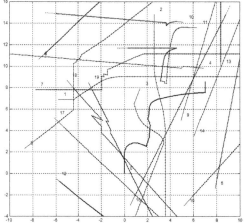

Figure 15. Computer simulation of multi-stage non-cooperative positional game algorithm *mpgame_nc* for safe own ship control in situation of passing 19 encountered ships in restricted visibility at sea, D_s=2.5 nm, $d(t_k)$=7.34 nm (nautical mile).

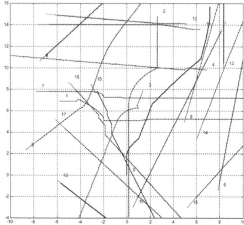

Figure 17. Computer simulation of multi-stage cooperative positional game algorithm *mpgame_c* for safe own ship control in situation of passing 19 encountered ships in restricted visibility at sea, D_s=2.5 nm, $d(t_k)$=7.06 nm (nautical mile).

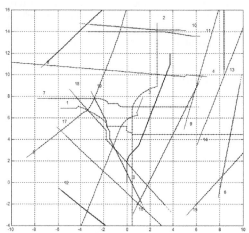

Figure 16. Computer simulation of multi-stage cooperative positional game algorithm *mpgame_c* for safe own ship control in situation of passing 19 encountered ships in restricted visibility at sea, D_s=1.5 nm, $d(t_k)$=3.75 nm (nautical mile).

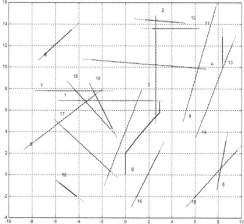

Figure 18. Computer simulation of non-game optimal control algorithm *mpngame_oc* for safe own ship control in situation of passing 19 encountered ships in restricted visibility at sea, D_s=1.5 nm, $d(t_k)$=2.92 nm (nautical mile).

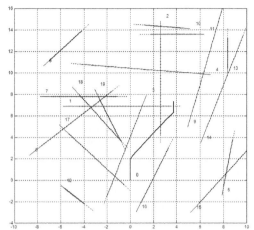

Figure 19. Computer simulation of non-game optimal control algorithm *mpngame_oc* for safe own ship control in situation of passing 19 encountered ships in restricted visibility at sea, D_s=2.5 nm, $d(t_k)$=3.76 nm (nautical mile).

On the above figures show that the value of a safe distance in conditions of a restricted visibility at sea, and the degree of cooperation between the own ship and the encountered ships have significantly affect to the value of the final deviation designated safe trajectory of its reference form.

5 CONCLUSIONS

The synthesis of an optimal on-line control on the model of a multi-stage positional game makes it possible to determine the safe game trajectory of the own ship in situations when she passes a greater *j* number of the encountered objects.

The trajectory has been described as a certain sequence of manoeuvres with the course and speed.

The computer programs designed in the Matlab also takes into consideration the following: regulations of the Convention on the International Regulations for Preventing Collisions at Sea, advance time for a manoeuvre calculated with regard to the ship's dynamic features and the assessment of the final deflection between the real trajectory and its assumed values.

The essential influence to form of safe and optimal trajectory and value of deflection between game and reference trajectories has a degree of cooperation between own and encountered ships.

REFERENCES

Baker G. 2016. Differential equations as models in science and engineering. New Jersey – Tokyo: Word Scientific.
Basar, T. & Olsder, G.J. 1982. Dynamic non-cooperative game theory. New York: Academic Press.

Bist, D.S. 2000. Safety and security at sea. Oxford-New Delhi: Butter Heinemann.
Bole, A., Dineley, B. & Wall, A. 2006. Radar and ARPA manual. Amsterdam-Tokyo: Elsevier.
Cahill, R.A. 2002. Collisions and thair causes. London: The Nautical Institute.
Clarke, D. 2003. The foundations of steering and manoeuvering, Proc. of the IFAC Int. Conf. on Manoeuvering and Control Marine Crafts, Girona: 10-25.
Cockcroft, A.N. & Lameijer, J.N.F. 2006. The collision avoidance rules. Amsterdam-Tokyo: Elsevier.
Engwerda, J.C. 2005. LQ dynamic optimization and differential games. West Sussex: John Wiley & Sons.
Fadali, M.S. & Visioli, A. 2009. Digital control engineering. Amsterdam-Tokyo: Elsevier.
Fang, M.C. & Luo, J.H. 2005. The nonlinear hydrodynamic model for simulating a ship steering in waves with autopilot system. Ocean Engineering 11-12(32):1486-1502.
Fletcher, R. 1987. Practical methods of optimization. New York: John Wiley and Sons.
Fossen, T.I. 2011. Marine craft hydrodynamics and motion control. Trondheim: Wiley.
Gałuszka, A. & Świerniak, A. 2005. Non-cooperative game approach to multi-robot planning. Int. Journal of Applied Mathematics and Computer Science 15(3):359-367.
Gierusz, W. & Lebkowski A. 2012. The researching ship "Gdynia". Polish Maritime Research 1(19): 11-18.
Gluver, H. & Olsen, D. 1998. Ship collision analysis. Rotterdam-Brookfield: A.A. Balkema.
Isaacs, R. 1965. Differential games. New York: John Wiley & Sons.
Isil Bozma, H. & Koditschek, D.E. 2001. Assembly as a non-cooperative game of its pieces: Analysis of ID sphere assemblies. Robotica 19: 93-108.
Jaworski B., Kuczkowski L., Smierzchalski R. 2012. Przeglad Elektrotechniczny 10B(88): 252-255.
Keesman, K.J. 2011. System identification. London-NewYork: Springer.
Kula, K. 2015. Model-based controller for ship track-keeping using Neural Network. Proc. of IEEE 2nd Int. Conf. on Cybernetics, CYBCONF, Gdynia, Poland: 178-183.
Landau, I.D., Lozano, R., M'Saad, M. & Karimi, A. 2011. Adapive control. London-New York: Springer.
Lazarowska, A. 2012. Decision support system for collision avoidance at sea. Polish Maritime Research 1(19):19-24.
Lisowski J. 2012. Game control methods in avoidance of ships collisions. Polish Maritime Research 1(19):3-10.
Lisowski J. & Lazarowska A. 2013. The radar data transmission to computer support system of ship safety. Edited by Garus J., Szymak p., Zak B.: Mechatronic Systems, Mechanics and Materials, Book Series: Solid State Phenomena, Vol. 180: 64-69.
Lisowski J. 2014. Computational intelligence methods of a safe ship control. Edited by Jedrzejowicz P., Czarnowski I., Howlett R.J.: Knowledge-based and Intelligent information & Engineering systems 18th Annual Conference KES 2014, Book Series: Procedia Computer Science, Vol. 35: 634-643.
Lisowski J. 2014. Optimization-supported decision-making in the marine game environment. Edited by Garus J., Szymak p.: Mechatronic Systems, Mechanics and Materials, Book Series: Solid State Phenomena, Vol. 210: 215-222.
Lisowski J. 2015. Determination and display of safe ship trajectories in collision situations at sea. Solid State Phenomena, Vol. 236: 128-133.
Lisowski J.: Sensitivity of the game control of ship in collision situations. Polish Maritime Research, 4(22): 27-33.
Lisowski J. 2015. Multi-stage safe ship control with neural state constraints. 2nd IEEE International Conference on Cybernetics, CYBCONF, Gdynia, Poland: 56-61.

Lisowski J. 2015. Comparison of anti-collision game trajectories of ship in good and restricted visibility at sea. In: A. Weintrit (ed.): Activities in Navigation - Marine Navigation and Safety of Sea Transportation, CRC Press/ Balkema, Taylor & Francis Group, London: 201-210.

Lisowski J. 2016. The sensitivity of state differential game vessel traffic model. Polish Maritime Research 2(23):14-18.

Lisowski J. 2016. Analysis of methods of determining the safe ship trajectory. TransNav, the International Journal on Marine Navigation and Safety of Sea Transportation 2(10): 223-228.

Lebkowski A. 2015. Evolutionary Methods in the Management of Vessel Traffic. In: A. Weintrit & T. Neumann (eds): Information, Communication and Environment, Marine Navigation and Safety at Sea Transportation: 259-266.

Luus, R. (2000). Iterative dynamic programming, CRC Press, Boca Raton.

Malecki J. 2013. Fuzzy track-keeping steering design for a precise control of the ship. Solid State Phenomena, Vol. 196: 140-147.

Mehrotra, S. 1992. On the implementation of a primal-dual interior point method. SIAM Journal on Optimization 4(2): 575-601.

Mesterton-Gibbons, M. 2001. An introduction to game theoretic modeling. Providence: American Mathematical Society.

Millington, I. & Funge, J. 2009. Artificial intelligence for games. Amsterdam-Tokyo: Elsevier.

Modarres, M. 2006. Risk analysis in engineering. Boca Raton: Taylor & Francis Group.

Mohamed-Seghir, M. 2016. Computational intelligence method for ship trajectory planning. Proc. XXI Conference Methods and Models in Automation and Robotics, MMAR, Miedzyzdroje: 636-640.

Nisan, N., Roughgarden, T., Tardos, E. & Vazirani, V.V. 2007. Algorithmic game theory. New York: Cambridge University Press.

Osborne, M.J. 2004. An introduction to game theory. New York: Oxford University Press.

Pantoja, J.F.A. 1988. Differential dynamic programming and Newton's method. International Journal of Control 5(47): 1539-1553.

Perez, T. 2005. Ship motion control. London: Springer.

Straffin, P.D. 2001. Game theory and strategy. Warszawa: Scholar (in polish).

Szlapczynski R. & Szlapczynska J. 2012. Customized crossover in evolutionary sets of safe ship trajectories. Int. Journal of Applied Mathematics and Computer Science 4(22): 999-1009.

Tomera, M. 2015. A multivariable low speed controller for a ship autopilot with experimental results. Proc. XXth Int. Conf. Methods and Models in Automation and Robotics, MMAR, Miedzyzdroje: 17-22.

Zak A. 2013. Trajectory tracking control of underwater vehicles. Solid State Phenomena, Vol. 196: 156-166.

Zio, E. 2009. Computational methods for reliability and risk analysis. Series on Quality, Reliability and Engineering Statistics 14: 295-334.

Mathematical Principles for Vessel's Movement Prediction

E. Kulbiej
Maritime University of Szczecin, Szczecin, Poland

ABSTRACT: The following paper presents novel approach to mathematical vessel's movement prediction and incoming manoeuvre estimation by proposing non-linear approximation. The stated calculus aims to derive formulas allowing any tracking and traffic systems to estimate position of a nautical vessel in a short-term future, which bases on data analysis via polynomial equation approximation technique. A numerical example is intended to prove possibilities of potential application in navigational decision support systems and in manoeuvring operations either in port or in open sea. Ultimately the calculus is compared to other parallel mathematical tools, widely used for proposition of vessel's position behaviour as a function of time, in order to contemplate the constatation of positive application's value.

1 INTRODUCTION

In order to provide and then develop maritime technology upgrades, a comprehensive mathematical model of a vessel is required. It stems from the fact that numerical calculations need a specific assumptions and algorithms in order to be effective. By ship's mathematical model, set of rules describing her movement is meant. One of such models is trivially described in (Lenart, 2012). The object is shown as a point-blank kinematic object, moreover regarded as if the mass was concentrated at a point. The motion parameters are ground or sea referenced as a drift angle is assumed to be zero. However, purely kinematic manoeuvres tend to misshape the image of the situation as they only propose linear solutions in regards to movement predictions, albeit in some particular cases it may be proven to be sufficient, for instance if the velocity and course over ground is not altering in time.

Last moment manoeuvre of a vessel in collision scenario is an example, when quality and precision of movement description is required. With stationary means of mathematical determination it turns out to be all but useful. Yet another activity where movement determination may prove to be convenient, is the storage of data, possibly regarding past voyages of vessels in coastal area. In that scenario data might be stored only partially. Neglecting most of the strings from AIS does not damage informative function required by authorised institutions since the whole track of a vessel, with particular accuracy, may be re-generated basing on a base of sample points (aforementioned partial information). Such approach could reduce the time of transmission of these data and will provide an opportunity to reduce physical space needed for binary storage (Czapiewska, 2015).

Commencing with these premises, this paper intends to provide a resourceful method of determining future part of the ship's movement, basing on values of motion parameters rather than analysing following positions of the object of narrative and determining probable next position. The main principle of this method is to determine an non-linear approximation of a function describing change of particular movement traits in time and then implement them into a calculus designed for finding sought position. Due to its analytical origin, the method can be directly compared to other methods of determining objects position: linear and circular movement estimation. The source of numerical data for every estimation attempt shall be the means of navigational equipment, Automatic Identification System (AIS) and Global Positioning System (GPS) or any other equative to those mentioned. It is not crucial to decide which equipment items are to be used, the case is to assure accuracy of the provided information. However, in the proposed schematics, even if the data is supposedly encumbered with an omni-present and constant error, such as relativistic effects in satellite systems described in Kulbiej (2016), it is not taken into account in the course of calculation, as the main numerical value that is being examined is a difference. It naturally is derived of any constant errors since they are simply reduced (subtracted).

2 MATHEMATICAL NAVIGATION

The following paper shall introduce the term mathematical navigation (MN). It is the general calculus and all sort of numeral operation and mathematical means undertaken in order to determine sea-borne vessel's position in particular point of time, both in past and in short-term future; as well as establishing set equations that can act as a mathematical model of analysed vessel's movement. By these means, MN appears to be broader than commonly used dead reckoning (DR). The latter is a tool of navigational bridge team for route calculation and position archiving, but it does not include dynamics of the ships and possibility that pre-set assumption may alter in time. It is specified in IMO regulations (IMO, 1978) that an officer of the navigational watch ought to be able to calculate vessel's position advance after taking into account simple cases of wind and current impact; by simple it is meant stationary. This inconvenience is dealt with by mathematical navigation.

Many works describe vessel movement as purely kinematic (Lenart, 2012). This approach has its disadvantages, for instance inaccuracies caused by either low speed of manoeuvre reaction or major size of a vessel in comparison to waterway length and width (especially in last moment manoeuvres). That is why a dynamic approach seems to bring positive balance into MN. It comprises Newton's laws of classical physics as well as comprehensive derivative and integral calculus.

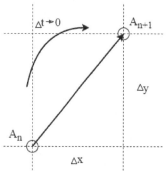

Figure 1. Position change after infinitesimally small time offset.

Mathematical navigation bases its calculation on a basic assumption, that every ship's movement can be divided into infinite number of minute position advances. It shall be labeled as a differential of vessel's positon (DVP). DVP can be described graphically on the provision that time offset between two successful positions is infinitesimally small (figure 1).

Vessel's velocity is a vector described by its magnitude, i.e. the linear speed (which may or may not be constant) and direction, i.e. course over ground (COG) or heading, depending on the data input. In two basic dimensions, OX axis and OY axis, it can be decomposed into two component vectors parallel to aforementioned axes, whose magnitudes depend on the value of φ. This action is shown in figure 2.

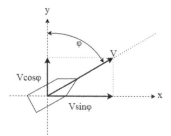

Figure 2. Velocity vector decomposition.

It is to be underlined that actual movement between two adjacent successful positions of a vessel in normal time offset (by normal it is meant non-infinitesimally small but a real value) is in mathematical means an integral of DVP over that time. In terms of factual equations it can be described as follows:

$$x(t_{n+1}) = x(t_n) + \int_{t_n}^{t_{n+1}} V(t) \cdot \sin[\varphi(t)] \cdot dt \qquad (1)$$

$$y(t_{n+1}) = y(t_n) + \int_{t_n}^{t_{n+1}} V(t) \cdot \cos[\varphi(t)] \cdot dt \qquad (2)$$

where Δt is the time offset between analysed points of time t_n and t_{n+1}.

Yet it has to be underlined that maritime navigation employs system of coordinates basing on latitudinal and longitudinal degrees as a unit. However latitude is described by units that are trivially recalculated for nautical miles (Nm) in case of longitudinal differences, if distance is to be measured, a concept of departure requires employment. It can be described as:

$$a = \Delta\lambda \cdot cos\varphi \qquad (3)$$

where a is the sought departure, $\Delta\lambda$ is the distance travelled in longitudinal degrees, φ is the latitude of the departure. Consequently (1) and (2) can be rearranged in order to meet Cartesian projection measurements:

$$\varphi(t_{n+1}) = \varphi(t_n) + \int_{t_n}^{t_{n+1}} V(t) \cdot \cos\left[COG(t)\right] dt \qquad (4)$$

$$\lambda(t_{n+1}) = \lambda(t_n) + \int_{t_n}^{t_{n+1}} V(t) \cdot \frac{\sin\left[COG(t)\right]}{\cos\left[\varphi(t)\right]} dt \qquad (5)$$

where position's coordinates are now found deliberately in degree units.

The accuracy of position calculation depends here on the accuracy of derivation of functions of speed and course. The method proposed in this paper bases on this premise and thus it is intended to employ an empirically derived third-degree polynomial as a function describing the alteration schema of ship's velocity and course over ground. For every vessel's position P_n, and consequently, for a specific time t_n, a matrix v_n may be introduced that will store all necessary information about ships position:

$$v_n = \begin{bmatrix} x(t_n) \\ y(t_n) \\ V(t_n) \\ COG(t_n) \end{bmatrix} \qquad (6)$$

These are vessel's motion numeral particulars for moment of t_n. Further crucial traits are acceleration a (change of linear velocity over time), angular speed ω (change of course over time) and angular acceleration ε (change of angular speed over time). Every of mentioned surplus parameter is a derivative over time of a function mentioned in particular bracket. However, if no exact formula for a, ω, ε can be found, then approximation is to be used that declares those values as equal to mean values, namely:

$$\bar{a} = \frac{v(t_{n+1}) - v(t_n)}{t_{n+1} - t_n} = \frac{\Delta v(\Delta t)}{\Delta t} \qquad (7)$$

$$\bar{\omega} = \frac{\varphi(t_{n+1}) - \varphi(t_n)}{t_{n+1} - t_n} = \frac{\Delta \varphi(\Delta t)}{\Delta t} \qquad (8)$$

$$\bar{\varepsilon} = \frac{\omega(t_{n+1}) - \omega(t_n)}{t_{n+1} - t_n} = \frac{\Delta \omega(\Delta t)}{\Delta t} \qquad (9)$$

If proposed linear approximation is all but sufficient then the course of the function can be estimated through means of cubic polynomial (10). Then, consequently, its derivative over time shall be a square polynomial, and another successive derivative shall be a linear function. For if instance course of time is a third-degree polynomial, then angular velocity is a second-degree polynomial and

angular acceleration is a first-degree polynomial. In following paper an arithmetic formula of a third-degree polynomial is derived. The formula looks followingly:

$$y = Ax^3 + Bx^2 + Cx + D \qquad (10)$$

where A, B, C, D are yet unknown constants determining the shape of y=f(x) curve.

3 FORMULA DERIVATION

This paragraph aims to picture a derivation process of course and speed function in time, both as third-degree polynomials. Both can be mathematically described as follows:

$$V(t) = A_V t^3 + B_V t^2 + C_V t + D_V \qquad (11)$$

$$COG(t) = A_C t^3 + B_C t^2 + C_C t + D_C \qquad (12)$$

If such equations are written for four consecutive moments in time, every parameter in (11) or (12) may be derived:

$$\begin{cases} y_k = Ax_k^3 + Bx_k^2 + Cx_k + D \\ y_{k+1} = Ax_{k+1}^3 + Bx_{k+1}^2 + Cx_{k+1} + D \\ y_{k+2} = Ax_{k+2}^3 + Bx_{k+2}^2 + Cx_{k+2} + D \\ y_{k+3} = Ax_{k+3}^3 + Bx_{k+3}^2 + Cx_{k+3} + D \end{cases} \qquad (13)$$

Then it is necessary to solve this set of equations. If by any chance they are not solvable for third-degree polynomial then it is valid to leave the answer in a way it has been obtained (a special case of for instance constant function– then the solution is y(t) = D). Many various derivation methods exist, one of which is to solve this set by pure, step-by-step transformation of equations and consequent elimination of unknowns Skipping through this comprehensive equation transformation, the sought function is obtained.

4 MATHEMATICAL MODEL

Eventually when sought equations (11) and (12) has been obtained, they are to be inserted into (2) and (3):

$$\varphi(t_{n+1}) = \varphi(t_n) + \int_{t_n}^{t_{n+1}} \left[A_V t^3 + B_V t^2 + C_V t + D_V\right] \cdot \cos\left[A_C t^3 + B_C t^2 + C_C t + D_C\right] dt \qquad (14)$$

$$\lambda(t_{n+1}) = \lambda(t_n) + \int_{t_n}^{t_{n+1}} \left[A_V t^3 + B_V t^2 + C_V t + D_V\right] \cdot \frac{\sin\left[A_C t^3 + B_C t^2 + C_C t + D_C\right]}{\cos\left[\varphi(t)\right]} dt \qquad (15)$$

The positon of a vessel in means of ship's information vector can be calculated for every point in time between the two extreme time moments that were taken into consideration in (13). After sufficient amount of time, in the same time offset a new information matrix will be obtained and thus the polynomials will need recalculation. However, in short term prediction schema, it is possible to declare vessel's future position basing on the already derived functions. The premise for this action is that in small neighbourhood of the last point in time the formulas shan't differ greatly. The dynamics of physics forbids any system to alter significantly in a short amount of time. In nature, phenomena such as wind are continuous functions and thus consistent in time. That means for instance wind or current speed will alter with some pace that will be taken into mathematical account while deriving the (11) and (12) equations. Naturally, anomalies of weather and other interferences are not taken into scope of the research as they are all but predictable.

5 REAL CONDITION ANALYSIS

In order to prove aforementioned MN effectiveness, actual situations involving real-condition journey of a vessel were examined. Two particular ships were taken in consideration while doing the research, one being mounted ferry Wawel and the other being container vessel Msc Paris. Their particular parameters are compared in a table 1 below. Figure 3 presents simulated overview of two movement prediction methods for m/v Msc Paris, while figure 4 presents the same scope of research for vessel m/f Wawel.

The data was originally recorded during commercial voyages of both ships, however for the purposes of this paper, they have been simulated using advanced navigational decision support system (NDSS) Navdec, which has already been presented in scientific literature (Pietrzykowski et.al., 2016) and interestingly used even in preventing historical collision analysis scenario (Kulbiej & Wołejsza, 2016). Situation of both vessels at the point of commencement of the experiment can be spotted in figure 5 for vessel Msc Paris and in figure 6 for vessel Wawel.

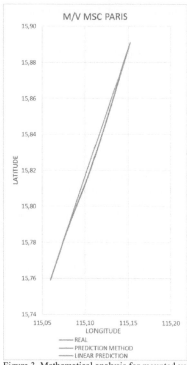

Figure 3. Mathematical analysis for mounted vessel Msc Paris.

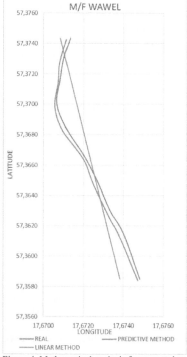

Figure 4. Mathematical analysis for mounted vessel Wawel.

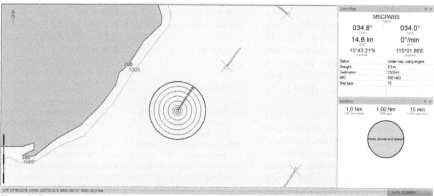

Figure 5. Container vessel Msc Paris at the point of commencement of the experiment.

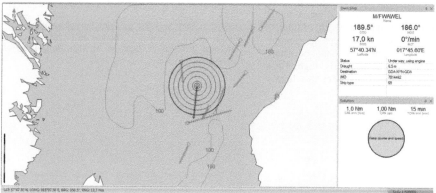

Figure 5. Ro-ro ferry Wawel at the point of commencement of the experiment.

Closer look at the outcome of the calculation and its graphic representation provides several conclusions:

− in case of m/v Msc Paris, where the track was nearly exactly linear, with small cross track errors due to COLREG specified anti-collision manoeuvres, all three routes are extremely closely plotted. This obviously stems from the fact, that through the journey, voyage parameters had been altering in only minute values.

− as far as case of m/f Wawel is concerned, a distinctive waypoint may be spotted, where the ferry altered hers course from 189° to 165°. Because of that distinctive shape of route (actual route, not the planned one), linear method of movement prediction appears to be useless. On the other hand, calculus derived in this paper has found itself accurate in terms of real route repetition.

− mathematical navigation means are in route determining purposes useful, as they can recreate travelled shape with high resemblance. It proves, that it is obviously unneeded to store all the information about the journey, since only partial

amount of data is sufficient to plot historical route of any ship.

Table 1. Vessels' parameters.

VESSELS NAME	WAWEL	MSC PARIS
TYPE	Ro-ro ferry	Container vessel
LOA	151 m	334 m
AVG SPEED	17 kn.	16 kn.
AREA	Baltic Sea	South China Sea

6 CONCLUSIONS AND FUTURE WORK

In the following paper, means of mathematical navigation has been introduced in regards to methods of analytical determination of vessel's past and short-term future positions. Such a calculus has been derived and its outcome scrupulously evaluated. Concerning the data field, it is shown that indeed, it may be used for purposes of movement prediction as in many cases it provides the user with appropriate approximation of ship's localisation. However, as mentioned before, the accuracy of movement prediction bases on the accuracy of provision of key factors such as course over ground and velocity of a ship. Namely if the time offsets, that is to say reporting intervals received from AIS or acquisition's and plotting's alteration delay in ARPA, is to large then the solution received from MN may be proven unreliable. In order to prevent this negative effect, a suitable frequency of real data (used for calculations) acquisition should be decided.

Compared to others means of mathematical position prediction it is shown that only in particular special cases the solutions are of similar accuracy. However, as stems from real-condition analysis, most of high-sea voyages are in majority straight-line legs of route. The resolution's inconsistency increases greatly for the advantage of MN in either close-quarter situations, where the manoeuvres must be taken rapidly so they can provide positive outcome or in coastal traffic scenarios, where many course alterations are needed in short span of time.

The paper analysed situations where information needed was provided in surplus amounts and where the redundancy level was particularly high. Case whereas the real-conditions shall meet this highly expected criteria is not discussed, but is a subject for further works of author. Moreover future works are to maintain the broad topic of mathematical navigation and its anti-collision purposes.

REFERENCES

Czapiejewska A. & Sadowski J. 2015. Algorithms for ship movement prediction for location data compression. *Transnav* 9(1): 75-81.

IMO International Convention on Standards of Training, Certification and Watchkeeping for Seafarers, 1978.

Kulbiej E. 2016. Relevance of relativistic effects in satellite navigation. *Scientific Journals of Maritime University in Szczecin*, 47: 85-90.

Kulbiej E. & Wołejsza P. 2016. An analysis of possibilities how the collision between m/v 'Baltic Ace' and m/v 'Corvus J' could have been avoided. *Annual of Navigation* 23: 121-135.

Lenart A. S. 2012. Approach parameters in marine navigation.

Pietrzykowski Z., Wołejsza P., Borkowski P. 2016. Decision Support in Collision Situations at Sea. *The Journal of Navigation* 0, 1–18.

Ships Manoeuvring - Practical Aspects

Ship Course Planning and Course Keeping in Close Proximity to Banks Based on Optimal Control Theory

H. Liu, C. Shao, N. Ma & X.C. Gu
Shanghai Jiao Tong University, Shanghai, China

ABSTRACT: Ship navigation safety in restricted water areas is of great concern to crew members, because ships proceeding in close proximity to banks will undertake a powerful influence from the so-called ship-bank interaction. The purpose of this paper is to apply the optimal control theory to help helmsmen plan ships' trajectory and maintain the expected course in restricted waters. To achieve this objective, the motion of a very large crude carrier (VLCC) close to a bank is modeled with the linear equations of manoeuvring and the influence of bank effect on the ship hydrodynamic force is considered in the model. State-space framework is cast in a Multiple-Input Multiple-Output (MIMO) system, where the scheme of model predictive control (MPC) is designed for course planning and the linear quadratic regulator (LQR) is used for course keeping. Simulation results show that the control methods effectively work in ship trajectory planning and course keeping with varying ship-bank distances. And the advantage of adopting speed variation as the second control input is obvious.

1 INTRODUCTION

Nowadays the ports around the world witness larger size and higher speed ships. The safe navigation of ships, especially in narrow shipping waterways, is of the most concern to the maritime authorities (Li et al. 2012). The higher risk of accident in restricted waters like ports, straits and channels activates the study on the course control when proceeding in restricted waters.

Course control in restricted waters is usually for ship berthing to dock or avoiding certain obstacles like other ships or constructions (as shown in Figure 1) in the waterways. The problem involves not only the trajectory plan, but also the course keeping in the designed route with the disturbance of the ship-bank interaction. When a ship moves in proximity to a bank, it will experience a suction force towards the bank and a yawing moment, which is called ship-bank interaction or bank effect. Many published studies presented the way to estimate the bank induced forces (Norrbin 1974, Ch'ng et al. 1993), and also pointed that the ship-bank interaction can be quite huge to influence the ship's course keeping ability as well as directional stability (Fujino 1968, Sano et al. 2014, Liu et al. 2016).

The under-actuated nature of the ship manoeuvring problems, namely with more variables to be controlled than the number of control actuators, makes the control problem quite challenging, because some commonly used control techniques are not suitable to this nature (Fossen 2003). Such problem has received a lot of attention from the control community. Pettersen & Nijmeijer (2001) provided a high-gain, local exponential tracking result. By applying a cascade approach, a global tracking result was obtained in Lefeber et al. (2003). Path following approach based on the line-of-sight method was proposed by Moreira et al (2007) and then presented in more researchers' works (Børhaug et al. 2008, Skjetne et al. 2011). However, the performance of control systems is in fact limited by the constraints on the control inputs. The aforementioned works has not taken the limitations into account.

To tackle this problem, the optimal control theory considering physical constraints was put forward to solve the problems of ship path planning (Djouani & Hamam 1995). Many researches based on that to design the ship path controller with various models like linear quadratic regulator (LQR) (Thomas & Sclavounos 2006, Mucha & el Moctar 2013) and evolutionary algorithms (Szłapczyński 2013). The model predictive control (MPC) approach, which allows multiple control inputs to response to the ship motion in the under-actuated system, has been used for path following (Oh & Sun 2009, Li et al. 2010), and improved with disturbance compensating unit to apply to ship heading control (Li & Sun 2012). A significant work by Feng et al. (2013) introduced the LQR to the MPC scheme to improve the robust control quality, so that the system can achieve the offset-free path following under external disturbances.

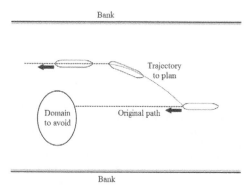

Figure 1. Trajectory control for avoidance in a waterway.

Enlightened by the previous work, this paper proposed a modified MPC scheme for the ship course planning in close proximity to a bank, which combines the LQR to overcome the bank induced forces and maintain the ship course. The rationale of the MPC scheme aiming at trajectory planning and the introduced unit of LQR for course keeping and system stabilization is elaborated in section 2. The illustration of bank induced forces and the corresponding hydrodynamic derivatives is given in section 3. In section 4, the performance of the controller is evaluated on the condition of different ship-bank distances, and the conclusion is drawn in the section 5.

2 NUMERICAL MODEL

2.1 Maneuvering theory

Two coordinate systems are used as shown in Figure 2. The space-fixed coordinate system $O_0\xi\eta\zeta$ is the frame with the positive ζ axis pointing into the page. The ship-fixed coordinate system $Oxyz$ is defined with the origin O at the mid-ship and positive z axis directing downwards. The ship is expected to move in the direction of ξ axis at a speed U. The bank induced forces will change ship motion which generates velocities u, v and turning rate r as well as heading angle ψ. δ denotes the rudder angle and y_{bank} denotes the separation distance between the ship and the bank.

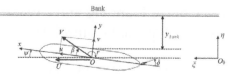

Figure 2. Coordinate systems.

Therefore, the 2-DoF linear maneuvering model of ship horizontal motion has been augmented with the components of hydrodynamic force in terms of ship-bank interaction, which is shown in the following equations:

$$\left(m+m_y\right)\dot{v}+\left(m+m_x\right)ur+mx_G\dot{r}=Y \tag{1}$$

$$\left(I_z+J_z\right)\dot{r}+mx_G\left(\dot{v}+ur\right)=N \tag{2}$$

where m is the ship mass and I_z is the yaw moment of inertia. m_x, m_y and J_z are the added mass and added moment of inertia for the surge, sway and yaw motion. Y and N represent the hydrodynamic sway force and yaw moment acting on the ship, which include hydrodynamic inertia terms and will be expressed by the following polynomial equations.

$$Y=Y_{\dot{r}}\dot{r}+Y_v v+Y_r r+Y_{\eta=0}+Y_\eta\eta+Y_\delta\delta \tag{3}$$

$$N=N_{\dot{v}}\dot{v}+N_v v+N_r r+N_{\eta=0}+N_\eta\eta+N_\delta\delta \tag{4}$$

Herein the force and moment caused by the bank effect are expressed as $Y_{\eta=0}+Y_\eta\eta$ and $N_{\eta=0}+N_\eta\eta$. The subscript "$\eta=0$" means the constant bank induced forces on the initial lateral position, and $Y_\eta\eta$ and $N_\eta\eta$ means the force due to the ship's lateral displacement η. The four items are called asymmetric derivatives.

All variables in the Equations 1 and 2 above are nondimensionalized in terms of ship length L, draft d, speed U and water density ρ through the equations as follows.

$$m'=\frac{m}{0.5\rho L^2 d},\quad I'_z=\frac{I_z}{0.5\rho L^4 d},\quad \dot{v}'=\frac{\dot{v}L}{U^2},\quad \dot{r}'=\frac{\dot{r}L^2}{U^2}$$

$$u'=\frac{u}{U},\quad v'=\frac{v}{U},\quad r'=\frac{rL}{U},\quad \eta'=\frac{\eta}{L},\quad x'_G=\frac{x_G}{L} \tag{5}$$

$$Y'=\frac{Y}{0.5\rho LdU^2},\quad N'=\frac{N}{0.5\rho L^2 dU^2}$$

With the substitution of Equations 3, 4 into 1, 2 and nondimensionalization, equations of ship maneuvering is written as:

$$\mathbf{M}\begin{bmatrix}\dot{v}'\\\dot{r}'\end{bmatrix}=\mathbf{N}\begin{bmatrix}v'\\r'\end{bmatrix}+\mathbf{L}[\eta']+\mathbf{F}_R[\delta]+\mathbf{F}_B \tag{6}$$

where the denotations are as follows.

$$\mathbf{M} = \begin{bmatrix} -Y_v' + m' & -Y_r' + m'x_G' \\ -N_v' + m'x_G' & -N_r' + I_z' \end{bmatrix} \tag{7}$$

$$\mathbf{N} = \begin{bmatrix} -Y_v' & -Y_r' + m' \\ -N_v' & -N_r + m'x_G' \end{bmatrix} \tag{8}$$

$$\mathbf{L} = \begin{bmatrix} Y_\eta' \\ N_\eta' \end{bmatrix} \quad \mathbf{F}_R = \begin{bmatrix} Y_\delta' \\ N_\delta' \end{bmatrix} \quad \mathbf{F}_B = \begin{bmatrix} Y_{\eta=0}' \\ N_{\eta=0}' \end{bmatrix} \tag{9}$$

With variables v, r and η existing in (6), the heading angle ψ could be added and the following expressions can be deduced.

$$\begin{bmatrix} \dot{v}' \\ \dot{r}' \\ \dot{\psi}' \\ \dot{\eta}' \end{bmatrix} = \begin{bmatrix} -\mathbf{M}^{-1}\mathbf{N} & \begin{matrix} 0 \\ 0 \end{matrix} & \mathbf{M}^{-1}\mathbf{L} \\ 0 & 1 & 0 & 0 \\ 1 & 0 & 0 & 0 \end{bmatrix} \begin{bmatrix} v' \\ r' \\ \psi \\ \eta' \end{bmatrix} + \mathbf{M}^{-1}\left(\mathbf{F}_R[\delta] + \mathbf{F}_B\right) \tag{10}$$

Since this dynamic system accepts a small scale variation of speed that is defined as Δu, the bank induced suction force and yaw moment in relation to forward speed can be expressed in the following form.

$$F_{bank} = 0.5\rho L d Y_{\eta=0}' \left(U + \Delta u\right)^2$$
$$= 0.5\rho L d Y_{\eta=0}' \left(U^2 + 2U\Delta u + \Delta u^2\right) \tag{11}$$

$$F_{bank} = 0.5\rho L^2 d M_{\eta=0}' \left(U + \Delta u\right)^2$$
$$= 0.5\rho L^2 d M_{\eta=0}' \left(U^2 + 2U\Delta u + \Delta u^2\right) \tag{12}$$

The last term in the equations above is of second order and negligible, while the second term directly shows the contribution of the perturbation velocity Δu to the bank induced forces. Keeping only leading order terms in the perturbation, the matrix of ship-bank hydrodynamics is rewritten as:

$$\mathbf{F}_B = \begin{bmatrix} \dfrac{F_{bank}}{0.5\rho L d U^2} \\ \dfrac{M_{bank}}{0.5\rho L^2 d U^2} \end{bmatrix} = \begin{bmatrix} Y_{\eta=0}' \\ N_{\eta=0}' \end{bmatrix} \cdot \left(\dfrac{U^2 + 2U\Delta u}{U^2}\right)$$
$$= \begin{bmatrix} Y_{\eta=0}' \\ N_{\eta=0}' \end{bmatrix} \left(1 + 2\Delta u'\right) \tag{13}$$

Now the linearized ship maneuvering model can be written into the state equation with matrix form:

$$\dot{\mathbf{x}} = \mathbf{A}\mathbf{x} + \mathbf{B}\mathbf{u} + \mathbf{E} \tag{14}$$

where

$$\mathbf{x} = \begin{bmatrix} v' & r' & \psi & \eta' \end{bmatrix}^T, \quad \mathbf{u} = \begin{bmatrix} \delta & \Delta u \end{bmatrix}^T \tag{15}$$

$$\mathbf{A} = \begin{bmatrix} -\mathbf{M}^{-1}\mathbf{N} & \mathbf{M}^{-1}\mathbf{L} & \begin{matrix} 0 \\ 0 \end{matrix} \\ 0 & 1 & 0 & 0 \\ 1 & 0 & 0 & 1 \end{bmatrix}, \quad \mathbf{B} = \begin{bmatrix} \mathbf{M}^{-1}[\mathbf{F}_R & 2\mathbf{F}_B] \\ 0 & 0 \\ 0 & 0 \end{bmatrix}$$

$$\mathbf{E} = \begin{bmatrix} \mathbf{M}^{-1}\mathbf{F}_B \\ 0 \\ 0 \end{bmatrix} \tag{16}$$

2.2 Course planning control

The continuous-time model of Equation 14 can be transformed into discrete-time scheme given a specific sampling time, which is:

$$\mathbf{x}_{k+1} = \mathbf{A}_d\mathbf{x}_k + \mathbf{B}_d\mathbf{u}_k + \mathbf{E}_d \tag{17}$$

The standard MPC scheme can be formulated based on the discrete model but unable to eliminate the steady cross-track error due to the existing disturbance of bank effect. So the MPC scheme proposed here is derived from the steady state system equations under the steady disturbance $\mathbf{E}_d$,

$$\begin{cases} \mathbf{x}_\infty = \mathbf{A}_d\mathbf{x}_\infty + \mathbf{B}_d\mathbf{u}_\infty + \mathbf{E}_d \\ \mathbf{y}_\infty = \mathbf{C}_d\mathbf{x}_\infty \end{cases} \tag{18}$$

where $\mathbf{u}_\infty$ is the steady state control; $\mathbf{y}_\infty$ is the output that is controlled to approach the desired value. Considering the following equation:

$$\mathbf{u}_\infty = -\left[\mathbf{C}_d\left(I - \mathbf{A}_d\right)^{-1}\mathbf{B}_d\right]^{-1}\mathbf{C}_d\left(I - \mathbf{A}_d\right)^{-1}\mathbf{E}_d \tag{19}$$

However, the current error dynamic formulation, which means that $\mathbf{A}_d$ has two eigenvalues at 1, leads to a singular $(I-\mathbf{A}_d)$ matrix. Therefore the precise control response to balance the bank induced forces cannot be obtained. To resolve this problem, the original system needs to be stabilized first through feedback while the MPC scheme determines the optimal control input at each time step. In this study, the stabilization is achieved by an LQR controller. The cost function and weighting matrices for the LQR controller are chosen to be the same as for the MPC. More specifically, the control input at each time step is divided into two parts:

$$\mathbf{u}_k = \mathbf{u}_k^{MPC} + \mathbf{u}_k^{LQR} = \mathbf{u}_k^{MPC} - \mathbf{K}_{LQR}\mathbf{x}_k \tag{20}$$

And the modified system dynamics will be:

$$\mathbf{x}_{k+1} = \mathbf{A}_d\mathbf{x}_k + \mathbf{B}_d\left(\mathbf{u}_k^{MPC} - \mathbf{K}_{LQR}\mathbf{x}_k\right)$$
$$= \left(\mathbf{A}_d - \mathbf{B}_d\mathbf{K}_{LQR}\right)\mathbf{x}_k + \mathbf{B}_d\mathbf{u}_k^{MPC} = \overline{\mathbf{A}}_d\mathbf{x}_k + \mathbf{B}_d\mathbf{u}_k^{MPC} \tag{21}$$

With the stable system, the corresponding control input to overcome the ship-bank interaction forces can be calculated by:

$$\mathbf{u}_\infty^{MPC} = -\left[\mathbf{C}_d\left(I-\overline{\mathbf{A}}_d\right)^{-1}\mathbf{B}_d\right]^{-1}\mathbf{C}_d\left(I-\overline{\mathbf{A}}_d\right)^{-1}\mathbf{E}_d \quad (22)$$

It can be seen from Equation 21 and Equation 22 that the function of LQR includes: 1) to avoid the singular matrix $(I-\mathbf{A}_d)$ as the system varies at each time step; 2) to stabilize the part of course keeping in the control system so that the MPC algorithm can be run in an offset-free situation.

As to the standard MPC scheme, first we define the cost function to be minimized as:

$$J\left(\mathbf{U}_k^{MPC};\mathbf{x}_k\right) = \sum_{j=0}^{N_p-1}\left[\mathbf{x}_{k+j}^T\mathbf{Q}\mathbf{x}_{k+j} + \right.$$
$$\left. \left(\mathbf{u}_{k+j}^{MPC}-\mathbf{u}_\infty^{MPC}\right)^T\mathbf{R}\left(\mathbf{u}_{k+j}^{MPC}-\mathbf{u}_\infty^{MPC}\right)\right] \quad (23)$$

$$\mathbf{U}_k^{MPC} = \left[\mathbf{u}_k^{MPC},\mathbf{u}_{k+1}^{MPC},\cdots,\mathbf{u}_{k+N_p-1}^{MPC}\right] \quad (24)$$

N_P is the prediction horizon; $\mathbf{Q}$ and $\mathbf{R}$ are the weighting matrices for the states and control inputs, respectively. $\mathbf{U}_k^{MPC}$ is the optimal control sequence, in which the $\mathbf{u}_k^{MPC}$ only is remained by the MPC scheme as the actual control input to be implemented. Notice that $\mathbf{u}_\infty^{MPC}$ is included in the cost function to achieve offset-free path planning. The optimal control sequence is subject to:

$$-\mathbf{u}_{max}-\mathbf{u}_{k+j}^{LQR} \leq \mathbf{u}_{k+j}^{MPC} \leq \mathbf{u}_{max}-\mathbf{u}_{k+j}^{LQR} \quad (25)$$

$$\begin{cases}\mathbf{u}_{k+j-1}^{MPC}-\mathbf{u}_{k+j}^{LQR} \leq \Delta\mathbf{u}_{max}T_s+\left(\mathbf{u}_{k+j}^{LQR}-\mathbf{u}_{k+j-1}^{LQR}\right)\\\mathbf{u}_{k+j}^{MPC}-\mathbf{u}_{k+j-1}^{LQR} \leq \Delta\mathbf{u}_{max}T_s-\left(\mathbf{u}_{k+j}^{LQR}-\mathbf{u}_{k+j-1}^{LQR}\right)\end{cases} \quad (26)$$

where $j=0, 1 ,...,N_P$; $\mathbf{u}_{max}$ is the rudder saturation deflection; $\Delta\mathbf{u}_{max}$ is the maximum rudder turning rate; T_s is the sampling time.

3 ASYMMETRIC HYDRODYNAMIC FORCES

Table 1 lists the principal dimensions of the test model KVLCC2, which is a crude oil tanker as a benchmark for study of ship hydrodynamics. The PMM test was conducted in the CWC at Shanghai Jiao Tong University. The dimensions of measuring section are 8.0m×3.0m (water width) ×1.6m (water depth). The ship-bank interaction force and moment is measured by off-centerline straight towing test. The velocity of water flow was set at 0.703m/s, which corresponds to the Froude number $F_n=0.142$. As Figure 3 shows, the straight towing tests for determining the asymmetric hydrodynamic derivatives were conducted by locating the ship off the centerline with the distance $b=0$m to $b=0.95$m, which is from $y_{bank}=2.8B$ to $y_{bank}=0.8B$ accordingly.

Table 1. Principle dimensions of the KVLCC2

Parameters		Full scale	Model
Length between perpendiculars	L m	320.00	2.4850
Breadth	B m	58.00	0.4504
Design draft	d m	20.80	0.1615
Displacement	Δ m^3	312540	0.1464
LCB from Mid-ship	x_B m	11.04	0.086
Scale		128.77:1	

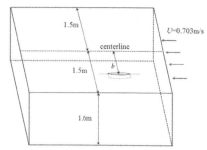

Figure 3. The arrangement of model ship in testing the asymmetric hydrodynamics.

Figure 4 plots the nondimensional hydrodynamic forces of ship-bank interaction. As the sign of the values indicate, the suction force to the bank wall and the bow-out moment were acting on the ship hull and they are increasing as the ship approaches the bankside. The value of Y' and N' also equals $Y'_{\eta=0}$ and $N'_{\eta=0}$ in the current position. In this paper the case of $b=0$m, 0.5m and 0.65m, corresponding to $y_{bank}=2.8B$, $1.7B$ and $1.35B$, are selected as $\eta=0$, respectively. The points around each selected case are used to fit a curve by the polynomial regression. Then the first-order coefficient for the polynomial is the asymmetric derivative Y'_η or N'_η. The curves of case $b=0$m, $b=0.5$m and $b=0.65$m are shown in Figure 5. The coefficients Y'_η and N'_η in the three locations are presented in Table 2.

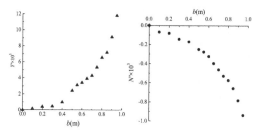

Figure 4. Asymmetric hydrodynamic force and moment.

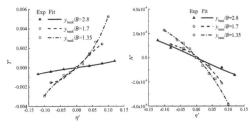

Figure 5. Asymmetric hydrodynamic force and moment versus η'.

Table 2. Asymmetric derivatives

	$y_{bank}=2.8B$	$y_{bank}=1.7B$	$y_{bank}=1.35B$
Y'_η	0.00105	0.0255	0.0237
N'_η	-0.00112	-0.00174	-0.00309

4 SIMULATION OF COURSE CONTROL

The controller focuses on the trajectory planning problem illustrated in Figure 1. Firstly the MPC scheme is tuned in the condition of $h/L=0.5$ where the bank effect can be ignored. The aim of the scheme is to make the ship proceed on a parallel course to the bank with a lateral distance of from the ship's initial course. The actuator constraints are $\mathbf{u}_{max}=[0.524\ 0.08]^T$ and $\Delta\mathbf{u}_{max}=[0.21\ 0.03]^T$; the sampling time $T_s=1$sec.

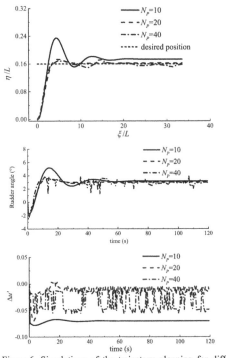

Figure 6. Simulations of the trajectory planning for different prediction horizons.

4.1 Controller tuning

The length of prediction horizon is first tuned by the simulations as shown in Figure 6. It reveals that the simulation results converge to the disired lateral value when N_P approaches to 20. And this achieves a good balance between the course control performance and the time consumption. So the length of prediction horizon is set as $N_P =20$ for further tuning of the weighting matrices.

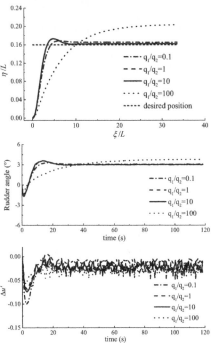

Figure 7. Simulations of the trajectory planning for different ratios between q1 and q2.

The weighting matrices $\mathbf{Q}$ and $\mathbf{R}$ shape the closed loop response to achieve the desired performance in the form of $\mathbf{Q} =\{0,0,\ q_1,q_2\}$ and $\mathbf{R} =\{r_1,r_2\}$. The tuning for the matrices is actually the trial-and-error program to find the optimum ratio between q_1 and q_2 as well as the ratio between r_1 and r_2. The ratio q_1/q_2 sets the preference for the controller to eliminate the cross-track errors while the ratio r_2/r_1 sets the preference for the controller to use one actuator over the other. Firstly $q_1=10$ and $r_1=1$, $r_2=2$ are selected while the value of q_2 is varied to examine the simulation results. As shown in Figure 7, the control scheme eliminates the cross-track error effectively with the range of $q_1/q_2=1$ to 10, but the path following performance deteriorate when q_1/q_2 reach 100. The amplitude of speed reduction also increases as the value of q_2 grows. Compared to the ratio $q_1/q_2=10$, the system succeeds in yielding a smaller fluctuation of the speed reduction during the whole process of the ship course adjustment and stabilization when $q_1/q_2=1$. So $q_1=q_2=10$ is chosen as the value of the **Q** matrix.

Next, the tuning of r_2/r_1 is conducted by varying r_2 with the value of r_1 kept at 1. The results are presented in Figure 8. In the range from $r_2/r_1=0.5$ to $r_2/r_1=10$, the simulation results of ship trajectory and the rudder reflection do not change quite large, which proves the finding by Feng et al. (2013) that the **R** matrix has less influence on the course changing performance than the **Q** matrix. But too small r2 will put larger weight on the usage of ship speed change and cause the failure of the speed variation constraints. As the ship approaches to the desired path with the least overshoot rudder angle at the ratio $r_2/r_1=2$, the value $r_1=1$ and $r_2=2$ is selected for the **R** matrix in the following simulations.

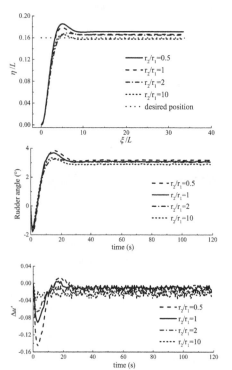

Figure 8. Simulations of the trajectory planning for different ratios between r1 and r2.

4.2 *Simulation in close proximity to banks*

The tuned scheme is further used for the course planning of KVLCC2 ship with two different ship-bank distances, i.e. $y_{bank}=1.7B$ and $y_{bank}=1.35B$. The combination of rudder deflection and speed variation as Multiple-Input Multiple-Output control (which is denoted as MIMO in Figure 9, 10) is compared with the performance of rudder-only controller (as Rudder-only in Figure 9, 10). The results show that

the favourable prediction horizon and weighting matrices chosen for multiple input system also work for the single rudder case. Comparison of the steady state error of η/L indicates that the MIMO control scheme can eleminate the cross-track error more effectively than the scheme that uses the rudder only. This can prove that the speed variation input has a positive effect on improving the control process. It is also noticeable in Figure 9, 10 that due to the offset-free scheme, there will be nonzero steady rudder angle as well as a certain value of speed reduction to overcome the bank induced forces.

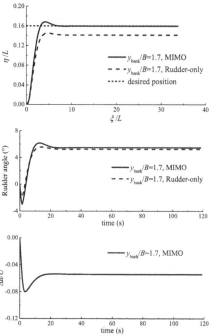

Figure 9. Simulations of the trajectory planning in close proximity to a bank ($y_{bank}=1.7B$)

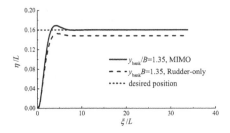

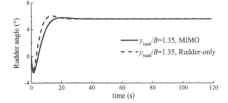

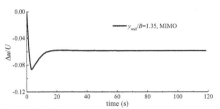

Figure 10. Simulations of the trajectory planning in close proximity to a bank (y_{bank}=1.35B).

5 CONCLUSIONS

The paper and proposes the modified MPC scheme to solve ship course planning in close proximity to the bank as well as counteracting bank effect. A linear maneuvering model including the hydrodynamic components of bank induced forces is introduced and transformed to state equations for the design of control system. The experimental approach to obtain the asymmetric derivatives shows that the bank induced hydrodynamic force acting on the ship is the suction force and bow-out moment. To stabilize the error dynamics formulation and to get a reliable solution of the control response to bank effect, the LQR unit is introduced into the system. And then the steady control input to overcome the disturbance due to bank effect is implemented into the MPC scheme. The controller is tuned first to obtain the suitable prediction horizon and weighting matrices for improving performance. Then, simulations of varying ship-bank distances are conducted. The feasibility of the offset-free MPC scheme in trajectory planning for ship proceeding close to a bank is proved. The advantage of adopting speed variation as the second control input is also demonstrated in that the steady state errors under ship-bank interactions can be reduced.

ACKNOWLEDGEMENT

The research is supported by the National Key Basic Research Program of China: No. 2014CB046804, and the China Ministry of Education Key Research Project "KSHIP-II Project" (Knowledge-based Ship Design Hyper-Integrated Platform): No. GKZY010004. Authors also thank Mr. Yi Dai and Dr. Fei Wang for their contribution to the model tests of this project.

REFERENCES

Børhaug, E., Pavlov, A., Pettersen, K. Y. 2008. Integral LOS control for path following of underactuated marine surface vessels in the presence of constant ocean currents. *Proc. 47th IEEE Conference on Decision and Control*: 4984-4991.

Ch'ng, P. W., Doctors, L. J., Renilson, M. R.1993. A method of calculating the ship-bank interaction forces and moments in restricted water. *International Shipbuilding Progress* 40(421): 7–23.

Djouani, K. & Hamam, Y. 1995. Minimum time-energy trajectory planning for automatic ship berthing. *IEEE Journal of Oceanic Engineering* 20(1): 4-12.

Feng, P. Y., Sun, J., Ma, N. 2013. Path following of marine surface vessels using bow and aft rudders in wave fields. *Control Applications in Marine Systems* 9(1): 120-125.

Fossen, T. I., Breivik, M., Skjetne, R. 2003. Line-of-sight path following of underactuated marine craft. *Proceedings of 6th IFAC conference on manoeuvring and control of marine craft*. Girona, Spain.

Fujino, M. 1968. Studies on manoeuvrability of ships in restricted waters. *Selected Papers Journal of Society of Naval Architecture of Japan* 124: 51-72.

Lefeber, E., Pettersen, K. Y., Nijmeijer, H. 2003. Tracking control of an underactuated ship. *IEEE Transactions on Control Systems Technology* 11: 52–61.

Li, S.Y., Meng, Q., Qu, X.B. 2012. An overview of maritime waterway quantitative risk assessment models. *Risk Analysis* 32(3): 496-512.

Li, Z., Sun, J., Oh, S. 2010. Handling roll constraints for path following of marine surface vessels using coordinated rudder and propulsion control. *American Control Conference* 6010-6015.

Li, Z. & Sun, J. 2012. Disturbance compensating model predictive control with application to ship heading control. *Control Systems Technology IEEE Transactions* 20(1): 257-265.

Liu, H., Ma, N., Gu, X. C. 2016. Numerical simulation of PMM tests for a ship in close proximity to sidewall and maneuvering stability analysis. *China Ocean Engineering* 30(6):884-897.

Moreira, L, Fossen, T I, Soares, C G. 2007. Path following control system for a tanker ship model. *Ocean Engineering* 34(14-15): 2074-2085.

Mucha, P., el Moctar, O. 2013. Ship-Bank interaction of a large tanker and related control problems. *Proc. of the 32nd ASME International Conference on Ocean, Offshore and Arctic Engineering (OMAE 2013)*. Nantes, France.

Norrbin, N. H. 1974. Bank effects on a ship moving through a short dredged channel. *Proceedings of the 10th Symposium on Naval Hydrodynamics*: 71–87. Cambridge, USA.

Oh, S. R. & Sun, J. 2010. Path following of underactuated marine surface vessels using line-of-sight based model predictive control. *Ocean Engineering* 37(2):289-295.

Pettersen, K. Y. & Nijmeijer, H. 2001. Underactuated ship tracking control: Theory and experiments. *International Journal of Control* 74(14): 1435–1446.

Sano, M., Yasukawa, H., Hata, H. 2014. Directional stability of a ship in close proximity to channel wall. *Journal of Marine Science and Technology* 19(4): 376-393.

Skjetne, R., Jørgensen, U., Teel, A. R. 2011. Line-of-sight path-following along regularly parametrized curves solved as a generic maneuvering problem. *Proc. 50th IEEE Conference on Decision and Control*: 2467-2474.

Szłapczyński, R. 2013. Evolutionary sets of safe ship trajectories with speed reduction manoeuvres within traffic separation schemes. *Polish Maritime Research* 21(1): 20-27.

Thomas, B.S. & Sclavounos, P.D. 2007. Optimal control theory applied to ship maneuvering in restricted waters. *Journal of Engineering Mathematics* 58(1): 301-315.

Fuzzy Self-tuning PID Controller for a Ship Autopilot

M. Tomera
Gdynia Maritime University, Gdynia, Poland

ABSTRACT: The paper examines the application of a fuzzy self-tuning controller for ship course steering. The controller is composed of fuzzy and linear PID controllers. Following the control system heading error and its change combined with fuzzy control rules, the fuzzy controller can adjust parameters of PID controller. The initial values of PID gains were calculated with the use of the classical linear control theory and the placement method. The PID controller was synthesized using the Nomoto model of ship dynamics identified on the basis of standard Kempf's zigzag manoeuvre and the Ant Colony Optimization algorithm (ACO). The ACO algorithm is a bio-inspired optimization method that has proven its success through various combinatorial optimization problems. The research of the designed control system was carried out on the training ship Blue Lady in the Ship Handling Research and Training Centre on the lake Silm in Ilawa/Kamionka. The results of full-scale trials have revealed that the proposed scheme has smaller overshoot and shorter settling time.

1 INTRODUCTION

The automatic steering device commonly known as the ship steering autopilot controls the motion of the vessel in accordance with passed course changes, or a set track. The development of ship steering autopilots is the field where control theories were applied very early and good results were achieved. In the early 1920s Minorsky (1922) published the theoretical analysis of automatic steering and the specification of the three-term or proportional-integral-derivative (PID) controller for course-keeping control. Whilst Minorsky developed the PID controller for automatic ship steering, PID controllers have largely remained an industrial standard in automatic control systems (Roberts 2008). Today, conventional PID controllers are extensively used for dynamic process control due to their simplicity of operation, ease of design, low cost, and effectiveness in the majority of linear systems. However it has been known that conventional PID controllers generally do not work well in nonlinear systems, higher order and time-delayed linear systems, and vague systems that have no precise mathematical model.

Ship autopilots designed based on proportional-integral-differential (PID) controllers are simple, reliable and easy to construct. However, dynamic characteristics of the ship change in the navigation process, following changes of ship's speed, load, sea conditions, and other factors. Consequently, their performance in various conditions is not as good as desired (Le et al. 2004). To overcome these difficulties, various modified types of conventional PID controllers, such as auto tuning and adaptive controllers, have been developed recently (Astrom et al. 1992). Moreover, some sophisticated autopilots are proposed which are based on advanced control engineering concepts, where the gain settings for proportional, derivative and integral terms of heading are adjusted automatically to suit the dynamics of the ship and environmental conditions. The list of such autopilots includes model reference adaptive control (Amerongen 1982), self-tuning (Mort & Linkens 1980), optimal (Zuidweg 1981), H_∞ theory (Fairbairn & Grimble 1990), model reference adaptive robust fuzzy control (Yang et al. 2003).

Nowadays, a tendency which gains much interest is the combination of various methodologies taking advantage of each one. One of most effective solutions here is a combination of the linear control theory and fuzzy logic control (FLC). He et al. (1993) applied the fuzzy self-tuning method to tune the gains of PID controller. The fuzzy self-tuning controller combining the fuzzy system with the traditional adaptive control has the same linear structure as the conventional PID controller but has self-tuned proportional, integral and derivative gains, which are nonlinear functions of the inputs signals.

Fuzzy control has appeared to be one of the most active and important applications of fuzzy sets theory since the first realization of the fuzzy controller using fuzzy logic by Mamdani (Mamdani 1974).

The essential part of the fuzzy logic controller is a set of linguistic control rules. The resultant control law has an analytical PID controller form and a linguistic form. Different applications of the fuzzy self-tuning PID controller can be found in the literature. This controller was examined in simulation tests oriented on controlling dynamic processes (Chen et al. 2009, Yang & Bian 2012), aircraft pitch control (Nurbaiti & Nurhaffizah 2012), and ship steering control (Shen & Guo 2008, Liu et al. 2010, Zhao et al. 2011).

Recently, several nonlinear controllers have also been proposed in the literature to deal with nonlinear steering conditions. However, attempts to use this method in real marine applications of ship steering have revealed numerous problems which needed solving. In order to obtain optimal quality of control for the designed nonlinear course controller, its parameters need tuning. Well known design methods presented in the literature make use of the backstepping method (Fossen & Strand 1998, Witkowska et al. 2007, Witkowska & Śmierzchalski 2012, Tomera 2010, Yoazhen et al. 2010, Perera & Soares 2012) and the sliding mode control (Tomera 2010, Yoazhen et al. 2010, Perera & Soares 2012).

The article presents results of examination and development of a course changing control system which controls the heading angle of a ship with the aid of a fuzzy self-tuning PID controller. The fuzzy self-tuning PID controller is combination of conventional PID and fuzzy logic control scheme. The performance of PID control strategy and fuzzy self-tuning with respect to the heading angle of ship

yaw dynamics was examined. The tests were performed on a training ship Blue Lady sailing on the lake Silm in the Ship Handling Research and Training Centre in Ilawa/Kamionka (Foundation 2012).

2 MATHEMATICAL MODEL OF SHIP DYNAMICS

The equations describing the horizontal motion of the ship (u, v, r) are well established. These equations can be derived from Newton's laws expressing conservation of linear and angular momentum, in the form given by Clarke (2003)

$$m(\dot{u} - vr - x_G r^2) = X \tag{1}$$

$$m(\dot{v} + ur + x_G \dot{r}) = Y \tag{2}$$

$$I_z \dot{r} + m x_G (\dot{v} + ru) = N \tag{3}$$

Here x_G is the distance from the centre of gravity to the midship point, m is the constant mass of the ship, I_z is the moment of inertia about the z-axis and u, v and r denote surge, sway and yaw angular velocity, respectively X and Y are the components of the hydrodynamic forces on the x- and y-axes, and N is the yaw moment about the z-axis. Abkowitz (1964) suggested the following functional form for X, Y and N

$$X = f(u, v, r, \delta, \dot{u}, \dot{v}, \dot{r}) \tag{4}$$

$$Y = f(u, v, r, \delta, \dot{u}, \dot{v}, \dot{r}) \tag{5}$$

$$N = f(u, v, r, \delta, \dot{u}, \dot{v}, \dot{r}) \tag{6}$$

He approximated the functions (4-6) with Taylor series expansions about the steady state condition $u = u_0$, $v = r = \delta = \dot{u} = \dot{v} = \dot{r} = 0$, where δ is the rudder angle.

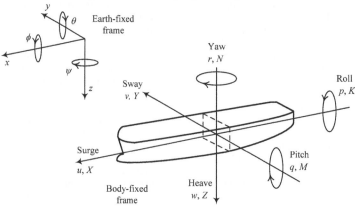

Figure 1. The coordinate systems for ship steering.

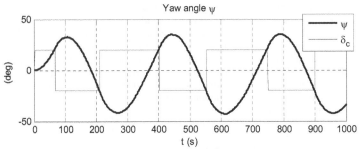

Figure 2. Time responses of zigzag manoeuvre (ψ – ship heading, δc – command rudder angle).

2.1 *Linear ship models*

For surface vessel moving at constant (or at least slowly varying) forward velocity

$$U = \sqrt{u^2 + v^2} \approx u \qquad (7)$$

the ship control problem is usually solved by applying the classical method which takes into account only two coupled movements, which are yaw and sway. Linearising the equations of motion (1-3) about $v = r = 0$ and $u = u_0$ gives the Davidson and Schiff (1946) model

$$\mathbf{M}\dot{\mathbf{v}} + \mathbf{N}(u_0)\mathbf{v} = \mathbf{B}\delta \qquad (8)$$

where $\mathbf{v} = [v \quad r]^T$. The equation (8) bases on the assumption that the cruise speed

$$u = u_0 \approx \text{constant} \qquad (9)$$

and that v and r are small. The matrices $\mathbf{M}$, $\mathbf{N}(u_0)$ and $\mathbf{B}$ in equation (8) are defined as

$$\mathbf{M} = \begin{bmatrix} m - Y_{\dot{v}} & mx_G - Y_{\dot{r}} \\ mx_G - N_{\dot{v}} & I_z - N_{\dot{r}} \end{bmatrix} \qquad (10)$$

$$\mathbf{N}(u_0) = \begin{bmatrix} -Y_v & mu_0 - Y_r \\ -N_v & mx_G u_0 - N_r \end{bmatrix} \qquad (11)$$

$$\mathbf{B} = \begin{bmatrix} Y_\delta \\ N_\delta \end{bmatrix} \qquad (12)$$

where Y_v -force against sway, Y_r - sway force due to yaw velocity, N_v - yaw moment due to sway, Y_δ, N_δ - hydrodynamic coefficients due to rudder, N_r - moment against yaw, moment due to rudder, $Y_{\dot{v}}$ - added mass due to sway acceleration, $Y_{\dot{r}}$ -sway force due to yaw acceleration, $N_{\dot{v}}$ - yaw moment due to sway acceleration, $N_{\dot{r}}$ -added mass due to yaw (SNAME 1950).

The second-order Nomoto model (Nomoto et al. 1957) can be derived by eliminating the sway velocity v from the earlier model derived by Davidson and Schiff (1946) and described by equation (4). This way we get the second order model:

$$\frac{r(s)}{\delta(s)} = \frac{K(sT_3 + 1)}{(sT_1 + 1)(sT_2 + 1)} \qquad (13)$$

where K is the static yaw rate gain, and T_1, T_2 and T_3 are time constants. The sea trial data-based identification indicates that the values of parameters T_2 and T_3 in (13) are not very different (Saari & Djemai 2012). This suggest that further simplification of (13) is possible, after which the first Nomoto model takes the form:

$$\frac{r(s)}{\delta(s)} = \frac{K}{sT + 1} \qquad (14)$$

where $T = T_1 + T_2 - T_3$ is the effective yaw rate time constant. Since the yaw rate is actually the time derivative of the ship heading angle $\dot{\psi} = r$, the model can be written in time domain as

$$T\ddot{\psi} + \dot{\psi} = K\delta \qquad (15)$$

The first Nomoto model defined by (9) is widely used in the ship steering autopilot design. The yaw dynamic is described by parameters K and T which can be easily identified from standard ship manoeuvring tests. In practice, ship steering autopilots are designed for heading angle control. Hence it is the transfer function relating the heading angle ψ and the rudder angle δ which is needed in autopilot design

$$\frac{\psi(s)}{\delta(s)} = \frac{K}{s(sT + 1)} \qquad (16)$$

2.2 *Identifying the Nomoto model for Blue Lady from zigzag manoeuvre*

A few ship manoeuvres have been proposed for testing the manoeuvre ability and identification of manoeuvring characteristics (Kempf 1932; Nomoto et al. 1957; Nomoto 1960). To determine the Nomoto model for the training ship Blue Lady the zigzag manoeuvre, which is most often used in such cases, has been applied. The zigzag time response is obtained by moving the rudder angle by 20o from an initially straight course. The rudder angle is held constant until the heading is changed to 20 degrees,

95

then the rudder is reversed. This process continues until a total of five rudder step responses have been completed. Common values for the rudder angle are 20/20 and 10/10, although other combinations can be applied as well. For larger ships the rudder angle of 10 degrees is recommended to reduce the time and waterspace required. The zigzag manoeuvre was first proposed by Kempf (1932). Hence, the name Kempf's zigzag manoeuvre also is used in the literature.

A widely used technique determines the parameters in the Nomoto model from the zigzag manoeuvre using an index estimator published in (Nomoto 1960). Further work in the field of parameters identification from zigzag trials was accomplished by Norrbin (1963). Journee (1970) developed a method to deal with overshoot and transient effects caused by rudder delay and limitations. Using Nomoto's first order model, a large number of zigzag manoeuvres have been calculated within a practical range of K and T values. These data were analyzed and the relation between the zigzag manoeuvring characteristics and the Nomoto parameters were presented in graphs. Based on the results of towing tank tests and sea-trials data, Azarsina & Williams (2013) used computer simulations to predict the Nomoto indices for Autonomous Underwater Vehicle during constant-depth zigzag manoeuvres. The Nomoto's first-order model for the rate of turn of the vehicle during horizontal zigzag manoeuvres in response to a square-wave input for the rudder deflection angle was solved analytically.

The well-known methods for determining the parameters of the Nomoto model (15) from the zigzag manoeuvre are quite complex and, therefore, in order to facilitate the process of finding the parameters K and T, in the present study ant colony optimization were used (Dorigo et al. 1996).

2.3 Ant colony optimization algorithm (ACO)

A colony of artificial ants cooperates to find good solutions, which are an emergent property of the ant's co-operative interaction. Based on their similarities with ant colonies in nature, ant algorithms are adaptive and robust and can be applied to different optimization problems.

The ant colony optimization algorithm (Fig. 4) was applied to estimate the manoeuvring indices K and T of the Nomoto's first order model from full-scale zigzag trial data, described as follows

$$T\ddot{\psi} + \dot{\psi} = K(\delta - \delta_r) \qquad (17)$$

where T is the time constant (s), K is the proportionality constant (1/s), δ is the level of the rudder angle (deg) and δ_r is the rudder angle at which the ship sails at a straight course (deg). The rudder servomechanism has not to be taken into

account, so in this case the rudder angle is equal to the command rudder angle ($\delta = \delta_c$).

For each parameter of Nomoto model ($K^1 = K$, $K^2 = T$, $K^3 = \delta_r$) was created a set of possible solutions between the minimum K^i_{min} and the maximum K^i_{max} values. To simplify, the proposed values are equally distributed between these bounds (Bououden et al. 2011).

$$K_{i1} = K^i_{min}, K_{i2} = K_{i1} + \frac{K^i_{max} - K^i_{min}}{J-1}, ..., K_{iJ} = K^i_{max}, \qquad (18)$$

where J is the number of possible values of identified parameter K^i.

Therefore, for each parameter K^i obtains the following set of potential values of the optimal solution

$$K^i \in \{K_{i1}, K_{i2}, ..., K_{iJ}\} \qquad (19)$$

Limits K^i_{min} and K^i_{max} for each searched parameters are contained in Table 1. This method allows to assign the distribution of each parameter of the vector $\boldsymbol{K} = [K^1, K^2, K^3]^T$, J possible values. Graphical representation of the optimized problem is shown in Figure 3, where searched parameters are distributed in three vectors. Each of the possible values is represented by one node. The problem is to find the best parameter combination that minimizes the cost function (20).

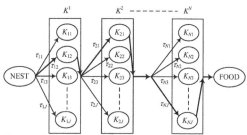

Figure 3. Graphical representation of the optimization problem by ACO.

Table 1. Bounded values of parameters in estimated Nomoto model.

Parameter	Min	Max
$K^1 = K$	0.0	1.0
$K^2 = T$	0.0	1000.0
$K^3 = \delta_r$	−2.0	0.0

The cost function used as the identification criterion is a discrete version of the integral of the absolute error (IAE)

$$J_c = \frac{1}{N}\sum_{i=1}^{N}|\Delta\psi_i| \qquad (20)$$

where N is the total number of iterations in the zigzag test simulations, $\Delta\psi_i$ is the i-th heading angle error between the measured heading and that obtained in the simulation making use of the Nomoto model. The ant optimization algorithm tend to minimise the value of function (20) so $\Delta\psi_i$ will be minimised too.

For each set of parameters, the node visited by the ant is selected as the value of parameter. Selection of a parameter value is based on pheromone trails between parameter vectors. The size of global pheromone matrix τ_{ij} is 3xJ, i =1, 2, 3 and j = 1, 2, ..., J.

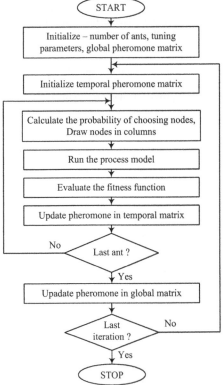

Figure 4. Graphical representation of the optimization problem by ACO.

The general steps of ACO (Fig. 4) are the following:

Step 1. Place a number of ants M in the nest. Set a maximum number of iterations. Initialize a global pheromone concentration τ_{ij} to each link (i,j)

$$\tau_{ij} = \frac{1}{J_c^0} \qquad (21)$$

where J^0_c is random obtained the expected value of the performance index.

Step 2. Initialize a temporal pheromone concentration τ_{ij}^k

$$\tau_{ij}^k = 0 \qquad (22)$$

Matrix τ_{ij}^k has the same size as the global pheromone matrix (21) and are stored the pheromone updates for the whole population of ants M.

Step 3. For every ant, iteratively is built a path to the food source using the following equation

$$p_{ij}^k(t) = \frac{\tau_{ij}^\alpha(t)}{\sum_{j=1}^{N}\tau_{ij}^\alpha(t)} \qquad (23)$$

Equation (23) represents the probability for an ant k located on a node i selects the next node denoted by j. N is the set of feasible nodes connected to node i with respect to ant k, τ_{ij} is the total pheromone concentration of link (i, j), and α is a positive constant used as a gain for the pheromone influence. After determining the transition probabilities for each node j, takes the draw method of the roulette wheel, the next node in the next vector. The greater the value of the probability of connection p_{ij}^k for the ant k, gives the greater chance of selection of a particular node.

Step 4. Run the simulation zigzag manoeuvre with Nomoto model (17) contained the values of parameters selected by the ant k.

Step 5. For the obtained results, compute the cost function J_c^k (20) and check if it is better then the best solution J_c^* obtained so far.

Step 6. Each ant after passing its path saves in the temporal matrix your pheromone trail to the formula

$$\tau_{ij}^k(t) \leftarrow \tau_{ij}^{k-1}(t) + \frac{1}{J_c^k}, \qquad k = 1,2,...,M \qquad (24)$$

Step 7. Update of the pheromone concentration using the following formula

$$\tau_{ij}(t+1) = (1-\rho)\tau_{ij}(t) + \tau_{ij}^k(t) + \rho\cdot\Delta\tau_{ij}(t) \qquad (25)$$

where $\Delta\tau_{ij}(t) = \frac{1}{J_c^*}$.

3 FUZZY SELF-TUNING CONTROLLER

The fuzzy self-tuning controller used for course changing concerns the change of ship heading as a response to a step command passed from the autopilot (Fig. 5). These commands are usually step changes in the heading reference which are then used by the controller to change the heading of the ship ψ by manipulating the rudder deflection angle δ. The value of the heading angle change is determined by the amplitude of the step command ψ_{ref} (McGookin et al. 2000).

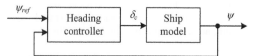

Figure 5. Ship course changing control system.

The course steering controller made use of fuzzy self-tuning PID digital controller (Fig. 6). The main idea of this control is to adjust PID controller parameters in order to improve the quality control of the entire system.

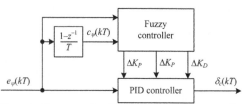

Figure 6. Structure of self-tuning fuzzy PID controller parameters.

The fuzzy PID controller has the following special properties:

1 It has the same linear structure as the conventional PID controller, but has self-tuned control gains, among which the proportional K_P, integral K_I and the derivative K_D are nonlinear functions of the input signals.

2 The controller design bases on the theory of classical discrete PID controller.

3 The fuzzy controller consists of the fuzzy rule base, the fuzzy inference system, and the fuzzification and defuzzification interfaces. The membership functions are simple triangular functions.

The inputs to the fuzzy controller are the error $e_\psi(kT)$ and the error derivative approximated by first-order difference $c_\psi(kT) = [e_\psi(kT) - e_\psi(kT-T)]/T$, while its outputs are ΔK_P, ΔK_I, ΔK_D changes in the PID controller (Fig. 6). This allows to modify the controller parameters according to the following relationships

$$K_P = K'_P + \Delta K_P \qquad (26)$$

$$K_I = K'_I + \Delta K_I \qquad (27)$$

$$K_D = K'_D + \Delta K_D \qquad (28)$$

where K_P, K_I, K_D are the PID controller parameters after modification, ΔK_P, ΔK_I, ΔK_D are the values of changes, and K_P, K_I, K_D are the values of controller parameters before modification. The initial values of the parameters used by the linear PID controller were determined by standard methods.

3.1 Digital PID controller

A classical conventional PID controller used to control ship's course changes is described by the following control rule

$$\delta_c(t) = K_P \left[e_\psi(t) + \frac{1}{T_i} \int_0^t e_\psi(\tau)d\tau + T_D \frac{de_\psi(t)}{dt} \right] \qquad (29)$$

where $e_\psi(t) = \psi_{ref}(t) - \psi(t)$ is the heading angle error between the desired and obtained heading, and δ_c is the set rudder deflection. A discrete version of this controller was implemented in the ship steering control system after replacing the continuous time by a series of discrete sampling points

$$t \approx kT_s, \quad k = 0,1,2,...$$

where T_s is the sampling period. Then, the continuous integral part of controller can be replaced by following approximate numerical integration

$$\int_0^t e_\psi(\tau)d\tau \approx \sum_{j=0}^{k} e_\psi(kT_s) = T_s \sum_{j=0}^{k} e_\psi(k) \qquad (30)$$

and the differential part can be substituted by the subtraction of the neighbouring error

$$\frac{de_\psi(t)}{dt} \approx \frac{e_\psi(kT_s) - e_\psi[(k-1)T_s]}{T_s} = \frac{e_\psi(k) - e_\psi(k-1)}{T_s} \qquad (31)$$

By substituting (31) and (30) to equation (29), the control law is formed as

$$\delta_c(k) = K_P e_\psi(k) + K_I \sum_{j=0}^{k} e_\psi(k) + K_D \left[e_\psi(k) - e_\psi(k-1) \right] \qquad (32)$$

where

$$K_I = \frac{K_P T_s}{T_I}, \quad K_D = \frac{K_P T_D}{T_s}$$

To avoid the sum in equation (32), we can get

$$\delta_c(k-1) = K_P e_\psi(k-1) + K_I \sum_{j=0}^{k-1} e_\psi(k)$$

$$+ K_D \left[e_\psi(k-1) - e_\psi(k-2) \right] \qquad (33)$$

By substracting (27) and (29) the output of the controller is obtained as

$$\delta_c(k) = \delta_c(k-1) + (K_P + K_I + K_D)e_\psi(k)$$

$$+ (-K_P - 2K_D)e_\psi(k-1) + K_D e_\psi(k-2) \qquad (34)$$

In the block diagram schematically shown in Figure 5, a mathematical model of ship dynamics described by equation (34) was used to control the ship on the course. The parameters of the linear PID controller were selected using the pole placement method with Nomoto model of the ship (16), in details described by Fossen (2002) where

$$K_P = \frac{\omega_n^2 T}{K} \qquad (35)$$

$$T_D = \frac{2\zeta\omega_n T - 1}{Kpk} \tag{36}$$

$$T_I = \frac{10}{\omega_n} \tag{37}$$

In equations (35-37) ω_n is the natural frequency computing by the formula

$$\omega_n = \frac{\omega_b}{\sqrt{1 - 2\zeta^2 + \sqrt{4\zeta^4 - 4\zeta^2 + 2}}} \tag{38}$$

where ω_b is the bandwidth and ζ is the relative damping ratio of the designed control system which was shown in Figure 4.

3.2 Fuzzy tuner

The parameters K_P, K_I and K_D which have to be tuned by the fuzzy tuner depend on the configuration of the fuzzy controller. The fuzzy tuner has two inputs: the course error (e_ψ) and the course error change (c_ψ), and three outputs: ΔK_P, ΔK_I and ΔK_D, as shown in Figure 7. The membership functions of these inputs and outputs are shown in Figures 8 and 9. The fuzzy sets for each input and output variables consist of seven linguistic values {NB, NM, NS, ZE, PS, PM, PB} representing the levels described as "negative big", "negative medium" and so on. The same set of linguistic values can be shown in the numeric form as {−3, −2, −1, 0, 1, 2, 3}, more convenient to write in computer software.

Let us notice in Figures 8 and 9 that the widths of the membership functions are parameterized by g_e and g_{uP} and will be adjusted by changing these values. The membership functions on the universe of discourse and the linguistic values for the second inputs $c_\psi(kT)$ are the same as for $e_\psi(kT)$, with the exception that the adjustment parameter is denoted by g_c. Respectively, the same situation is for the next outputs ΔK_I and ΔK_D, which have the same membership functions as for ΔK_P with the exception that the adjustment parameters are denoted by g_{uI} and g_{uD}. Therefore, there are five parameters for changing the fuzzy controller g_e, g_c, g_{uP}, g_{uI} and g_{uD}.

Assuming that there are seven membership functions on each input universe of discourse, there are 49 possible rules that can be put in the rule base. For the considered fuzzy controller consisting of two inputs and two outputs, each fuzzy rule takes the following form

$$IF \left(e_\psi \text{ is } E_i \right) AND \left(c_\psi \text{ is } C_j \right) THEN \tag{39}$$
$$\left(\Delta K_P \text{ is } \Delta K_{Pk} \right) \left(\Delta K_I \text{ is } \Delta K_{Ik} \right) \left(\Delta K_D \text{ is } \Delta K_{Dk} \right)$$

Generally, fuzzy rules are dependent on the plant to be controlled and the scope of designer's knowledge and experience.

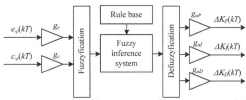

Figure 7. Structure of self-tuning fuzzy PID controller parameters.

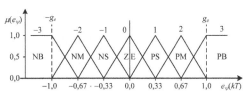

Figure 8. Membership functions for input $e_\psi(kT)$.

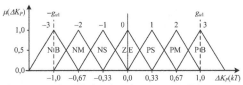

Figure 9. Membership functions for output $\Delta K_P(kT)$.

4 COURSE CHANGING COST FUNCTION

Evaluating the quality of operation of the examined controllers makes use of the cost function used as the design criterion in course steering and defined by equation (36). This function is a discrete version of the integral least squares criterion.

$$J_E = \psi_E + \lambda \delta_E \tag{40}$$

$$\psi_E = \frac{1}{N} \sum_{i=0}^{N} (\Delta\psi_i)^2, \quad \delta_E = \lambda \frac{1}{N} \sum_{i=0}^{N} \delta_i^2$$

$\lambda = 0.1$. The first term of this equation refers to the course change, where N is the total number of iterations in the control system process, λ is the weight coefficient, $\Delta\psi_i$ is the i-th heading error between the desired and obtained heading, δ_i is the i-th rudder deflection. The quantity $\Delta\psi$ gives an indication of how close the actual heading is to the desired heading, thus showing how well the controller is operating. The component δ is used to keep the magnitude of the rudder angle to a minimum. The main task of the weight coefficient λ is to amplify the course error term to the same level as the term representing the rudder deflection.

In addition, the overshoot M_p and the settling time t_s values are given. The overshoot M_p is the maximum value to which the ship heading overshoots its final value divided by its final value, expressed as a percentage. The settling time t_s is the time required for the transient of ship heading to decay to a small value and to be almost in the steady state. Here, the measure of smallness amounting to 5% is used.

5 RESULTS

In order to evaluate the quality of the derived algorithm of fuzzy self-tuning controller, simulation tests were performed using Matlab/Simulink and full-scale experiments using the training ship Blue Lady which was navigated on the lake Silm in Ilawa/Kamionka in the Ship Handling Research and Training Centre (Foundation 2012). Moreover, to compare the obtained results, additional tests were performed on a classical PID controller (34). The PID controller parameters were determined within the aid of the first Nomoto model identified from full-scale zigzag manoeuvre.

First, the ant colony optimization algorithm (Fig. 4) was applied to estimate the manoeuvring indices K, T and δ_r of the Nomoto's first order model (17) from full-scale zigzag trial data. The tests of parameter's estimation of the first order Nomoto model were performed and their results are collected in Table 2. The minimum value of the performance index (20) was obtained for Test no. 4.

Figure 10 presents the results of zigzag manoeuvres obtained in the full-scale test and in the simulation test making use of the identified first Nomoto model (17).

Table 2. Results of estimation the parameters for the first order Nomoto model using the ant colony optimization algorithm $\alpha = 3.0$, $\rho = 0.2$.

Test no.	K	T	δ_r	J_c
1	0.074	339.3	−1.565	12.7450
2	0.110	641.6	−1.469	14.4752
3	0.161	677.7	−1.455	16.6680
4	0.041	158.1	−1.620	9.7040
5	0.059	425.4	−1.485	14.4461
6	0.095	765.8	−1.315	16.4155
7	0.035	192.1	−1.729	11.2180
8	0.125	888.8	−1.691	16.8357
9	0.038	178.2	−1.167	15.2787
10	0.059	908.9	−0.941	20.2486

The best values of parameters of ten Nomoto model were collected in Table 3 and in this model the rudder angle δ_r at which the ship sails a straight course was omitted.

The parameters K_P, K_I and K_D of the PID controller with constant coefficients were calculated from equations (35-37) and are collected in Table 4. These parameters are also the initial values of the fuzzy self-tuning PID controller.

Table 3. Parameters estimated for the first Nomoto model (16).

	K	T
First order Nomoto model	0.041	158.1

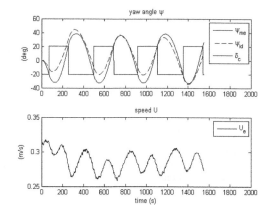

Figure 10. Zigzag test for identification of the Nomoto model (ψ_{me} – measured heading, ψ_{id} – heading from the identified first Nomoto model, δ_c – command rudder angle, U_e – ship velocity).

Table 4. Parameters calculated for the PID controller.

	K_P	K_I	K_D
PID controller (PID)	1.88	0.0048	92.1

The full control system with the fuzzy self-tuning controller shown in Figure 6 was modeled in Matlab/Simulink used for performing simulation tests. In the simulation tests the complex mathematical model of the training ship Blue Lady, described in details by Gierusz (2005), was applied. To obtain perfect performance of the whole control system, it is necessary for fuzzy tuner to follow fuzzy control rules, which is difficult to obtain. The knowledge of the functions performed by the parameters K_P, K_I and K_D along with observations of dynamic characteristics of the process being controlled and analyzes of numerous simulations, have made the basis for formulation of the fuzzy rule base presented in tabular form in Tables 5-7.

Table 5. Rule table for ΔK_P (Fig. 5)

		E_i						
		−3	−2	−1	0	1	2	3
	−3	3	3	2	1	1	1	0
	−2	3	3	1	1	1	0	1
	−1	2	1	2	1	0	1	1
Cj	0	−1	0	0	0	0	0	−1
	1	1	1	0	1	2	1	2
	2	1	0	1	1	1	3	3
	3	0	1	1	1	2	3	3

In these tables the premises for the input e_ψ are represented by the the linguistic values found in the top row, the premises for the input c_ψ are represented by the linguistic values in the leftmost column, and the linguistic values representing the consequents for each of 49 rules can be found at the

100

intersections of the row and column of the appropriate premises.

Table 6. Rule table for ΔK_I (Fig. 5)

		E_i					
	−3	−2	−1	0	1	2	3
−3	0	0	0	0	0	0	1
−2	0	0	−2	−1	0	0	0
−1	0	−1	−3	0	0	1	0
Cj 0	0	−1	−1	0	1	1	0
1	0	−1	0	0	3	1	0
2	0	0	0	0	2	0	0
3	1	0	0	0	0	0	0

Table 7. Rule table for ΔK_D (Fig. 5)

		E_i					
	−3	−2	−1	0	1	2	3
−3	0	0	−2	−3	−2	−1	0
−2	0	0	0	−2	−1	−1	−1
−1	−1	0	−1	−1	−1	−1	−1
Cj 0	−1	0	0	0	0	0	−1
1	−1	−1	−1	−1	−1	0	−1
2	−1	−1	−1	−2	0	0	0
3	0	−1	−2	−3	−2	0	0

During the simulation tests, the scaling gains for fuzzy tuner inputs and outputs (Fig. 7) were selected as

$$g_e = 40,0;\ g_c = 0,8;\ g_{uP} = 0,6;\ g_{uI} = 0,001;\ g_{uD} = 50,0; \qquad (41)$$

After completing the simulation tests, the proposed algorithms were tested in full-scale trials performed on the training ship Blue Lady on the lake Silm in Ilawa/Kamionka. For this purpose the control and measurement system was used which had been worked out in the Department of Ship Automation, Gdynia Maritime and described in detail in (Pomirski et al. 2012). Figure 11 presents the experimental results of full-scale tests of two

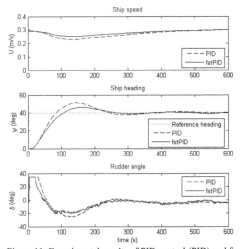

Figure 11. Experimental results of PID control (PID) and fuzzy self-tuning PID control (fstPID) to step reference input.

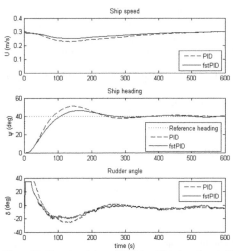

Figure 12. Experimental results of time-histories of controllers' setting values: PID controller (PID), fuzzy self-tuning PID controller (fstPID).

controllers, which were the conventional linear PID controller (34), the results for which are marked with a dashed line, and the fuzzy self-tuning PID controller (Fig. 6), marked with a solid line.

The experimental results of the time-histories of parameters of the examined controllers are shown in Figure 12. The exact values of the time performance indices, determined from the step response of these controllers are collected in table 8. The step responses shown in Figure 11 reveal that the use of the fuzzy self-tuning PID controller gave better results.

Table 8. Values of time performance indices obtained in experimental results with PID controller (PID) and fuzzy self-tuning PID controller (fstPID)

	ψ_E	δ_E	J_E	M_p	t_s
PID	112.8	166.9	129.5	27.9	309.0
fstPID	109.4	199.2	121.4	15.4	235.0

6 REMARKS AND CONCLUSIONS

Two points of interest have arisen from this study. First it has been shown that the ant colony based optimisation technique can be successfully used to obtain the parameters of the first Nomoto model from the full-scale zigzag manoeuvre data.

Another interesting point arises from the application of the fuzzy self-tuning PID controller in course changing operations. However, course changing is effective in manoeuvres in which the positional course is not of particular importance (e.g. in open waters). The proposed control system cooperating with the fuzzy self-tuning PID controller reveals better performance than the conventional PID controller, which manifests itself in smaller overshoot and shorter settling time, as was demonstrated by the full-scale trials results. Both controllers worked with sampling time $T_s = 1$ (s).

Further studies are planned on the floating model of the real vessel, which is building in Ship Automation Department at Gdynia Maritime University (Gierusz, 2012).

REFERENCES

Abkowitz, M. A. 1964. *Lectures on ship hydrodynamics – steering and manoeuvrability*, Technical report Hy-5, Hydro- and Aerodynamics Laboratory, Lyngby, Denmark.

Amerongen, J. Van. 1982. *Adaptive steering of ships: a model-reference approach to improved manoeuvring and economical course keeping*, PhD thesis, Delft University of Technology, The Netherlands. URL: http://www.ce.utwente.nl/amn.

Aström, K. J., Hang, C. C., Persson, P. & Ho, W. K. 1992. Toward intelligent PID control, *Automatica*, 28(1): 1–9. doi: 10.1016/0005-1098(92)90002-W.

Azarsina, F. & Williams, C. D. 2013. Nomoto Indices for Constant-Depth Zigzag Manoeuvres of an Autonomous Underwater Vehicle, *SRN Oceanography*, pp. 1-8. http://www.hindawi.com/isrn/oceanography/2013/219545/.

Bououden, S., Filali, S., Chadli, M. & Allouani, F. 2011. Using Ant Colony Optimization Algorithm for Enhancing an Optimal PI Controller, *International Journal of Sciences and Techniques of Automatic Control & Computing Engineering*, 5(2): 1648-1659.

Chen, W., Yuan, H.-M. & Wang, Y. 2009. Design and implementation of digital fuzzy-PID controller based on FPGA, *Proceedings of the 4th IEEE Conference on Industrial Electronics and Applications*, Beijing 25-27 May 2009, China, pp. 393-397.

Clarke, D. 2003. The foundations of steering and manoeuvring, *Proceedings of IFAC Conference Manoeuvering and Control Marine Crafts*, IFAC, Girona, Spain, pp. 10-25.

Davidson, K. S. M. & Schiff, L. I. 1946. Turning and Course Keeping Qualities, *Transactions of Society of Naval Architects Marine Engineers*, 54:152-200.

Dorigo, M., Maniezzo, V., Colorni, A. 1996. The ant system: Optimization by a colony of cooperating agents, *IEEE Transactions on Systems, Man, and Cybernetics-Part B*, 26(1):1-13. doi: 10.1109/3477.484436.

Fairbairn, N. A. & Grimble, M. J. 1990. H∞ marine autopilot design for course-keeping and course-changing, *Proceedings of the 9th Ship Control System Symposium*, September 10-14, Bethesda, Maryland, USA, 3: 311-336.

Fossen, T. I. & Strand, J. P. 1998. Nonlinear ship control (Tutorial paper), *Proceedings of IFAC Conference on Control Applications in Marine Systems*, (CAMS-98), Fukuoka, Japan, pp. 1-75.

Fossen, T. I. 2002. *Marine Control Systems: Guidance, Navigation, and Control of Ships, Rigs and Underwater Vehicles*, Marine Cybernetics, Trondheim, Norway.

Gierusz, W. 2005. *Synthesis of multivariable systems of precise ship motion control with the aid of selected resistant system design methods*, Publishing House of Gdynia Maritime University (in Polish).

Gierusz, W. & Łebkowski, A. 2012. The researching ship "Gdynia", *Polish Maritime Research*, 19(S1):11-18. doi: 10.2478/v10012-012-0017-3.

He, S., Tan, S. & Xu, F. 1993. Fuzzy self-tuning of PID controllers, *Fuzzy Sets and Systems*, 56(1):37–46. doi: 10.1016/0165-0114(93)90183-I.

Foundation. 2012. *Foundation for Safety of Navigation and Environment Research*, URL: www.ilawaship handling. com.pl.

Journée, J. M. J. 1970. A simple method for determining the manoeuvring indices K and T from zigzag trial data, *Report 267*, Delft University of Technology, Ship Hydromechanics Laboratory.

Kempf, G. (1932). Measurements of the Propulsive and Structural Characteristics of Ships, *Transactions of Society of Naval Architects Marine Engineers*, 40:42-57.

Kula, K. S. (2016). Heading control system with limited turning radius based on IMC structure, *Proceedings of the 21st International Conference on Methods and Models in Automation and Robotics (MMAR)*, pp. 134-139. doi: 10.1109/MMAR.2016.7575121.

Mamdani, E. H. 1974. Applications of fuzzy algorithm of simple dynamic plant, *Proceedings IEE*, 121(12):1585-1588. doi: 10.1049/piee.1974.0328.

Le, M., Nguyen, S. & Nguyen, L. 2004. Study on a new and effective fuzzy PID ship autopilot, *Artificial Life and Robotics*, 8(2):197-201. doi: 10.1007/s10015-004-0313-9.

Lisowski, J. 2013. Sensitivity of computer support game algorithms of safe ship control, *International Journal of Applied Mathematics and Computer Science*, 23(2):439-446. doi: 10.2478/amcs-2013-0033.

Liu, Y., Mi, W. & Guo, C. 2010. Study of fuzzy self-tuning steering controller for ship course, *Navigation of China*, 33(1): 11-16 (in Chinese).

McGookin, E.W., Murray-Smith, D. J. Li, Y. & Fossen, T. I. 2000. Ship Steering Control System Optimisation Using Genetic Algorithms, *IFAC Journal of Control Engineering Practice*, 8(4):429-443.

Minorsky, N. 1922. Directional Stability of Automatice Steered Bodies, *Journal of the American Society for Naval Engineers*, 34(2): 280-309, doi: 10.1111/j.1559-3584.1922.tb04958.x.

Mort, N. & Linkens, D. A. 1980. Self-tuning controllers steering automatic control. *Proceedings of the Symposium on Ship Steering Automatic Control*, June 25-27, Genova, Italy, pp. 225-243.

Nomoto, K., Taguchi, T., Honda, K. & Hirano, S. 1957. On the steering Qualities of Ships. *Technical Report*. International Shipbuilding Progress, 4:354-370.

Nomoto, K. 1960. Analysis of Kempf's standard maneuver test and proposed steering quality indices, *Proceedings of the 1st Symposium in Ship Maneuverability*, DTRC Report 1461.

Norrbin, N. H. 1963. On the design and analyses of the zigzag test on base of quasi-linear frequency response. *Technical report*, Technical Report B 104-3,Gothenburg, Sweden.

Nurbaiti, W. & Nurhaffizah, H. 2012. Self-tuning Fuzzy-PID Controller Design for Aircraft Pitch Control, *Proceedings of the Third International Conference on Intelligent Systems Modelling and Simulation*, 08-10 February, Kota Kinabalu, Malaysia, pp. 19-24.

Perera, L. P. & Soares, C. G. 2012. Pre-filtered sliding mode control for nonlinear ship steering associated with disturbances, *Ocean Engineering*, 51(9): 49-62. doi: 10.1016/j.oceaneng.2012.04.014.

Pomirski, J., Rak, A. & Gierusz, W. 2012. Control system for trials on material ship model, *Polish Maritime Research*, 19(S1):25-30. doi: 10.2478/v10012-012-0019-1.

Roberts, G. N. 2008. Trends in marine control systems, *Annual Reviews in Control*, 32(2): 263-269. doi: 10.1016/j.arcontrol.2008.08.002.

Saari, H. & Djemai, M. 2012. Ship motion control using multi-controller structure, *Ocean Engineering*, 55(12): 184-190. doi: 10.1016/j.oceaneng. 2012.07.028.

Shen, Z. & Guo, C. 2008. Fuzzy self-tuning PID steering control for ultra large container ship, *Proceedings of the 7th World Congress on Intelligent Control and Automation*, 25-27 June, Chongqing, China, pp. 7661-7666, doi: 10.1109/WCICA.2008.4594119, (in Chinese).

SNAME. 1950. Nomenclature for Treating the Motion of a Submerged Body through a Fluid. *Technical Report Bulletin* 1-5. Society of Naval Architects and Marine Engineers, New York, USA.

Tomera, M. 2010. Nonlinear controller design of a ship autopilot, *International Journal of Applied Mathematics and Computer Science*, 20(2): 271-280. doi: 10.2478/v10006-010-0020-8.

Witkowska, A., Tomera, M. & Śmierzchalski, R. (2007). A backstepping approach to ship course control, *International Journal of Applied Mathematics and Computer Science*, 17(1):73-85. doi: 10.2478/v10006-007-0007-2.

Witkowska, A. & Śmierzchalski, R. 2012. Designing a ship course controller by applying the adaptive backstepping method, *International Journal of Applied Mathematics and Computer Science*, 22(4): 985-997. doi: 10.2478/v10006-012-0073-y.

Yang, Y., Zhou, C. & Ren, J. 2003. Model reference adaptive robust fuzzy control for ship steering autopilot with uncertain nonlinear system, *Applied Soft Computing*, 3(4): 305-316. doi: 10.1016/j.asoc.2003.05.001.

Yang, Y. & Bian, H. 2012. Design and realization of fuzzy self-tuning PID water temperature controller based on PLC, *Proceedings of the 4th International Conference on Intelligent Human-Machine Systems and Cybernetics*, Nanchang, Jiangxi 26-27 August 2012, China, 2:3-6.

Yoazhen, H., Hairong, X., Weigang, P. & Changshun, W. 2010. A fuzzy sliding mode controller and its application on ship course control, *Proceedings of the 7th International Conference on Fuzzy Systems and Knowledge Discovery*, pp. 635-638.

Zhao, Q., Li, L. & Chen, G. 2011. The research of fuzzy self-tuning of PID autopilot, *Proceedings of 3rd International Conference on Transportation Engineering*, Chengdu, China, pp. 985-990.

Zuidweg, J. K. 1981. Optimal and sub-optimal feedback in automatic track-keeping system. *Proceedings of the 6th Ship Control Systems Symposium*, October 26-30, Ottawa, Canada, vol. 3, pp. G1.1.1-G1.1.18.

Navigational Risk

Estimated Risks of Navigation of LNG Vessels through the Ob River Bay and Kara Sea

D. Ivanišević & A. Gundić
University of Zadar, Zadar, Croatia

Đ. Mohović
University of Rijeka, Rijeka, Croatia

ABSTRACT: The project of LNG exploitation in the area of the Yamal peninsula and the Ob Bay in the Russian Federation combines all the elements of the LNG Value/Supply Chain of LNG exploitation and transport by sea. LNG ARC7 Ice Class vessels, that in many ways surpass common standards in construction and exploitation of LNG vessels, have been designed and constructed specifically for the purposes of this project. Carriage of LNG by ARC7 Ice Class LNG vessels is very demanding, especially in the conditions of polar night and permanent ice of various thickness, and very low atmospheric temperatures. The complexity of the endeavour is particularly emphasized while manoeuvring in the port, transiting narrow approach channels, and while navigating through the area of Northern Sea Route or just part of it, between the Arctic and Russian Continent. Successful and profitable projects of LNG exploitation depend on planned and uninterrupted ex-port of LNG as well as its safe and reliable delivery to the buyers. Therefore, the emphasis of the paper will be the identification of risks for the vessel and the environment as well as measures taken to mitigate the risks to the acceptable minimum.

1 INTRODUCTION

Natural gas has become the third source of energy in the last decade, after oil and coal. It is expected that it will become the second source of energy by 2025 (International Energy Outlook 2016). Approximately one third of natural gas has been shipped by LNG carriers as a Liquefied Natural Gas (hereinafter, LNG). It is estimated that, by 2035, the transport of LNG by LNG Carriers will surpass the transport of gas by pipelines (BP Energy Outlook 2016). LNG carriers have existed in the shipping industry for less than 50 years. The carriers usually operate in accordance with the long-term time charter party between the exploitation areas and final buyers and predetermined (un)loading terminals. The key factor for a successful and profitable project of LNG exploitation is planned and uninterrupted export of LNG as well as its safe and reliable delivery to the buyers. Transporting LNG by carriers in the area north of the Arctic Circle is unique since it is the first time that LNG is transported to the final buyers at very low temperatures and during the periods of polar night regardless of the fact that these areas are ice covered for nine or more months a year. Such an "endeavour" is far more difficult than previously experienced in LNG transportation. This fact refers primarily to the navigation through hydrographically poor chartered areas, manoeuvring in the ports covered by ice of various thicknesses, transits through narrow approach channels and navigation with icebreaker escort or icebreaker escort and tow. All the above-mentioned segments of navigation will take place with predominant conditions of ice-covered waters, polar nights and extremely low atmospheric temperatures.

2 THE EXPLOITATION AND LNG TRANSPORT IN THE AREA NORTH OF THE ARCTIC CIRCLE

In the industry of exploitation and seaborne transportation of LNG the current experiences have been limited to relatively smooth conditions of navigation in ice covered waters in the south of Sakhalin in the Russian Federation, the Prigorodnoye terminal, and the Melkøya terminal on the island of Hammerfest in Norway. In regards to the ice concentration, Sakhalin terminal is having mild ice conditions in winter months, resulting with less demanding requirements for the construction of ships and on-board equipment.

The Melkøya terminal on the island of Hammerfest in the Norwegian Sea is an ice-free port that does not freeze due to the effect of the Gulf Stream, so there are no special requirements for the construction of ships and on-board equipment. Given the recently available technology and adverse weather conditions, natural gas resources in the Russian Arctic have been insufficiently exploited. So far most of the exploration has taken place in the continental part of Siberia and transported to the market by pipelines.

The aim of the ongoing Yamal LNG Project is to utilize natural gas resources close to the Ob Bay in the Yamal peninsula permafrost, and further LNG transport by carriers. The natural gas exploitation in the area of the Yamal peninsula and the Ob Bay has consolidated all the elements of the so called LNG Value/Supply Chain in the process of production and transportation of LNG. Namely; research and development, production, purification and liquefaction of LNG, shipping to the end destination, discharging, storage and regasification at the receiving terminal (Figure 1).

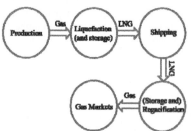

Figure 1. LNG Value Chain (Weems,P. & Howell, N. 2014)

The results of geological exploration near the port of Sabetta, have confirmed that the field's "proven" natural gas reserves are sufficient to produce up to 16.5 million tons of LNG per year (Yamal LNG Project 2013). Taking into consideration these results and the market analysis, the contract was signed for the construction of 15 ARC7 Ice-Class LNG vessels and about 10 conventional LNG carriers necessary for the carriage of contracted LNG volumes. Ice Class defines construction standards for hull and ship propulsion and propulsion power suitable for safe ice navigation. ARC classification has been established to the requirements of the RMRS (Russian Maritime Register of Shipping).

3 ARC7 ICE-CLASS LNG TANKERS

In order to ensure safe and successful LNG transport from the Yamal project region through the Ob Bay, the Kara Sea and other waterways north of the Arctic Circle (Figure 2), it has been necessary to adjust and optimize the design of LNG carriers. This includes a strengthened construction of LNG carriers in order to enable navigation in ice conditions and "winterization". The term "winterization" refers to the preparation of the equipment as well as living and working quarters to ensure efficient operation at extremely low temperatures and during the polar night.

Figure 2 Chart of planned routes north of the Arctic Circle from Sabetta port (MAIRES 2011)

The outline design of LNG carriers for required navigation has been the result of a comprehensive program of model testing and studies conducted in collaboration with the leading technical institutes and classification societies. A further detailed ship design optimization, particularly with respect to the shape and structure of the hull, cargo containment system and propulsion systems with azimuth thrusters, has been completed by the shipbuilders chosen on the basis of their long experience and expertise in the construction of LNG carriers. The final result has been an LNG carrier with the net capacity of 170,000 cubic meters, that is, 98% of the total tank capacity.

ARC7 ice-class LNG carrier, built to the requirements of the Russian Maritime Register of Shipping, is capable of operating in the first-year ice up to 2.1 m thick and penetrating the first-year ice ridges up to 15 metres high, which represent the extreme conditions anticipated on the planned navigation routes. ARC7 LNG carrier has a moderate ice-breaking bow for ice navigation, strengthened bottom and stern for optimal ice breaking, and three azimuth thrusters, which enables the ship to proceed headway through ice up to 1.5m thick and sternway up to 2.1 m thick, maintaining continuous minimum speed. In addition to this, the ship maintains her capability of maintaining 19.5 knots speed in the open water with 21% sea margin.

The ice-class ships of this size and operational capabilities represent a significant technological improvement over the existing standards in the construction of LNG carriers. The propulsion power and the configuration of azimuth thrusters have surpassed present experiences, requiring the upgrade of the existing technology.

Table 1. ARC7 LNG Basic specifications

Length overall	299.0m
Breadth	50.0m
Depth	26.5m
Draught, loaded	11.7m
LNG cargo capacity	172,600m3
Cargo tank type	GTT membrane
Power/propulsion system	Dual-fuel diesel-electric
Main generator engines	4 x 11,700kW + 2 x 8,775kW
Total main engine power	64,350kW
Speed	19.5knots
Builder	Daewoo Shipbuilding & Marine Engineering Co., Ltd.
Ice Class Notation	ARC7
Class	RMRS/BV

This type of concept, enabling the year-round operations, has been tested on smaller oil and product tankers as part of the Lukoil Varandey Project in the Pechora Sea as well as on the Norilsk Nickel company's ships transporting minerals from the Yenisei Gulf in the Kara Sea.

Larger tankers, in size and design similar to Yamal LNG tankers, have been successfully operating for years in much more moderate Baltic Sea ice conditions. They have been using the same azimuth propulsion principle and breaking ice both forward and astern and are referred to as "Aker - Double Acting Tanker ".

Taking into consideration the above-mentioned developments in the LNG sea transport industry, authors will analyse the possible risks during ARC7 LNG navigation through the Ob Bay and the Kara Sea.

4 RISK IDENTIFICATION AND ASSESSMENT

Risk is the possibility that a certain action will lead to an undesirable outcome (Stošić 2017). Risk assessment is a process that includes hazard identification, estimation of consequences, a characteristic of risk and an evaluation of the results (De Bollmann 2002).

In the initial phase of the risk assessment, it is necessary to determine and identify the potential risks and hazards that may occur while navigating certain region.

4.1 HAZID & HAZOP methods

As a "rule of the thumb", the first step is HAZard IDentification Study Method (hereinafter HAZID study) which identifies potential risks and hazards that might affect people, the environment, vessel and cargo equipment, installations and company reputation. HAZARD identification is a crucial activity for reducing accidents and other operability related losses (Seligmann, Benjamin J., et al. 2012). HAZID study is used to determine more precisely finances required for the reduction or elimination of identified risks and hazards.

The next step is a Hazard and Operability Studies Method (hereinafter HAZOP method) which identifies risks and hazards that might endanger normal functioning of equipment and on board operations regardless whether the vessel is berthed, loading/discharging cargo or underway. It is an important technique to evaluate system safety and risks (Ahn & Chang 2016). HAZOP Method identifies possible deviations from normal operating conditions which could lead to harmful consequences (Mohović 2011.). HAZOP enables duly determination and formulation of efficient procedures that will enhance safety when operating on-board equipment and ship operation as well.

4.2 Potential risks for navigation through the Ob River Bay and Kara Sea

When considering navigation area north of the Arctic Circle with predominant conditions of extremely low temperatures and the polar night, potential risks for navigation are numerous and very specific. The authors will analyse and divide identified risks into three categories:

- the navigational risks in the Arctic Circle in the conditions of low temperatures and polar night,
- the risks associated to on-board living and working in the conditions of low temperatures and polar night,
- the risks of navigation through the ice of various thickness (Table 2).

The risks relate primarily to navigation through the area north of the Arctic Circle, where due to the small number of ships sailing the area, investments by countries bordering the Arctic Sea and its marginal seas in the navigational infrastructure and hydrographic survey, are significantly lower. The risks associated to on-board living and working in the conditions of low temperatures and polar night can reduce the working efficiency and the ability of making optimal decisions. A special attention has to be given to the risks of navigation through the areas covered permanently with ice, since such a navigation requires higher standards in the construction of ships, experience and training of the crew.

Table 2. The risks of operating in low temperatures and polar night in the area north of the Arctic Circle.

The risks of navigation through low-density shipping areas north of the Arctic Circle

no Risk	Consequences	Risk reducing measures
1 The lack of reliable hydrographic data and nautical charts.	Possible grounding with casualties and material losses.	Development of the recommended shipping routes through areas with chart giving accurate representations. Translation of the existing charts and publications into English.
2 Not able to receive the information on ice condition and ice movement while navigating in high latitudes.	The loss of time and inability of planning the optimal ice route due to lack of data.	The reporting system on ice condition and synopsis must be capable of sending and receiving small data files.
3 Grounding while transiting channels and passages constrained by shallow water.	Grounding and damage to the hull or part of the hull. Transit suspended for other ships.	Installing the additional system for determining ship's position, especially during the season of ice. Additional crew simulator training.
4 Insufficient number of Search Rescue Coordination Centres in the area north of the Arctic Circle.	The loss of valuable time in cases of emergencies and/or marine accidents during navigation north of the Arctic Circle.	Additional crew training. Ship design with enhanced and characteristics appropriate for the navigation north of the Arctic Circle.
5 Cargo handling equipment and other equipment icing due to weather conditions.	Problems during cargo handling, loss of time and possible damage to the certain parts of equipment.	Assess which equipment needs extra protection from low temperatures and ice in order to prevent icing and reduce the time needed to remove the ice prior to equipment handling.

The risks associated to on-board living and working in the conditions of low temperatures and polar night

no Risk	Consequences	Risk reducing measures
1 Insufficient crew training for the navigation in low temperature conditions and polar night.	Possible grounding or damage to the equipment, hull or screws.	Early training of all members of the crew.
2 Crew fatigue due to adverse usual working conditions, days,	The crew fatigue has an impact on decision making process which might result in unintended consequences.	Consider a different scheme of crew changes from the one, taking into account the periods of polar nights and the experience of the crew and the requirements set by the industry. The increased number of crew members compared to conventional LNG tanker.

The risks of navigation in different ice conditions

no Risk	Consequences	Risk reducing measures
1 «Man overboard» during crew navigation through area with different ice conditions.	A fall on ice might result in serious injuries and death. In the case of falling into water, in the pieces of ice, there is a danger of severe injuries and death as a result of crush injuries.	Handling aft gantry for fast platform lowering and the trained for this activity. Verify that the thermal protective clothing is sufficiently buoyant to keep the person float. Establish the deck in pairs operations system when operating in ice conditions an low temperatures.
2 Collision between assistant icebreaker and ARC7 LNG.	Damage to the ARC7 LNG tanker and/or icebreaker.	Develop the specific procedures for the vessel escort, the towing operations and icebreaker assistance in conditions where ice thickness is such that ARC7 LNG tanker's own propulsion is not sufficient to continue navigation. Additional crew simulator training.
3 Damage to the screw and/or azimuth thrusters during ice Simulator navigation.	The possible damage to the hull and azimuth propulsion unit and consequently delays and financial losses.	Precisely defined procedures for ice navigation. Assistance of ice- until the crew is sufficiently experienced. training with an emphasis on ship handling with one or two damaged azipods.

5 CONCLUSION

The navigation north of the Arctic Circle, as far as LNG Carriers are concerned, has been almost non-existent. The exception was the transit of two LNG carriers through the Northern Sea Route from Europe to Japan and backwards. In reference to estimated amounts of natural gas in this area, the new projects are expected to be carried out, which will, consequently, lead to greater number of carriers and larger traffic as well. Geographical and climatic features of the area north of the Arctic Circle make the navigation more difficult, which has consequently led to stricter demands regarding design, construction, and equipment as well as the education and experience of the crew.

The risks, that may occur while navigating this area are numerous and new in the LNG transport industry. They primarily refer to low temperatures, ice of various thickness that covers navigational routes and the periods of polar night.

An example of how to reduce the risks is the construction of LNG ARC7 vessels for Yamal LNG project. Some of the risks have been reduced to the acceptable level, and now, a special attention has to be given to the risks that refer to additional and early crew training, fatigue reduction as a result of adverse working conditions, and definition of rigid procedures for ice navigation. That is the only way which will enable the safe and reliable transportation of the cargo to the customers.

REFERENCES

Ahn, J. & Chang, D. 2016. "Fuzzy-based HAZOP study for process industry." *Journal of Hazardous Materials* 317: 303-311.

BP Energy Outlook 2016 edition, available at: http://www.bp.com/content/dam/bp/pdf/energy-economics/energy-outlook-2016/bp-energy-outlook-2016.pdf

De Bollmann, L. C. M. 2002. "Hazard identification, risk assesment and risk evaluation in the laboratory". *ACTA BIOQUIMICA CLINICA LATINOAMERICANA 36.4*: 547-561.

International Energy Outlook 2016. U.S Energy Information Administration, available at: http://www.eia.gov/outlooks/ieo/

Mohović, Đ. 2011. *Upravljanje rizikom u pomorstvu.* Rijeka: Pomorski fakultet u Rijeci.

Monitoring Arctic Land and Sea Ice from Russian and European Satellites / MAIRES / Collaborative Project (SICA), 2011., available at: http://dib.joanneum.at/maires/index.php?page=objects

Seligmann, Benjamin J., et al. 2012. "A blended hazard identification methodology to support process diagnosis." *Journal of Loss Prevention in the Process Industries* 25.4: 746-759.

Stosic, Biljana, et al. 2016. "Risk identification in product innovation projects: new perspectives and lessons learned." *Technology Analysis & Strategic Management* : 1-16.

Yamal LNG Project 2013., available at: http://yamallng.ru/en/project/about/

Weems, P. & Howell, N. 2014. Japan's Pivotal Role in the Global LNG Industry's 50-Year History. *Energy Newsletter, available at:* http://www.kslaw.com/library/newsletters/EnergyNewsletter/2014/August/article2.html

A Research on Concept of Ship Safety Domain

A. Baran, R. Fışkın & H. Kişi

Dokuz Eylül University, İzmir, Turkey

ABSTRACT: This study presents the results of the content analysis of articles related to "Ship Safety Domain". The content and statistical analysis of 44 articles that published in 21 different journals between 1970 and 2016 were conducted. This study includes various data types such as publishing years of articles, obtaining the type of data, the contribution of authors, the contribution of countries, the contribution of the institution, the rank of the journal, keywords of articles and etc. The content analysis aims to provide necessary indicators for readers, followers, and contributors of relevant discipline and a glimpse of lots of articles. The published articles about ship safety domain are discovered by this study so as to contribute lecturers and researchers interested in navigation especially ship domain concept. The results of the research were revealed that Po-land and China are the most contribution countries, Dalian Maritime University and The Maritime University of Szczecin are the most contribution institutions and Ning Wang and Rafal Szlapczynski are the most contribution researchers.

1 INTRODUCTION

Safety is a basic priority of waterway transportation (Liu et al, 2016). The determination of clear area around the ship is mainly important for safe navigation (Wang and Chin, 2015). Safe navigation is an ever-growing issue, especially due to the rapidly increasing numbers of ships, and improving maritime industry. Even though statistics display developed levels of safety in the industry, which carries 90% of the world trade, the risk of navigational incidents remains a major concern and priority (Baldauf et al., 2015).

The concept of collision avoidance helps to eliminate these risks. It is important the analysis of ship collision risk, probability of occurrence, the understanding of collision probability and possible results of collisions (Tang et al., 2013). Making collision risk assessment is the main issue in the decision systems, and it is a major concept in navigation (Xu and Wang, 2014). A research prepared by the Nautical Institute (NI) demonstrated that 60% of collision and grounding cases are caused by direct human error (Gale and Patraiko, 2007) on the other hand, the statistics shows that human errors have caused 80% of marine incidents for which the basic reason is improper determinations of the navigational case and the result incorrect decisions

(Wang et al., 2009). The two major human involved causes are "inadequate determination of the situation" (24%) and "ineffectual look out" (23%) navigation (Wang and Chin, 2015).

Most of the ship domain models are defined in a geometrical manner which is easy to grasp however not conducive for application to practices or simulations. Therefore, there is a need for a uniform analytical framework to describe ship domain models in order that these models could play a powerful role in marine traffic engineering (Wang et al., 2009).

During the ship navigation, the safety zone is referred as the ship domain. Any infringing of the ship domain is evaluated as a danger to navigational safety (Pietrzykowski and Uriasz, 2009). A ship domain is commonly thought as the space around the ship, which the OOW (Officer of the Watch) wants to keep clear of other objects. (Szlapczynski and Szlapczynska, 2015). Ship domains are powerful and primary accesses to dealing with ship navigation risk determination, with collision avoidance, with marine traffic simulations and with optimal trajectory, etc. Many researchers have submitted ship domains type with various dimensions and shapes in the past decades (Wang, 2013). Ship domain capacity (shape and size) generally depends on ship's speed and length although parameters of other ships may also be considered (Szlapczynski and Szlapczynska, 2015).

The concept of ship domain is crucial for safety navigation so, in this context, the aim of this study is to make a content analysis of articles related to ship safety domain in related literature to guide the academicians and researchers study in the related field.

2 A CONCEPT OF SHIP DOMAIN

Various ship domains have been presented by many researchers having taken into account different shapes and sizes. The determination of ship domains strongly depends on the statistical data and operator experience. It is found that the ship domains described by geometrical figures including ellipse, polygon, circle and other complex figures rather than analytical manner. The circular ship domain proposed by Goodwin (1975) which is divided into three sectors. The model is also obtained from statistical methods from a great number of records and simulator data. The other type of ship domain called as elliptical ship domain is proposed by Fuji and Tanaka (1971). The domain is created by using statistical methods. Coldwell (1983) proposed another elliptical ship domain by similar statistical methods for head on and overtaking situations. Polygonal ship domain proposed by Pietrzykowski (2004, 2006) is also another important type of ship domain. The domain depends on the determination of dynamic functions of ship dimensions and speed (Wang et al., 2009).

3 METHODOLOGY

In this study, 44 articles that were published in various journals between 1970 and 2016, were performed to make a content analysis. This analysis was separated various categories such as the contribution of authors, the rank of the journal, keywords of articles and etc. When applying statistical analysis, we utilized from computer-assisted Statistical Package for the Social Sciences (SPSS) software.

We adopted content and statistical analysis for 44 articles based on ship safety domain. First of all, we determined categories as authors, countries, institutions, journals, keywords, words in the articles names and years to apply analysis.

The basic aim of this study is to guide researchers who will study about ship safety domain and assist further studies that will utilize to the ship safety domain research area.

4 FINDINGS

4.1 *Authors statistics*

Articles in the related field prepared by commonly double and more authors. As seen in Figure 1 and Table 1, 18,2% of articles prepared by double authors and 40,9% of articles prepared by multiple authors.

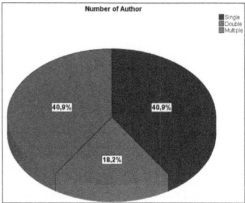

Figure 1. Distribution of the articles according to authors group.

Table 1. The number of authors in articles.

The Number of Authors in Articles	(f)	(%)	C (%)
Single	18	40,9	40,9
Double	8	18,2	69,1
Multiple	18	40,9	100,0
Total	44	100,0	

The most contributing ten authors was listed in Table 2 and Figure 2. 'Ning Wang' is the most contributing author within a total of 80 authors with 6 articles. Followed by Rafal Szlapczynski with 5 articles and Zbigniew Pietrzykowski with 4 articles.

Table 2. The most contributing authors.

Author	Rank	(f)	(%)	C(%)
Ning Wang	1	6	5,8	5,8
Rafal Szlapczynski	2	5	4,8	10,6
Zbigniew Pietrzykowski	3	4	3,8	14,4
Krzysztof Marcjan	4	3	2,9	17,3
Feng Zhou	5	2	1,9	19,2
Jingxian Liu	6	2	1,9	21,2
Lucjan Gucma	7	2	1,9	23,1
Mirosław Wielgosz	8	2	1,9	25,0
Paul Vernon Davis	9	2	1,9	26,9
Q.Y. Hu	10	2	1,9	28,8
Others	74	7	1,2	100,0
Total	80	104	100,0	

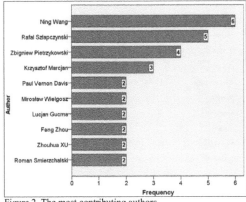

Figure 2. The most contributing authors.

4.2 Bibliometric statistics

A total of 44 selected articles were published in 21 different journals. "The Journal of Navigation" is the most contributing journal with a total of 17 articles.

Table 3. The most contributing journals.

Journal	Rank	(f)	(%)	C(%)
The Journal of Navigation	1	17	38,6	38,6
Journal of Dalian Maritime University	2	3	6,8	54,5
Ship & Ocean Engineering	3	2	4,5	65,9
Activities in Navigation	4	1	2,3	68,2
Annual of Navigation	5	1	2,3	70,5
Information Processing and Security Systems	6	1	2,3	75,0
Journal of Wuhan University of Technology	7	1	2,3	77,3
Missing value		1	2,3	79,6
Others		17	38,6	100
Total	21	44	100	

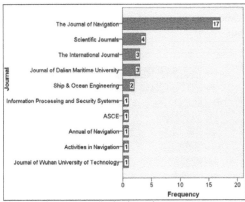

Figure 3. The most contributing journals.

All articles are separated according to 5-years period to simplify cluster. According to analysis results, Studies in related field showed a rapid increase since 2000. While a total number of studies that published between 1970-2000 is equal to 6, 35 articles were published after 2000. The most contributing period is 2011-2016 with the total of 24 studies. On the other hand, no studies have been conducted between 1991 and 1995. Following the recession, a large increase was observed.

Table 4. Distribution of articles (5-year period).

Year	(f)	(%)	C(%)
1970-1975	1	2,2	2,2
1976-1980	1	2,2	4,4
1981-1985	1	2,2	6,7
1986-1990	1	2,2	8,9
1991-1995	0	0,0	8,9
1996-2000	2	4,4	13,3
2001-2005	3	6,7	20,0
2006-2010	11	26,7	46,7
2011-2016	24	53,3	100,0
Total	44	100	

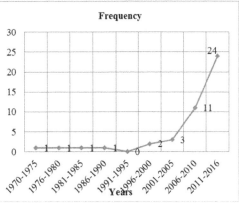

Figure 4. Distribution of articles (5-year period).

4.3 Affiliation statistics

Geographical locations of contributing institutions are shown in Figure 5. The most contributing institutions are in China and Poland. Dalian Maritime University is the most contributing institution followed by Maritime University of Szczecin. Both of them accounted for 34,1% of all articles.

China is the most contributing country with the total of 17 articles and approximately half of them were conducted by Dalian Maritime University. Followed by Poland with the total of 16 articles and approximately one third of them were realized by the Maritime University of Szczecin. China and Poland accounted for 75% of all countries.

Figure 5. The geographical distribution of contributing countries.

Table 5. The most contributing institutions.

Country	Rank	(f)	(%)	C(%)
China	1	17	38,6	38,6
Poland	2	16	36,4	75,0
Sweden	3	2	4,5	79,5
UK	4	2	4,5	84,1
USA	5	2	4,5	88,6
Denmark	6	1	2,3	90,9
Japan	7	1	2,3	93,2
Singapore	8	1	2,3	95,5
Slovenia	9	1	2,3	97,7
Taiwan	10	1	2,3	100,0
Total	10	44	100,0	

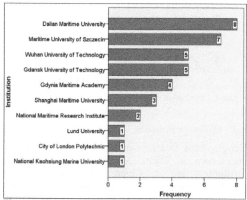

Figure 6. The most contributing institutions.

Table 6. The most contributing countries.

Institution	Rank	(f)	(%)	C(%)
Dalian Maritime University	1	8	18,2	18,2
Maritime University of Szczecin	2	7	15,9	34,1
Wuhan University of Technology	3	5	11,4	45,5
Gdansk University of Technology	4	5	11,4	56,8
Gdynia Maritime Academy	5	4	9,1	65,9
Shanghai Maritime University	6	3	6,8	72,7
National Maritime Research Institute	7	2	4,5	77,3
World Maritime University	8	1	2,3	79,5
National University of Singapore	9	1	2,3	81,8
National Kaohsiung Maritime University	10	1	2,3	84,1
Missing Value		1	2,3	86,4
Others		6	13,6	100,0
Total	17	44	100,0	

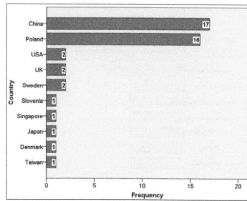

Figure 7. The most contributing countries.

4.4 Keyword statistics

The words used in article titles are analyzed in order to reveal the most used words. 'Domain' (37) and 'Ship' (34) is revealed the most utilized words as shown in Table 7. and Figure 8. According to keywords analysis, 'Collision avoidance' (8) and 'ship domain' (7) is the most used keywords within 70 keywords as shown in Table 8. and Figure 9.

Table 7. The words used in article titles.

The words in the article titles	Rank	(f)	(%)	C(%)
Domain	1	37	11,6	11,6
Ship	2	34	10,7	22,3
Collision	3	12	3,8	26,0
Area	4	9	2,8	28,8
Safety	5	9	2,8	31,7
Model	6	8	2,5	34,2
Risk	7	7	2,2	36,4
Based	8	6	1,9	38,2
Water	9	6	1,9	40,1
Assessment	10	5	1,6	41,7
Others		286	58,3	100,0
Total	133	319	100,0	

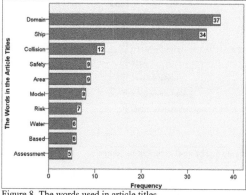

Figure 8. The words used in article titles.

116

Table 8. The keywords used in article.

Keywords	Rank	(f)	(%)	C%
Collision avoidance	1	8	9,3	9,3
Ship domain	2	7	8,1	17,4
Collision risk	3	3	3,5	20,9
Restricted area	4	2	2,3	23,3
DCPA	5	1	1,2	24,4
3D Model	6	1	1,2	25,6
Accident prevention	7	1	1,2	26,7
AIS data	8	1	1,2	27,9
Analytical model	9	1	1,2	29,1
Approach parameters	10	1	1,2	30,2
Others		60	69,8	100,0
Total		70	86	100

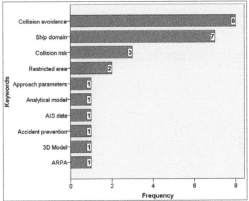

Figure 9. The keywords used in article.

5 CONCLUSION

In the literature related to 'Ship Safety Domain', Poland and China is revealed the most contributing countries. On the other hand, there is no study prepared by major maritime countries such as Turkey, Netherland, Deutschland and etc. The Journal of Navigation is the most contributing journal as expected because the journal is based on navigation. The most productive periods are revealed as '2011-2016' and '2006-2010'. This shows that the topic has become a popular topic in recent years. Dalian Maritime University and Maritime University of Szczecin are the most contributing institutions. China is intensely interested in the topic and Poland is the most contributing country in EU region. 'Collision avoidance' and 'Ship domain' are the most used keywords as expected. Similarly, the words that most used in article titles are 'Domain', 'Ship' and Collision'. It is revealed that the most contributing authors are 'Ning Wang', 'Rafal Szlapczynski' and ' Zbigniew Pietrzykowski', respectively. Especially, researchers from China and Poland are mostly interested in the topic.

As a result, there is very few study related field. It is thought that new researchers can focus the topic. This is because the ship navigation can be safer if more researchers are interested in the topic.

REFERENCES

Baldauf, M., Mehdi, R., Deeb, H., Schroder-Hinrichs, J.U., Benedict, K., Kruger, C., Fischer, S. & Gluch, M. 2015. Maneuvering areas to adapt ACAS for the maritime domain. *Scientific Journals* 43(115): 39-47.

Coldwell, T. G. 1983. Marine traffic behavior in restricted waters. *The Journal of Navigation* 36: 431–444.

Fuji, J. & Tanaka, K. 1971. Traffic capacity. *The Journal of Navigation* 24: 543–552.

Goodwin E. M. 1975. A statistical study of ship domains. *The Journal of Navigation* 28: 329–341.

Gale, H. & Patraiko, D. 2007. Improving navigational safety. *Seaways*, pp. 4–8.

Liu, J., Zhou, F., Li, Z., Wang, M. & Liu, R.W. 2016. Dynamic ship domain models for capacity analysis of restricted water channels. *The Journal of Navigation* 69(3): 481-503.

Pietrzykowski, Z. & Uriasz, J. 2009. The ship domain – a criterion of navigational safety assessment in an open sea area. *The Journal of Navigation* 62(1): 93-108.

Pietrzykowski, Z. & Uriasz, J. 2004. The ship domain in a deep-sea area. *Proceeding of the 3rd International Conference on Computer and IT Applications in the Maritime Industries*, Siguenza, Spain.

Pietrzykowski, Z. & Uriasz, J. 2006. Ship domain in navigational situation assessment in an open sea area. *Proceeding of the 5th International Conference on Computer and IT Applications in the Maritime Industries*, Oegstgeest, Netherlands.

Szlapczynski, R. & Szlapczynska, J. 2015. A simulative comparison of ship domains and their polygonal approximations, *TransNav, the International Journal on Marine Navigation and Safety of Sea Transportation*, 9(1): 135-141.

Tang, C., Xu, Z., Liu, Z. & Liu, J. 2013. Research on collision probability model based on ship domain, ICTIS & ASCE, pp. 2309-2310.

Wang, Y. & Chin, H.C. 2015. An empirically-calibrated ship domain as a safety criterion for navigation in confined waters, *The Journal of Navigation* 69(2): 257-276.

Wang, N., Meng, X., Xu, Q. & Wang, Z. 2009. A unified analytical framework for ship domains, *The Journal of Navigation* 62(4): 643-655.

Wang, N. 2013. A novel analytical framework for dynamic quaternion ship domains, *The Journal of Navigation* 66(2): 265-281.

Xu, Q & Wang, N. 2014. A survey on ship collision risk evaluation. *Traffic and Transportation* 26(6): 475-486.

Global Navigation Satellite Systems (GNSS)

Evaluation of the Influence of Atmospheric Conditions on the Quality of Satellite Signal

M. Siergiejczyk, K. Krzykowska & A. Rosiński
Warsaw University of Technology, Warsaw, Poland

ABSTRACT: Limitation of satellite systems' operating use is caused by different errors that inevitably accompany these systems' operation. The satellite system, in all its complexity, can be regarded as a complex measurement system. Operational accuracy of such a system is one of its most important characteristics. Therefore, as soon as during the design stage, it is important to minimize the error related to the measurement accuracy. Errors of this type are specified as systematic and in the case of the satellite systems, they apply to the receiver clock, and their elimination is possible by increasing the number of measured distances from the satellites by one. Passing of the radio waves through the medium may change their speed and frequency. On the way from the satellite to the receiver's antenna, the signal passes through the atmosphere, whose properties change dynamically in a stochastic way. Two atmosphere layers have the greatest impact on the signal propagation – ionosphere and troposphere. The article presents basics researches on the evaluation of the influence of atmospheric conditions on the quality of satellite signal using fuzzy sets.

1 INTRODUCTION

Telecommunications challenges are an inherent aspect of modern transportation. Due to economics aspects – the growth of new technologies is truly important all over Europe (Garcia – Pereiro & Dileo 2015). In telecommunication, they may concern on transport telematics (Sumiła 2014), intelligent transport systems (Zając & Świeboda 2015), or the IT application for different use (Laskowski & Łubkowski 2013) (Piro et al. 2015). To be able to deploy advanced technologies - it is necessary to conduct research in which important step is modeling. The model is a representation of reality - its essential characteristics relevant to the objective of pursued research. The quality of this representation of reality depend on various factors, including the capabilities of a research tool and the purpose of taken research (Jacyna 2008). Nevertheless, models are reflection of systems that we submit to research to get to know the phenomena taking place within them (Stawowy & Dziula 2015). A particularly important and applicable are decision models, in which the essence of modeling is to extract new information about the model for decision-making to drive, control and manage system (Kierzkowski & Kisiel 2016). Their research can be carried out using mathematical methods

(Stawowy 2015), including artificial neural networks, fuzzy sets and simulation methods (Kaniewski et al. 2015).

In air transport, telecommunications equipment are the CNS (Communication, Navigation, Surveillance). Their development determines the increase of efficiency of air traffic management. However, the correct operating of this type of system is possible after verification of their functioning in a complex external environment.

The analysis of the phenomena associated with the satellite signal used for operational use in civil aviation is not possible without model analysis. Experimenting on actual flight operations may in fact lead to dramatic effects (Stanczyk & Stelmach 2015). Numerous efforts are taken to study signal interference and different processes inside satellite systems while operations having a potential impact on its quality.

2 SIGNIFICANCE OF TOPIC

Research topic including the impact of meteorological phenomena on the satellite signal parameters is widely described in the literature, e.g. in publications of (Januszewski 2010) and (Narkiewicz 2007), but also in professional journals scored by the Ministry of Science and Higher Education. Numerous research units, including Rzeszów University of Technology and the University of Warmia and Mazury, conduct research projects focusing on the satellite signals accuracy. It is worth indicating here publications of the team of Oszczak, Ciećko, Grzegorzewski, Ćwiklak of 2010-2014 and the team of Professor Fellner and their publications of 2011-2014. In the first case – the problem of determining the aircraft's position and this position's accuracy using EGNOS, particularly in the north-eastern Poland was undertaken. Whereas in the second on – focus was put on verification methods for testing satellite systems in various flight operations. The issue is also known in the world. Habarulema & McKinell (2009) present the artificial neural networks as the best, in their opinion, method of monitoring the signal, but they use the TEC (total electron content) indicator in the prediction. Researches on impact of atmospheric conditions on satellite signal were held also in (Yin et al. 2011) or (Warnant et al. 2007).

The literature review performed for the purpose of this study also points to a number of other research conducted in the field of modelling and predicting the signals with the artificial neural networks. Oronsay, McKinnell and Habarulema prove additionally possibilities for prediction using the neural networks with available data on the latitude and longitude of points. The paper (Habarulema et al. 2009), which use the ionospheric error in modelling and predicting the satellite signal, seems to be particularly close to the described issue. The indicators prediction in the air transport using the neural networks was also undertaken in the work of Chen T. (2012) and Alekseev, Seixas (2009). So far, however, detailed research that would include a tropospheric error (precipitation, cloudiness, air humidity), the ionospheric error (solar activity) and the historical accuracy and availability data (HNSE, VNSE) for the prediction of the satellite signal parameter values has not been found. Performance of the research will, therefore, have a significant impact on the development of science in this area. This is particularly important from the point of view of the dynamic growth of research works on satellite systems and, in particular, on improving the signal parameters' accuracy.

3 LIMITATION OF SATELLITE SYSTEMS' OPERATING USE

Limitation of satellite systems' operating use is caused by different errors that inevitably accompany these systems' operation. The satellite system, in all its complexity, can be regarded as a complex measurement system. Operational accuracy of such a system is one of its most important characteristics

(Krzykowska & Siergiejczyk 2014). Therefore, as soon as during the design stage, it is important to minimise the error related to the measurement accuracy. Errors of this type are specified as systematic and in the case of the satellite systems, they apply to the receiver clock, and their elimination is possible by increasing the number of measured distances from the satellites by one.

Determination of position and time in the satellite navigation systems is governed by the distance of the satellite from the receiver's antenna measured as the time, in which the radio signal travels from the satellite to the receiver's antenna. This phenomenon is called often the signal delay. Generally, the satellite system errors can be divided into four groups:
– signal propagation errors, including:
  – ionospheric errors,
  – tropospheric errors,
  – multipath errors,
– relativistic errors
– system operation errors, including:
  – satellite ephemeris errors,
  – satellite clock errors,
– receiver errors, including the DOP (Dilution of Precision) error.

Passing of the radio waves through the medium may change their speed and frequency. On the way from the satellite to the receiver's antenna, the signal passes through the atmosphere, whose properties change dynamically in a stochastic way. Two atmosphere layers have the greatest impact on the signal propagation – ionosphere and tropospheres. Without a doubt, Marconi contributed to the discovery, who in 1901, as the first, sent the radio signals across the Atlantic. The receiving station location beyond the reach of the broadcast station's visibility link gave Marconi reasons to believe that the electromagnetic wave is reflected from the atmosphere. The next step in the discoveries was an indication that in the atmosphere, there is an electrically conductive layer. This discovery was made in 1902 by Oliver Heaviside and Arthur Kennelly. In the twenties of the 20th century, its height was already known and it was discovered that it is of a layered nature.

The ionosphere begins at 50 km and ends (according to various sources of literature) in the 500-1000th kilometre. The solar radiation causes ionisation of gases occurring in this layer, resulting in a cloud of free electrons. These electrons have an impact on broadcast on the signals in the satellite system. Interferences taking place in the ionosphere cause the absorption and polarisation of electromagnetic waves, which leads to radio connectivity interferences. The ionosphere itself is also divided into layers, which are marked by letters of the alphabet (D, E, F). The division is independent from the variable content of electrons in a volume unit, in different parts of the ionosphere. The ionosphere, in particular the F layer, is important for the data transmitted from the satellite to the receiver because it causes the greatest change in speed of the transmitted signals. The ionosphere condition in GPS is specified by the TEC coefficient (Total Electron Content), whose unit is the number of electrons per m2. With the help of this coefficient, the number of particles present in the cuboid with the basis of 1m2 connecting the receiver antenna with the satellite is set. In the world literature, one can meet publications using the TEC coefficient to model and even predict the satellite signal. Ionospheric error occurs due to the delay of the satellite signal when passing through the ionosphere. It is called interchangeably ionospheric delay and ionospheric refraction. The occurring delay results from changes in the signal way, in addition also the change in the group and phase velocity of the wave (signal), as well as the carrier wave frequency. This leads to change in the signal movement time.

The ionosphere condition also is impacted by the solar activity, whose manifestation are the changes taking place on the Sun surface and atmosphere. These changes cause fluctuations in the radiation that reaches the Earth in the form of electromagnetic waves, including light, and a stream of particles emitted by the Sun (solar wind). The solar activity also includes changes in the number and distribution of the sunspots. The basic cycle of changes in the solar activity is c. 11 years. In one such cycle, the Sun changes the activity level from the minimum to the next minimum. The measure of solar activity is the Wolf number.

The troposphere is the lowest and also the thinnest atmosphere layer. It extends from the Earth's surface to an altitude of c. 10 km. The troposphere's property, which is important for the development of the processes occurring here, is a drop in pressure and temperature with increase of altitude. Temperature decreases from, on average, (under normal conditions) +15 degrees Celsius at sea level to -50 degrees Celsius in the vicinity of the poles and to -80 degrees Celsius in the equatorial regions. Similarly, atmospheric pressure decreases from (under normal conditions) about 1013 hPa at the Earth's surface to even 100-200 hPa (depending on season and latitude). In the troposphere, there occur all the most important processes influencing weather and climate on the Earth. Meteorological conditions prevailing here cause a curvature of the satellite signal and its delay, extending the way to the receiver. Tropospheric error, also known as tropospheric delay, is caused both by the dry and wet part of the troposphere. The dry part must be understood as atmospheric pressure, air density and its temperature changing with altitude. It is assumed that the dry factor causes almost 90% of the tropospheric error (delay). The wet part of the troposphere is closely connected with a very high content of water vapour in the atmosphere layer. This content results from the formation of the weather in the troposphere the presence of clouds and precipitation cause that in the lowest part of the atmosphere, there is almost 99% of moisture (counting also the Earth's surface waters). The water vapour content depends, therefore, also on altitude and atmospheric phenomena (storms, weather fronts).

The signals sent by the satellites may encounter obstacles such as buildings, trees, hills, and others in their path. This causes that, apart from the signals arriving directly from the satellites, the signals reflected from various obstacles, as well as from the ground and water, also reach the receiver. This phenomenon introduces the multipath error, which may affect the correct operation of the system, especially in far built-up areas. The multipath effect can, for example, be detected through analysing the signal-to-noise ratio in a function of time.

The correct time measurement in the satellite system is obtained through determining the time difference in the satellites and the receiver's clocks. According to Einstein's theory, this difference is affected by both the gravity and the movement of satellites. Moreover, this issue is further described in the special theory of relativity. Basically, it is assumed that smaller gravity force accelerates the clock run. However, according to the special theory of relativity, high velocity of satellites orbiting the Earth causes a delay in the satellite clocks in relation to the clocks on Earth. This is called relativistic error. It can be reduced by reducing the satellites' time pattern frequencies before placing them in the orbit, so that this frequency is equal to the nominal one already on it.

Ephemeris error can be explained as the difference between the actual parameters, i.e. the orbits parameters, and the ones emitted towards the Earth by the satellite. On the example of the system supporting GPS – EGNOS, it is sought to minimise the differences between the actual and transmitted status through the developed tracking stations, which, in small intervals, examine the actual status and correct information forecast and transmitted to the satellites by the ground segment.

Time pattern error of the satellite affects not only the accuracy of the user's position (in the case of satellite navigation systems), but also the data up-to-dateness. The satellite time pattern deviation affects directly the occurred error's magnitude. The time pattern for GPS is atomic time scale in accordance with UTC. On the satellites, there are atomic clocks that must have long useful life because of their rare replacement related with the satellite operation. The time measurement error equal to 1 ns causes the error of distance measurement from the satellite of 0.3 m.

The object position determined by the satellite systems is at the intersection of the surface with radii equal to the measured distances of the satellites from the receiver antenna. However, the distance measurements are burdened with errors and at the intersection of the surface, there is not a point but an area (this phenomenon is called dilution). The error resulting from the specified dilution is called Dilution of Precision (DOP). The error value depends on the shape of the planes intersection area. The satellites proper configuration (e.g. low on the horizon) reduces the DOP value. Several types of DOP are distinguished:

- GDOP – geometric dilution of precision (the position accuracy in four dimensions – space and time);
- PDOP – position dilution of precision (three-dimensional position coordinates);
- HDOP – horizontal dilution of precision (two-dimensional position coordinates);
- VDOP – vertical dilution of precision (height);
- TDOP – time dilution of precision.

The analysis of errors in the satellite systems is a major aspect limiting the operational functioning of such systems in the air transport. From the point of view of this work, the tropospheric and ionospheric errors, especially solar activity, deserve for special attention. It turns out that the time of year and even time of day can have a significant impact on the quality of the satellite signal and, therefore, on the operation safety of aircrafts. The high safety level in aviation is placed on top of the pyramid of modern operators and service providers' challenges in the industry.

4 IMPORTANCE OF SATELLITE SYSTEMS IN AIR NAVIGATION

The rationale for the selection of the research problem is the fact that the satellite systems are considered to be the future of navigation and surveillance in the aviation. Failure to meet the requirements set to the satellite signals prevents their operational use. The satellite systems play a significant role in programmes relating to the development of aviation technology, including in the SESAR programme (Single European Sky ATM Research), which is a technological component of the SES (Single European Sky) project implemented in the EU (Siergiejczyk & Krzykowska 2013). The concept of the Single European Sky was founded in 1999, when the need to adjust the EU's air navigation system to the growth of air traffic was pointed. The SESAR program, however, created in 2004, is a response to the need to support this traffic through specified, innovative ATM/CNS systems (Air Traffic Management/Communication Navigation Surveillance). Thanks to the SESAR

programme, the key elements affecting the efficiency of the European air navigation system are to be improved. In recent years, special attention is paid to the CNS systems in technical projects, in which solutions taking into account the implementation of 4D trajectory, reduction of air traffic delays, and increase in the airspace capacity using, among others, these systems, are proposed.

CNS consists of communication, navigation and surveillance. Navigation in other words – leading, pointing the right direction. This is where the implementation of the GNSS system and its development seem to be reasonable. In the era of programmes such as SESAR, it becomes obvious that the use of the satellite navigation may lead to a more flexible use of the airspace, and thus its higher throughput, shorter time of waiting for an aircraft (lower delays) and other benefits (Siergiejczyk & Rosiński & Krzykowska 2013).

Surveillance in the aviation can be defined as a set of technical measures, whose task is to provide information about the position of the aircraft in the airspace, in other words – to monitor this object. The selected technique for surveillance determines the amount of information transmitted to the air traffic control (information about the current situation both in the airspace and at the airport, as well as weather information). These data are used to ensure safe and smooth air traffic, including separation minima. In terms of surveillance, solutions using the satellite signal develop as dynamically as in navigation. One example may be ADS-B – Automatic Dependant Surveillance – Broadcast considered to be a key element of the air traffic management systems, including – the European SESAR and the US NextGen (Siergiejczyk et al. 2015). The ADS – B system is dependent in terms of reliance on the source and method of transmitting information about the location of the aircraft (in this case, GNSS).

ICAO emphasises that the use of GNSS in the air transport as a basic navigation method will be likely only when a signal meets the strict requirements. The existence of GPS and Galileo separately, but compatibly and interoperably to each other will ensure the signal's reliability and high quality. In addition, service providers are obliged to provide the GNSS service in the form of a continuous and accurate signal. Continuity in this case was considered in two ways. In a technical sense, it was about minimising the impact of system failure causing discontinuity events. In a legal sense – the functionality failures connected with military and political activities should have been excluded.

5 RESEARCH METHODOLOGY

Descriptions of external factors, including the weather conditions, is possible by using mathematical methods. However, is seems that one of the best method and tool to assess the impact of weather conditions on the interference of satellite signals tend to be fuzzy sets. Fuzzy set is a mathematical object with a defined membership function which takes values in the range [0, 1]. The theory of fuzzy sets was introduced by Lotfi A. Zadeh in 1965. As an extension of the classical theory. This reasoning is applied in the case of inaccurate or incomplete information, which makes it necessary to take decisions under conditions of uncertainty. Knowledge of the phenomenon is subjective and is expressed by the opinion or descriptive expert studies. It is determined by an informal basis as linguistic uncertainty. In the literature there are numerous references to the use of fuzzy sets, also in the context of air transport. Their big advantage is a graphical presentation of the results of research, modeling uncertainties and researches considering the effect of multiple factors on the different phenomenon (Petry et al. 2005).

In research works, it is planned to use the Mamdani fuzzy model. This type of model, to solve the problem, uses a linguistic approach, as well as the fuzzy conditional sentences and an assembly inference rule (Skorupski & Uchroński 2016). It is planned to introduce standard linguistic values in the range from large positive to large negative into the model. They all are identified with the fuzzy sets. The rule base can be built on the basis of the analysis of the literature and available results of research conducted in the similar field. In the process of building the model, two stages should be distinguished:

1 indication of the input variables and definition of their terms (based on the literature research);
2 assigning ranges of variable linguistic values.

It is extremely important to assign membership function to each of the variables. Its task, in fuzzy reasoning, is to express the extent, to which each element x of the universe of discourse X={x} belongs to the fuzzy set A – from $\mu\_A\_(x)=0$ for total non-membership to $\mu\_A (x)=1$ for full membership though all intermediate degrees ($0<\mu\_A (x)<1$).

After analysis of the existing literature, MatLab's Fuzzy Logic Tool was chosen, in which a trapezoidal function was applied to describe the input data. Currently, thanks to the data obtained from the Institute of Meteorology and Water Management, three groups of meteorological factors which are input data have been distinguished in the model, including: precipitation, air humidity and cloudiness. In addition, the following factor was added: solar energy to the tests. The measurements include 2014 and were executed in the station in Rzeszów once a day. Then, 274 rules specifying dependencies between particular factors in the satellite signal interferences were set.

Table 1. Linguistic values of input and output variables ranges

I/O variable	Linguistic values				
Cloudiness	-	low	medium	high	-
Precipitation	very low	low	medium	high	very high
Humidity	very low	low	medium	high	very high
Solar energy	very low	low	medium	high	very high
Interferences	very low	low	medium	high	very high

Table 2 shows the numbers reflecting the total membership of $\mu\_A(x)=1$ to the fuzzy set for the linguistic variables of input data.

Table 2. Function of linguistic variable of input data's membership

Input data	Linguistic variable	Measure (day by day)	Values for total membership to a fuzzy set
Cloudiness	Low	Sky coverage degree (in oktas)	2-3
	Medium		4-5
	High		6-8
Precipitation	Very low	(mm of water)	0-1
	Low		5-8
	Medium		10-14
	High		20-24
	Very high		29-43
Humidity	Very low	%	33-42
	Low		49-55
	Medium		62-71
	High		78-87
	Very high		93-100
Solar energy	Very small	Sunspots number	0-10
	Low		36-53
	Medium		80-113
	High		142-174
	Very high		205-234

The Mamdani's type was used in the model construction. The standard linguistic values in the range from large positive to large negative were introduced into the model. The rule base was built on the basis of the analysis of the literature and available results of research conducted in the similar field (Lower & Magott & Skorupski 2013).

6 RESEARCH RESULTS

The Mamdani model to simulate the satellite signal interferences under the influence of external factors is presented below.

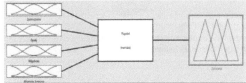

Figure 1. Mamdani model to simulate the satellite signal interferences

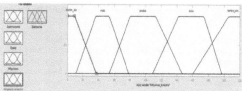

Figure 2. Adopted membership functions for fuzzy sets describing an input datum: solar activity

As an example, Figure 2 shows the adopted membership functions for fuzzy sets describing an input datum: Solar activity (according to Table 2 and available measurements). In the further part of the description, the results of the simulation, which define the relationships between the particular external factors and their influence on the satellite signal, were presented.

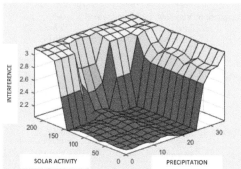

Figure 3. Graphic results for signal interferences depending on solar energy and precipitation

On Figure 3, one can clearly observe that both solar energy and precipitation significantly influence the satellite signal interferences but only when they take very high values.

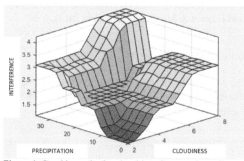

Figure 4. Graphic results for signal interferences depending on precipitation and cloudiness

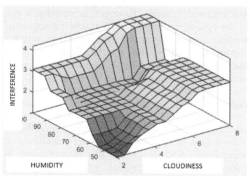

Figure 5. Graphic results for signal interferences depending on humidity and cloudiness

The graph on Fig. 4 is a premise to say that the cloudiness influences the signal interferences to a larger extend than precipitation. Fig. 5 presents a very similar situation. Therefore, an impact of weather conditions on the satellite signal quality is proven. Preliminary tests conducted for the needs of this work confirm this fact.

7 SUMMARY

It should be noted that the causes of the limited operating application of the satellite systems derive from various technical, organisational and administrative aspects. Some of them are associated with satellite connectivity. They require very precise legal regulations and a wide international cooperation. One cannot forget the errors occurring in the satellite systems. They are a big limitation, the ionospheric and tropospheric errors. It turns out that the time of year, and even the time of day can have a significant impact on the quality of the satellite signal and, therefore, on the operation safety of aircrafts. The high safety level in aviation is placed on top of the pyramid of modern operators and service providers' challenges in the industry. The growth rate of the satellite market in the navigation sector is 25% per annum. By 2025, the demand for almost 18 thousand new aircrafts will have been caused by threefold growth in passenger traffic and similarly rapid increase in air cargo transport. For Europe, the development and implementation of the Galileo system remains the most important goal. Accuracy and correctness, which will be ensured by this system, will enable greater use of existing airports that currently cannot work in bad weather conditions and low visibility. It cannot forgot that the development of GNSS services represents an important point of the European programme of SESAR (Krzykowska & Siergiejczyk & Rosiński 2014).

The satellite systems are considered to be the future of navigation. They constantly are the subject of projects, and even entire research programmes, which aim primarily at the product development. The similarity of the navigation satellite systems structure makes it possible to extract some general lines of action by 2020. The main technical changes in space and ground segments, whose aims is as follows, are distinguished:

- improving the stability of the clocks on the satellites,
- increasing the broadcast signals strength,
- extension of the satellite lifetime,
- broadcasting new signals on additional frequencies,
- increasing the measurements accuracy (especially under conditions of high city housing and other ones restricting signal transmission),
- inclusion of transmission ionoshperical fixes in the primary navigation signal,
- increasing the signal's resistance to interferences,
- providing communication between the satellites within a given system.

REFERENCES

Alekseev, K.P.G. & Seixas, J.M.. 2009. A multivariate neural forecasting modeling for air transport e preprocessed by decomposition: a Brazilian application. J. Air Transp. Manag. 15, 212-216.

Chen, T. 2012. A collaborative fuzzy-neural approach for long-term load forecasting in Taiwan. Comput. Ind. Eng. 63, 663-670.

Habarulema J.B. & McKinnell L-A. & Opperman B.D.L. 2009. Towards a GPS-based TEC prediction model for Southern Africa with feed forward networks, Advances in Space Research 44, pp. 82-92.

García-Pereiro T. & Dileo I. 2015. Determinants of nascent entrepreneurial activities: the Italian case. RIVISTA ITALIANA DI ECONOMIA DEMOGRAFIA E STATISTICA. VOLUME LXIX – N. 4. Roma, pp. 5-16.

Jacyna M. 2008. Wybrane zagadnienia modelowania systemów transportowych, Oficyna Wydawnicza Politechniki Warszawskiej.

Januszewski J. 2010. Systemy satelitarne GPS Galileo i inne. Warszawa.

Kaniewski, P. & Lesnik, C. & Susek, W. & Serafin, P. 2015. Airborne radar terrain imaging system. In 16th International Radar Symposium (IRS), Dresden, Germany, pp. 248-253.

Kierzkowski A. & Kisiel T. 2016. Simulation model of security control system functioning: A case study of the Wroclaw Airport terminal, Journal of Air Transport Management, http://dx.doi.org/10.1016/j.jairtraman.2016.09.008

Krzykowska K., Siergiejczyk M. 2014. Signal monitoring as a part of maintenance of navigation support system in civil aviation. Archives of Transport System Telematics ISSN 1899-8208.

Krzykowska, K., Siergiejczyk, M., Rosiński A. 2014. Parameters analysis of satellite support system in air navigation. 23rd International Conference on Systems Engineering, ICSEng 2014, Advances in Intelligent Systems and Computing, Springer, Las Vegas, USA, August 19 – 21.

Laskowski D. & Łubkowski P. 2013. The end-to-end rate adaptation application for real-time video monitoring, Advances in Intelligent Systems and Computing, Springer International Publishing AG, Switzerland, Advances in Intelligent Systems and Computing, pp. 295-305.

Lower M., Magott J., Skorupski J. 2013. Analiza incydentów lotniczych z zastosowaniem zbiorów rozmytych. Warszawa.

Narkiewicz J. 2007. GPS i inne satelitarne systemy nawigacyjne. Warszawa, pp. 87-91.

Petry F.E. 2005. Fuzzy Modeling with Spacial Information for Geographic Problems. Springer.

Piro G. & Yang K. & Boggia G. & Chopra N. & Grieco L.A. & Alomainy A. 2015. Terahertz communications in human tissues at the nano-scale for healthcare applications. IEEE Transactions on Nanotechnology 14(3), pp. 404-406.

Siergiejczyk M. & Krzykowska K. 2014. Some issues of data quality analysis of automatic surveillance at the airport. Diagnostyka 1/2014 ISSN 1641-6414.

Siergiejczyk, M. & Krzykowska, K. & Rosiński, A. 2015b. Reliability assessment of integrated airport surface surveillance system. In W. Zamojski, J. Mazurkiewicz, J. Sugier, T. Walkowiak, J. Kacprzyk (ed.), Proceedings of the Tenth International Conference on Dependability and Complex Systems DepCoS-RELCOMEX, given as the monographic publishing series – „Advances in intelligent systems and computing", Vol. 365: 435-443. Springer.

Siergiejczyk M. & Rosiński A. & Krzykowska K. 2013. Reliability assessment of supporting satellite system EGNOS. The monograph New results in dependability and computer systems editors: Wojciech Zamojski, Jacek Mazurkiewicz, JarosławSugier, Tomasz Walkowiak, Janusz Kacprzyk, given as the monographic publishing series – „Advances in intelligent and soft computing", Vol. 224. Wydawca: Springer, pp: 353 – 364.

Skorupski J. & Uchroński P. 2016. A fuzzy system to support the configuration of baggage screening devices at an airport. Expert Systems With Applications 44, pp. 114–125.

Stańczyk P. & Stelmach A. 2015. Selected aspects of modeling the movement of aircraft in the vicinity of the airport with regard to emergency situations, Theory and Engineering of Complex Systems and Dependability, Springer International Publishing, pp. 465-475.

Stawowy M. 2015. Comparison of uncertainty models of impact of teleinformation devices reliability on information quality, Proceedings of the European Safety and Reliability Conference ESREL 2014, CRC Press/Balkema, pp. 2329-2333.

Stawowy, M. & Dziula, P. 2015. Comparison of uncertainty multilayer models of impact of teleinformation devices reliability on information quality. In Proceedings of the European Safety and Reliability Conference ESREL 2015: 2685-2691. Zurich.

Sumiła M. 2014. Evaluation of the drivers distraction caused by dashboard MMI interface, Monografia Telematics support for transport. J. Mikulski (ed.), Communications in Computer and Information Science: pp. 396-403. Springer, Berlin Heidelberg.

Warnant R. & Kutiev I. & Marinov P. & Bavier M. 2007. Ionospheric and geomagnetic conditions during periods of degraded GPS position accuracy: 1. Monitoring variability in TEC which degrades the accuracy of Real-Time Kinematic GPS applications. Advances in Space Research nr 37(2007): pp. 875 – 880.

Yin P. & Mitchell C. 2011. Demonstration of the use of the Doppler Orbitography and Radio positioning Integrated by Satellite (DORIS) measurements to validate GPS ionospheric imaging. Advances in Space Research nr 48 (2011): pp. 500 – 506.

Zajac M. & Świeboda J. 2015. Process hazard analysis of the selected process in intermodal transport. In Military Technologies (ICMT), International Conference on (pp. 1-7). IEEE. DOI: 10.1109/MILTECHS.2015.7153698

Reliable Vessel Navigation System Based on Multi-GNSS

A. Angrisano
G. Fortunato University, Benevento, Italy

S. Gaglione, S. Del Pizzo, G. Castaldo & S. Troisi
Parthenope University of Naples, Napoli, Italy

ABSTRACT: One of the most recent aims established by IMO (International Maritime Organization) is the improved reliability and integrity of navigation information, in order to enhance navigation safety and security services at sea. Actually, the GPS (Global Positioning Systems) and AIS (Automatic Identification System) provide ship position, speed and course to VMS (Vessel Management System) platforms, without any consideration about the safety and security.
The reliable navigation system proposed is based on a multi – GNSS satellite receiver and on vessel speed log and direction finder (magnetic or gyro compass). The robustness of navigation solution in ensured by: RAIM (Receiver Autonomous Integrity Monitoring) algorithms able to avoid (or limit) the consequences of gross errors between measurements, and by a coherence check between GNSS and position solution provided by a Dead Reckoning algorithm, based on log and compass measurement, able to highlight drift in position domain, caused for example by spoofing on GNSS receiver. In order to validate the algorithms, several tests have been carried out in different scenario, in static and kinematic condition.

1 INTRODUCTION

When a vessel is in open sea, the accurate knowledge of its position, velocity and course will ensure to reach its destination in the safest, most economical and timely way. The need for accurate position information becomes even more critical as the vessel departs from or arrives to port. Vessel traffic and other potential waterway hazards make maneuvering more difficult and the risk of accidents becomes greater, therefore the need of an accurate and reliable navigation become a primary aim of IMO. Furthermore, GNSS positioning is critical when satellite signals are blocked or strongly degraded by natural or artificial obstacles (generating the multipath effect), when an intentional interference occurs causing the unavailability of the system (e.g. a jamming attack) and when GNSS receiver is forced to provide a fake position/time (Spoofing) (*Ruegamer et al. 2015*).

For these reasons, in order to provide a reliable navigation information, in this paper, two interference mitigation algorithms and a coherence check are implemented for multi-constellation receiver, improving the accuracy of the positioning in case of satellite signals interferences. These are the Observation Subset Testing and the Forward-

Backward algorithms (Angrisano et al. 2013, Castaldo et al. 2014), which have the purpose of detecting and excluding measurements affected by gross errors. In the coherence check, the positions provided by GNSS are compared with the navigation solutions provided by Dead Reckoning (DR) algorithm, obtained by vessel instruments which are different from GNSS and, for this reason, free by a potential spoofing. A complete description of DR algorithm can be found in (Kaplan et al. 2006).

The performances of the proposed algorithms have been evaluated in terms of position accuracy and integrity. In order to validate the algorithms, several tests have been carried out, including static and kinematic data collections.

The reminder of the paper is organized as follows. An overview of the estimation technique is provided in Section 2; in Section 3 RAIM algorithms are detailed. In Section 4 Reliable navigation method are introduced. Test and experimental results are analyzed in Section 5 and, finally, Section 6 concludes the paper.

2 ESTIMATION TECHNIQUE

Estimation is the process of obtaining a set of unknowns from a set of uncertain measurements, according to a definite optimization criterion (Brown et al. 1997a).

The most common adjustment method employed in geomatics is the Least Squares (LS), that uses only a measurement model to estimate the state vector:

$$\underline{z} = H\underline{x} + \underline{e} \tag{1}$$

where $\underline{z}$ represents the measurement vector, H is the design matrix, $\underline{x}$ is the state vector and $\underline{e}$ is the measurement noise vector.

The optimization criterion adopted by the LS method is the minimization of the squared residuals defined as:

$$\underline{r} = \underline{z} - H\hat{\underline{x}} \tag{2}$$

In this work the adopted estimation technique is a Weighted LS (WLS), whose solution is:

$$\hat{\underline{x}} = \left(H^T W H\right)^{-1} H^T W \underline{z} \tag{3}$$

The used weights matrix (W) is related to satellite elevation for GNSS measurements, as shown in (Kaplan et al. 2006).

3 RAIM ALGORITHMS

In reliability testing, it is assumed that the measurements follow a predefined statistical distribution and the presence of an outlier cause the measure no longer belongs to that (Brown et al 1997b). Actually, it is not possible to undoubtedly know if a measure belongs or not to a predefined distribution (Mezentsev. 2005). In order to get that information with a good chance, a statistical approach is adopted, defining a decisional variable and a threshold wherewith performing a comparison. The situation where outliers are not encountered is called Null Hypothesis (H_0) and the situation where outliers are present is called Alternative Hypothesis (H_a); if the decisional variable D is lower than the threshold T, the H_0 event is assumed, while if D is higher than T, the H_a event is considered and the presence of outliers is assumed. This procedure is performed using two different tests: a global and a local test. The first one is carried out to verify the measurement set consistency and is called Global Test (GT), if such test fails, i.e. the measurement set is declared inconsistent. The second one is a test to identify the outliers, so called Local Test (LT).

3.1 Global Test

A common reliability test is the GT, based on least squares residual analysis, that determines the goodness of the used model or highlights the presence of outliers. It defines a decision variable D, based on the quadratic form of the residual $\overline{r}$, weighted through the weighting matrix W, and compares it with a threshold. The decision variable is defined as:

$$D = \overline{r}^T W \overline{r} \tag{4}$$

It is assumed that the decision variable follows a central chi-square (χ^2) distribution with (m-n) degrees of freedom (DoF) (Mezentsev 2005); this is, in turn, based on the assumption that the observation errors follow a standard normal distribution and on the relationship between the normal and chi-square distributions (Brown et al. 1997b).

The threshold is defined by:

$$T_G = \chi^2_{1-\alpha,DoF} \tag{5}$$

with α is the false alarm probability and $\chi^2_{1-\alpha,DoF}$ is the abscissa value corresponding to a false alarm probability $(1-\alpha)$ of a chi-square distribution with DoF degrees of freedom.

The hypothesis testing in the GT is:

$$H_0 : D < T_G$$
$$H_a : D > T_G \tag{6}$$

If H_0 is rejected and H_a accepted, an inconsistency in the measurement set is assumed, and the blunder should be identified and mitigated. The GT is applied to the whole set of measurements, while to identify an outlier, a Local Test has to be carried out or GT should be applied to several measurement subsets.

3.2 Local Test

Local Test is performed analysing standardized residual w_i for each satellite

$$w_i = \left| \frac{r_i}{\sqrt{(C_r)_{ii}}} \right|, i = 1 : m \tag{7}$$

where r_i is the ith element of residual vector $\overline{r}$, $(C_r)_{ii}$ the ith diagonal element of the residual covariance matrix C_r.

$$C_r = W^{-1} - H\left(H^T W H\right)^{-1} H^T \tag{8}$$

The standardized residuals are assumed to be normally distributed; each w_i that exceeds the local threshold T_L is flagged as blunder. Threshold is defined as:

$$T_L = N_{1-\alpha_0/2} \tag{9}$$

where T_L is the abscissa corresponding to the probability value $(1 - \alpha_0 / 2)$ of a normal distribution (N).

The measurement corresponding to the largest standardized residual exceeding T_L is excluded after a separability check. Separability refers to the ability to separate any two measurements from one another (Hewitson et al. 2006); this concept is primary to avoid the incorrect exclusion of blunder-free measurements. A large blunder could cause multiple local test failures and therefore an erroneous measurement rejection. In RAIM techniques, a parameter properly representing the separability is the correlation coefficient, shown in the following:

$$\gamma_{i,j} = \frac{(C_r)_{ij}}{\sqrt{(C_r)_{ii} (C_r)_{jj}}} \qquad (10)$$

In the separability check, if $\gamma_{i,j}$ exceeds a threshold, the measurement suspected to be a blunder is not rejected or deweighted, because it is strictly correlated to another measurement.

3.3 Preliminary check

In case of poor satellite geometry, the navigation accuracy degrades, integrity monitoring algorithms performance is degraded and large errors can occur before the outliers are detected (Kaplan et al. 2006); hence a preliminary check has to be performed to screen out bad geometries, which could imply erroneous detections.

The method proposed for this scope is based on Approximate Radial-error Protected (ARP) defined as:

$$ARP = slope_{max} \sqrt{T_g} \qquad (11)$$

where SLOPE parameter is the ratio between the radial error (horizontal HRE or vertical VRE) and the statistic test when a deterministic error is considered on a single measurement and stochastic perturbations are omitted.

The satellite whose bias error causes the largest position error is the most difficult to detect and it is associated to the $slope_{max}$. The classical parameter, ARP, considers all measurements with the same accuracy. In order to include a different weights for each satellite, in this work, a generalization of the ARP (Parkinson et al. 2006) is studied; it is called Weighted ARP (WARP) and its expression is:

$$WARP = Wslope_{max} \sqrt{T_g} \qquad (12)$$

3.4 Observation Subset Testing Algorithm

Subset testing is an FDE technique based only on GT, it can be adopted to localize gross errors by assessing the LS residuals (Angrisano et al. 2013). LS technique may spread multiple gross errors across the whole measurement set. Hence localization of the blunders based on statistical rejection of residuals could be very difficult. A possible solution is to perform several LS adjustments using subsets of the measurements, in order to find a subset from which the supposed blunders are excluded. If a measurements set is stated inconsistent, all the possible combinations of measurements are checked, i.e. all the possible subsets including n+1 to m−1 measurements, where m is the number of measurements and the n is the number of unknowns. Only the subset that satisfies the GT is declared consistent and is used to compute the navigation solution; if more subsets pass the GT, the set with the minimum statistic variable and the largest number of measurements is chosen. In this technique, the separability check is not performed, because standardized residuals are not analyzed. A complete scheme of the algorithm is shown in Figure 1.

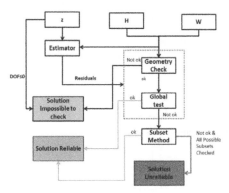

Figure 1. Subset Testing Workflow

3.5 Forward Backward Algorithm

Forward-Backward is a Fault Detection and Exclusion (FDE) technique, involving the use of both GT and LT to identify and exclude the outliers. This method consists of two different steps (Kuusniemi. 2005). The first algorithm section, called Forward, is carried out to identify and exclude erroneous measurements. In the Forward phase, preliminary check, GT, LT and separability check are used. This part of the algorithm contemplates the following steps:
- the measurement set is preliminary tested for the integrity geometry to screen out bad geometries which could imply erroneous detections;
- if the system has strong geometry, the GT is carried out in order to verify measurements consistency;

131

- in case of GT failure, the LT is performed to identify erroneous measurements;
- in case of failure of the LT, the separability check is carried out to avoid erroneous rejection of a good observation due to the mutual influence of the observations. If no outliers are identified, the solution is declared unreliable due to contradictory results between GT and LT;
- the measurement flagged, in the LT, as possible blunder is excluded only if it is not strongly correlated with other measures.

The forward process is performed recursively until no more erroneous measurements are found and the solution is declared reliable or unreliable.

If the solution is declared reliable and k measurements are excluded (with k≥2), the Backward scheme is applied to reintroduce observations wrongly excluded. The Backward procedure use only the GT. Rejected measurements are iteratively re-introduced and the GT is performed to verify the consistency of the measurement set. The observation set which passes the GT is used to compute the navigation solution to ensure that the order of the excluded measurements does not cause a wrong exclusion and to minimize the number of unnecessary exclusions. A complete scheme of the Forward-Backward technique is shown in Figure 2

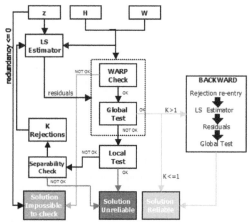

Figure 2. Forward-Backward workflow

4 CONSISTENT NAVIGATION METHOD

The reliable solution obtained by RAIM is compared with the position provided by dead reckoning, computed starting from the last available reliable position and considering the vessel velocity and course given by speed log and gyrocompass respectively. If the distance between the position given by RAIM and the position given by dead reckoning is smaller than a threshold, the solution is

classified as reliable and spoofing free, otherwise it is considered unreliable. A complete scheme of the reliable navigation method is show on Figure 3.

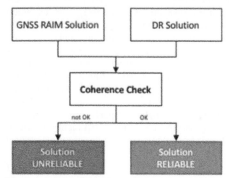

Figure 3. Reliable Navigation method workflow

5 TEST AND RESULTS

In order to validate the proposed approach two different data collections have been performed. The first one in static conditions and the second one in kinematic environments using GPS/GALILEO constellations, in order to study the potentiality of the European satellite system.

In order to stress the developed RAIM algorithms for the static test, an urban canyon has been chosen as scenario; it is a very noisy environment where the skyscrapers may reflect the signals, introducing gross errors among measurements and leading to a wrong estimate of the position.

Moreover, in order to simulate a ship in a port during the maneuvering operations, a test in kinematic conditions has been done using a properly equipped boat; the traveled path was characterized by entry and exit maneuvers from port.

In next sections the tests and the obtained results are described.

5.1 Static Test

5.1.1 Data collection

The measurement session has been collected, using low cost U-blox single frequency receiver (U-blox Neo M8N), in 8th July 2016 for about 2 hours and 30 minutes with a one second data rate. The antenna was placed on the roof of the PANG laboratory building at Parthenope University in Centro Direzionale of Naples (Italy) (Fig. 4); in this environment many GNSS signals are blocked by skyscrapers or are strongly degraded because of the multipath and non-line-of-sight phenomena.

The reference solution is computed by a post-processing geodetic method, guaranteeing a position accuracy of mm order.

Figure 4. Antenna Position.

The reference solution is computed by a post-processing geodetic method, guaranteeing a position accuracy of mm order.

5.1.2 Accuracy and precision analysis

The RAIM algorithm performance are analyzed in terms of:

- Accuracy, using as figure of merit RMS and maximum errors for horizontal and vertical components in the position domains;
- Integrity, using as figure of merit the reliable availability, defined as the time percentage when solution is classified as reliable.

Herein three different GNSS configurations are compared, combining the two considered systems and the different developed RAIM algorithms:

- GPS only without RAIM application (briefly indicated as GPS ONLY);
- GPS/GALILEO with Forward-Backward RAIM application (GPS+GALILEO FB);
- GPS/GALILEO with Observation Subset Testing RAIM application (GPS+GALILEO SUB);

The performance of the developed RAIM algorithms is compared with GPS only configuration, using the classical representation for the horizontal component, i.e. East and North coordinates, in order to have a clear amplitude of the point clouds of the different configurations.

The qualitative analysis of the figure 5 shows that the cloud relative to "GPS+GALILEO FB" (blue dots) and "GPS+GALILEO SUB" (black dots) configurations are more concentrated with respect to the "GPS Only" one (green dots). So, also in this case, the configurations that use RAIM algorithms are characterized by smaller errors.

Figure 5. Horizontal Scatter.

Figure 6 shows horizontal and vertical position errors obtained using the different configurations during the static session. It can be noted that for the first 20 minutes only the GPS solution is available, due to the low number of visible satellites. Then, GPS+GALILEO FB provides the best behavior compared to the other configurations for horizontal and vertical components.

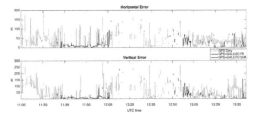

Figure 6. Horizontal and vertical error considering all reliable epochs for each configuration.

The comparison shows that for both considered RAIM techniques, the errors decrease using RAIM algorithms with respect to "GPS only case", characterized by horizontal and vertical RMS errors amounting, respectively, to 42,70 m and 65,02 m and by horizontal and vertical maximum errors, respectively, of 245,93 m and 291,19 m.

The horizontal and vertical RMS errors obtained using GPS/GALILEO with Forward-Backward RAIM configuration are, respectively, 23,47 m and 32,55 m; the maximum horizontal and vertical error amounts respectively to 166,61 m and 227,42 m.

Using the GPS+GALILEO with Observation Subset testing RAIM configuration the horizontal RMS error is 24,73 m and the vertical RMS error 38,94 m; the maximum horizontal and vertical error are, respectively, 149,96 m and 233,84 m.

The reliable availability is about 28,3% for the Forward-Backward RAIM algorithm and 26,5% for the Observation Subset Testing algorithm.

Static test results obtained considering only reliable epochs for each configuration, are summarized in Table 1.

Table 1. Statistical position error parameter in static test considering only reliable epochs for each configuration.

	Hor. RMS [m]	Ver. RMS [m]	Hor. MAX [m]	Ver. MAX [m]	Reliable Availability [%]
GPS only	42.70	65.02	245.93	291.19	/
GPS/ GALILEO FB	23.47	32.55	166.61	227.42	28.26
GPS/ GALILEO SUB	24.73	38.94	149.96	233.84	26.56

The obtained results show the effectiveness of the adopted algorithms in terms of reliable availability and of RMS and maximum errors. The reliable availability is the percentage of time mission when the solution is declared reliable by the adopted RAIM algorithm; the highest value of this parameter is obtained with the Forward-Backward method, which provides also the smallest errors proving the validity of the separability check module (which cannot be applied to Observation Subset testing method). The Observation Subset testing is instead characterized by similar accuracy performances and by smallest reliable availability.

5.2 Kinematic Test

5.2.1 Data collection

The kinematic session was performed by a boat test and was taken place on 10th September 2016 in the Gulf of Naples (Italy). In this session about 45 minutes of data at 1 Hz rate were collected. The vehicle left Massa Lubrense port and first headed towards the north to reach the port of Sorrento, by performing the entry and exit maneuvers from port. Then the course has been inverted until Massa Lubrense port. Also in this case the entry and exit procedures from the port have been repeated, during the final part of the measurement session, the boat was headed south along the Sorrento Peninsula. The path is shown in Figure 7.

Figure 7. Reference trajectory for the kinematic test.

The vehicle was equipped with a U-blox receiver and a geodetic double frequency Novatel receiver (Fig. 8) that was used to generate a precise reference trajectory computed by a post-processing geodetic method.

5.3 Accuracy and precision analysis

In order to validate the proposed approaches, the same analysis carried out for the static test is performed for the kinematic test.

From Figure 9, it can be noted that the clouds of the considered configurations are nearly coincident. This result is due to the low presence of blunder during the whole measurement session.

Horizontal and vertical errors obtained in kinematic test are shown in Figure 10.

Figure 8. Kinematic test equipment.

Figure 9. Horizontal Scatter

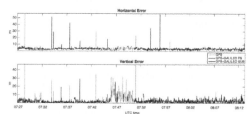

Figure 10. Horizontal and vertical error considering all reliable epochs for each configuration.

From the comparison, for the horizontal error, it can be noted that the three strategies proposed are similar, whereas the inclusion of RAIM algorithms provides significant improvement for the vertical performance with respect to the GPS-only case.

The obtained vertical RMS errors decrease, respectively, of 29,2% for GPS+GALILEO FB and 30% for GPS+GALILEO SUB with respect to the GPS-only configuration.

The maximum vertical error for both RAIM configuration decrease from 46,82 m, obtained with GPS only, to 29,48 m, with an enhancement of 37%.

From 7:45 to 7:50 (UTM) it can be noted an evident decreasing of GPS performance, in fact errors of stand-alone configuration are considerable. This state happens during the maneuvers in the port of Sorrento (yellow dots trajectory in Fig. 11).

Figure 11. Trajectory with higher vertical errors.

By analyzing the PDOP values recorded (Fig.12), for the considered period, they are significantly higher than the rest of the session.

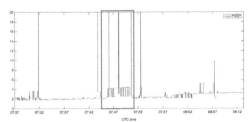

Figure 12. GPS/GALILEO PDOP parameters

This is due to the decrease of visible satellites and the presence of a blunder in measurements, that has been properly identified by both RAIM algorithms. This is evident in Figure 13 which shows the total number of visible satellites (green line) and the number of satellites used after the application of RAIM algorithms (blue line).

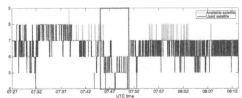

Figure 13. Available and reliable satellite for kinematic test

Results obtained during maneuvers in the port of Sorrento are shown in table 2.

In table 2 it could be noted that the application of RAIM algorithms enhance GNSS performance in terms of vertical RMS, horizontal and vertical maximum errors, identifying the blunder and classifying as unreliable the epochs with a lower visible satellite number.

Kinematic test results are summarized in Table 3.

Table 2. Statistical position error parameter obtained during maneuvers in the port of Sorrento.

	Hor. RMS [m]	Ver. RMS [m]	Hor. MAX [m]	Ver. MAX [m]	Reliable Availability [%]
GPS only	4.51	10.72	23.21	46.81	/
GPS/ GALILEO RAIM	4.59	8.88	8.23	26.12	28.62

Table 3. Statistical position error parameter in kinematic test considering only reliable epochs for each configuration.

	Hor. RMS [m]	Ver. RMS [m]	Hor. MAX [m]	Ver. MAX [m]	Reliable Availability [%]
GPS only	4.21	4.62	56.73	46.82	/
GPS/ GALILEO FB	4.27	3.27	55.69	29.48	71.18
GPS/ GALILEO SUB	4.20	3.23	55.69	29.48	72.08

From table 3 it is evident that, in this case, Observation Subset Testing RAIM algorithm provides slightly better performances in terms of reliable availability, horizontal and vertical RMS errors with respect to GPS+GALILEO FB configuration.

Lastly the RAIM reliable solutions and the one obtained by dead reckoning are compared. In this case, the speed log and the gyrocompass are not available, hence the vessel course and velocity information have been simulated by using the Position and Velocity estimation provided by the double frequency GNSS receiver. Figure 14 shows the number of solutions resulting reliable from the comparison.

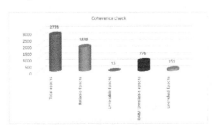

Figure 14. Coherence check results

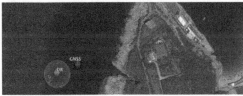

Figure 15. Unreliable epoch

The total number of solutions is 2778, 776 of them are considered unreliable by RAIM, 151 epochs are unchecked since the velocity of the vessel was insufficient to simulate course information with GNSS. In the 99.50% of verifiable epochs, the solution is classified reliable.

Considering one of the 13 epochs marked unreliable, from figure 15 it can be noted that the distance between the vessel position obtained by GNSS and the dead reckoning solution exceeds the threshold (T_d), hence the solution is declared unreliable by the algorithm. In this case, the RAIM algorithm cannot detect the presence of blunder, consequently an error on the GNSS estimated position is deduced.

6 CONCLUSIONS

In this paper, two different RAIM techniques (Observation Subset testing and Forward-Backward algorithms) have been developed and tested using a multi-constellation (GPS/GALILEO) receiver to improve the robustness of the GNSS solution and to satisfy IMO's objectives. Furthermore, a coherence check between GNSS and Dead Reckoning position solution has been performed in order to identify potential spoofing attack on GNSS receiver.

RAIM algorithm performances have been evaluated in real condition using two different data collection: the first one in static condition and the second one in kinematic mode. Static data set has been collected in urban canyon, that is the most challenging scenario to validate the integrity requirements.

The results obtained show the effectiveness of the adopted algorithms in terms of reliable availability, RMS and maximum errors. The highest value of reliable availability is obtained with the Forward-Backward method, which provides also the smallest errors, demonstrating the validity of the separability check module (which cannot be applied to Observation Subset testing method). The Observation Subset testing is instead characterized by similar accuracy performances and by smallest reliable availability.

Moreover, in order to simulate a ship in port during the maneuvering operations, a test in kinematic conditions has been done using a properly equipped boat, the path traveled was characterized by entry and exit maneuvers in the port.

In this case, because of the low presence of blunder during the whole measurement session, it can be noted that the three proposed strategies results (GPS only, GPS+GALILEO FB and GPS+GALILEO SUB) are similar for the horizontal error, whereas the inclusion of RAIM algorithms provides significant improvement for the vertical performance with respect to the GPS-only case.

Furthermore, during the maneuvers in the port, an evident decreasing of GPS performance has been noted, this is due to the decrease of visible satellites and the presence of a blunder in measurements, that have been properly identified by both RAIM algorithms.

Also in this case both RAIM algorithms provide better performances for all the considered figures of merit with respect to GPS only configurations.

Finally, the coherence check between the position obtained by GNSS and the position estimated by dead reckoning was carried out. From this check, only 13 solutions are classified as unreliable, since RAIM algorithms incorrectly detect blunders and they bring to an error on the GNSS estimated position.

ACKNOWLEDGEMENT

This research is supported by ASI (Agenzia Spaziale Italiana), Italy, and is included in SMILE (Satellite Multi-constellation Identification techniques for Liable Enhanced applications) project.

REFERENCE

Angrisano A., C. Gioia, S. Gaglione, and G. del Core. 2013. "GNSS Reliability Testing in Signal-Degraded Scenario," International Journal of Navigation and Observation, vol. 2013, Article ID 870365, 12 pages.

Brown, R. G., and Hwang, P.Y.C. 1997a. Introduction to Random Signals and Applied Kalman Filtering. John Wiley & Sons.

Brown, R. G., Chin G. Y. 1997b. GPS RAIM: Calculation of Threshold and Protection Radius Using Chi-Square Methods-A Geometric Approach in Global Positioning System: Inst Navigation, vol. V, pp. 155–179

Castaldo, G.; Angrisano, A.; Gaglione, S.; Troisi. 2014. P-RANSAC: An Integrity Monitoring Approach for GNSS Signal Degraded Scenario. International Journal of Navigation and Observation.

Hewitsonand S., Wang J. 2006. "GNSS receiver autonomous integrity monitoring (RAIM) performance analysis," GPS Solutions, vol. 10, no. 3, pp. 155–170.

Kaplan, E.D.; Hegarty, J. 2006. "Fundamentals of Satellite Navigation" in "Understanding GPS: Principles and Application", Ed., Artech House Mobile Communications Series, 2nd edition.

Kuusniemi H. 2005. User-level reliability and quality monitoring in satellite-based personal navigation [Ph.D. thesis], Tampere University of Technology, Tampere, Finland.

Mezentsev, O. 2005. Sensor Aiding of HSGPS Pedestrian Navigation. PhD Thesis, published as UCGE Report No. 20121, Department of Geomatics Engineering, University of Calgary

Parkinson B. and Spilker J. J. 1996. Global Positioning System: Theory And Applications, vol. 1-2, American Institute of Aeronautics and Astronautics, Washington, DC, USA.

Ruegamer, Alexander, and Dirk Kowalewski 2015. "Jamming and Spoofing of GNSS Signals–An Underestimated Risk?!.", the Wisdom of the Ages to the Challenges of the Modern World Sofia, Bulgaria, 17-21 May 2015: 1-21.

The SMILE Project: Satellite Multi-Constellation Identification Techniques for Liable Enhanced Applications

S. Gaglione, S. Del Pizzo, A. Innac & S. Troisi
Parthenope University of Naples, Napoli, Italy

N. Marchese, G. Pellecchia, A. Gentile & A. Amatruda
ITSLAB S.r.l., Torre Annunziata (NA), Italy

G. Mangani, G. Cecilia, V. Fontana & M. Lombardi
Blue Thread S.r.l., Roma, Italy

ABSTRACT: The SMILE (Satellite Multi-constellation Identification techniques for Liable Enhanced applications) project is supported by the Italian space agency, ASI (Agenzia Spaziale Italiana); it started in November 2015 and will last in October 2017. The SMILE partnership, including ITSLAB S.r.l.(Prime Contractor), University of Naples Parthenope and Blue Thread S.r.l., covers the different aspects of the project, starting from the analysis of commercial potential in terms of market and business development. Based on current maritime market, identifying the additional necessities of the final users, SMILE project aims to enhance and extend the performance of Vessel Management System (VMS) providing an advanced application solution for the maritime context. The main goal of SMILE is to assure a reliable, not repudiable, robust and accurate navigation system through GNSS (Global Navigation Satellite Systems) and AIS (Automatic Identification System) equipment that are the basis of Vessel Management System platforms. In particular, in the present paper two main technical aspects of SMILE system will be described: the Receiver Autonomous Integrity Monitoring (RAIM) algorithm used to guarantee a robust navigation system, and an innovative vessel authentication technique based on a dynamic masking of information received by onboard GNSS receivers.

1 INTRODUCTION

The SMILE (Satellite Multi-constellation Identification techniques for Liable Enhanced applications) project is oriented to offer an innovative application solution within the maritime sector, aiming to improve and extend the functionality of Vessel Management System (VMS). Actually, these platforms are based on the AIS (Automatic Identification System), that is an equipment used to detect and transmit the position of the ship determined by a GNSS receiver (Tetreault, 2005), together with the vessel ID and other ancillary information such as course, velocity, destination, ETA (Estimated Time of Arrival) etc. On the other hand, such system does not provide any guarantee about the security, especially in terms of countermeasures to the spoofing attacks. The SMILE project is going to improve and extend the features of the VMS platform.

The project partners are 3: ITSLAB S.p.a. (prime), University of Naples Parthenope and Blue Thread S.r.l.. The partnership consists of one university and two small – medium enterprises.

The SMILE project is divided in 6 work-packages, every partner manages two nodes in according to the scheme shown in Figure 1.

Project activities are distributed in 15 tasks as the breakdown structure shows; basically every partner has all needed facilities and know how to develop the assigned work-packages

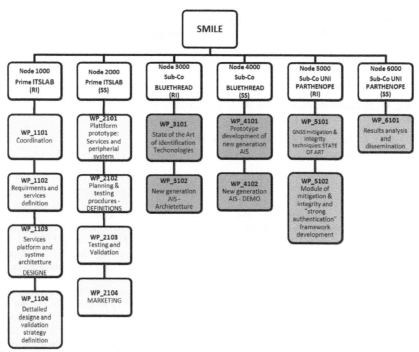

Figure 1. Work Breakdown Structure.

1.1 *Project aims*

SMILE aims to enhance the above mentioned technologies to offer more reliable services. The project idea is moving, therefore, on two technological lines:

– the former is the time to strengthen the on board equipment, the GNSS receiver and the AIS, in order to make them more robust to the noise and to the interferences,
– the second one is to develop a novel technique for the identification and non-repudiation of the vessel based on the geo-location.

The technical objectives of SMILE system could be summarized in the following:

– Make sure of the authenticity of the vessel where the transmitter is installed: the transmitter (GNSS receiver + AIS) and the vessel are loosely matched because it is possible to unmount the transmitter and transfer it on board of another entity or to masquerade directly for another entity;
– Make sure of the authenticity of the stated position of the transmitter: the transmitter could be interfered (intentionally or accidentally) and the position achieved by it could be not reliable;
– Ensuring the reception of the authentic data transmitted: the security cycle should be closed, informing the transmitter that the information have been received by the true receiver.

Each point above mentioned is related with one or more security threats and neglecting one or more of them could cause a lack of reliability, making the GNSS based services built on these solutions pretty unreliable.

Two main on-board equipment are involved in these security issues: the GNSS receiver and the AIS.

The solution suggested from SMILE project are:

– Enhance the GNSS receiver by a multi constellation. The GNSS information, provided through AIS, will be based on a multi GNSS (GPS, GLONASS and Galileo) positioning with a particular attention to the European Satellite Navigation Systems.
– Provide a reliable navigation information through the application of new RAIM techniques for multi-constellation receiver improving also the accuracy of the positioning in case of satellite signals interferences.
– Provide an innovative vessel authentication, the so called strong authentication, through GNSS information. In particular, a coherence check on the GNSS receiver position and the in-view satellites (GNSS sky) was implemented. In order to provide a reliable identification, the list of satellites in-view is opportunely modified by the on-board system, ciphered and transmitted on-shore. The kind of modification is specific for

each vessel and can be changed on configuration basis.

1.2 *Motivations*

The key aims of this project are founded on an analysis of the maritime market, the main needs highlighted include:
- Improvement of the positioning information of the ship, especially in terms of Course Over Ground (COG) and Speed Over Ground (SOG), it is useful analyze the exact route of the ship and therefore generate alarms if the XTE (cross track error, otherwise known as abandonment of the intended course) is too high, which can be a symptom of an anomalies. Improving speed information can also be kept up to provide information on ETA and generate alerts of "under speed" and "over speed";
- Unique and certificate identification of the naval unit, which broadcasts the own position; it is required a robust system (not spoofed) able to authenticate the received data (name of the vessel, position);
- Position, speed and time certifiable and non-discarded of the naval unit.

This latter feature is extremely useful, in the services of VMS, in respect of any accidents (collisions with other units or with port facilities) for taking responsibility away from the owner. Such functionality is also required as part of the passenger transport services, especially those in regional convention, to certify the arrival and departure from each port and to certify the actual miles travelled naval unit. But it is also required by vessel owners operating in the cruise sector for accounting related to the time spent in the territorial waters, as well as the service providers accounting for tugs of their support services / areas maneuvering in port.

2 TECHNICAL ASPECTS OF SMILE

One of SMILE project goal is to provide a reliable and accurate navigation system through GNSS and AIS equipment. The information on ship position is of great interest for seafarers in different operational scenarios such as open sea, congested harbors and waterways. When a vessel is in open sea, the accurate knowledge on its position, velocity and course will ensure to reach its destination in the safest, most economical and timely way.

Vessel traffic and other waterway hazards make maneuverings more difficult, and the risk of accidents becomes greater. So, there is the need of an accurate and reliable navigation.

GNSS positioning is critical when lack of available signals occurs. This could happen:

- when satellite signals are blocked or strongly degraded by natural or artificial obstacles (generating the multipath effect)
- when an intentional interference occurs targeting the unavailability of the system (a jamming attack);
- in case of faking of a false position/time towards a target GNSS receiver (spoofing).

The multi-GNSS navigation, which consists in combining different GNSS, is the approach adopted in SMILE to improve the available measurements, guarantying more continuous and accurate solution (Gaglione et al, 2015).

The follows sub-paragraphs provide an overview about the adopted integrity monitoring technique and on the basis of GNSS Authentication.

2.1 *RAIM*

As described previously, SMILE project proposes to enhance the performance of on board GNSS receivers to give information on ship position, reducing the risk of getting unreliable information because of noise, interference or spoofing attacks that can produce a completely wrong position.

The integrity is the most critical performance requirement that must be satisfied by a navigation system (Kaplan, 2006; Parkinson and Spilker, 1996). In fact, even if GNSSs provide integrity information to users into the navigation message, this may be available with a time delay and it will be ineffective for real-time applications (Angrisano et al, 2013). So, GNSS stand-alone needs to be integrated with integrity monitoring algorithm. Such algorithms have been adopted to provide a measure of the trust that can be placed in the correctness of the solution; this concept in the GNSS field is represented by the reliability which is related to the ability to identify and exclude measures affected by large errors (blunders) that, if not rejected, could corrupt the final output of the system. The method adopted in SMILE is a Fault Detection and Exclusion (FDE) algorithms able to detect gross errors caused by different sources such as multipath, non-line-of-sight (NLOS) reception (if only reflected signals arrive at the receiver), receiver failures, unusual atmospheric conditions, interference or system failure.

In GNSS context, RAIM is used for the reliability testing of redundant measurements. RAIM algorithm is based on the investigation about the coherence between the measurements used for the navigation solution computation. RAIM main goal is to protect against excessive horizontal position error defined as the difference between the estimated position and the true position (Angrisano et al, 2013; Hewitson and Wang, 2006; Kuusniemi, 2005).

The core of the adopted procedure is the assessment of single-epoch solutions with only current redundant measurements being used in the self-consistency check to improve the reliability and accuracy of navigational solution for SMILE.

FDE RAIM algorithm is developed in MatLab® environment to detect and exclude the blunders present in the dataset that can degrade the accuracy of navigation solution. The inputs to the algorithm are the raw GNSS measurements and GNSS broadcast ephemeris while the output is an indicator about the integrity level of GNSS navigation signals.

In detail, RAIM algorithm provides to the users information on the reliability of the obtained navigational solutions that can be:
- Reliable if the faulty measurement is detected and excluded, and there is consistency between remaining observations that are used to compute the solution;
- Unreliable if the algorithm is not able to detect the blunders;
- Unverifiable when the redundancy (defined as the difference between the measurements and unknowns number) is not enough for the identification of the outliers.

A detailed description of the integrity monitoring performed by SMILE system is shown in Figure 2.

2.2 Identification Technique

One of SMILE project goal is to provide a strong authentication based on GNSS satellite position. This information is actually provided by the AIS, the GNSS navigation system delivers the ship position toward the AIS, which transmits to other ships or on-shore via standardized VHF transceiver.

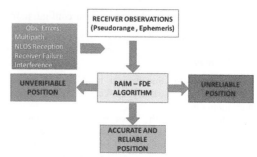

Figure 2. RAIM scheme

The AIS system is quite weak in terms of security and the position can be easily forge, actually no checking system is enable to detect false broadcasting position. Nevertheless, several scenarios need to authenticate the position information transmitted by AIS, both for safety purpose and for economical one (Robards, 2016).

In this project a strong identification system was designed. Such system is composed by two subsystems:
- On-board subsystem, equipped with an AIS (class A or B), opportunely strengthened with anti-tampering measures and a software component named GAA (GNSS Anti-spoofing Agent), which was developed during the project.
- On-shore subsystem, also called SMILE service center, which allows to verify and certificate the broadcasted position of the ship by means the GAM (GNSS Anti-spoofing Manager).

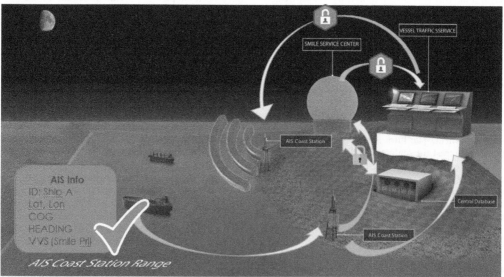

Figure 3. SMILE System Overview. The figure shows the workflow adopted to certificate the position of the ship

Basically, the list of receiver in-view satellites, masked in accordance to a specific strategy adopted, could permit to the SMILE service center to verify and authenticate the position information received through the AIS system by the vessel (Figure 3). The end-to-end ship-to-shore service chain guarantees that the ship ID and position information, received by the VMS, is true or wrong due to technical failure or alteration from malicious intents.

A validation process has been developed using the Matlab environment. The simulation uses a virtual environment model to reproduce behavior of the system, in controlled environment in order to easily detect critical issue system.

The final architecture is composed by three fundamental parts:

1 A SMILE box mounted on board, that include the GAA
2 The SMILE service center, which verify the identity of the vessel transmitter by means the GAM and integrate others functions ancillary to the fleet monitoring
3 The communication subsystem, which allows a linking between the ship and SMILE center.

ACKNOWLEDGEMENT

This research is supported by ASI (Agenzia Spaziale Italiana), Italy, and is included in SMILE (Satellite Multi-constellation Identification techniques for Liable Enhanced applications) project.

REFERENCE

Angrisano A., C. Gioia, S. Gaglione, and G. del Core. 2013. "GNSS Reliability Testing in Signal-Degraded Scenario," International Journal of Navigation and Observation, vol. 2013, Article ID 870365, 12 pages.

Gaglione S., A. Angrisano, P. Freda, A Innac, M. Vultaggio, and N. Crocetto. 2015. "Benefit of GNSS multiconstellation in position and velocity domain". Paper presented at the 2nd IEEE International Workshop on Metrology for Aerospace, MetroAeroSpace 2015 - Proceedings, 9-14. doi:10.1109/MetroAeroSpace.2015.7180618

Hewitson S., Wang J. 2006. "GNSS receiver autonomous integrity monitoring (RAIM) performance analysis," GPS Solutions, vol. 10, no. 3, pp. 155–170.

Kaplan, E.D.; Hegarty, J. 2006. "Fundamentals of Satellite Navigation" in "Understanding GPS: Principles and Application", Ed., Artech House Mobile Communications Series, 2nd edition.

Kuusniemi H. 2005. "User-level reliability and quality monitoring in satellite-based personal navigation" [Ph.D. thesis], Tampere University of Technology, Tampere, Finland.

Parkinson B. and Spilker J. J. 1996. "Global Positioning System: Theory And Applications", vol. 1-2, American Institute of Aeronautics and Astronautics, Washington, DC, USA.

Robards, M. D., et al. 2016. "Conservation science and policy applications of the marine vessel Automatic Identification System (AIS)—a review." Bulletin of Marine Science 92.1 (2016): 75-103.

Tetreault, B. J.2005. "Use of the Automatic Identification System (AIS) for maritime domain awareness (MDA)". In OCEANS, 2005. Proceedings of MTS/IEEE (pp. 1590-1594). IEEE.

EGNOS Poland Market Analysis in SHERPA Project

A. Fellner

Silesian University of Technology, Katowice, Poland

ABSTRACT: The "EGNOS Poland Market Analysis " was prepared as part of the SHERPA Project and is based on research conducted by airports and aircraft's operators. The part of results from these research were described in European document "EGNOS National Market Analysis". They were the first in Poland conducted examinations through Polish Air Navigation Services Agency, in frames of which was made implementation of LPV GNSS procedures at airports in order to assure the High Quality and the Operating Safety. Furthermore, these research was consistent with the expectations of aircraft operators and user of the EGNOS and was conducted in relation to signed by Poland ICSO Resolution A-36/37 concerning the Implementation of PBN (Performance Based Navigation). The assessment of some tangible criteria will allow the best selection after the complete process. The paper presenting these issues and has been divided into following parts: the introduction presents purpose, the scope of research, methodology and applied during the identification, selection and the evaluation, conclusion. The paper was drawn up on the basis of examinations and results achieved as part of the SHERPA project under the Agreement Grant No. 287246 with the GSA (European GNSS Agency).

1 INTRODUCTION

The proposed common methodology for the airports and aircraft operator selection (namely "scenarios") is a rational and simple 3-phases process (fig. 1):

1 Identification and description of potential scenario candidates;
2 SELECTION CRITERIA targeting a clear definition of the criteria used to assess each of the potential scenarios identified, based on the criteria proposed in this document or including any additional aspect under PANSA consideration for the final scenario selection;
3 ASSESSMENT based on the list parameters, criteria selected. Each potential scenario shall be assessed justifying the selection of the best or most beneficial "final scenario" where LPV procedures publication will be committed, as included in the corresponding.

During the assessment of the different "scenarios" following the proposed methodology, the PANSA will be able to evaluate the implications and benefits derived from the implementation of LPV procedures in each of them such as minima reduction, increased accessibility, continuous horizontal or/and vertical guidance, noise and emissions reduction, etc. This evaluation process will be the basis for the justification of the final selected scenario. These activities must be performed in close co-operation with all the stakeholders such as Aircraft Operators, PANSA, Regulators and Airports.

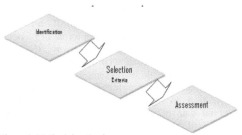

Figure 1. Methodology's phases.

2 PHASE OF IDENTIFICATION AND DESCRIPTION OF POTENTIAL SCENARIO CENDIDATES

Identification as one of the main objectives of SHERPA project is to support Eastern European countries through the setup of a regional working group and to understand the actions to be undertaken by their relevant stakeholders (ANSPs, Regulators and Aircraft Operators) in support to EGNOS adoption, the first step in the process is the identification of some State airports and aircraft operators where LPV approaches implementation and its later operation will represent tangible benefits. They assumed that PANSA should generate the list of potential candidates of the script (airports and operators of the plane) to be subsequently assessed. In principle, pairs of aircraft operators and airports where the first ones operate in would be the preference. The identification of candidates should be done taking into account organizational, technical and institutional requirements similar, but not limited to, the following ones:

- National strategic objectives,
- Aircraft operator requests,
- PBN implementation plans and airspace concept, ATM operational requirements,
- Environmental policy directives.

As part of the SHERPA project was appointed Polish Team – National Implementation Team.

2.1 *Airports*

PANSA identified and analyzed most appropriate airports, landing field in Poland, which were included in the list of candidates. This preliminary identification was based on mentioned above parameters, specific domestic criteria and is accepted in the Polish Team (National Implementation Team). For the purposes of the PANSA identification he is provided with all essential information of the pre-selected airports. The information for every airport was presented in the assumed form - below specified tables for six airports. The candidate airports (Katowice EPKT, Kraków EPKK, Rzeszów EPRZ, Warszawa-Chopina EPWA, Wrocław EPWR, Warszawa- Modlin EPMO, Mielec EPML) shall fulfill minimum technical requirements, at least in terms of physical characteristics (e.g. table 1. runway shall be classified as instrumental according to ICAO Annex 14, ATS/AFIS should be present, etc.). This form of the examination is aspiring for summarizing the information, to relieve the process and to collect essential data of airports of the candidate in the coordinated way for the entire SHERPA ANSP partners.

Table 1. Identification chart for airports – example of Katowice Airport (ICAO: EPKT, IATA: KTW).

		Name	KATOWICE - PYRZOWICE		
		ICAO code	EPKT		
		IATA code	KTW		
Airport's overview					
International airport, located in Pyrzowice, 30 km north of center of Katowice. The airport has third biggest passenger flow in Poland.					
Location	PYRZOWICE				
Traffic	SCHEDULE, NONSCHEDULE, GENERAL AVIATION				
RWYs layout	RWY 09-27				
Taxiway	9 TAXIWAY, CONCRETE SURFACE				
RWY		09	27	...	...
Main parameters	Magnetic [deg]	90.81°	270.36°	...	...
	Dimensions [m]	2800 x 60	2800 x 60	...	...
	IFR/VFR	VFR	IFR/VFR	...	...
	Lightning	THR, REDL, RENL	THR, REDL, RENL	...	...
Navigation	PA	NIL	ILS Cat I	...	...
	NPA	DVOR/DME	LOC, NDB, DVOR/DME	...	...
Equipment	App lightning system	NALS PAPI 3°	PALS Cat I PAPI 3°	...	...
	Approach aids	DVOR/DME	ILS, NDB, DVOR/DME	...	...
	RVR measurement available	Yes	Yes	...	...
Traffic 2011	Movements	-5	30 664	...	...
	IFR/VFR [%]		93.7	...	...
	Commercial flights		27 628	...	...
	People/Load %		90,569,44	...	...
	A/C types used for schedule flights		ATR CRJ AJ20 B737	...	...
ATS	Approach control	APP			
	Aerodrome Control	TWR			
	TWR Opening hours	H24			
Obstacle Clearance Height (OCH)		Cat. Of ACFT (AD 2 EPKT 6-1-1)	A	B	C	D
		Straight-in	Cat I	65	68	71	74
			Loc- DME	105	105	105	105
			Loc- DME when stepdown fix not received	130	130	130	130
		Circling OCH ALL		145	190	260	260
Noise footprint	NIL				
Notes relevant data	AIP POLAND AD 2 EPKT 1-6 (AD 2.11)				
Visual Segment Surface (VSS)	AERODROME OBSTACLE CHART (AIP POLAND AD 2 EPKT 2-1-1)				
Expected APV benefits	Approach with Vertical Guidance - it's an instrument approach based on a navigation system, that is not required to meet the precision approach standards of ICAO annex 10, but				

2.2 *Aircraft operators*

Similarly to section 2.1 to make a good identification of aircraft operators, it is essential to collect enough information of the ones with potential interest in EGNOS based operations. Appointing the Polish team we know, that is very important it aware regarding current and projected RNAV capabilities onboard the aircraft operating at the airport of interest. Including, though not restricted, to the following: aircraft equipment and navigation capabilities, airworthiness and operational approval, current experience with RNP APCH procedures, operator requirements and preferences for RNP APCH procedures, plans in terms of future equipage and operational approval e.g. Aircraft operator Royal Star Aero (table 2).

Table 2. Identification chart for aircraft operator – example of Royal Star Aero.

Royal Star aero	Name	P.P.H.U. ROYAL-STAR KRZYSZTOF PAWELEK
	Location	DROGOWCOW 7, 39-200 DEBICA POLAND
	Area of business	TRAINING, AEROTAXI, REPAIR
	Contact point	UL. LOTNISKOWA 16, 39-300 MIELEC POLAND
Operator's overview		
Business profile: • Manufacture of aircraft components and engine parts. • Design and construction of flight simulators. • Major repair and overhaul of engines for Lycoming, Continental. • Major repair and overhaul of propellers for McCauley, Hartzell, Sensenich. • Production of refrigeration equipment and components for aviation industry. • Training of pilots: Private Pilot License PPL (A), Commercial Pilot License CPL (A), Air Transport Pilot License ATPL (A). • PART -147: Training for mechanic's license: A A1 - Airplane - turbine engines, A A2 - Airplane - piston engines, A A4 – Helicopter - piston engines, B1 B1.2 - Airplane - piston engines		
Fleet (number and type)	4 x Cessna 152	PZL M20	PA34-200T	PA28R-201T
Company size	LARGE			
Movements	+30			
National airports operated	EPML, EPWA, EPKT, EPKK, EPRZ			
Questionnaire				
Q1	How many aircraft of your fleet are equipped for RNAV NPA, APV Baro or APV SBAS approach? 2			
Q2	Do you have certified GPS receivers onboard the aircraft? List the types (e.g. TSO-C129a, TSO-C145a, TSO-C146a) M20 - TSO-C129a , PA-34 - TSO-C146a			
Q3	Does the aircraft have an airworthiness approval for the use of GPS and/or EGNOS in the approach phase of flight? Which type of operation? Yes. Enroute, terminal and NPA and APV approaches			
Q4	Do you intend to upgrade the navigation equipment of aircraft which are not equipped for any type of RNAV approach mentioned in the first question? Yes			
Q5	Do you have any plans to sign new a/c purchases? If yes, are these new a/c suitably equipped for RNAV approaches? Which ones (NPA, APV Baro, APV SBAS)? APV SBAS			
Q6	Which type of approach are you most interested in with respect to your present or future navigation equipment – APV Baro or APV SBAS ? APV SBAS			
Q7	Do you have already any experiences with RNAV approaches? At which airports and what are your experiences? Yes. During APV approaches evaluation at EPML, EPKT airports			
Q8	Which national airport (or runway) would you consider to have the highest priority for RNAV approach implementation? All certified			
Q9	Do you consider having sufficient information and documentation about RNAV approaches operations and aircraft certification? Yes			

3 SELECTION PHASE

This is the core section of this document due to the direct application of the contents included here. It describes the main criteria to be taken into account by the participating PANSA, when evaluating the identified scenarios (airports +aircraft operators) and the selection of the "best" one.

After the identification and presentation, through the specific forms, of the potential scenario candidates ("airports" where LPV approaches bring tangible benefits and "aircraft operators" with potential interests on EGNOS based operations), next step is the definition of the scenario selection criteria.

The proposed criteria to be used by the participating PANSA come from the well-known key benefits that the EGNOS adoption brings for aviation in operational, safety, economical and environmental aspects within the Performance Based Navigation (PBN) concept.

Airport capabilities shall be studied to determine whether APV SBAS operations can be implemented on specific aerodromes. These criteria are detailed in the following subsection. Five areas are considered to group the proposed criteria based on EGNOS benefits for the assessment of the candidate airports: operational, safety, economical, environmental, capabilities

Some operational criteria that would represent benefits in the adoption of LPV approaches are:
- LPV is particularly attractive to runways not equipped with ILS, although also it could;
- be used as back-up of ILS. EGNOS provides lower operational minima on non ILS;
- runways and one achievable minima estimation is suggested;
- EGNOS allows to perform instrument approaches with vertical guidance (APV) based on SBAS down to LPV minima to airports which currently only provide NPA or visual approaches;
- a minimum of physical aerodrome infrastructure (runway, taxiway, approach lighting etc.) and CNS Systems are required.;
- it provides increased accessibility at airports with weather/terrain constraints. Improving lateral guidance and proposing vertical guidance, creating a direct approach that does not currently exist with standard navigation resources;
- meteorological data such as wind statistics, cloud ceiling and RVR per runway end are required;
- the airports with existing high OCH (over 500ft) are specially preferred;
- the existence of ATC/ATS services and the airport traffic and number and distribution of flight operations, must also be studied;
- the Visual Segment Surface (VSS) has to be assessed since its penetration may represent an

obstacle for the publication of the RNP APCH procedures;
- a possible reduction in the decision height and lowering the slope on the final approach;
- the possibility to implement advanced procedures (e.g. curved approaches) and the integration of the new procedure into the terminal area impact.

Of relevant interest are the criteria regarding safety, for example:

LPV is able to reduce Controlled Flights Into Terrain (CFIT) accidents, because it provides vertical guidance and situational awareness to pilots.

It also provides better precision in low altitude routes such as for helicopters.

From an economical point of view, the criteria could be:
- LPV improves the attractiveness of airports not equipped with ILS to new airlines (e.g. major airlines, regional aviation, business aviation, general aviation, cargo aviation, aerial works, helicopters, etc).
- APV/SBAS allows enhancing accessibility (% of avoidable disruptions) flights that can land at the intended destination.
- EGNOS could reduce and rationalize ground navigation infrastructure with cost reduction in maintenance of ground infrastructure and conventional navigation aids (e.g. NDB, VOR and ILS).

Finally, there are environmental parameters than can be included for the selection process:
- noise reduction in populated areas.
- LPV provides more efficient approaches and time and fuel saving.

The proposed criteria based on EGNOS benefits for the assessment of the candidate aircraft operators are listed hereafter:

It will be very positive assessed if the aircraft operator operates at the airport under study.
- fleet composition of aircraft operator will inform about availability of a target type of aircraft to be selected for SHERPA project.
- it will be taken into account if there is any LPV equipped aircraft.
- the aircraft operator investment plan is important due to several costs the process involves: equipment upgrade, certification, procedure design, training, manuals update, etc.
- the traffic at a specific airport giving detailed information of movements and composition (people/load).
- if the aircraft operator is flying advanced procedures.
- time saving estimation after adoption of APV/SBAS approach procedures.
- fuel saving estimation after adoption of APV/SBAS approach procedures.

- the operation of LPV approaches is done through low cost and high performance avionics available for all users.

4 ASSESMENT PHASE

The analysis consists of using the forms presented in section and applying the corresponding marks and weights for all the criteria. Crossing marks with the assigned weights will result in a specific figure to assess the feasibility of the LPV implementation for each individual airport and aircraft operator. As explained before, this analysis shall be completed with a rationale justification of criteria, marks and weights selected in each case. This process shall be repeated for all the airport and aircraft operator candidates, paying special attention in crossing data, due to the aircraft operator criteria form shall analyses the corresponding aircraft operator together with a specific airport (table 3 and table 4).

Table 3. Airport criteria for Katowice Airport EPKT

Table 4. Aircraft Operator criteria for AIRCOM.

CRITERIA	WEIGHT	AIRPORT NAME						
		EPRZ	EPKK	EPKT	EPWA	EPMO	EPML	EPWR
Operation at proposed airport?	100	5	13	10	10	4	6	5
Fleet composition?	60	4	6	8	8	3	5	4
LPV equipped A/C?	90	4	6	8	10	3	8	4
Upgrade investment foreseen?	80	10	2	10	10	1	5	5
Cost SBAS avionics	70	4	4	10	8	3	6	2
Movement?	30	7	7	9	8	3	5	2
People/Load (%)	20	3	3	8	8	3	3	1
Advanced procedures?	70	3	8	8	8	1	5	4
Time saving estimated?	10	5	5	8	9	3	5	4
Fuel saving estimated?	90	9	9	9	9	3	4	4
Strategic plan?	40	4	6	9	9	3	5	1
Investment local actors	55	5	4	10	10	4	7	1
Other user's interest	30	1	1	9	9	6	6	5
TOTAL		43,9	49,2	71,8	71,4	27,12	52,69	38,00

5 FINAL SCENARIO

Final scenario shall be selected as the best choice of "airport + aircraft operator" after crossing all information forms (table 5.).

146

Table 5. Final scenario summary.

AIRPORT	RUNWAY				AIRCRAFT OPERATOR	
					ROYAL STAR AERO	AIRCOM
EPRZ	9	27				
	77,9	62,9			44,0	41,9
EPKK	7	25				
	87,8	72,8			38,5	49,2
EPKT	9	27				
	80,3	65,3			54,7	71,8
EPWA	11	29	15	33		
	69,5	84,5	84,5	69,5	46,5	71,4
EPMO		25				
	60,77	60,77			27,12	27,12
EPML	9	27				
	46,12	51,29			51,00	51,00
EPWR	11	29				
	75,34	62,41			38,50	38,50

This table presents the outcome of the study and states that RWY 29 of EPWA Airport and RWY 09 of EPKT Airport, are the best option for implementing a new LPV approach. Including above analyses, they distinguished one pair: EPKT airport – AIRCOM.

6 CONCLUSION

Conducted examinations showed, that:
- EGNOS must assure required by ICAO: accuracy, integrity availability, continuity in Poland;
- GNSS is lacking domestic regulations concerning the application;
- All responsible institutions must in the employed and coordinated way join in into the process of implementing GNSS;
- The part of the fleet only has an essential avionics.

REFERENCES

Draft Guidance Material for the Implementation of RNP APCH Operations PBN TF6 WP06 Rev 1 05/01/2012
SHERPA Grant Agreement Grant number 287246
EASA - AMC 20-26 : Airworthiness Approval and Operational Criteria for RNP AR Operations;
EASA - AMC 20-27: Airworthiness Approval and Operational Criteria for RNP APPROACH (RNP APCH) Operations Including APV BARO VNAV Operations;
EASA - Helicopters Deploy GNSS in Europe (HEDGE) project documentation,
EATMP Navigation Strategy for ECAC;
EGNOS Introduction in European Eastern Region MIELEC project documentation,
EUR Document 001/RNAV/5 Guidance Material Relating to the Implemen-tation of European Air Traffic Management Programme;
FAA - AC 20-105: Approval Guidance for RNP Operations and Barometric Vertical Navigation in the U.S. National Airspace System;
FAA - AC 20-129: Airworthiness Approval for Vertical Navigation (VNAV) Systems for Use in the U.S. National Airspace System (NAS) and Alaska;
FAA - TSO C146A: Stand-Alone Airborne Navigation Equipment Using the Global Positioning System Augmented by the Wide Area Augmentation System (WAAS);
FAA: TSO C145A: Airborne Navigation Sensors Using the Global Positioning System (GPS) Augmented by the Wide Area Augmentation System (WAAS);
Fellner A. SHERPA-PANSA-NMA-D11EP Issue: 01-00 EGNOS Poland Marked Analysis, 2012
ICAO Annex 10,
ICAO Doc 8168 – PANS-OPS,
ICAO Doc 9613 – PBN Manual,
ICAO Doc 9905 – RNP AR Procedure Design Manual
ICAO Doc. 7754 European Region Air Navigation Plan;
ICAO European Region Transition Plan to CNS/ATM;
ICAO Global Air Navigation Plan for CNS/ATM Systems. Doc 9750;

Automatic Identification System (AIS)

Automatic Identification System (AIS) as a Tool to Study Maritime Traffic: the Case of the Baltic Sea

A. Serry
University Le Havre Normandie, Le Havre, France

ABSTRACT: The Automatic Identification System (AIS) is an automatic tracking system used on as a tool to increase navigation safety and efficiency as well as vessel traffic management. It enhances maritime safety and security. AIS' contributions are undeniable in spite of some deficiencies and technical restrictions. This article presents the impacts and uses of AIS technology that can provide useful information to study maritime traffic, especially for the scientific community and port authorities. This desktop study is carried out in the framework of the implementation of a platform to reconstruct shipping routes using AIS data. It also focuses on the Baltic Sea as a case study.

1 INTRODUCTION

Maritime transportation is protected by several safety devices such as the development of maritime surveillance systems. Currently, vessels take on board more and more aid to navigation systems. AIS is a system of data exchange between ships that was made obligatory by the International Maritime Organisation (IMO) in 2004. AIS presents advantages for maritime transportation actors: improvements in safety, progresses in the management of fleets and navigation. Its distribution also presents numerous advantages in seaway management. Nevertheless, the generalisation of AIS poses problems of confidentiality for ship-owners, indeed for safety. In effect, the data transmitted by AIS are available to all, including the scientific community.

This paper is a synthesis of a reflection conducted during the development of a research platform for the analysis of maritime traffic and maritime transportation, the CIRMAR project. This tool makes it possible to envisage multiple operational applications that concern navigational safety as well as maritime economy, analysis of the strategies used by maritime actors or the environmental impact of maritime traffic.

The article is based essentially on a documentary analysis of existing studies but also on in-depth bibliographical research. The paper proceeds as follows. Section 2 focuses on a definition of AIS including its characteristics and objectives. The third part intends to highlight the contribution made by this system and its use. The fourth section focuses on the potential use of the data produced by AIS when the last one focuses on the case of the Baltic Sea. The case of the Baltic Sea is very interesting because it is one of the world's busiest seas and the transportation volumes in the Baltic Sea have increased significantly in recent years.

By means initially of an empirical approach, the focus of this article, therefore, is to identify the relevance of AIS for the maritime and scientific communities.

2 THE AUTOMATIC IDENTIFICATION SYSTEM

2.1 *A tool for navigation security*

So as to increase maritime safety, the IMO has adopted mandatory regulations concerning the installation of automatic identification systems capable of providing information from one vessel to another as well as to on-shore authorities. These regulations form part of Chapter V of the SOLAS convention. They concern principally vessels of a gross tonnage equal to or exceeding 300 tonnes making international voyages.

2.2 Main features of AIS

The AIS system uses a transponder which transmits and receives automatically in VHF. It also includes a GPS receiver which records the position and details of movement.

Broadcast and reception is carried out continuously and autonomously. AIS data is the realistic data of vessel traffic including dynamic information (position, speed, course, etc.) and static information (ship type, dimensions, name, etc.) (Yao, 2016). Generally, ships receive information in a radius of 15-20 nautical miles.

Introduction of AIS satellites have been carried out since 2009 therefore considerably reducing the number of white areas (Chen, 2013). So, the latest change to the AIS technical standard includes a message specifically designed for AIS reception from satellite (AIS SAT). Any vessel equipped with AIS today is easily trackable, and this at any moment wherever it may be. The generalization of AIS does not entail removing the use of pre-existing systems and they are complementary. The objective is to combine the pre-existing data coming from these sub-systems with an integrated system commonly called *Vessel Traffic Monitoring Information System*.

2.3 AIS main objective: maritime security

The management of maritime traffic has become a major current issue. In fact, one of the challenges of the maritime community now is to conciliate the surge in marine shipping while at the same time guaranteeing the protection of marine resources in a context of climatic change. In highly frequented waters, active surveillance of maritime traffic has taken on an even greater meaning.

Maritime security and safety cover a wide and increasing area: from the management of commercial traffic to the fight against piracy, including sea rescue, counter-terrorism and the protection of port infrastructures. Recent AIS development aims to reinforce maritime safety with reference to the protection of life at sea, the preservation of transported goods, and the protection of the vessel and prevention of collision.

3 CONTRIBUTIONS OF AIS AND USE OF DATA

AIS is part of e-Navigation which is expected to provide digital information and infrastructure for the benefit of maritime safety, security and protection of the marine environment (Weintrit, 2016).

3.1 An efficient operational tool

"AIS was initially intended to assist ships in avoiding collisions, and the port and maritime authorities in monitoring traffic and ensuring better surveillance of the sea" (Thery, 2012). This system enables vessels to be traced but also to anticipate their movements. Availability of precise data on the position of ships in real time renders it possible to manage traffic efficiently, to react more swiftly in the event of an accident or incident.

The use of AIS as an aid to navigation is a precious source of information not only with regard to ships but also with regard to all the navigational aid beacons (Świerczyński, Czaplewski, 2013). AIS is placed notably as a relevant tool for the protection of the marine environment (Serry, 2013). AIS systems have a potential ability to reduce the frequency of polluting accidents linked to navigation by simply supplying an update of information concerning ships. Similarly, they can shorten the response time in the face of accidents by supplying information about the situation in near-real-time. AIS is therefore an important asset for the protection of the marine environment. (Schwehr, Mc Gillivary, 2007).

Illegal discharges are the second cause of pollution of the marine environment. This type of pollution could be substantially reduced and AIS technology can make a contribution. For instance, in the Baltic Sea, the Helsinki Commission has been using AIS data since 2005 (HELCOM AIS) to assess the risks of oil spills associated with specific vessels (Figure 1). Today, Automatic Identification System also be able used to estimate ships emissions.

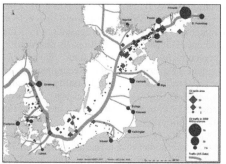

Figure 1. Survey of maritime traffic and oil spills using AIS.

AIS is expected to become an important element in the fight against marine pollution caused by maritime traffic. For example, Lloyds of London has already used AIS data from the AISLive service in legal proceedings involving accidents of ships.

3.2 Obstacles and limits of AIS

"If the advantages of the new technologies are undeniable, as long as these advances form part of economic and social life, they run up against the risk of violations of privacy and individual liberties" (Deboosere, Dessouroux, 2012).

3.2.1 AIS' technical limits

Merchant ships under 300 tonnes are exempt from the system which limits AIS's capacity with regard to maritime surveillance. Sot it does not make it possible to detect every ship. As a matter of fact, AIS reliability is far from perfect. The captain can cut off the system. VHF links can decline in certain conditions. Information relative to the voyage on the nature of the cargo and the ports of departure and destination is entered manually on board. It could be erroneous, voluntarily or not.

So, AIS and ARPA are complementary and should be used in conjunction with one another, even if AIS provides more complete information than shipboard radars (Serry, 2016). AIS shows data often more stable and precisely (Wawruch, 2015).

Meanwhile, the system is potentially vulnerable to more sophisticated attacks:
- The system is vulnerable to intentional or unintentional interference;
- Intentional broadcast of erroneous information;
- Transmission of computer viruses.

The situation of a maritime area can therefore not be controlled exhaustively with AIS.

3.2.2 The case of piracy

Maritime piracy is not a new phenomenon but, it is increasing, especially in the Gulf of Aden. AIS system is progressively frequently used by modern-day pirates, in order to locate their potential targets. The pirate mother ships are consequently equipped with the latest technology which enables them to target and organise an attack with great accuracy by taking a targeted ship by surprise and despatching speed boats which are sometimes undetectable.

Another simple solution exists, that of an autonomous terminal, marketed freely and designed for amateur yachtsmen. All that is needed is for pirates to be in the right position to cover the usual itineraries of maritime traffic and choose their prey according to the name of the ship, its cargo or its destination.

Furthermore, the AIS system can also be used to broadcast false information that can be fabricated with relative ease. The aim of these misleading messages (distress signal, wrong locality of the ship, etc.) is essentially to attract attention and lead the ships targeted into a trap.

Yet, even when navigating without lights, ships remain detectable through their VHF transmissions linked to AIS. In this case, the solution is to deactivate the AIS system of vessels entering at-risk areas such as the Gulf of Aden which are often also areas of heavy traffic in which recourse to AIS is primordial in order to reduce risks of collision.

3.2.3 Socio-economic activities and significances

AIS is an efficient tracking tool. Together with information agencies, ship-owners were the first to take possession of it because it enables them to track their fleet from land and ensure that the logistics are optimum. For instance, recordings of arrival and departure are also provided by the *Lloyd's Register Fairplay* for commercial purposes in the framework of their Sea-web database (Kaluza & Al, 2010).

Furthermore, open AIS information generates fears of commercial espionage. Maritime companies and shippers in fact wish to remain as discreet as possible regarding commercial data and information. Imparting information other than the automatic transmission of certain data such as the ship's destination or its ETA makes them fear the risk of commercial espionage.

4 AIS IN SCIENTIFIC RESEARCH

The data acquired from AIS systems constitute a new means of information not only for the maritime community and the wider public but for research scientists as well. They comprise a potential source of information on maritime traffic, essentially commercial traffic. Consequently, broadcasting it in real-time makes a tangible contribution especially to the scientific community.

4.1 A recent source with a yet restrictive use

The state of the art, primarily founded on Francophone literature, brings to the fore works based principally on the subject of security (Fournier, 2012) or on the uses of specific, mainly coastal spaces (Bay of Brest, marine coastal areas and insular areas). Moreover, the analysis of works, reports, academic and research studies confirm the relative scarcity of works in Human Sciences and highlights a fragmented literature which approaches subjects like international maritime law, physics, signal processing, geopolitics and many others.

As for the far more extensive literature in English, it is a relatively different situation. More research has been done on AIS on different spatial scales both global (Shelmerdine, 2015), regional (Cairns, 2005) and local (Perkovic & Al., 2012). As appears in studies done by Richard L. Shelmerdine (Shelmerdine, 2015), the research focuses on surveillance of itineraries taken by maritime transport and the intensity of shipping traffic (Eriksen, Høye, Narheim, Meland, 2006), the prevention of maritime accidents and the detection of unusual situations and on the environmental impacts of maritime traffic.

Thus, the Eastern Research Group (ERG) used a Geographic Information System (GIS) to map and analyze both individual vessel movements and general traffic patterns on inland waterways and within 9 miles of the Texas coastline. ERG then linked the vessel tracking data to individual vessel characteristics from Lloyd's Register of Ships, American Bureau of Shipping, and Bureau Veritas to match vessels to fuel and engine data, which were then applied to the latest emission factors to quantify criteria and hazardous air pollutant emissions from these vessels. The use of AIS data provides the opportunity for highly refined vessel movement and improved emissions estimation, however, such a novel and detailed data set also provides singular challenges in data management, analysis, and gap filling, which are examined in depth in this paper along with potential methods for addressing limitations (Perez, 2009).

Some researchers also use AIS to study ships comportments due to meteorological circumstances. For instance, the Baltic Sea is a seasonally ice-covered sea located in a densely populated area in northern Europe. Severe sea ice conditions have the potential to hinder the intense ship traffic considerably. Thus, sea ice fore- and nowcasts are regularly provided by the national weather services. In their study, Löptien and Axell provide an approach by comparing the ship speeds, obtained by the AIS, with the respective forecasted ice conditions. They find that, despite an unavoidable random component, this information is useful to constrain and rate fore and nowcasts. More precisely, 62–67% of ship speed variations can be explained by the forecasted ice properties when fitting a mixed-effect model (Löptien, Axell, 2014).

The website *Marine Traffic* (http://www.marinetraffic.com) is a good example of the dissemination of information. It provides information, partially free of charge and in real-time, on the movement of ships in an almost global area of coverage (Thery, *2012*). It is part of an academic project whose objective is to gather and disseminate these data with a view to exploiting them in various domains. This is an open project and the organisers are constantly looking for partners prepared to share data from their region so that they can cover more maritime areas and ports worldwide. Everyone can explore at leisure each of the areas for which the information is available.

The AIS system being an open one, it has given rise to other sites like *AISHub* (http://www.aishub.net/). The sites broadcasting AIS information, therefore, have a great advantage, that of making it possible to visualise maritime traffic in real-time free of charge. In terms of research, the interest of these sites of visualisation of AIS data is certainly more important than making archived databases available but it makes it possible namely to compare the reality of marine traffic with the rhetoric coming from shipping companies by checking, for example, the vessels operating on regular lines. By linking this information with a shipping database, it is additionally possible to determine the capacities offered by these same maritime lines.

4.2 *An evident potential*

The usage of AIS is based on a multi-scaled character; spatial scales (local/global) and temporal (short term/long term) of the information produced by AIS signals linked to other bases. This makes operating functions possible in a variety of areas. A potential application of archived AIS data, therefore, consists in extracting statistics of the voyage time for a population of ships (Mitchell et al, 2014).

AIS data makes possible to improve a network of stations covering more and more coastal areas, providing new possibilities to the mapping of transport activity. *"Several studies carried out are based on the exploitation of AIS data with the aim of detecting unusual situations (risks of collision) and of qualifying the behaviour of vessels in real time. Thanks to the availability of AIS data, it is possible to identify, quantify and map navigation lanes of vessels"* (Le Guyader, Brosset, Gourmelon, 2011). The method, founded on a spatial analysis within a geographical information system (GIS) combined with a database server, makes it possible to reconstruct each vessel's trajectory in such a way as to identify the navigation lanes then to match the daily traffic in its temporal and quantitative dimensions. It is therefore possible to complete the maritime transport map which has been traditionally directed towards analysing maritime networks and flows globally, towards recording the departure and arrival ports or analysing the spatial influence of maritime transport. *"In a global approach to running maritime activities, this information can be analysed with other information describing the operation of nautical and fishing activities, in order to characterise their interaction and bring to light potential conflicts. In the medium term, applying maritime traffic tracking systems to the entirety of activities involving different types of vessels, as intended in the framework of E-Navigation, will no doubt represent a precious data source as an aid for navigation in real- time, fishing management and contribute to the integrated management of the sea and coast"* (Le Guyader, Brosset, Gourmelon, 2011).

The availability of a reliable and consistent data source has proved difficult in order to make it possible to build a picture of flows of exchange in the short, medium and long terms, at both regional and global scales. Maritime companies' schedules are very heterogeneous and subject to the above-mentioned vagaries, and movements recorded by the port authorities' harbour master's offices are very difficult to gather without considerable means. The availability of archived AIS data opens interesting perspectives for the characterisation of maritime activities on spatial, temporal and quantitative levels. There is great potential of AIS to contribute to scientific research: analysis of the maritime itineraries taken by vessels, estimation of vessel discharges, identification of port calls and duration, analysis of maritime companies' strategies, mapping vessel flows, analysis of interactions with the vessel's environmental elements such as climatic conditions, state of the sea or density of traffic (Lévêque, 2016).

The automatic character of transmitting vessel positioning signals and its generalisation provide an opportunity to track and analyse the vessels' itineraries. Once this source of information has been properly checked through matching it with external data with regard to vessels and ports, it opens the way to reasoning on a global scale as well as on the scale of port approaches, in real-time as well as long term. With regard to scientific research, it represents, for example, the possibility to test traffic models, predictive or dynamic, long or short term, which could also be of interest to port authorities. As for the professional world of shippers and logistics providers, it represents the possibility of better apprehending the vagaries of maritime transport which, by comparison with terrestrial logistics, is often seen as a "black box". It is also the opportunity to study the positioning of ports in a global network of port calls.

So, using AIS data is a solution to perform multiscale, diachronic and synchronic analyses multi scales (Lévêque, 2016).

4.3 CIRMAR: Acquisition and treatments of AIS signals

Geo-economic entry in the use of AIS signals request a number of methodological and technical issues. Some related to the duration of observation which must be long enough to account for seasonal trends, others related to the geographic scope, which must be comprehensive if you want to account for the economic gap between different regions of the globe. All this requires the accumulation of large masses of data. Arises also the question of the intrinsic quality of the data transmitted by the AIS signals.

The CIRMAR device is based on affiliation to a collaborative network. It allows to develop and test processing of analysis tools and representation of the data.

Therefore, a proper platform is needed to receive, decode, clean, store and analyze AIS messages. Our project is made of some different phases. During the first phase, AIS messages are received. The received messages of each day are stored. As the received messages are coded, a Java application using an open source library decodes the messages and make them ready to be stored. We ignore messages which includes errors like checksum error. For this project, only AIS message types 1, 2, 3, 4, 5, 18 and 19 are interesting, so we ignore the other ones. The next step is to store AIS messages in the database.

Speaking about data treatment, the two major obstacles to produce information for geo-economic issues are the amount of data to process the quality of data and particularly static data. Even if they are of small size (39 characters for example) the number of AIS messages received, led to the formation of large amounts of data. A relevant analysis of marine traffic must also consider to the best of the seasonality of flows (Peak season, Chinese new year) and at least the duration of the transit times for the longest roads. The volume of data to decode and then analyze is therefore considerable.

Regarding data quality, most of the problems relate to information that are populated on each trip by an operator on the ship. They concern the reference to the port of destination, the navigation status, the draught of the ship and ETA (estimated time of arrival). Positioning and road information transmitted automatically, entering the static information in the transponder is a subsidiary operation compared to other more crucial operations for navigation, especially during the approach or departure from the port.

So, use of aggregated material from AIS messages to analyze maritime trade flows must be based preferably on dynamic data (positioning and road), which are more reliable than the static ones. That information must be validated by external data: therefore we use a database of ships via the MMSI number and a referenced geo database on ports (Lévêque, 2016).

So, during the treatment, the first step is to decode all the messages. The amount of data to process may require to allocate treatment on several sessions and machines to avoid excessive wait times. The results serve as a base for different types of analyses: port performance analysis (duration of call and ship size...), shipping companies' strategies, maritime network study, regional markets analysis.

5 BALTIC CASE STUDY

The Baltic Sea is very transport-intense. The maritime traffic is relatively diffused on the whole of the Baltic Sea, despite a distinction of maritime and port activities within the Baltic Sea, mostly between the south and the north shores. Baltic Sea traffic growth is particularly important in the field of containerization which can be investigated applying traditional geographical tools but also using AIS data.

5.1 *Baltic containerization network*

Containerized cargo is transported via shipping lines, operating on a regular basis. It is characteristic of the integration of Baltic ports in the regular maritime lines and in world trade.

The amount of containers shipped in the Baltic Sea is determined by the proximity of consumer markets with Russia being the key destination point of containerized cargo. Only a few ports handle large quantities of containers. The largest regional container port, Saint-Petersburg, stands only 10th in Europe, the second largest (Gothenburg) 20th. Due to geography, Baltic ports cannot reach the volumes of large Western Europe ports, which act as transhipment hubs.

In 2014, the number of containers handled among Baltic Sea ports amounted to 9.4 million TEUs, which equals an annual growth of 7%. The composition of the 20 largest container ports remained stable: St. Petersburg is clearly the undisputable leader in this segment, while Gdansk recorded considerable and continued growth in container numbers handled.

The analysis of the containerized traffic in the ports of the Baltic Sea distinguishes four types of ports in the region:
- Traditional regional ports as Gothenburg whose location and the early opening of containerized goods explain the contemporary importance;
- Regional or national ports, mainly located on the western shore of the Baltic Sea. The often modest traffic of these ports must not minimalize their role in the regional economies;
- Ports of "Russia" are themselves made up of Russian, Baltic and Finnish (Hamina/Kotka) ports. This port's range is also the most dynamic one in the BSR.
- Emerging regional hubs like the port of Gdansk and Klaipeda as we can see in Figure 2 which is, in part, is elaborated by analysing AIS data in the Baltic Sea ports.

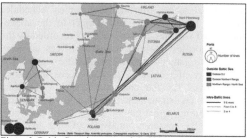

Figure 2. Baltic Container network in 2015.

The analysis of Baltic Sea ports cannot be done without a multi-scale methodology. On a global scale, the reorganization of traffic generated standardization on all shores of the Baltic Sea, responding to the hub and spoke model.

5.2 *AIS data exploitation: first results for the Baltic Sea*

AIS data allows some different geographical approaches. In the case of the Baltic Sea Region (BSR), we can develop some examples of treatment and uses (Note that this work is in progress and therefore the results are still incomplete).

Thanks to the AIS data and in relationship with external databases, we determined all container ships that called at a BSR port between 1 November 2015 and 1 November 2016. We kept the first 12 ports by their container traffic to refine the analysis (Figure 5).

At first, we can use AIS data to obtain quite generic information like the number of port calls in a certain period and/or a specific area (Figure 3) but also the number of operators involved in maritime transportation. It is also interesting to focus on ships to monitor maritime roads, their adaptations to weather or economic imperatives.

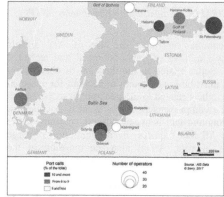

Figure 3. Container ships' port calls and operators from January 1st to November 1st 2016.

According to our data, we can establish that in the studied period, 60 different operators have provided containerized services to ports. From our geo-economic point of view we can go further in analyzing operators, their strategies or their networks. Indeed, the choice of companies, their presence (or not) in ports are good indicators to understand the functioning of the port system and to evaluate its potential.

Considering this, some companies like *Containerships* have quite local strategies with ships calling only in 4 ports (Helsinki, St-Petersburg and Riga) when others like *MSC* propose services to almost all the ports. In addition, it is also possible to determinate the capacity offered by each company in each port. For instance, during the period of the study, *Maersk Line* offered a capacity of 966336 TEU in the port of Gdansk but only 40233 TEU in Kaliningrad. It is an interesting way to focus on the three different types of actors present in the region: some major ones in the feeder field like *Unifeeder* or *Team Lines*, some global carriers (*MSC* or *CMA-CGM*) and regional ones (*Seagoline, Containerships, Mannlines...*). Furthermore, the mapping of maritime lines' networks is another opportunity offered when data on flows are available: we can map the real network of services proposed by each operator in the BSR. It clearly appears that companies have different strategies: global carriers concentrate their flows on some ports like Gdansk hub while companies specialized in feeder services have a more decentralized network (Figure 4).

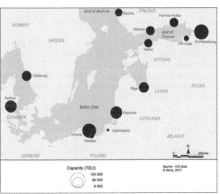

Figure 4. *Unifeeder* offered capacity from January 1st to November 1st 2016.

Nodaway, the question of port competitiveness is central for port authorities and operators. It especially includes port operation efficiency level, handling charges, reliability or landside accessibility. Regarding the Baltic ports, we can analyze operation efficiency using the duration of port calls given by AIS data. The results must be apprehended with caution, and we need remove

from the study all data that appear abnormal. For instance, discarding prolonged stops (one week or more), dedicated to the maintenance of ships. The critical examination of results is imperative, especially if one wants to start in evaluating port performance (Lévêque, 2016).

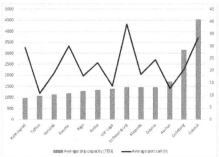

Figure 5. Comparison of container ships' capacity and duration of port call in some Baltic Sea ports.

Baltic container ports appear very different (Figure 5): two ports in particular, Goteborg and Gdansk, are served by ships offer a bigger capacity than in the other ports.

Combining this analyses to operator's strategies is also interesting. In the case of Gdansk, the situation is clearly the result of *Maersk Line* choice to make the polish port its Baltic hub. AIS data shows that 35% of all port calls are made by *Maersk Line*, with an average capacity of 4530 TEU.

By integrating the port traffic in the research process, it is possible to estimate the average length of handling of a TEU in each port. In that case study, container traffic has been weighted by the share of traffic RoRo but it should be borne in mind that these are estimations and that finer port statistics would be required to obtain a more accurate result.

It appears that container terminal efficiency is very variable in the BSR ports (Table 1): in the port of Gdansk, it takes three times less time than in Kaliningrad to operate one TEU. Average operational speed is 1.56 min per TEU and only five ports offer a better efficiency. Of course, such analysis could be more precise with the number of cranes used in each terminal for instance.

Table 1. Results of AIS data exploitation for some Baltic ports

	Gdansk	Goteborg	Klaipeda	St-Petersburg	Gdynia	Helsinki	Kaliningrad
Average ship capacity (TEU)	4529	3159	1459	1456	1462	1137	958
Average port call (h)	33,37	20,55	18,5	38,9	24,35	19,03	29,47
Number of calls	412	593	508	1032	616	640	229
Weighted traffic	971457	544680	359880	1421755	507834	320081	152684
Average TEU's per call	2358	919	708	1378	824	500	667
Average speed per TEU (minute)	0,85	1,34	1,57	1,69	1,77	2,28	2,65

157

Using AIS data, we can also study the relations between the ports and for instance the major European ports. It confirms that BSR is mainly connected to Europe with less than 10 regular container lines to the rest of the world in 2015. It is also a proof the concentration of transshipments in the Northern Range using a feedering system. The Port of Hamburg specializes in container handling, in particular in Asia-Europe and Baltic feeder traffic. In 2015, Hamburg was connected to 34 Baltic ports offering a yearly theoretical capacity of 5.1 M TEU). At the same time, Rotterdam had 17 connections and Antwerp only 15.

So the Baltic case study shows some possibilities of using AIS in the field of geo-economic research. Much more possibilities are offered by AIS data exploitation like considering connections between Baltic ports and north European ports, impact of ice conditions on navigation or maritime network seasonal disparities.

6 CONCLUSION

Satellite Automatic Identification System (AIS) technology has profoundly changed the landscape for monitoring the maritime domain. Improving upon existing AIS technology already deployed aboard all large vessels and many smaller vessels around the globe, satellite AIS is truly revolutionary in providing a complete and global picture of the world's maritime shipping.

AIS has rapidly become an operational tool exploited by a large number of actors. In effect, it provides precious information, not only to crews but also to terrestrial regulatory authorities, not forgetting individuals and research scientists. On-board security and safety for ships at sea are topical subjects owing to the growing number of acts of piracy.

The opportunities for exploiting information from AIS signals gives this device a character of global information. It is, in effect:
- Multi-scaled, temporal and spatial,
- Multi-purpose: an aid to navigation, following of global economic flows, analysis of the behaviour of economic players, behaviour of sailors, inter-action with the environment, etc.
- Multi-use: management of maritime lines, of traffic, of port calls, construction of indicators of re-liability, performance, impacts on logistics chains, etc.
- Wealth of opportunity for theoretical develop-ments in a large number of disciplines since it is, together with its air traffic counterpart, the only source of continuous tracking of moving objects on a planetary scale.

This huge potential is being exploited to set up a platform to integrate the data and for application development founded on the use of AIS signals. This poses scientific challenges and results in the requirement of an interdisciplinary methodology. First of all, to construct the platform for the acquisition of processing and availability of useable data according to the various ultimate aims and uses. The scientific validating of AIS data involves the implementation of new tools in close relation to computer processing specialists. At the same time, it is necessary to apprehend as widely as possible the different types of exploitation that will be required for this platform and consequently, collaboration is indispensable with all the different disciplines and professions concerned: geography, economics, statistics, engineering sciences, logistics providers, seagoing personnel, etc. It is preferable that this collaboration be done at the earliest possible stage so as to determine specifications for each development envisaged as this will help to improve the services provided by the platform. Lastly, even if the results are immediately available, this is also a project built on the medium to long term together with archiving the data.

REFERENCES

Cairns William R. 2005. AIS and Long Range Identification & Tracking. *Journal of Navigation* 58: 181-189.

Chen Y. *2013, Will Satellite-based AIS Supersede LRIT?* In: A. Weintrit (ed.), Advances in Marine Navigation, Marine Navigation and Safety of Sea Transportation: *CRC Press/ Balkema, London, UK: 91-94.*

Deboosere P., Dessouroux C. 2012. Le contrôle de l'espace et de ses usage(r)s : avancées technologiques et défis sociaux. *Espace populations sociétés* 2012/3: 3-11.

Dujardin B. 2004. L'AIS et ses capacités de surveillance maritime. *La revue maritime*, n°467.

Eriksen T., Høye G., Narheim B., Meland B. J. 2006. Maritime traffic monitoring using a space-based AIS receiver. *ActaAstronaut*, 58:5: 37–49.

Kaluza P., Kölzsch A., Gastner M.T. & Blasius B. 2010. The complex network of global cargo ship movements. *Journal of the Royal Society Interface*, vol.7, 48, 1093.

Le Guyader D., Brosset D., Gourmelon F. 2011. Exploitation de données AIS (Automatic Identification System) pour la cartographie du transport maritime. *Mappemonde*, 104.

Lévêque L. 2016. Les signaux AIS pour la recherché géoéconomique sur la circulation maritime. *Short Sea Shipping, myth or future of regional Transport*, Ed Serry A. & Lévêque L., pp.189-210.

Löptien U. & Axell L. 2014. Ice and AIS: ship speed data and sea ice forecasts in the Baltic Sea. *The Cryosphere*, 8: 2409-2418.

Maimun A. & Al. 2013. Estimation and Distribution of Exhaust Ship Emission from Marine Traffic in the Straits of Malacca and Singapore using Automatic Identification System (AIS) Data. *Jurnal Mekanikal 86:* 86-10.

Mitchell K.N. & Al. 2014. Waterway Performance Monitoring via Automatic Identification System (AIS) Data. *Transportation Research Board (TRB) 93rd Annual Meeting*, Chicago, 12-16 January 2014.

Perez H.M. 2009. Automatic Identification Systems (AIS) Data Use in Marine Vessel Emission Estimation. *18th Annual International Emissions Inventory Conference*, Baltimore.

Perkovic M., Gucma L., Przywarty M., Gucma M., Petelin S. , Vidmar P. 2012. Nautical risk assessment for LNG operations at the Port of Koper. *StrojniskiVestnik-J Mech Eng,* 58: 607-613.

Schwehr K., Mc Gillivary P. 2007. Marine Ship Automatic Identification System (AIS) for Enhanced Coastal Security Capabilities: An Oil Spill Tracking Application. *Oceans07 MTS/IEEE,* Vancouver.

Serry A. 2013. Le transport maritime en mer Baltique, entre enjeu économique majeur et approche durable. *Revue d'études comparatives Est-Ouest,* 44 : 89-123.

Serry A. 2016. The automatic identification system (AIS) : a data source for studying maritime traffic. *7th International Conference on Maritime Transport,* Barcelona, 27-29 June 2016.

Shelmerdine R.L. 2015. Teasing out the detail: How our understanding of marine AIS data can better inform industries, developments, and planning. *Marine Policy,* 54: 17–25.

Świerczyński S., Czaplewski K. 2013. The Automatic Identification System operating jointly with radar as the aid to navigation. *Zeszyty Naukowe Akademia Morska w Szczecinie,* 36: 156-161.

Thery H. 2012. Marine Traffic Project, un outil d'observation des routes et des ports maritimes. *Mappemonde,* 104.

Wawruch R. 2015. Study Reliability of the Information About the CPA and TCPA Indicated by the Ship's AIS. *TransNav, the International Journal on Marine Navigation and Safety of Sea Transportation,* Vol.10, N°3: 417-424.

Weintrit A. 2016. E-Nav, Is It Enough?. *TransNav, the International Journal on Marine Navigation and Safety of Sea Transportation.* Vol. 10, No. 4: 567-574.

Yao X., Mou J., Chen P., Zhang X. 2016. Ship Emission Inventories in Estuary of the Yangtze River Using Terrestrial AIS Data. *TransNav, the International Journal on Marine Navigation and Safety of Sea Transportation,* Vol. 10, No. 4: 633-640.

Proceedings of 12ᵗʰ International Conference on Marine Navigation and Safety of Sea Transportation, TransNav 2017
21-23 June 2017, Gdynia, Poland

Evaluation Method of Collision Risk Based on Actual Ship Behaviours Extracted from AIS Data

R. Miyake & J. Fukuto
National Maritime Research Institute, Tokyo, Japan

K. Hasegawa
Osaka University, Osaka, Japan

ABSTRACT: Unmanned ships have recently been studied mainly by Europe. For realization of such ship, a number of issues will be considered and one of the most important issues would be to avoid collisions with ships and obstructions floating at sea. It is essential to properly estimate a risk of collision for prevention of the collisions. There are a lot of indices for estimating risk of collision between two ships. However, it has not been validated whether such indices correspond to actual collision avoidance manoeuvres at sea. Therefore, the authors have analysed actual collision avoidance manoeuvres extracted from the recorded AIS data. Then, the authors propose a new method for evaluating collision risks based on two parameters, i.e. changing rate of relative bearing to a stand-on ship and distance between a give-way ship and a stand-on ship. In this paper, the method is described and validated based on comparison of two-ship encounter situations extracted from the AIS data and several simulation runs.

1 INTRODUCTION

Collisions are the most frequently occurred marine accidents. For safe and secured navigation at sea, one of the most simple and effective collision prevention measures is to alert operators to risk of collision. As a support system for collision prevention, ARPA (Automatic Radar Plotting Aids) is installed to radar display. However its alert sometimes does not correspond to an operators' judgment.

On the other hand, unmanned ships have recently been studied mainly by Europe from a view point of prevention of accidents caused by human error and labour-saving on-board. For realization of such ships, it is essential to estimate a collision risk to other ships in order to enable appropriate collision avoidance manoeuvres.

Many researchers have studied collision risk evaluation and proposed evaluation indices (Hara 1991, Hasegawa 2012). These indices were based on an interview survey or behaviour analysis of collision avoidance where two ships, i.e. a give-way and a stand-on ship avoided the collision in an extremely congested area in a ship handling simulator. Therefore, it has not been validated whether such indices correspond to actual collision avoidance manoeuvres at sea.

The purpose of this study is to provide a method for estimating collision risks between two ships. For this purpose, the authors analysed actual collision avoidance manoeuvres extracted from AIS (Automatic Identification System) data. In this paper, the authors provided the method of collision risk evaluation with two parameters, i.e. the distance between two ships and the rate of relative bearing change.

2 RELATION BETWEEN COLLISION RISK AND COLLISION AVOIDANCE MANOEUVRES

Collision avoidance manoeuvres can generally be described as below. In the simplest case that two ships encounter, a ship's operator of a give-way ship estimates collision risk against a stand-on ship based on the distance between the two ships and the rate of relative bearing change. When the collision risk is deemed high, the operator starts collision avoidance manoeuvres at a suitable timing in the proper manner taking into account a deviation from the planned route, approaching velocity of two ships, manoeuvrability of give-way ship and so on. The operator seeks to manoeuvre keeping a moderate distance between the two ships during the collision avoidance manoeuvre. Then, the operator stops the collision avoidance manoeuvre and returns to the original route or goes to a next way point when the operator estimates that collision risk is enough low.

It can namely be said that respective timings of changing course, such as starting collision avoidance manoeuvre or returning to the original course, are related to collision risks estimated by operators. Therefore, the authors consider that it is possible to estimate the collision risk by analysis of the state quantity of the give-way ship at each timing, especially the start time of the collision avoidance. In a previous research (Nagahata 1980), the distance between two ships at the start time of the collision avoidance are represented by two parameters, i.e. the distance between two ships and the rate of relative bearing change. Therefore, the authors also focused on the two parameters in order to estimate the collision risk.

3 ANALYSIS DATA

The authors extracted situations that give-way ships avoided stand-on ships from the AIS data. Hereafter, each extracted situation of the two ships is called *avoidance case*. At first, the authors extracted 199 avoidance cases from the AIS data recorded from 1st to 31st of August in 2013 in the western area of Seto Inland Sea as illustrated in Figure 1, and the AIS data recorded from 1st to 30th of June in 2013 in an area between Izu O-shima and Tokyo bay as illustrated in Figure 2. Next, the authors eliminated avoidance cases in which intention of changing course was deemed mainly due to navigation planning. Then the authors analysed 166 avoidance cases out of 199 cases (Miyake 2015). The avoidance cases are categorized into two groups, which are a case where a give-way ship passed behind of a stand-on ship, and a case where a give-way ship passed ahead of a stand-on ship. The number of cases of respective groups were 132 and 34.

A sequence during collision avoidance manoeuvre can be separated into the following four modes (Hasegawa 1978), because a collision risk will be related to each action of changing course angle, as mentioned in chapter 2:

1. WP (way point) mode before collision avoidance: action according to an originally planned route before collision avoidance;
2. avoiding mode: action avoiding a target ship;
3. returning mode: action of returning the planned route or moving to a next WP; and
4. WP mode after collision avoidance: action returned route after collision avoidance.

Taking into account the abovementioned modes, the following four timings of changing course angle are identified based on a time variation of heading of the give-way ship in each avoidance case as illustrated in Figure 3.
- Start time of collision avoidance (T_1): timing of change from WP mode (before collision avoidance) to WP mode (after collision avoidance);
- Start time of keeping course (T_2): timing of keeping steady course angle during avoiding mode;
- End time of collision avoidance (T_3): timing of change from avoiding mode to returning mode; and
- Returned time of planned course (T_4): timing of change from returning mode to WP mode (after collision avoidance).

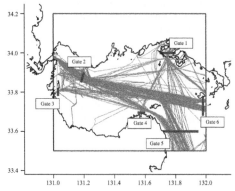

Figure 1. Target area and trajectories on 1st of August 2013 in the western area of Seto Inland Sea.

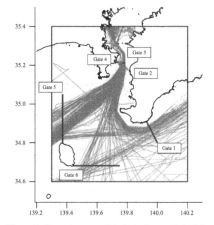

Figure 2. Target area and trajectories on 1st of June 2013 in area between Izu O-shima and Tokyo bay.

162

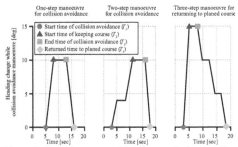

Figure 3. Typical patterns of heading angle in an action of collision avoidance on course recorder (drawn based on reference Fujii 1981).

4 COLLISION RISK EVALUATION BASED ON ACTUAL COLLISION AVOIDANCE AT SEA

4.1 Previous research of collision risk

In a previous research (Hara 1990), a questionnaire survey was conducted on 300 seafarers regarding the assessment of collision risks evaluated by ships' trajectories displayed on a radar in the range of 2 NM. Relationships between collision risk and the two parameters were analysed and the collision risks were categorized into three areas, i.e. "danger area", "caution area" and "safety area" as illustrated in Figure 4.

The ordinate axis in Figure 4 indicates the rate of relative bearing change against a stand-on ship and abscissa axis indicates the distance between give-way and stand-on ships. In Figure 4, there are four quadrants which are determined based on the relative position of the two ships illustrated in the circles. In each circle, the ships painted in black and the arrows denotes a give-way ship and vector speed, respectively.

However, it is not sure whether the evaluation of collision risks could be applied to the distance between two ships, which is greater than 2 NM. Therefore, the authors sought to analyse the avoidance cases and evaluate the collision risk between two ships whose distance is greater than 2 NM based on the previous research shown in Figure 4.

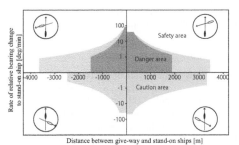

Figure 4. Subjective assessment of risk (drawn based on reference (Hara 1990)).

4.2 Collision risk evaluation based on actual manoeuvres Extracted from AIS Data

The authors analysed a relationship of the distance between two ships and the rate of relative bearing change at start time of collision avoidance (T1) (Miyake 2015, Miyake 2016).

Figure 5 shows the collision risk and the relationship of the two parameters at T_1. In the figure, two hatched areas are collision risk defined by the previous study, the dashed curve is an extended curve of boundaries of "safety area" and "caution area" in the previous study, the open circles denote each avoidance case in a situation where a stand-on ship passes ahead of a give-way ship, the crosses denote each avoidance case in a situation where a stand-on ship passes behind of a give-way ship, and the blue curve is an approximated curve estimated by the method of least squares. It is noticed that the ordinate of Figure 5 is the absolute value of the rate of relative bearing change.

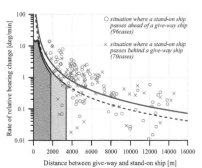

Figure 5. Collision risk defined by previous research and relationship of two parameters at start time of collision avoidance.

The avoidance cases are vertically distributed around the approximated curve. The authors deemed that collision risk is high when most of the operators started to take an action for collision avoidance. Therefore, we estimated two confidence intervals around the approximated curve under the assumption that the distributed around the approximated curve is equivalent to normal distribution. The two confidence intervals are 1σ and 2σ corresponding to the 68% and 95% confidence intervals respectively. Based on the confidence intervals, the authors distinguished four degrees of collision risk as illustrated in Figure 6. In this paper, areas of each collision risk in the figure are called the region of "Not dangerous", "To be noticed", "Cautious" and "Dangerous". Furthermore, the definitions of collision risk are shown in Table 1. Additionally, the boundaries of the regions is generally represented by Equation 1.

$$\dot{\theta} = a \cdot D^b \quad (a, b = const.) \tag{1}$$

where, D [m] is the distance between two ships and $\dot{\theta}$ [deg/min] is the rate of relative bearing change against the stand-on ship.

In Figure 6, a collision risk is estimated by the two parameters, i.e. the distance between two ships represented by the abscissa axis and the rate of relative bearing change represented by the ordinate axis. It should be noted that both axes are drawn through logarithmic graduation. In the figure, Equation 1 is given on Equation 2. Besides, the dotted line in Figure 6 is equivalent to the blue curve in Figure 5.

$$\log \dot{\theta} = \log a + b \cdot \log D \quad (a, b = const.) \qquad (2)$$

Incidentally the four regions of collision risk were distinguished based on the two confidence intervals around the approximated curve. It is said that the collision risk varies continuously in each region. Namely, collision risk corresponds to Equation 2.

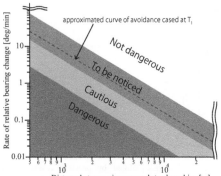

Figure 6. Collision risk evaluation between two ships

5 VERIFICATION OF COLLISION RISK EVALUATION

In order to confirm validity of the evaluation for collision risks, collision risks under the extracted two situations in which a give-way ship properly avoided a stand-on ship and the collision accident occurred are evaluated. Both situations were newly extracted from the AIS data and were not included in the avoidance cases described in chapter 3.

5.1 Collision avoidance manoeuvres in crossing situation extracted from AIS data

Figure 7 shows trajectories of the two ships recorded every three minutes. Both arrows are directions of movement of them. The position of T_1 corresponds to the start time of the collision avoidance.

Figure 8 shows the collision risk superimposed on the time variation of state quantity of the give-way ship. The time variation is separated into

different colures by the timings of changing course angle, i.e. T_1 - T_4, and a point turned blue from red corresponds to the start time of a collision avoidance. A point at U-turned corresponds to the moment when the distance between two ships was the smallest in this encounter situation. Each dotted line is the boundary of the regions in Figure 6.

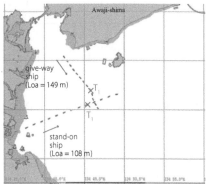

Figure 7. Trajectories of collision avoidance observed in Osaka bay.

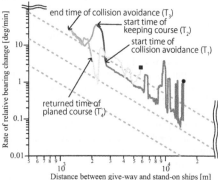

Figure 8. Collision risk and time variation of ship state during collision avoidance.

As shown in Figure 8, after collision avoidance, it is confirmed that the give-way ship keeps the collision risk in the region of "Not dangerous". In general, the situation of two ships after the collision avoidance maneuverer should be safety. Therefore, the collision risk of this situation shows that the evaluation method is proper.

5.2 Situation of collision accident

Figure 9 shows trajectories of the two ships recorded every two minutes when a collision accident occurred at around 7:09 on 7 June 2016 in an entrance to Kobe harbour. Figure 10 shows the collision risk superimposed on the time variation of state quantity of a ship whose length is 397 m.

Table 1. Definition of collision risk.

Region	Action of collision avoidance	Evaluation equation
Not dangerous	Not necessary	$\dot{\theta} > 3.0 \cdot 10^{6} \cdot D^{-1.7}$
To be noticed	To prepare for the action	$\dot{\theta} \leq 3.0 \cdot 10^{6} \cdot D^{-1.7}$ and $\dot{\theta} > 5.8 \cdot 10^{6} \cdot D^{-1.7}$
Cautious	Strongly recommended	$\dot{\theta} \leq 5.8 \cdot 10^{6} \cdot D^{-1.7}$ and $\dot{\theta} > 1.1 \cdot 10^{5} \cdot D^{-1.7}$
Dangerous	Immediately	$\dot{\theta} \leq 1.1 \cdot 10^{5} \cdot D^{-1.7}$

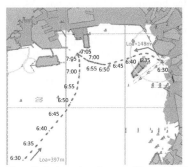

Figure 9. Trajectories when two ships crashed in Osaka bay.

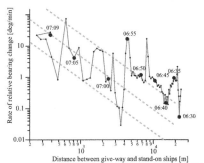

Figure 10. Collision risk and time variation of ship state when two ships crashed.

In general, a situation where the collision accident occurs is deemed that collision risk is extremely high. The collision risk after 06:57 is almost in the region of "Cautious" or "Dangerous" in Figure 10. Therefore, the collision risk of this situation also shows that the evaluation method is proper.

5.3 Time variation of distance and the rate of relative bearing change in encounter situation with different encounter angle

In Figures 8 and 10, it is confirmed that each time variation of state quantity basically proceeds in parallel with the boundaries of the regions. The authors, thus, sought to confirm whether the time variation of state quantity in any encounter situation also proceeds in parallel with the boundaries of the regions.

In this study, that is confirmed by simulation runs based on four scenarios which a give-way ship met a stand-on ship with a different encounter angle

supposed the following encounter situations: crossing situation similar to head-on situation; crossing situation; crossing situation similar to sail with the same course; and situation similar to parallel sailing. Figure 11 shows trajectories of the give-way ship and the stand-on ship in each scenario. The initial position of the give-way ship for all scenarios is on the origin. The initial position of the stand-on ship in each scenario is as shown in Figure 11 and the initial positions of the give-way ship and all stand-on ships are on a radius of 5.0 NM. The arrows show the direction of movement of each ship. Table 2 shows the initial conditions of each stand-on ship in the scenarios.

Figure 12 shows the collision risk superimposed on the time variation of state quantity of the give-way ship in each scenario. It is confirmed that each time variation basically proceeds along with the boundaries of the regions. The situation of the two ships in any encounter situation could be represented by the relationship between the distance and the rate of relative bearing change. Namely, the collision risk could be represented by Equation 2 and could be almost constant in any situation when give-way passed through the stand-on ship keeping above certain DCPA between them.

Besides, both DCPAs (Distances of Closest Point of Approach) in scenarios 3 and 4 are around 0.3 NM as shown in Table 2. However, the collision risk in scenario 4 is higher than the one of scenario 3. The difference is due to the relationship between the distance of two ships and the rate of relative bearing change. As an example, in a situation where the distance on the abscissa axis in Figure 12 is around 1,800 m, the rate of relative bearing change in scenario 4 is smaller than the one in scenario 3 because the difference between the course angle of the give-way ship and the stand-on ship in scenario 4 is smaller than the one in scenario 3 as shown in Figure 11, i.e. relative speed in scenario 4 is smaller than the one scenario 3. From the collision risks evaluated in scenario 3 and 4, we found that the collision risk could be estimated in consideration of the encounter situations.

Table 2. Initial conditions of stand-on ship in scenarios.

Scenario ID	HDG [deg]	SPD [kt]	Distance [NM]	DCPA [NM]
Scenario 1	210	16.5	abt. 9.1	abt. 0.1
Scenario 2	270	16.5	abt. 6.7	abt. 0.2
Scenario 3	320	16.5	abt. 3.2	abt. 0.3
Scenario 4	350	16.5	abt. 0.8	abt. 0.3

165

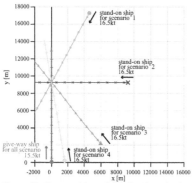

Figure 11. Superposition of ships' trajectories in scenarios.

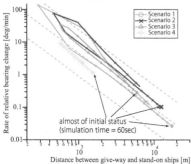

Figure 12. Collision risk and time variation of ship state in the scenarios.

6 CONCLUSIONS

The authors proposed a method of collision risk evaluation between two ships based on the actual collision avoidance manoeuvres extracted from the AIS data. In the evaluation, the collision risk is distinguished four regions on the plane of the distance between two ships and the rate of relative bearing change. Additionally, it is described that the collision risk continuously varies in each region and corresponds to the relationship of the two parameters.

In order to confirm validity of the evaluation for collision risks, the collision risks in two situations extracted from the AIS data were evaluated and it was confirmed as below. In the situation where the give-way ship accordingly avoided the stand-on ship, the collision risk after the start time of collision avoidance is in the region of "Not dangerous". In another situation where collision accidents occurred,

the collision risk of right before collision is in the region of "Cautious" or "Dangerous". These evaluation results of collision risk show that the evaluation method is almost proper.

Besides, the authors sought to confirm by simulation runs whether the time variation of state quantity in any encounter situation proceeds in parallel with the boundaries of the regions. As the results, the collision risk could be represented based on the distance and the rate of relative bearing change. Additionally, it was confirmed that collision risk in the method is estimated in consideration of the encounter situation.

Further study for an evaluation method of generalized collision risk is ongoing taking into account the ship length and relative speed.

ACKNOWLEDGMENT

This work was supported by JSPS KAKENHI Grant Number JP15K18297.

REFERENCES

Hasegawa, K. & Kouzuki, A. 1978. Automatic Collision Avoidance System for Ships Using Fuzzy Control, *The Journal of Japan Society of Naval Architects and Ocean Engineers*, vol. 205, 1-10 (in Japanese).

Nagahata, T. 1980. The Characteristics of Navigators on Maneuvers for Avoiding Collision-III: Modeling of the Relative Distance by Catastrophe, *The Journal of Japan Institute of Navigation*, vol. 63, 19-28 (in Japanese).

Fujii, Y., Makishima, T. & Hara, K. 1981, *Kaijo Kotsu Kogaku*: 86, Japan: Kaibundo, ISBN4-303-23401-X (in Japanese).

Hara, K., Nagasawa, A. & Nakamura, S. 1990, The Subjective Risk Assessment on ships Collisions, *The Journal of Japan Institute of Navigation*, vol. 83, 71-80 (in Japanese).

Hara, K. 1991. Proposal of Maneuvering Standard to Avoid Collision in Congested Sea Area, *The Journal of Japan Institute of Navigation*, vol. 85, 33-40 (in Japanese).

Hasegawa, K. & et al. 2012. An Intelligent Ship Handling Simulator with Automatic Collision Avoidance Function of Target Ships, *Proc. Of International Navigation Simulator Lecturer's Conference*, INSLC17, F23-1-10.

Miyake, R., Fukuto, J. & Hasegawa, K. 2015. Analyses of the Collision Avoidance Behaviours Based on AIS Data, *The Journal of Japan Institute of Navigation*, vol. 133, 66-74 (in Japanese).

Miyake, R., Hasegawa, K. & Fukuto, J. 2016. Quantitative Assessment of Collision Risk Based on Actual Behaviour Analysis of Collision Avoidance Manoeuvre, *The Journal of the Japan Society of Naval Architects and Ocean Engineers*, Vol.24, 283-290, (in Japanese).

Marine Radar

The Joint Waveform and Filter Design for Marine Radar Tasks

V.M. Koshevyy & I.Y. Gorishna

National University "Odessa Maritime Academy", Odessa, Ukraine

ABSTRACT: Waveforms, constructed on the base of m-Sequences, have good perspectives for being used in continuous wave radar of new generation. However, the side-lobe level of their ambiguity function does not always satisfy the requirements for separation of target vessels from other reflective objects. Therefore, along with m-Sequences it is necessary to consider other classes of waveforms with discrete modulation having the better properties of side-lobes suppression. In this paper has presented the basis of a continuous waveform with discrete modulation and filter design capable of achieving range side-lobe levels of better than in the case of m-Sequences with additional ability of clutter rejection for a given area on the Doppler-range plan, suitable for use in marine radar. Possibilities of such kind of joint waveform and filter design have been investigated.

1 INTRODUCTION

The existing model of marine radars has a high peak power of the radiation, which worsens the quality of the main lobe and leads to harmful emission into the environment and worsens the property of the electromagnetic compatibility. To improve the effectiveness of the radar must be developed radar with a reduced pulse radiation power, which is actual topic nowadays.

Further we consider it with more details. The Ambiguity Function (AF) corresponds to the time-frequency response function that is observed on the output of the filter. One of the most important characteristics of the AF is the level of side lobes, which in most cases are trying to reduce. Phase-manipulated (PM) signals are sequence of radio pulses, phases of which vary according to a given law. The complex envelope of such PM signals is a sequence of positive and negative pulses [1, 2]. Almost always is the same shape of pulses and in most cases it is rectangular. A rectangular pulse with unit amplitude and duration written as:

$$p_0(t) = 1 \text{ , at } 0 \le t \le \tau_0. \tag{1}$$

Let the amplitude of the n-th pulse in the video PM signal is equal to +1 or -1, which corresponds to the initial phases of 0 or π in radio PM signal. With this definition the PM signal is written as follows:

$$S(t) = \sum_{n-1}^{N} S_n p_0[t-(n-1)\tau_0]. \tag{2}$$

One of the important characteristics of the cross ambiguity function (CAF) is the level of side lobes, which in most cases are trying to reduce. Choice of form of CAF, that is actually a form of probing signal depends primarily on the purpose of radar station, interference environment, the form and nature of the objectives, parameters of movement, etc.

In general, the expression for the CAF phase-manipulated radio pulse can be written:

$$\chi_{sw}(\tau_0, f) = \sum_{n-1}^{N} W_n^* S_{n-k} e^{-j2\pi n f T_0}, \tag{3}$$

where
$S_{n-k} = e^{j\varphi_{n-k}}$ — the complex amplitude of signal;
$W_n^* = e^{-\varphi_n}$ — the complex amplitude of filter;

The analysis of CAF phase-manipulated signals with different phase modulation law leads to the conclusion that the level of side lobes of CAF of many peaks or many crest structure without the use of special measures has significant value, commensurable, sometimes, with a maximum value of CAF. Applying of weight processing with the use of quasi optimal weighting coefficients allows to reduce the side lobes level between peaks, but inevitably is the expansion of peaks, this, in its turn, affects the performance of accuracy and ambiguity of target coordinates estimation and their reliability.

Substituting into the expression (3) $\Delta f = \frac{1}{4NT_0}$, then the expression for the CAF phase-manipulated radio pulse can be written:

$$\chi_{sw}(\tau_0, f) = \sum_{n-1}^{N} W_n^* S_{n-k} e^{j2\pi \frac{nl}{4N}}. \tag{4}$$

Using formula (5) were calculated optimal filter weighting coefficients:

$$W = R^{-1}S, \tag{5}$$

where R - the correlation matrix of similar to signal obstacle.

More details in the case of a periodic signal:

$$S = \begin{bmatrix} s_0 \\ s_1 \\ s_2 \\ \vdots \\ s_n \end{bmatrix}$$

$$R = \underbrace{\begin{bmatrix} s_0 \\ s_1 \\ s_2 \\ \vdots \\ s_n \end{bmatrix} \begin{bmatrix} s_0 & s_1 & s_2 & \cdots & s_n \end{bmatrix}}_{R_0} + \underbrace{\begin{bmatrix} s_n \\ s_0 \\ s_1 \\ \vdots \\ s_{n-1} \end{bmatrix} \begin{bmatrix} s_n & s_0 & s_1 & \cdots & s_{n-1} \end{bmatrix}}_{R_1} + \cdots$$

$$\cdots + \underbrace{\begin{bmatrix} s_1 \\ s_2 \\ s_3 \\ \vdots \\ s_0 \end{bmatrix} \begin{bmatrix} s_1 & s_2 & s_3 & \cdots & s_0 \end{bmatrix}}_{R_n}; \tag{5`}$$

And in case of aperiodic signal:

$$R = \underbrace{\begin{bmatrix} s_0 \\ s_1 \\ s_2 \\ \vdots \\ s_n \end{bmatrix} \begin{bmatrix} s_0 & s_1 & s_2 & \cdots & s_n \end{bmatrix}}_{R_0} + \underbrace{\begin{bmatrix} 0 \\ s_0 \\ s_1 \\ \vdots \\ s_{n-1} \end{bmatrix} \begin{bmatrix} 0 & s_0 & s_1 & \cdots & s_{n-1} \end{bmatrix}}_{R_1} + \cdots$$

$$\cdots + \underbrace{\begin{bmatrix} 0 \\ 0 \\ 0 \\ \vdots \\ s_0 \end{bmatrix} \begin{bmatrix} 0 & 0 & 0 & \cdots & s_0 \end{bmatrix}}_{R_n} + \underbrace{\begin{bmatrix} s_1 \\ s_2 \\ s_3 \\ \vdots \\ 0 \end{bmatrix} \begin{bmatrix} s_1 & s_2 & s_3 & \cdots & 0 \end{bmatrix}}_{R_{n+1}} +$$

$$+ \underbrace{\begin{bmatrix} s_2 \\ s_3 \\ s_4 \\ \vdots \\ 0 \end{bmatrix} \begin{bmatrix} s_2 & s_3 & s_4 & \cdots & 0 \end{bmatrix}}_{R_{n+2}} + \cdots + \underbrace{\begin{bmatrix} s_n \\ 0 \\ 0 \\ \vdots \\ 0 \end{bmatrix} \begin{bmatrix} s_n & 0 & 0 & \cdots & 0 \end{bmatrix}}_{R_{n+k}}; \tag{5``}$$

2 ITERATION METHOD

The iteration procedure of optimization of the signal and filter to find the maximum of signal/(noise + interference) ratio at their various dimensions lies in sequential solution of integral equations for the filter and the signal at the fixed norm of the signal and filter [4].

At the first step, for a given initial signal vector is being solved equation, determining the filter impulse response, then for the thus-obtained filter is being solved equation defining the signal, etc.

Thus, each value of the found filter or the signal passes normalization process. The methods of joined optimization of the signal and filter were considered earlier, taking into account additional restrictions on the permanent resolution of time, the amount of losses in the signal/noise ratio and methods of signal and filter optimization at a fixed amplitude signal modulation. The research of their efficiency is being considered in this paper

In the calculations we used the iteration procedure of maximizing the signal/noise ratio, taking into account the restrictions on losses in the signal/noise ratio and a constant resolution in time where the signal/noise ratio was considered as [3]:

$$\Sigma = \frac{\left| \int_{-\infty}^{\infty} W^*(t) S(t) \, dt \right|^2}{v \int_{-\infty}^{\infty} |W(t)|^2 \, dt + \sigma_0 \int_{-\infty}^{\infty} \sigma_\xi(\tau, 0) |\chi_{sw}(\tau, 0)|^2 \, d\tau}, \tag{6}$$

where $\chi_{sw}(\tau, f) = \chi_\rho(\tau, f) \sum_{}^{N-1} W_n^* S_{n+k} e^{i2\pi nf T_0}$ — ambiguity function, T_0 – elementary pulse duration in the signal, f – Doppler`s frequency, N – number of pulses in the signal, NT_0 – period of signal (in periodic mode of work), $s_{n+k} = [s_{n+k}] e^{i\varphi_{n+k}} - S_{n+k} = [S_{n+k}] e^{i\varphi_{n+k}}$ the complex amplitude of signal, delayed on k positions, φ_{n+k} – phase of signal, w_n^* – complex amplitude of bearing signal (filter), σ_0 – the coefficient, which describes reflected properties of interference $, \sigma_\xi(\tau, 0)$ – the range-velocity distribution of interfering reflections, e parameters that determine the restrictions on losses of signal/noise ratio (v) and a constant restriction on time resolution (ξ). Expression of constant time resolution is as follows [1, 3]:

$$\frac{\int_{-\infty}^{\infty} |\chi_{sw}(\tau, 0)|^2 \, d\tau}{|\chi_{sw}(0,0)|^2} = \frac{\int_{-\infty}^{\infty} \left| \int_{-\infty}^{\infty} W^*(t) S(t+\tau) \, dt \right|^2 \, d\tau}{\left| \int_{-\infty}^{\infty} W^*(t) S(t) \, dt \right|^2} = T_R. \tag{7}$$

The constant time resolution T_R has the best value, when all the side lobes of the cross-correlation functions are zero. The solution of task will be found for the values of parameters

$v = N_0 \xi = 1$. In fact, we consider the problem of maximizing the signal/noise ratio (the noise with spectral density N_0) with restriction on constant time resolution. On the first iteration, we are looking for a filter that provides for the chosen parameters v и $N_0 = 10^{-3}$ almost complete suppression of side lobes. This is connected to the fact that the problem of digital signals corresponds to suppress $N-1$ side lobes, which corresponds to the condition of the zero zone. Zones with complete suppression of side lobes.

The expression of losses in signal/noise ratio is as follows:

$$\rho = \frac{\left| W^* S \right|^2}{W^* \cdot W \cdot S^* S}. \tag{8}$$

In the first step may be a large losses in the signal / noise ratio. Therefore, we shall use the procedure above to select the signals and filters, which allow to obtain minimal losses in the signal / noise ratio.

3 PROGRAM DEVELOPMENT USING THE ITERATIVE METHOD

In the Matlab was developed a program that allows to realize the iteration process of joint optimization of filter and signal and receive graphics with CAF-sections $\Delta f = \frac{1}{4NT_0}$ given below (figures 1a, 1b, 2a, 2b, 3a, 3b). In this case, as the initial approximation we considered a discrete signal sequence with N=5 with the following form s= [1; -1; -1; -1; -1], In the result of the iteration process we received a couple of signal and filter, which ensures a constant value of time resolution $T_R = 1$ (it means complete suppression of the side lobes in L=0) and $\rho = 1$, which no losses in signal/noise ratio (figure 1б) and corresponds to the agreed treatment.

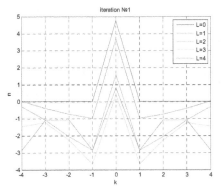

Figure 1a. The shape of CAF and her sections at l=0...4 of periodic signal s=[1; -1; -1; -1;- 1]. The value of optimal filter: Wn =[1.5811; -0.7906; -0.7906; -0.7906; -0.7906]. The value of losses in signal/noise ratio . = 0.9001.

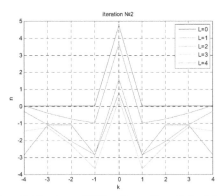

Figure 1b. The shape of CAF and her sections at l=0...4 for the optimal filter Wn=[1.3416; -0.8944; -0.8944; -0.8944; -0.8944] and optimal signal: Snorm=[1.3416; -0.8944; -0.8944; -0.8944; -0.8944]. The losses in signal/noise ratio $\rho = 1.0000$.

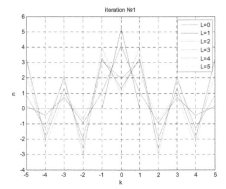

Figure 2a. The form of CAF and her sections at l=0...5 of periodic signal s=[1; 1; -1; 1; -1; 1]. The value of optimal filter Wn =[1.9639; 0.6547; -0.6547; 0.6547; -0.6547; 0.6547]. The value of losses in signal/noise ratio . = 0.7619.

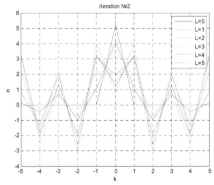

Figure 2b. The form of CAF and her sections at l=0...5 for the optimal filter Wn=[1.9638; 0.6547; -0.6547; 0.6547; -0.6547; 0.6547] and optimal signal: Snorm=[1.0004; 0.9999; -0.9999; 0.9999; -0.9999; 0.9999]. The value of losses in signal/noise ratio $\rho = 0.7621$.

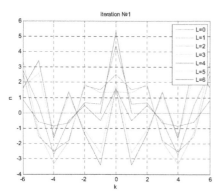

Figure 3a. The shape of CAF and her sections at l=0…6 of periodic signal s=[1; 1; 1; -1; -1; 1; -1]. The value of optimal filter Wn =[1.3229; 1.3229; 1.3229; -0.0007; -0.0007; 1.3229; -0.0007]. The value of losses in signal/noise ratio . = 0.5720.

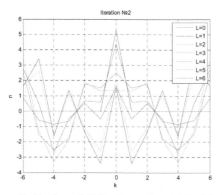

Figure 3b. The shape of CAF and her sections at l=0…6 for the optimal filter Wn=[1.3229; 1.3229; 1.3229; -0.0013; -0.0013; 1.3229; -0.0013] and optimal signal: Snorm=[1.0002; 1.0002; 1.0002; -0.9997; -0.9997; 1.0002; -0.9997]. The value of losses in signal/noise ratio ρ = 0.5725.

Furthermore, calculations were performed with N = 3, 8, 9 for the periodic case where similar results were obtained and also for aperiodic case. Using obtained signals for different N (N1, N2,…Np) new signals may be constructed with the method based on element-wise multiplication of signals with mutually prime periods [5]. In particular we can get resultant signal due to the product of two signals: N= N1N2 (for example N1=3, N2=4; N1=5, N2=4; N1=7, N2=9; and others). Also can be used products of three, four signals and so on.

Considered in this article method of signal-filter pair synthesis can also be used for range-velocity

distributions of the interfering reflections, which contain a few cross-sections CAF with different Doppler shifts.

4 CONCLUSIONS

The sidelobe suppression helps to reduce the level of harmful radiation to the surrounding space. Also an important part is to reduce the level of background noise in the antenna, as it is created due to the differences of the amplitudes and frequencies of side lobes from the main lobe.

In this paper we considered the the task of maximization of signal/noise ratio with additional restrictions in it and restrictions on constant time resolution. The results of calculations confirmed the effectiveness of the considered iteration procedure allowing at the appropriate choice of the initial signal to get known globally optimal solutions. Therefore, we will consider the tasks with the help of this procedure to suppress of interfering reflections with random range-velocity distribution of preventing reflections, the best solutions of which are unknown.

REFERENCES

[1] Cook C. E. and Bernfeld M. Radar signals. Iintroduction to theory and application. New York: Academic Press. 1967.
[2] Varakin L. E. Systems of communication with noise-like signals. Moscow: Radio and Communications. 1985.
[3] V.M. Koshevyy, M.B Sverdlik .'Synthesis of Signal – Filer pair under additional constraints'. Radio Engineering and Electronics. vol 21. no.6. pp. 1227-1234. June 1976.
[4] V.M. Koshevyy , M.B. Sverdlik. Joint Optimization of Signal and Filter in the Problems of Extraction of Signals from Interfering Reflections. Radio Engineering and Electronic Physics. Vol.20. N 10 pp.48-55.
[5] V.M. Koshevyy, I.V. Koshevyy, D.O. Dolzhenko. Synthesis of Composite Biphasic Signals for Continuos Wave Radar. In: A. Weinetrit & T. Neumann (eds.), Information, Communication and Enviroment, Marine Navigation and Safety of Sea Transplotation. CRC Press/Balkema, London, UK, 2015. pp. 55 – 59.
[6] V.M. Koshevyy, M.B. Sverdlyk. On the possibilities of complete suppression of the side lobes of the cross ambiguity function in a given region. Moscow: Radio Engineering and Electronics, 1974 - 8 p.
[7] V.M, Koshevyy, D.O. Dolzhenko. The synthesis of periodic sequences with given correlation properties – Proc. of IEEE East-West Design & Test Symposium (EWDTS'11), 2011 - pp. 341-344.

Improved Compound Multiphase Waveforms with Additional Amplitude Modulation for Marine Radars

V.M. Koshevyy & O. Pashenko
National University "Odesa Maritime Academy", Odessa, Ukraine

ABSTRACT: This paper has presented the basis of a compound multiphase waveform design with additional amplitude modulation, capable of controlling a waveform pick-factor, suitable for use with marine radar. The waveform shows good Doppler tolerance, with the low side-lobes performance maintained over the central zone of an ambiguity function. A clear waveform design procedure has been presented that does not require the implementation of numerical optimization procedures. It has been shown that a compromise between side-lobes suppression and the value of pick-factor can be found. These waveforms allow achieving better results as compared to compound multiphase waveforms without additional amplitude modulation under mismatched weighting filtering.

1 INTRODUCTION

The compound multiphase signal consists of multiplication of two sequences; each of them is a signal with a quadratic variation of phases. Expressions for complex amplitudes for the base (s_n^b) and external (s_n^v) signal and final multiphase compound signal's sequence (s_n) are follows [1,2]:

$$s_n^b = exp\left\{ j\frac{\pi}{4}\alpha'\left[2\left(n - N_B E\left[\frac{n}{N_B} \right] + 1 \right) - (N_B+1)\right]^2 \right\} \quad (1)$$

$$s_n^v = exp\left\{ j\frac{\pi}{4}\beta'\left[2\left(E\left[\frac{n}{N_{B_1}} \right] - N_V E\left[\frac{E\left[\frac{n}{N_{B_1}} \right]}{N_V} \right] + 1 \right) - (N_V+1)\right]^2 \right\}, n=\overline{0, N-1} \quad (2)$$

$$S_n = s_n^V \cdot s_n^B \quad (3)$$

where $\alpha'=\alpha T_0^2$; $\beta'=\beta(T_0 N_6)^2$; α, β, μ_0, N_{B1}– parameters of phase modulation; T_0 – duration of one pulse; N_B - period of the base sequence; N_V - period of the external sequence; $N = N_B N_V$- period of the sequence; $E[x]$ - integer part of $x[1]$.

Three types of aperiodic compound multiphase signals were considered. The length of sequence is N=324 (N_B=18, N_V=18) with parameters:

1 $\alpha'=-1/N_B$, $\beta' =1/N_B^2$. The signal with such parameters has a maximum ratio of the free zone (FZ) area around the central peak (CP) to the CP topographic section area (on the zero level) [2].

2 $\alpha'=-1/N_B$, $\beta' =1/N_B$. The side-lobes level is increased in the FZ region for these parameters in comparative to the case 1). But the Autocorrelation Function (ACF) has the lowest side-lobes level for its entire length. It should also be noted that FZ decreases around CP of Ambiguity Function (AF) [3].

3 $\alpha'=-1/N_B$, $\beta'=0$.

Signals with an additional amplitude modulation (AM) can be described by the following expression:

$$s_n^{am} = s_n * v_n \quad (4)$$

$$v_n = v_{n-E[n/N_B]N_B}^b * v_{E\left[\frac{n}{N_V}\right]+1}^v, n=\overline{0 \div N-1},$$

where s_n^{am}- compound multiphase signal with an additional AM; s_n - complex envelope of the signal; v_n^b, v_n^v- weighting coefficients for the base and external sequences.

The expression of the AF for the aperiodic compound multiphase signal with an additional AM has the form:

$$\chi_{ss}(k,l) = \sum_{n=0}^{N-1} s_n^{am*} \cdot s_{n+k}^{am} \cdot e^{j\frac{2\pi ln}{4N}} \quad (5)$$

where k - discrete values of time delay with the selected step T_0; l - discrete values of frequency.

It's convenient to use the pick-factor parameter for describing behavior of AM modulation [4]. It is defined as the ratio:

$$\xi = \frac{\left|s_{max}^{am}\right|^2 N}{\sum_{n=0}^{N-1}\left|s_n^{am}\right|^2} \qquad (6)$$

The law of the changing weighting function *sin* is considered in the paper. This function allows controlling the behavior of AF. Its advantage is possibility to control the signal peak-factor and the side-lobes level of AF.

$$v_n^B = sin\left[\pi\left(\frac{y_B}{N_B+1}+\frac{n}{N_B+z_B}\right)\right], n=0 \div N_B-1$$

$$v_n^V = sin\left[\pi\left(\frac{y_V}{N_V+1}+\frac{n}{N_V+z_V}\right)\right], n=0 \div N_V-1 \qquad (7)$$

where y_B, y_V - the positive numbers for the base and external sequences, respectively, varies in the range $1 \leq y_B < \frac{N_B+1}{2}$, $1 \leq y_V < \frac{N_V+1}{2}$; z_B, z_V - variables for the basic and external sequences are $z_B = \frac{2y_B N_B-(N_B+1)}{N_B+1-2y_B}$, $z_V = \frac{2y_V N_V-(N_V+1)}{N_V+1-2y_V}$.

The formula (8) describes one of the possible weighting functions for the signals (3):

$$v_n^B = \left(sin\left[\pi\left(\frac{y_B}{N_B+1}+\frac{n}{N_B+z_B}\right)\right]\right)^2, n=0 \div N_B-1$$

$$v_n^V = \left(sin\left[\pi\left(\frac{y_V}{N_V+1}+\frac{n}{N_V+z_V}\right)\right]\right)^2, n=0 \div N_V-1 \qquad (8)$$

The structure of AF of multiphase compound signals with an additional AM is compared and analyzed with Cross-ambiguity Function (CAF) compound multiphase signals without AM, but mismatched treatment with weighing functions *sin* (7) and *sin²* (8).

As a result (Fig.1), side-lobes level is significantly reduced after using the aperiodic compound multiphase signals from with an additional AM. CP area isn't expanded significantly. By increasing the y_B and y_V, the value of peak-factor is decreasing. By changing the parameters y_B and y_V we can control the peak-factor of the signal. Table 1 shows these values:

Table 1. The peak-factor dependence from parameters $y_V=y_B$ for aperiodic multiphase compound signal with an additional AM, $N=324$ ($N_B=18$, $N_V=18$) $\alpha'=-1/N_B$, $\beta'=1/N_B$, $\mu_0 = 0$, $N_{B1} = 1$, (matched processing) with the weighting coefficients *sin*

$y_V=y_B$	ξ
1	3.59
2	2.83
3	2.24
5	1.5
6	1.28
7	1.13
8	1.05
9	1,01

Table 2. The peak-factor dependence from parameters $y_V=y_B$ for aperiodic multiphase compound signal with an additional AM, $N=324$ ($N_B=18$, $N_V=18$) $\alpha'=-1/N_B$, $\beta'=1/N_B$, $\mu_0 = 0$, $N_{B1} = 1$, (matched processing) with the weighting coefficients *sin²*

$y_V=y_B$	ξ
1	6.38
2	4.97
3	3.77
5	2,09
6	1,6
7	1,28
8	1.09
9	1,01

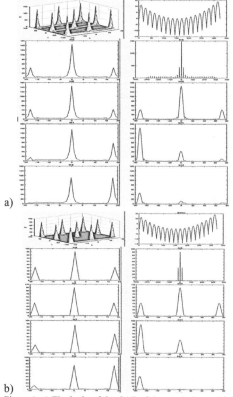

a)

b)

Figure 1. a) The body of the CAF of the aperiodic multiphase compound signal without an additional AM, $N=324$ ($N_B=18$, $N_V=18$) $\alpha'=-1/N_B$, $\beta'=1/N_B$, $\mu_0 = 0$, $N_{B1} = 1$, with the weighting function *sin* ($y_B = y_V = 1$) for (7), phases of the signal, sections of the body of the CAF; b) the body of the AF aperiodic multiphase compound signal with an additional AM, $N=324$ ($N_B=18$, $N_V=18$) $\alpha'=-1/N_B$, $\beta'=1/N_B$, $\mu_0 = 0$, $N_{B1} = 1$, with the weighting coefficients *sin* ($y_B =y_V =1$) for (7), phases of the signal, sections of the body of the AF

The calculations were made for three types of the signals, as indicated above:
1. $\alpha'=-1/N_B$, $\beta' =1/N_B^2$, $N=324$ ($N_B=18$, $N_V=18$).
2. $\alpha'=-1/N_B$, $\beta' =1/N_B$, $N=324$ ($N_B=18$, $N_V=18$).
3. $\alpha'=-1/N_B$, $\beta' =0$, $N=324$ ($N_B=18$, $N_V=18$).

The three kinds of signals with different sets of parameters were compared. The compound multiphase signal with an additional AM $N=324$ ($N_B=18$, $N_V=18$) $\alpha'=-1/N_B$, $\beta'=0$, $\mu_0=0$, $N_{B1}=N_B$, (matched processing) with the weighting function sin shows the highest ratio of the CP to the side-lobes level. The signal with $\beta'=1/N_B^2$ shows almost the same. For example, we demonstrated the cases of signals with parameters $\alpha'=-1/N_B$, $\beta'=1/N_B$, $N=324$ ($N_B=18$, $N_V=18$), $y_V=y_B=1,2$. They have the highest side-lobe level in the FZ region.

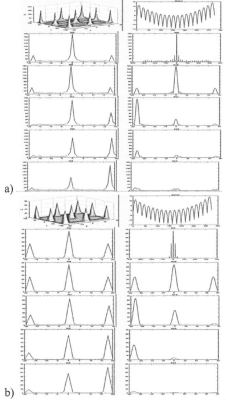

a)

b)

Figure 2. a) The body of the CAF of the aperiodic multiphase compound signal without an additional AM, $N=324$ ($N_B=18$, $N_V=18$) $\alpha'=-1/N_B$, $\beta'=1/N_B$, $\mu_0=0$, $N_{B1}=1$, with the weighting function sin^2 ($y_B=y_V=2$) for (8), phases of the signal, sections of the body of the CAF; b) the body of the AF aperiodic multiphase compound signal with an additional AM, $N=324$ ($N_B=18$, $N_V=18$) $\alpha'=-1/N_B$, $\beta'=1/N_B$, $\mu_0=0$, $N_{B1}=1$, with the weighting coefficients sin^2 ($y_B=y_V=2$) for (8), phases of the signal, sections of the body of the AF

Thus, the lowest side-lobes level is gave signals with peak-factor ($\xi = 3,59$ and $\xi = 2,83$), while $y_V=y_B = 1,2$. For comparison AF body for

rectangular shape of compound signals is presented on Figure 3.

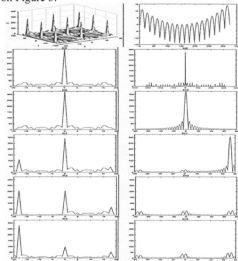

Figure 3. The body of the AF of the aperiodic multiphase compound signal, $N=324$ ($N_B=18$, $N_V=18$) $\alpha'=-1/N_B$, $\beta'=1/N_B$, $\mu_0=0$, $N_{B1}=N_B$, phases of the signal, sections of the body of the AF

These results can be used not only for the compound multiphase signals, but also for equivalent compound LFM signals [5, 6, 7].

REFERENCES

[1] V.M. Koshevyy.Synthesis of Waveform-Filter pairs under Additional Constrains with Group-Complementary Properties IEEE, Radar Conference 2015, May 2015, Arlington,VA (USA), pp 0616-0621.

[2] V.M. Koshevyy, Synthesis compound multiphase signals, Izvestiya VUZ. Radioelectronika (Radioelectronics and Communication Systems), vol. 31, N8, 1988, pp. 56-58.

[3] V.M. Koshevyy, V.I. Kuprovskyy. Investigation of properties of compound multiphase signals.Izvestiya VUZ. Radioelectronika (Radioelectronics and Communication Systems), N8, 1991, pp. 63-66.

[4] Ch.E. Cook, M. Bernfeld, Radar Signals. An Introduction to Theory and Application, Artech House, Inc., Boston 1993.

[5] B.L. Lewis, F.F. Kretschmer, Linear frequency modulation derived polyphase pulse compression codes, IEEE Trans. on Aerospace and Electronic Systems Vol. AES-18, N5, Sept. 1982, pp. 637-641.

[6] N. Levanon, E. Mozeson, Radar signals, J. Wiley, NJ, 2004.

[7] V. Koshevyy & O. Pashenko, Signal Processing Optimization in the FMCW Navigational Radars, Marin Navigation and Safety of Sea Transportation. Activities in Navigation. (edited) Adam Weintrit. CRC Press. 2015 pp. 55-60

Radar Radiation Pattern Linear Antennas Array with Controlling Value of Directivity Coefficient

V.M. Koshevyy & A.A. Shevchenko
National University 'Odessa Maritime Academy', Odessa, Ukraine

ABSTRACT: Improved linear antenna array design capable of obtain the given side-lobe suppression with controlled value of directivity coefficient is suggested. A clear linear antenna design procedure has been presented that does not require the implementation of numerical optimization procedures. The efficiency of suggested design has been investigated.

1 INTRODUCTION

The modern radars are high-powered technical means of navigation and take important part to ensure maritime safety. But they have high ship's antenna reception diagram side lobes level. It's lay to deterioration of radar angle and range resolution, especially when big and small vessel are situated on the same distance and have azimuth angle's close value. The solution of this problem can be received in ship radar antenna pattern formation with using linear antenna array with low side lobes level. The optimal methods of formation such diagrams are known and associated with the need of all array elements tuning by difficult algorithms [6]. At the same time the reaching of antenna pattern side lobes required suppression level may be with the simple methods by using the arrays with limited number of tunable weight coefficients of spatial filter (antenna array elements), for example, when there are only two of such tunable weight coefficients.

In this case all of the receiving antenna array of spatial filter's weights coefficients W_i of the processing, except two (first and last : W_1, W_N), are fixed (selected under the condition of providing the required antenna pattern side lobe's level) ($W_2; W_3; ...; W_{N-1}$). Value of the two tunable weights coefficients are selected for carried out the condition of providing zero values in two points (θ_1, θ_2) of the reception pattern. The expressions, which are describing the reception pattern of linear array antenna $G(\theta)$, may be written in the following form:

$$G(\theta) = \sum_{i=2}^{N} W_i \cdot e^{-j2\pi(i-1)\frac{d}{\lambda}\sin\theta} \tag{1}$$

The expression for tunable weight coefficients has the following form:

$$W_1 = \frac{G_{N-2}(\theta_2)*e^{j2\pi(N-1)\frac{d}{\lambda}\sin\theta_1} - G_{N-2}(\theta_1)*e^{j2\pi(N-1)\frac{d}{\lambda}\sin\theta_2}}{e^{j2\pi(N-1)\frac{d}{\lambda}\sin\theta_2} - e^{j2\pi(N-1)\frac{d}{\lambda}\sin\theta_1}}; \tag{2}$$

$$W_N = \frac{G_{N-2}(\theta_1) - G_{N-2}(\theta_2)}{e^{j2\pi(N-1)\frac{d}{\lambda}\sin\theta_2} - e^{j2\pi(N-1)\frac{d}{\lambda}\sin\theta_1}}. \tag{3}$$

where $G_{N-2}(\theta)$ – partial diagram of untunable (fixed) diagram, which has the following form:

$$G_{N-2}(\theta) = \sum_{i=2}^{N-1} W_i \cdot e^{-j2\pi(i-1)\frac{d}{\lambda}\sin\theta} \tag{4}$$

In this paper we consider the different weight functions effect on reception diagram, which allow to transform the reception diagram properties.

The expression for the tunable weight coefficient in this case has the next form:

$$W_n^{(1)} = sin\left[\pi\left(\frac{y}{N+1-2} + \frac{n-2}{N+z-2}\right)\right] \tag{5}$$

$$W_n^{(2)} = \left[sin\left[\pi\left(\frac{y}{N+1-2} + \frac{n-2}{N+z-2}\right)\right]\right]^2 \tag{6}$$

where: $n=2:N-1$;

$$y = 1 : \frac{N-2+1}{2}$$
$$z = \frac{2y(N-2)-(N-2+1)}{N-2+1-2y};$$

We can regulate the main lobe widening by parameter 'y'. The more value of parameter 'y', the less main lobe widening. For example, the reception diagrams calculated with (5) are shown on figures 1-2.

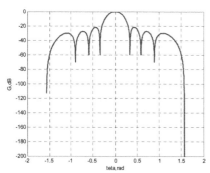

Figure 1. Partial reception diagram with y = 1($W_n^{(1)}$)

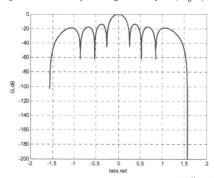

Figure 2. Partial reception diagram with y = 3($W_n^{(1)}$)

The fully coherent reception diagram $G_{N-2}(\theta)$ (equable correction), which corresponds the condition with absence of main lobe widening is shown on figure 3.

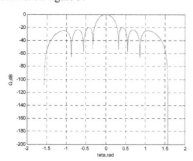

Figure 3. The fully coherent reception diagram $G_{N-2}(\theta)$

For compare the reception diagrams calculated by (6) are shown on figures 4-5.

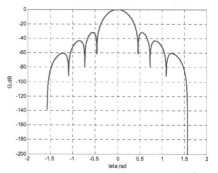

Figure 4. Partial reception diagram with y = 1($W_n^{(2)}$)

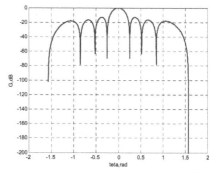

Figure 5. Partial reception diagram with y = 3($W_n^{(2)}$)

So, we can see that the weight function, described by (5) and (6), allows to correct the reception diagram properties widely.

The reception diagrams with suppression in given points, calculated with weight coefficients (2) and (3) are shown on figures 6-7.

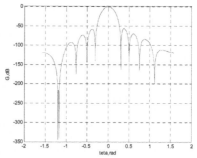

Figure 6. The reception diagram side lobes suppression in points: $\theta_1 = -1.2001, \theta_2 = -1.1718, \rho = -1,1138\,dB$

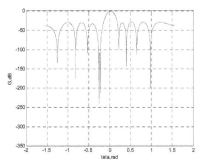

Figure 7. The reception diagram side lobes suppression in points: $\theta_1 = -0.2576, \theta_2 = -0.2293, \rho = -8,2945, dB$

The reception diagrams side lobes suppression in given points, calculated by (5) are shown on figures 8-9 .

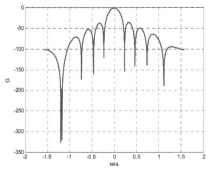

Figure 8. The reception diagram side lobes suppression in points: $\theta_1 = -1.2001 \theta_2 = -1.1718 \rho = -0.4235 \, dB$, y=3 ($W_n^{(1)}$)

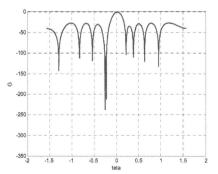

Figure 9. The reception diagram side lobes suppression in points: $\theta_1 = -0.2576 \theta_2 = -0.2293 \rho = -7.4690 \, dB$, y=3 ($W_n^{(1)}$)

Losses in antenna's directivity described by the next expression:

$$\rho = \frac{|G(0)|^2}{N_1 * \sum_{n=1}^{N} |W_n|^2} \qquad (7)$$

As we can see, the suppression in case of (5) is high enough, and the side lobes level between the suppressed points not high (about – 100 dB).

So, we can correct the regulated part of reception diagram by different weights functions. This approach doesn't require the implementation of numerical optimization procedures as were described in [1].

2 CONCLUSIONS

In this paper linear antenna array design capable of obtain the given side-lobe suppression with controlled value of directivity coefficient is suggested. The approach is simple enough for calculations and does not require the implementation of numerical optimization procedures. It's very useful for practical implementation, when it's necessary to get the given side lobes suppression.

REFERENCES

[1] V. Koshevyy, A. Shevchenko, 2016, The research of non-tunable part of antenna array amplitude distribution for side lobes suppression efficiency. 2016 International Conference Radio Electronics & Info Communications (UkrMiCo'2016), National Technical University of Ukraine "Kyiv Polytechnic Institute", Kyiv, Ukraine, pp. 156 – 160.

[2] V. Koshevyy, V. Lavrinenko, 1981, The target's selection on based on the discrete structure with a minimum quantity of controlled elements. «Izvestia VUZ. Radioelectronika», t. 24, №4, pp. 105 – 107.

[3] V. Koshevyy, A. Shershnova, 2013, The formation of zero levels of Radiation Pattern linear Antennas Array with minimum quantity of controlling elements, Proc. 9 Int. Conf. on Antenna Theory and Techniques (ICATT-13), Odessa, Ukraine, pp.264-265.

[4] V. Koshevyy , M. Sverdlik, 1974, About the possibility of full side lobes level suppression of ambiguity function in the given area. – « Radio Eng. Electron. Phys.», t. 19, № 9, pp. 1839 – 1846.

[5] V. Koshevyy, V. Lavrinenko , S. Chuprov, 1975, The efficiency of quasi-filter analysis. «RIPORT », VIMI. №2, – p. 7.

[6] Y. Shirman, V. Mandjos, 1981, Theory and technics of radar information processing under interferences. M. Radio I Svyaz, – 416c.

[7] V. Koshevyy, 1982, Moving target systems indication synthesis with the inverse matrix size restrictions. -«Izvestia VUZ. Radioelectronika», т.25, № 3, C. 84-86.

[8] V. Koshevoy, M. Sverdlik, 1973, About influence of memory and pass-band of generalized V-filter to efficiency of interference suppression. « Radio Eng. Electron. Phys.», t.18, №8, pp. 1618-1627.

Anti-Collision

A Framework of a Ship Domain-based Collision Alert System

R. Szlapczynski
Gdansk University of Technology, Gdańsk, Poland

J. Szlapczynska
Gdynia Maritime University, Gdynia, Poland

ABSTRACT: The paper presents a framework of a planned ship collision alert system. The envisaged system is compliant with IMO Resolution MSC.252 (83) "Adoption of the revised performance standards for Integrated Navigation Systems (INS)". It is based on a ship domain-oriented approach to collision risk and introduces a policy of alerts on dangerous situations. The alerts will include cautions, warnings and alarms, which will be triggered by predicted violation of a ship domain or predicted passage in the domain's proximity. The paper includes a description of the policy of alerts and examples of generating them for potentially threatening situations.

1 INTRODUCTION

Collision Alert Systems are decision-support tools, designed in maritime industry either for ship navigators or VTS center operators. They aim at alerting the users on possible vessel collisions via visual, sound or text signals. Unlike a typical collision avoidance system (also in some publications abbreviated as CAS, what might be misleading) a Collision Alert System (here and later referred to as CAS) is focused on collision risk assessment and decision making towards alerting on the potentially dangerous encounters rather than directly proposing collision avoidance maneuvers. The elementary CAS application utilizes Automatic Radar Plotting Aid (ARPA), a system that tracks via radar echo target vessels in proximity of the own ship. While it is possible to set up ARPA sound or visual alarms for indicating violations of Distance at the Closest Point of Approach (DCPA) and Time to the Closest Point of Approach (TCPA) threshold values, this approach is flawed. A combination of DCPA and TCPA parameters is not always adequate as collision risk indicator as it does not carry enough information (e.g. it ignores bearing of a target). Moreover, various ARPA sound or flashing alarms may be too frequent during regular navigation (Goerlandt et al. 2015) and in case of a real collision alert the operator may easily overlook the message.

Due to the abovementioned reasons, a number of more sophisticated maritime CAS solutions have been already proposed (Baldauf et al. 2011, Bukhari et al. 2013, Simsir et al. 2014, Goerlandt et al. 2015), developed in line with IMO Resolution MSC.252(83) "Adoption of the Revised Performance Standards for Integrated Navigation Systems (INS)" (IMO, 2007). Some of those works propose new indicators or methods of assessing collision risk, based on a set of well-known parameters. In this paper, instead of introducing such a new method, the concept of a ship domain is applied (Coldwell 1983). A default elliptic domain similar to the ones proposed in (Coldwell 1983, Hansen et al. 2013) is assumed, with the possibility of adjusting the domain according to navigator's preferences, conditions and ship's maneuvering abilities. Two domain-based parameters: degree of domain violation (DDV) and time to domain violation (TDV), proposed in (Szlapczynski & Szlapczynska 2016) are used as criteria for generating alerts.

The rest of the paper is organized as follows. In Section 2 examples of existing CAS solutions are briefly described. Following this, the proposed framework of a new CAS is presented in Section 3. Examples of alerts generated for given navigational situations are provided in Section 4 followed by a brief summary of the paper.

2 RELATTED WORKS

According to the IMO Resolution MSC.252(83) "Adoption of the Revised Performance Standards for

Integrated Navigation Systems (INS)" (IMO, 2007), the alert management should distinguish between the three priorities (alarms, warnings & cautions) as listed:

- alarms should indicate the need of immediate attention and action by the navigators;
- warnings should indicate a change in conditions and thus should be presented for precautionary reasons (to point out that dangerous situation may develop if no action is taken);
- cautions should indicate a condition which does not need an alarm or a warning, but requires attention and special consideration of the situation or of given information.

Some examples of existing CAS solutions, mostly in line with the above mentioned policy, are presented below in chronological order.

In (Baldauf et al. 2011) the authors propose a general CAS framework offering four output levels of a collision risk:

- risk of collision is developing,
- risk of collision exists,
- danger of collision is developing,
- danger of collision exists.

The research has been inspired by air traffic control and has been based on empirical field studies on-board ships. A general risk model for situation assessment has been based on:

- safe distance (DCPA and related parameters),
- COLREGS compliance,
- visibility type (good or restricted),
- basin type (open waters or restricted waters),
- wind and current conditions.

Similarly, in (Bukhari et al. 2013) an almost-CAS solution is proposed that calculates the degree of collision risk in range [-1.0, 1.0] based on radar data from VTS. The solution takes into account DCPA, TCPA and bearing-based variation of compass degree (VCD). It utilizes fuzzy inference rules to determine the degree of collision risk in terms a of linguistic values: positive small, positive medium small, positive big, negative small, negative medium small, negative big, positive medium big. The Authors have tested the proposed system in a series of experiments on a simulator fed with real radar data with very promising results.

Analogically, in (Simsir et al. 2014) a CAS-like framework has been presented. It is intended for reciprocally passing vessels in narrow straits and has been tested for the Istanbul Strait. It offers one output level of collision risk (either 'safe' or 'risk exists') reporting any existing risk of collision for this ship for her 3-min-onward positions. The system is based on Artificial Neural Networks (ANN):

- training of the ANN has been completed with the use of the Levenberg–Marquardt (LM) algorithm,
- ANN has been trained with data of four ships.

In (Goerlandt et al. 2015) a more advanced system has been proposed in the form of Risk-Informed CAS (RICAS). It is a compete CAS, with solid risk theoretical background, offering four output levels of a ship collision risk, in line with (IMO, 2007) recommendations:

- safe,
- caution,
- warning,
- alarm.

The theoretical framework of the research has been inspired by frameworks for road traffic encounters. An encounter is analyzed from four perspectives in relation to:

- COLREGS (eight zones around a ship are taken into account for overtaking, crossing and head-on encounters),
- imminence of accident occurrence (based on collision course, four types of imminence are distinguished: A, B, C and D),
- deviations from a reference level (violation of a safe distance and unexpected ship turns),
- ambiguous situations (e.g. a head-on encounter near crossing may result high risk of collision if classified differently by each of two ships).

Risk levels assigned to an encounter of two ships are based on data processed by a fuzzy expert system. The data consists of on a number of parameters, including:

- DCPA and TCPA,
- Bow Cross Range (BCR),
- Bow Cross Time (BCT),
- relative bearing,
- range,
- time of day (day/night),
- visibility (good or restricted).

For each parameter a trapezoidal membership function is defined by means of Transition Interval Estimation expert elicitation method. Membership functions are utilized by the fuzzy expert system. Eventually, for given parameter values, final risk levels are assigned by fuzzy rules.

3 A FRAMEWORK OF A PROPOSED CAS SYSTEM

As opposed to the systems presented in the previous section, the newly proposed framework does not introduce new methods of assessing ship collision risk. Instead, it uses an old, but continuously developed concept of a ship domain (Coldwell 1983, Pietrzykowski & Uriasz 2009). In general, the main advantage of a ship domain is that it is able to synthesize a number of parameters, including not only a combination of DCPA and TCPA but also target's bearing. Therefore ship safety domain models have long been proposed as safety criteria. Lately it has been argued that criteria of safety assessment are different for various water regions and consequently a number of region-specific domains have been presented (Hansen et al. 2013, Wang & Chin 2013). Some domain-related concepts have also been developed: in (Krata & Montewka 2015), ship's maneuvering characteristics is used to determine an area around the own ship, whose violation should be preceded by a collision avoidance maneuver.

3.1 A domain based approach

To apply ship domain directly in a system, some domain-based measures are needed. It is not enough to predict that a domain may be violated: equally important is to know when it will be violated and to what extent. This issue has been addressed in (Szlapczynski & Szlapczynska 2016), where two parameters have been presented: degree of domain violation (DDV) and time to domain violation (TDV).

TDV is the time remaining to entering other ship's domain assuming unchanged courses and speeds of both ships. It can be seen as a domain-based generalization of a time to violating the safe distance – a parameter that has been introduced and analyzed in (Lenart 2015). In practice TDV is almost always smaller than TCPA, because entering a ship's domain usually precedes reaching CPA by a few minutes (depending on the domain's size and relative speed of a target). As for DDV, it is given by:

$$DDV = \max\left(1 - f_{min}, 0\right) \qquad (1)$$

where f_{min} is the approach factor defined as follows. For an encounter of a target, the approach factor f_{min} is the scale factor by which the target's domain has to be multiplied so that the own ship passed on the boundary of the f_{min}-scaled target's domain (assuming unchanged courses and speeds of both ships). DDV and f_{min} are visualized in Figure 1.

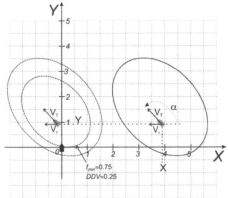

Figure 1. A predicted violation of a target's domain presented in the own ship's relative coordinate system (VT – true speed, Vr – relative speed), source: (Szlapczynski & Szlapczynska 2016)

Since f_{min} cannot be a negative number, DDV is always within a [0.0 ; 1.0] range, with 0.0 denoting safe distance throughout an encounter and values larger than 0.0 denoting domain violations. The time of reaching f_{min} (and consequently of reaching DDV) is denoted by t_{fmin}. It must be noted here that for most encounters TDV is significantly smaller than TCPA. Also, TDV is always smaller than t_{fmin} as domain violation always precedes the closest approach in the domain's sense.

A comparison of DCPA and TCPA parameters with DDV and TDV has been made in (Szlapczynski & Szlapczynska 2016). It has also been shown there that domains of various shapes, e.g. those according to (Pietrzykowski & Uriasz 2009, Wang & Chin 2015) can be approximated with a reasonable accuracy by an off-centered ellipse. Following that, detailed formulas for determining DDV and TDV analytically for any off-centered ship domain have been provided. An off-centered elliptic (Figure 2) is described by four length-dependent parameters:
- a – semi-major axis,
- b – semi-minor axis,
- Δa – a ship's displacement from the ellipse's center towards aft along the semi-major axis,
- Δb – a ship's displacement from the ellipse's center towards port along the semi-minor axis.

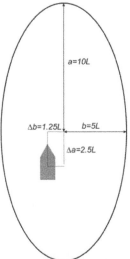

Figure 2. A decentralized elliptic ship domain (L – ship's length), source: (Szlapczynski & Szlapczynska 2016)

The first two variables make it possible to specify the domain's length and width, the other two – to vary the sizes of the fore, aft, starboard and port sectors. A domain with fore and starboard sectors larger than astern and port ones favors passing astern and manoeuvres to starboard, which is compliant with COLREGS (IMO 1972, Cockcroft & Lameijer 2011) by.

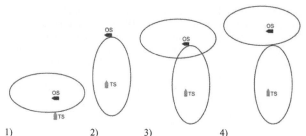

Figure 3. Different domain-based safety criteria: 1) OS domain is not violated, 2) TS domain is not violated, 3) neither OS nor TS domain is violated, 4) domains do not overlap. Source: (Szlapczynski & Szlapczynska, in review)

In practice, using a ship domain in an encounter situation may be combined with one of the following four safety criteria, as presented in Figure 3.
1 own ship's (OS) domain should not be violated by a target ship (TS),
2 a target ship's (TS) domain should not be violated by the own ship (OS),
3 neither of the ship domains should be violated (a conjunction of the first two conditions),
4 ship domains should not overlap - their areas should remain mutually exclusive (the effective spacing will be a sum of spacing resulting from each domain).

All of these approaches are represented by some researchers. However, it must be emphasized that the first two approaches may lead to largely different assessment of the same situation by two ships involved in an encounter. As for the fourth one, it is not complaint with classic domain definitions and may be misleading as it results in the actual spacing being a sum of spacing for each of the domains. Therefore, the third of the above listed approaches has been chosen here, namely the one according to which neither of the ships' domains should be violated (Figure 3c).

3.2 Alert levels and policy – criteria for generating different alerts

It has been decided that alert levels in line with (IMO, 2007) will be generated. According to this resolution, a collision warning should be sent for DCPA smaller than 2 nautical miles and TCPA shorter than 12 minutes (precise conditions for cautions and alarms are not given). In the current proposal the following conditions for all three categories of alerts have been chosen.
1 **Cautions**. It has been assumed that each ship passing in proximity of the own ship's domain in the nearest time should generate a caution. Such passing may theoretically cause a dangerous situation if the target suddenly changes motion parameters to the ones threatening the own ship. Thus, a caution is triggered even if there is no predicted domain violation (DDV=0), but if $f_{min} \leq 3$ and $t_{fmin} \leq 20$ min. or if DDV>0 and TDV≤ 20

min. The time limits given here are chosen so as to be significantly larger than a TCPA limit recommended in (IMO, 2007) for warning. It must be noted here, that depending on the ship domain's size and target's bearing, TDV of 20 minutes is an equivalent of TCPA and t_{fmin} being about 23-26 minutes As for the domain-oriented f_{min} parameter, the value of 3 means that a domain of a triple size (all dimensions multiplied by factor 3) would be violated.
2 **Warnings**. It has been assumed that a predicted domain violation or passing in a direct proximity of ship's domain, should result in a warning, if it is expected to happen within 15 minutes. Thus, a warning is triggered if: $f_{min} \leq 2$ and $t_{fmin} \leq 15$ min or DDV>0 and TDV<15 min. Depending on the ship domain's size and target's bearing, TDV of 15 minutes is an equivalent of TCPA and t_{fmin} being about 18-21 minutes.
3 **Alarms**. It has been assumed that an alarm is generated if an immediate action is necessary to avoid domain's violation. The conditions for that are: DDV>0 and TDV<10 min. Depending on the ship domain's size and target's bearing, TDV of 10 minutes is an equivalent of TCPA being about 13-16 minutes.

4 EXAMPLE SCENARIOS

Two scenarios (for crossing and head-on encounters) are presented in this section. For both of them an off-centred elliptic ship domain from Figure 1 has been assumed, with ships lengths ranging from 250 to 350 metres, which corresponds to minimal allowed spacing between ships being approximately:
– 0.6 nm, when passing port to another ship's port during head-on encounters,
– 0.85 nm, when passing starboard to another ship,
– 1.2 nm, when passing astern of another ship,
– 2.0 nm, when passing ahead of another ship.

4.1 Scenario 1: a crossing (perpendicular courses)

An example of crossing potentially leading to a collision is shown below. The own ship is in the middle of the coordinate system (the own speed vector marked with an arrow) and the stand-on target on starboard. In Figure 4 a caution is generated for a target on port (no predicted domain violation, DDV=0, f_{min}=1.97, t_{fmin}≈TCPA≈20 min.), which is approaching fast. At the same time a caution is generated for a target on starboard (predicted domain violation, DDV=0.39, f_{min}=0.61, TDV=20 min., also in Figure 5).

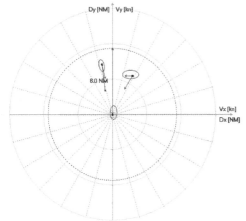

Figure 4. A crossing with a target on starboard. A caution is generated for a target on port, which is approaching fast (no predicted domain violation, DDV=0, f_{min}=1.97, t_{fmin}≈TCPA≈20 min.)

In Figure 5 a warning is generated for a target on port (t_{fmin}≈TCPA≈15 min.) because of f_{min} being smaller than 2.

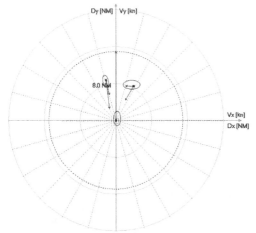

Figure 5. A crossing with a target on starboard. A warning is generated for a target on port (t_{fmin}≈TCPA≈15 min.) and a caution is generated for a target on starboard (predicted domain violation, DDV=0.39, f_{min}=0.61, TDV=20 min.)

Another warning is generated for target on starboard (TDV=15 min.) in Figure 6.

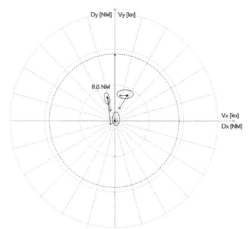

Figure 6. A crossing with a target on starboard. A warning is generated for the target on starboard (TDV=15 min.)

Finally, if the own ship has not performed a manoeuvre yet, an alarm for a target on starboard is generated in Figure 7 (TDV=10 min.). There is no alarm for a target on port because there is no predicted domain violation.

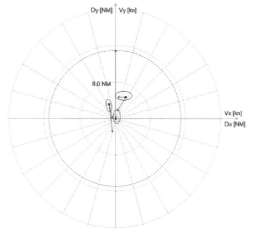

Figure 7. A crossing with a target on starboard. An alarm is generated for the target on starboard (TDV=10 min.)

4.2 Scenario 2: a head-on encounter

An example of a potentially dangerous head-on encounter is shown here. In Figure 8 a caution is generated for a target on port (no predicted domain violation, DDV=0, f_{min}=2.92, t_{fmin}≈TCPA≈20 min.), which is approaching fast.

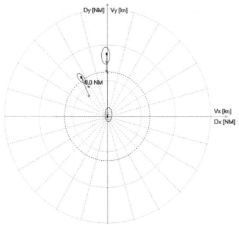

Figure 8. A head-on encounter with a target ahead. A caution is generated for a target on port, fast approaching (no predicted domain violation, DDV=0, f_{min}=2.92, t_{fmin}≈TCPA≈20 min.)

In Figure 9 a caution is generated for a target ahead (predicted domain violation, DDV=0.77, f_{min}=0.23, TDV=20 min.) There is no warning for a target on port, since f_{min} for this target is larger than 2.

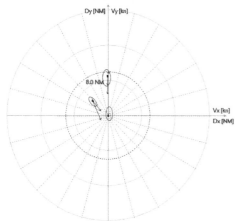

Figure 9. A head-on encounter with a target ahead. A caution is generated for the target ahead (predicted domain violation, DDV=0.77, f_{min}=0.23, TDV=20 min.)

Five minutes later a warning is generated for a target ahead (TDV=15 min.) shown in Figure 10.

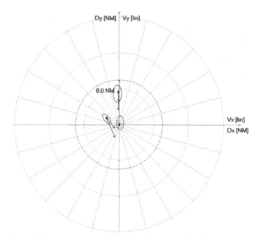

Figure 10. A head-on encounter with a target ahead. A warning is generated for the target ahead (TDV=15 min.)

A target on port is safely passing the own ship n Figure 11. However, an alarm is generated for a target ahead (TDV=10 min.) as it is high time to initiate a collision avoidance manoeuvre to starboard.

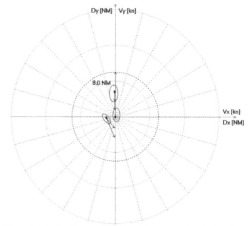

Figure 11. A head-on encounter with a target ahead. An alarm is generated for the target ahead (TDV=10 min.)

5 SUMMARY AND CONCLUSIONS

In the paper a new domain-based approach to Collision Alert Systems has been introduced. The main difference between this framework and existing solutions is that it does not involve additional collision risk analysis. Contemporary ship domain models already take into account major factors affecting collision risk so they can be directly incorporated into the framework instead of proposing new measures and risk estimators. A length-dependent off-centered ellipse has been used as a domain shape in the presented research and some basic rules for generating alerts have been proposed for this domain. In the accompanying example scenarios it has been shown how this approach can be applied in practice. It must be mentioned though, that the paper documents a work in progress. This includes the presented rules, which are a subject to ongoing research and may be modified in the final solution so as to address the IMO resolution (IMO, 2007) more accurately.

REFERENCES

Baldauf, M., Benedict, K., Fischer, S., Motz, F. & Schroeder-Hinrichs, J.-U. 2011. Collision avoidance systems in air and maritime traffic. *Proc. Inst. Mech. Eng. Part O J. Risk Reliab.* 225(3): 333-343.

Bukhari, A. C., Tusseyeva, I., Lee, B.-G. & Kim, Y.-G. 2013. An intelligent real-time multi-vessel collision risk assessment system from VTS view point based on fuzzy inference system. *Expert Syst. Appl.* 40: 1220-1230.

Cockcroft, A.N. & Lameijer, J.N.F. 2011. A Guide to Collision Avoidance Rules. Butterworth-Heinemann.

Coldwell, T.G. 1983. Marine Traffic Behaviour in Restricted Waters. *The Journal of Navigation* 36: 431-444.

Goerlandt, F., Montewka, J., Kuzmin, V. & Kujala, P. 2015. A risk-informed ship collision alert system: Framework and application. *Safety Science* 77: 182-204.

Hansen, M.G., Jensen, T.K., Lehn-Schiøler, T., Melchild, K., Rasmussen, F.M. & Ennemark, F. 2013. Empirical Ship Domain based on AIS Data. *Journal of Navigation* 66(6): 931-940.

IMO (1972). [with amendments adopted from December 2009]. Convention on the International Regulations for Preventing Collisions at Sea. International Maritime Organization.

IMO (2007). Adoption of the revised performance standards for integrated navigation systems (INS). Resolution MSC 83/23/Add.3-ANNEX 30.

Krata, P., Montewka, J. 2015. Assessment of a critical area for a give-way ship in a collision encounter. *Archives of Transport* 34(2): 51–60.

Lenart, A. 2015. Analysis of Collision Threat Parameters and Criteria. *The Journal of Navigation* 68: 887-896.

Pietrzykowski, Z. & Uriasz, J. 2009. The Ship Domain – A Criterion of Navigational Safety Assessment in an Open Sea Area. *The Journal of Navigation* 62: 93-108.

Simsir, U., Amasyali, M. F., Bal, M., Äelebi, U. B. & Ertugrul, S. 2014. Decision support system for collision avoidance of vessels. *Applied Soft Computation* 25: 369-378.

Szlapczynski, R. & Szlapczynska, J. 2016. An analysis of domain-based ship collision risk parameters. *Ocean Engineering* 126(1): 47–56.

Szlapczynski, R. & Szlapczynska, J. (In review). Review of ship safety domains: models and applications to collision avoidance and traffic engineering.

Wang, Y. & Chin H. C. 2016. An Empirically-Calibrated Ship Domain as a Safety Criterion for Navigation in Confined Waters. *The Journal of Navigation* 69: 257-276.

Proceedings of 12th International Conference on Marine Navigation and Safety of Sea Transportation, TransNav 2017
21-23 June 2017, Gdynia, Poland

Is COLREG enough? Interaction Between Manned and Unmanned Ships

T. Porathe
Norwegian University of Science and Technology, Trondheim, Norway

ABSTRACT: In recent years, there has been a growing interest for autonomous unmanned shipping. Both from research and industry. Between 2013-2015 the EU-project MUNIN investigated the feasibility of transocean unmanned shipping. While trans-ocean drones could be expected to seldom encounter other ships, short-sea shipping would mean intense interaction with other manned SOLAS and non-SOLAS vessels. This rises some serious questions. How can we expect watch keepers on other SOLAS vessels, fishing boats and unexperienced leisure boat skippers to react when they meet an unmanned vessel? Will they behave in the same way as with manned ships today? Is there a need for the unmanned vessel to communicate intentions in different ways than today? Is there a need for humans to know that they are detected by the autonomous vessel, or is enough that they know the drone will adhere to COLREG? Is there a need to designate separate fairways for unmanned vessels? This is a discussion paper pointing to pending research needs relating to interaction between manned and autonomous vessels.

1 INTRODUCTION

1.1 *The MUNIN project*

In recent years, there has been a growing interest in unmanned, autonomous shipping. A number of projects and conferences has attracted interests from stakeholder all over the world.

One such project was the EU 7th Framework project MUNIN (Maritime Unmanned Navigation through Intelligence in Networks, 2013-15).

The objective of the MUNIN project was to show the feasibility of unmanned, autonomous merchant shipping. The ships would be under control of on-board crew approaching and leaving a harbor, being autonomous and unmanned only from pilot drop-off point to pilot pick-up point. However, there might be maintenance teams on-board if necessary. The goal was also that the ship would be under autonomous control during the main part of the ocean voyage, remotely monitored from the Shore Control Center. Only in exceptional cases, the shore control center was expected to actually maneuver the ship remotely.

Limited tests were successfully conducted in a simulator environment before the project ended in 2015.

1.2 *Collision avoidance*

The proposal from the MUNIN project was that a ship during the major part of the unmanned deep sea voyage should proceed in an "autonomous execution" mode. This meant that the ship's autopilot should follow the pre-programmed voyage plan in track-following mode. This is just as ordinary ships do today: a voyage plan is programmed into the navigation system, which the autopilot follows. Sometimes the operator has to acknowledge a change of course at a waypoint, but the autopilot can also do this automatically.

The MUNIN project also designed an Autonomous Navigation System, which based on information from the onboard sensor system, the nautical chart, and the uploaded voyage plan automatically could detect obstacles and conduct evasive maneuvers if no intervention from the Shore Control Center was made. If the obstacles was identified as other ships, collision avoidance should be made according to the rules of the road, the International Regulations for Preventing Collisions at Sea (COLREGS). The ship could also be remotely maneuvered from the shore control center using radar and cameras. As a last resort, if e.g. radio communication were lost, the ship would go into a "fail to safe" mode, drifting or hovering on its station (assuming dynamic positioning capabilities).

An autonomous ship on the high seas, conducting a weather-routed voyage somewhat on the side of the shipping corridors would not meet many other vessels. The technical challenge as far as interaction with other ships would therefore be limited. Instead the problem would be testing on international waters in absence of legislation from the IMO.

However, both Norway, Finland (LVM, 2016) and the United Kingdom (DfT, 2016) has declared that they want to be the first nations conducting autonomous shipping. If approved by national authorities testing can be done on national waters, and e.g. in 2016 the Norwegian Maritime Authority and the Norwegian Coastal Administration signed an agreement, which allows for testing of autonomous ships in the Trondheim fjord in the middle of Norway (NMA, 2016). Tests with smaller crafts has already been conducted here and the first tests with larger ships are already being planned, see Figure 1.

Figure 1. Kongsberg will together with British Automated Ships Ltd build the first full scale autonomous ship to be tested in the Trondheims fjord in 2018 (Kongsberg, 2016).

Testing autonomous navigation in national waters means coastal and inshore navigation. This is an altogether different challenge that ocean navigation.

2 UNMANNED INSHORE NAVIGATION

2.1 *Challenges for coastal and inshore navigation*

Navigation in confined waters means increased difficulties that needs to be tackled by the autonomous systems onboard, even if the vessel is remotely monitored by a shore control center. I will in the following assume that autonomous systems will conduct all but exceptional emergency handling. I will discuss three types of problems: (a) navigation in narrows and in proximity of land and shallow water, (b) interaction with SOLAS vessels, and (c) interaction with non-SOLAS vessels, as leisure crafts, small fishing boats and carjacks.

2.2 *Navigation in confined waters*

Navigation in close proximity to land and shallow water will require good precision by the onboard

positioning systems, good nautical charts, and good maneuvering capabilities by the autonomous vessel.

Precise positioning will be fundamental and will be based on Global Navigation Satellite Systems (GNSS). Apart from the American GPS, the Russian GLONASS, the European Galilei and the Chinese BeiDou is under (re-)construction so in the future redundant satellite coverage of high precision can be expected. The passenger ferries on the Hurtigruten, which traffic the narrow inshore fairways of the Norway from Bergen to Kirkenes every day, the whole year around, already today conduct most of the voyage with autopilots in track-following mode based on GNSS data. However, GNSS is vulnerable to intentional interference due to weak signal strength. Thus, the GNSS signal must be crosschecked by independent systems such as radar and/or LIDAR systems comparing satellite position derived from radar maps and 3D terrain models.

One might also speculate on the needs to install new aids to navigation in narrow and tricky fairways. For instance, automatic positioning based bearing and distance to the new type of e-RACONS, which transmits a unique identification code. Maybe also electronic leading lines, which giver very high precise cross track position like an airport ILS.

Equally important will be good nautical charts based on high-resolution bathymetrical surveys. The resolutions must be good enough to allow back-up systems like radar, LIDAR and echo sounder to crosscheck and verify chart data from the satellite position with independent measures from the onboard instruments. For radar and LIDAR the nautical chart needs to have terrain elevation features also for the land areas. Many nations are already collecting this kind of data (e.g. Kartverket, 2015).

Together with good positioning and good maps, the maneuvering properties of the autonomous vessel needs to be good. Winds, waves and currents will pose a great challenge for autonomous navigation in areas like the Norwegian west coast. It will be difficult to replicate the ship handling skills of experienced mariners, instead an autonomous vessel will need dynamic positioning capabilities allowing it to hover on a set position and translate in any direction.

2.3 *Interaction with SOLAS vessels*

If close proximity to land and shallows constitute one problem of inshore navigation, high traffic density and interactions with other ships, constitute the other. In this text, I have assumed that the autonomous vessel is a SOLAS vessel carrying stipulated equipment like an AIS transponder, transmitting position, course and speed to other SOLAS vessels in the vicinity. Relying on radar, AIS and that all ships obey COLREG, one could assume that the problem of collision avoidance would be solved. However, COLREG does not unambiguously define which ship should give way and which should stand on. For instance, there are no precise definition of the terms "restricted visibility", "safe distance" or "safe speed". Ships should give way for other ships on their starboard side (rule 15), but also ships should cross a traffic separation scheme "shall cross on a heading as nearly as practicable at right angles to the general direction of traffic flow" (rule 10). Conflicting opinions on which of these two rules are most important has caused incidents in the past (Lee & Parker, 2007).

Ship Traffic Management is a new paradigm of route exchange that has been, and is, investigated by the MONALISA (SMA, 2014), the STM (STM, 2017) and SESAME Straight (Kongsberg, 2014) projects. Briefly, the meaning is that ships send their voyage plans to a coordination center that coordinates all plans to make sure no two ship are at the same place at the same time. The coordination center will also coordinate the arrival time to an available port slot by recommending speed changes. Because ships might not be able to precisely follow preset plans due to influence of currents and weather, the routes need to be updated at regular intervals and the automatic coordination mechanism needs to constantly update recommendations for course and speed to keep the separation between ships. For the shipping industry, the benefit is more efficient voyages arriving just in time to an allocated port slot with a minimum of fuel consumption. For the interaction between manned and autonomous vessels, this means that collision avoidance can be done in advance by the coordination center. (Although sensors and collision avoidance algorithms will still be needed, e.g. in cases where ships maneuverability breaks down.)

However, there remains an even bigger problem: the interaction between autonomous ships and non-SOLAS vessels.

2.4 Interaction with non-SOLAS vessels

The International Convention for the Safety of Life at Sea (SOLAS) is a maritime treaty, which requires flag states to ensure that ships flagged by them comply with minimum safety standards in construction, equipment and operation. Most bigger ships are SOLAS vessels while smaller crafts like leisure and small fishing boats are not. That also means that they mostly miss stipulated equipment like AIS transponders and receivers, ECDIS and radars displaying AIS targets, and for the future, abilities to do route exchange, participate in a Ship Traffic Management regime that coordinates traffic, and ensure separation.

One of the biggest challenges as far as the interaction between manned and autonomous ships will involve non-SOLAS crafts. Because they normally do not have an AIS transponder, they do not automatically exist in the internal representation of the world in the Automatic Navigation System of the autonomous ship. Instead, it has to be detected by its sensors (radar, infrared or daylight cameras or LIDAR). Tests done during the MUNIN project with fused radar and infrared data showed that objects with a size down to a bath-ball could be detected in calm weather with no sea state. In reality, small boats and kayaks will be difficult to detect among the waves of an open inlet – just as they might be for the naked eye as. Technology will no doubt improve within this field, but the problem remains: the sensor systems of the autonomous ship sees what it sees, and how can the fisherman or leisure boat skipper know that they are detected by the autonomous vessel? If a fisher, laying still, pulling up lobsterpots observes an autonomous ship coming in his direction, how can he know wheatear or not it will turn, or run maybe him down? Or a slender kayak crossing the fairway?

Probably there will be a need for new technical solutions here. With mobile coverage in the coastal areas, a smartphone could act as an AIS transponder/receiver (see Figure 2).

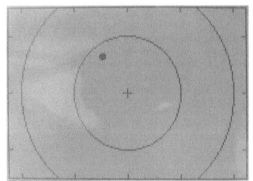

Figure 2. The interface of an AIS receiver. The own boat is in the center and the circles depicts a range of 2 and 4 nautical miles. In this case, an AIS target is approaching from the northwest. Source: ATSB (2013).

The application should also be able to verify that the phone (and the boat it is in) exists on the map of the autonomous vessel. In addition, it might be beneficial if the autonomous vessel was able to communicate its intentions to the small boat as to whether it will change course or if the small boat will have to move away, and when, and in what direction. The lobster fisher might get an alarm on his phone that an autonomous vessel will pass over his position in 5 minutes. Such an app would also be beneficial also in today's traffic environment. In the foggy waters of South Korea several small fishing boats is every year run down by commercial ships while fishing in the approach fairways to major ports (e.g. Maritime Herald, 2016).

In a transition period designated and marked fairways for autonomous vessels will probably be a way to communicate to other mariners where they might expect to meet drone ships, and where not. As well as evident and conspicuous making on the ships as well.

3 CONCLUSION

As the technical development of systems for autonomous ships ramps up, it will be necessary to find answers to the questions from the public on issues that may seem scary or dangerous. Therefore, research is needed in issues relating to maritime human factors and the man-machine interaction realized when we now will encounter and must interact with drone ships. In Scandinavian waters thousands of leisure crafts, every summer invades the fairways and port approaches. They are problems for manned commercial shipping today, what about tomorrow.

GLOSSARY

AIS, Automatic Identification System
COLREG, The International Regulations for Preventing Collisions at Sea
ILS, Instrument Landing System
LIDAR, Light detection and ranging
RACON, Radar beacon
SOLAS, The International Convention for the Safety of Life at Sea

REFERENCES

ATSB, Australian Transport Safety Bureau. 2013. Collision between the bulk carrier Furness Melbourne and the yacht Riga II north of Bowen Queensland 26 May 2012. ATSB report 295-MO-2012-006.

DfT, Department for Transport. 2016. https://www.gov.uk/government/speeches/maritime-critical-to-uks-future-success [Acc. 2017-03-02]

Kartverket. 2015. http://kartverket.no/Om-Kartverket/Nyheter/klarsignal-for-3d-modell-av-norge/ [Acc. 2017-03-02]

Kongsberg. 2014. https://kongsberg.com/en/kds/knc/news/2014/august/knc-to-revolutionise-ship-traffic-management-with-sesame-straits-e-navigation-project/ [Acc. 2017-03-02]

Kongsberg. 2016. https://www.tu.no/artikler/norsk-selskap-bak-verdens-forste-autonome-skip-til-kommersiell-drift/363811 [Acc. 2016-02-08]

Lee, G.W.U. & Parker, J. 2007. Managing Collision Avoidance at Sea. London: The Nautical Institute. p. 63

LVM (Minestry of Transport and Communication. 2016. https://www.lvm.fi/en/-/finland-to-take-the-lead-in-automation-experiments-in-the-maritime-sect-1 [Acc. 2017-03-02]

Maritime Herald. 2016. http://www.maritimeherald.com/2016/fishing-boat-sank-after-collision-with-cargo-ship-keum-yang-post-in-tsushima-strait/ [Acc. 2017-03-02]

MUNIN. 2013-2015. http://www.unmanned-ship.org/munin/ [Acc. 2017-02-08]

NMA, Norwegian Maritime Authority. 2016. https://www.sjofartsdir.no/en/news/news-from-the-nma/worlds-first-test-area-for-autonomous-ships-opened/ [Acc. 2017-02-08]"

SMA, Swedish Maritime Administration. 2014. http://www.sjofartsverket.se/en/MonaLisa/ [Acc. 2017-03-02]

STM. 2017. http://stmvalidation.eu/ [Acc. 2017-03-02]

Model Research of Navigational Support System Cooperation in Collision Scenario

P. Wołejsza & E. Kulbiej
Maritime University of Szczecin, Szczecin, Poland

ABSTRACT: The purpose of following paper is to analyse how a navigational support system would behave in a collision situation upon encountering another vessel also steered by a NSS. This case study of two AI-driven vessels is executed under laboratory circumstances within the use of Navdec System, a navigational decision support tool established as a direct aid to the vessel's bridge Officer team. Decisions undertaken by the Navdec are compared to the experience-based choices of the Master in the exact case. Latter analysis of experiment's outcome is given as a premise for the real introduction of a NSS into practical use.

1 INTRODUCTION

Human error is the main factor of collisions of ships. Depends on sources it is estimated from 75 to 96% (Antão and Guedes Soares, 2008; Celik and Cebi, 2009; Harrald et al., 1998). According HELCOM report (HELCOM, 2014), only in 2013 and only on the Baltic Sea, there were 150 accidents (figure 1). Out of this, 38% were contacts and collisions (figure 2). It means 57 collisions in one year. These figures steadily increasing after crisis in 2008-09. Actually groundings are also collisions (with fixed object), so the numbers increase rapidly to over a hundred per year. The Swedish Club (The Swedish Club, 2012, 2011) estimated that average cost of collision is more than 1,000,000 USD (figure 3). Currently there are 87.233 merchant vessels worldwide (Equasis, 2015). Around 2.5% of vessels are in collision every year i.e. over 2100. To reduce number of collision and thereby cost of collisions, decision support systems are developed. For the time being, they deliver to the navigator ready solutions, but in the future such solutions will be executed by autopilot and engine telegraph autonomously to lead finally to unmanned vessels.

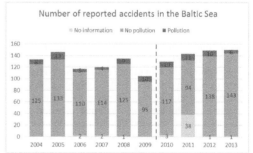

Figure 1. Number of reporting accidents in the Baltic Sea (HELCOM, 2014)

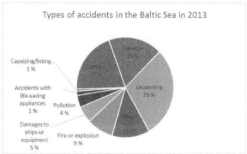

Figure 2. Type of accidents in the Baltic Sea in 2013 (HELCOM, 2014)

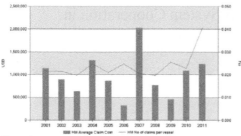

Figure 3. Average claim cost & frequency 2001 — 2011, (The Swedish Club, 2011)

2 MULTI-AGENT SYSTEM OF DECISION SUPPORTING IN COLLISION SITUATIONS

NAVDEC is by all means the very first tool that serves as a profound navigational decision support system (NDSS), namely navigational diagnosis, information and decision aid functions. Due to its novel functionalities, greatly extending the performance of devices generally carried by ships, the system possess currently a status of patent applications distinguished globally. NAVDEC has been created by research team of prof. Pietrzykowski from Maritime University Szczecin.

NAVDEC serves the navigator in the real time situations and under any circumstances in order to provide direct and accurate aid to decision handling problems regarding safe navigation of a ship. It comprises minor navigational equipment (commonly these are log, gyro, APRA, AIS, GNSS, ENC) and gathers the data into own database so that it is later successfully identifies a navigational situation in the area close to the ship and declares solution for own ship's movement that would ensure its safety and provide economical advantage via thoughtful expertise and optimization. In a way similar to those widely known from ECDIS (Electronic Chart Display and Information System) NAVDEC presents bathymetric data, tracking radar information (radar targets and their plotting), positioning information from GNSS (mainly GPS) and AIS. Ultimately, it calculates movement parameters of targets in vicinity, who are later presented to the user. Schema of data operation pyramid in Navdec is shown in fig.4.

Figure 4. Data operation pyramid in Navdec.

Since enormous amount of data is derived from numerous navigational devices it is NAVDEC's additional function to provide an effective mean of their filtration. Surplus data may render the decision aid untrue since the redundancy is not always a desirable trait. It is a case obviously when information gathered from distinctively different pieces of navigational equipment exclude each other. That is one of the premises why NAVDEC differs from other trivial decision support tools. It is namely the fact, that the aforementioned NDSS successfully and profoundly employs methods of artificial intelligence in a way of neural networks. Its graph model of numeral data filtration is schematically shown in figure 5.

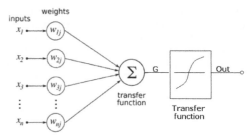

Figure 5. Intelligent data filtration algorithm (own study).

Consequently, such approach eliminate a risk of choosing a wrong data source but provides optimization and balanced decision support for already sorted and weighted intel.

3 MODEL AND ASSUMPTIONS FOR SIMULATION

All the information that was used for simulating an encounter situation of the ships is derived from a Denish Maritime Administration report available on the website www.dma.dk (Danish Maritime Administration, 2003).

From the positions of both ships and their movement parameters at 0900, the system qualified this encounter as crossing courses and pointed out Gotland Carolina as the give-way vessel, which is in compliance with Rule 15 of COLREGs. In this connection, a recommended manoeuvre was generated for the ship Gotland Carolina, with an assumption that the stand-on ship is actually maintaining its course and speed (Rule 17 of COLREGs).

The following parameters were adopted for simulations:
- Closest Point of Approach CPA = 1852 m (1 nautical mile),
- good visibility,
- length overall of the Gotland Carolina – 183 m,
- length overall of Conti Harmony – 210 m,

- position of AIS/GPS antenna on-board Gotland Carolina – 153 m from the bow,
- position of AIS/GPS antenna on-board Conti Harmony – 180 m from the bow.

Based on the data included in the report (DMA, 2003), a simulation was made to determine parameters of the encounter and to generate possible anti-collision manoeuvres at certain moments of time. Figure 6 presents a reconstructed situation at 0900 hrs. The range of courses that assure safe passing at the pre-set CPA or larger is marked yellow on the circle. The recommended manoeuvre is indicated as 'NEW COURSE' and enables the ships to pass each other at the assumed CPA. The speed range satisfying the assumed criteria is marked green, and proceeding at 'NEW SPEED' will result in the ships' distance during passing being equal the assumed CPA.

Following actions and manoeuvres which enable to pass on presumed CPA were presented in (Wołejsza P., Magaj J. 2010). On the next screen

shoots, authors presented results of simulation experiment which was conducted on NAVDEC system which was integrated with the Navi Trainer navigational manoeuvring simulator by Transas Marine, designed for training purposes resulting from STCW 78/95 Convention purposes. The Navi Trainer Professional software works in an integrated network environment based on Windows NT operational system. Radar work simulation appliances, ARPA, ECDIS system, the gyrocompass, the log, GPS receiver and other systems and navigational appliances fulfil all the functioning standards that find their application accepted by IMO (International Maritime Organization) and international conventions. The verification was carried out in the open sea, as the main task was to check the correctness of calculation made by the system of decision support without the restrictions of water area geometry and hydrometeorological conditions. The main goal was to simulate Last Moment Manoeuvre.

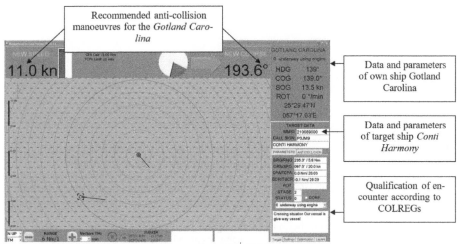

Figure 6. Classification of encounter situation according to COLREGs (own study).

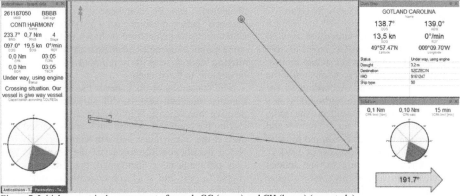

Figure 7. Initial geometrical arrangement of vessels GC (upper) and CH (lower) (own study).

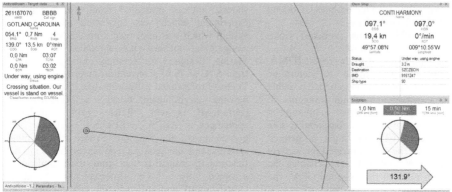

Fig. 8. Initial geometrical arrangement of vessels GC (upper) and CH (lower) (own study).

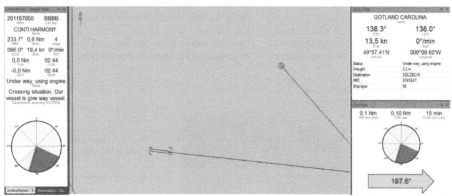

Figure 9. Moment when GC should commence course alteration to starboard if the manoeuvre is to be successful (own study).

4 SIMULATION NARRATIVE

For the sake of analysis, the northern vessel with course 139° and velocity 13,5 kn. is to be referred to as "GC" (Gotland Carolina), while the southern with course 97° and velocity 19,5 kn. as "CH" (Conti Harmony). The simulation of collision between ships is commenced in a point of time when distance between vessels equals 0,7 Nm and the angle of courses' transection is 42°. The initial geometrical configuration of both ships is shown in figures 7 and 8.

It is clearly visible from the screenshot as well after brief COLREG analysis that GC ought to give way, while CH should remain on the actual course. Nonetheless both vessels are proceeding until the danger of collision is imminent. According to the Rules, GC should alter its course to starboard, as it is shown in figure 7, while CH should do the same (figure 8). However, with every successive second of not taking any action, the course required for safe avoidance increases, namely instead of 191,7° course of 197,6° is required (figure 9). The actions triggers highly close approach of the vessels (figure 10), which is somewhat of dangerous, since the

success rate depends on the precise initial distance and due to possible data inaccuracies .

During the course of experiment it was found out that distance of 0,6 Nm between vessels is required in order to render the manoeuvre successful. But as it was later revealed, an approach, compatible with the good seamanship but on the other hand inconsistent with COLREG regulations, could be undertaken, namely GC's course alteration to port. The most significant difference is the fact, that the manoeuvre can be commenced even when the distance between ships is 0,32 Nm. Mid-time situation of the manoeuvre is shown in figure 11. At some point it becomes mathematically visible that altering course to port is required if the positive outcome is to be reached.

The consensus of the paper is the research on the minimal distance that is still sufficient to obtain positive outcome if both vessels are cooperating. During the course of experiment it was obtained that even if the linear range between ships was only 0,2 Nm it is still possible to "save" the situation. The only provision is, as mentioned, that GC alters her course to port and C H consequently to starboard. The initial sequence of the manoeuvre is shown in figures 12 and 13.

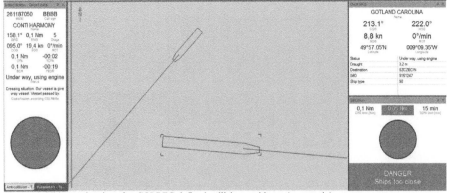

Figure 10. Ultimate situation after COLREG defined collision avoidance (own study).

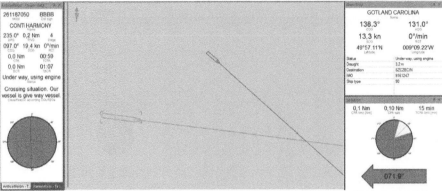

Figuer 11. Course alteration to port suggested by Navdec, which is the only way to avoid collision (own study).

It is to be noted that in this case automatic steering (by the means of autopilot) and course setting was rejected in favour of pure rudder management. In aforementioned situation both vessels executed consequently hard-a-port and hard-a-starboard. After a minute amount of time, a moment before the headings (approximately courses over ground) became parallel, both vessels executed rudder swing fully in the opposite direction. Thanks to first activity, bow close-quarter situation was prevented and due to the other – bow close-quarter. When the dynamic rotation of the vessels was relatively stable, the rudders were altered to midships. The result of the manoeuvre of vessels cooperation is shown in figures 14 and 15.

5 ANALYSIS OF THE SIMULATION'S RESULTS

Results of aforementioned simulation can be stated in form of conclusions that are as following:
– Undertaking a manoeuvre that is non-complying with COLREGS may prove effective enough to support the exception. During simulation a course

alteration to port triggered the positive outcome when the distance between vessels was 0,32 Nm, whereas alteration to starboard, which is as intended in the Rules of the Road, required as much as 0,6 Nm in order to omit collision.
– Undertaking the manoeuvre with two vessels proves to be over 50% more efficient in regards to space in the sea required to fulfil the action successfully.
– The course alterations required to be taken simultaneously and time-efficiently in order to obtain desired effect. In other words, employing system of artificial intelligence capable of determining and executing correct last moment manoeuvre could ensure safety of both vessels engaged in close-quarter situation. It is however on the very basis that each of the ships possesses such a system on- board. As far as human-human cooperation is concerned, it would require perfect synchronisation of steering decisions. Since either of the navigators may be not sufficiently informed of the other's ship manoeuvring capabilities, such cooperation is extremely difficult to achieve.

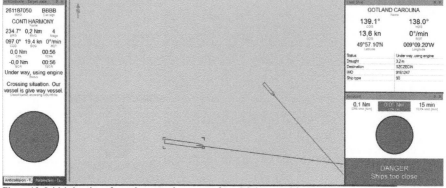

Figure 12. Initial situation of vessel cooperation as seen from GC (own study).

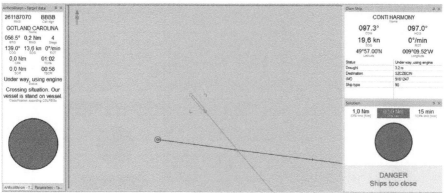

Figure 13. Initial situation of vessel cooperation as seen from CH (own study).

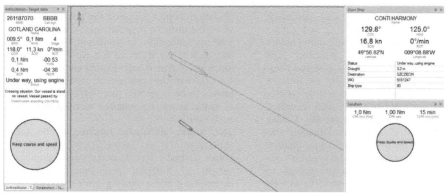

Figure 14. Final positions of ships after successful cooperation as seen from GC (own study).

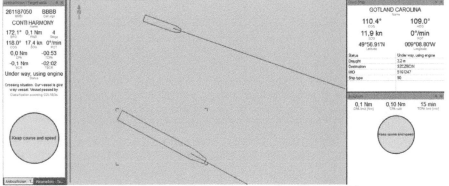

Figure 15. Final positions of ships after successful cooperation as seen from CH (own study).

6 FEASIBILITY OF AUTONOMOUS VESSELS

NAVDEC is a commercial system proven for more than a dozen ships. Navigators confirmed that the system properly qualifies the situation under COLREGs and worked out solutions useful in navigation and in avoiding collisions. NAVDEC can also be used to analyze collisions of ships, as was the case in this article. Indications that avoid collisions by altering course to starboard, and when this maneuver was no longer feasible, by altering course to the port, were correct and feasible, which was confirmed during the tests on the ECDIS simulator. The critical moment is when the navigator does not respond in time for the proposed solutions. So it was in this situation. Then the only solution is a coordinated maneuver of both ships, which has been successfully carried out at a distance of 0.2 Nm. However, these solutions were not suggested by the system, since the last moment maneuver (LMM) algorithm is currently not implemented in NAVDEC.

LMM is not fully examined and poorly described in the literature. At first glance it would seem that in the era of autonomous vessels, LMM will not be necessary. However, given that even the best systems can fail, we should be ready for emergency scenario, which in the case of collision avoidance maneuver is LMM. The article is part of the work leading to commercial implementation of LMM. Despite these imperfections, NAVDEC can now be used for autonomous vessels for collision avoidance. Work carried out in the IMO on Performance Standards for navigational decision support systems,

bringing much closer the moment when such vessels to be used in merchant shipping. The authors have no doubt that NAVDEC with the LMM algorithm will be a key element of autonomous navigation.

REFERENCES

Antão, P.,Guedes Soares, C. 2008. Causal factors in accidents of high-speed craft and conventional ocean-going vessels. *Reliab. Eng. Syst. Saf.* 93(9): 1292–1304.

Celik, M.,Cebi,S. 2009. Analytical HFACS for investigating human errors in shipping accidents. *Accid. Anal. Prev.* 41(1): 66–75.

IMO. 1972. Convention on the international regulations for preventing collisions at sea.

Equasis, The world merchant fleet in 2015.

Harrald, J.R., Mazzuchi,T.A., Spahn,J., et al. 1998. Using system simulation to model the impact of human error in a maritime system. *Saf. Sci.* 30(1): 235–247.

HELCOM. 2014. Annual report on shipping accidents in the Baltic Sea in 2013.

Kulbiej E. 2016. Relevance of the relativistic effects in satellite navigation. *Scientific Journal of Maritime University of Szczecin* 47: 85-90.

Kulbiej E. & Wołejsza P. 2016. An analysis of possibilities how the collision between m/v 'Baltic Ace' and m/v 'Corvus J' could have been avoided. *Annual of Navigation* 23: 121-135.

The Swedish Club. Basic facts 2012.

The Swedish Club. Collisions and groundings 2011.

Wołejsza P., Magaj J. 2010. Analysis of possible avoidance of the collision between m/v Gotland Carolina and m/v Conti Harmony, *Annual of Navigation* No 16: 165-172.

www.dma.dk, Danish Maritime Administration, CASUALTY REPORT, Collision between Chinese bulk carrier FU SHAN HAI and Cypriot container vessel GDYNIA, 2003.

An Analysis of Vessel Traffic Flow Before and After the Grounding of the MV Rena, 2011

A. Rawson

Marine and Risk Consultants Ltd. (Marico Marine), Southampton, United Kingdom

ABSTRACT: On the 5th October 2011, the 3,351 TEU containership MV Rena ran aground on the Astrolabe Reef whilst sailing to Tauranga, New Zealand. The break-up of the vessel, and subsequent oil spill, became one of New Zealand's most significant maritime accidents in recent years. Whilst the investigation detailed the factors which contributed to the grounding, no analysis was conducted into the disposition of navigation in the vicinity of the reef. This paper analyses historical AIS data both before and after the grounding and considers two issues; firstly, the extent to which the MV Rena's track deviated from established shipping routes and, therefore, whether a credible risk of grounding was evident. Secondly, the effectiveness of risk controls implemented in response by the local authorities to prevent the incident from reoccurring. The analysis in this case study will improve the understanding of the relationship between vessel traffic flow and grounding risk to aid in the prevention of similar incidents worldwide. Furthermore, the results provide important lessons to researchers in the uncertainties of building risk models from the relationship between vessel traffic analysis and historical incident data.

1 INTRODUCTION

The grounding of the MV Rena is the highest profile maritime accident in New Zealand in recent years. For an island nation dependent on seaborne trade, significant media attention was directed at the event and subsequent salvage, with many commentators describing the event as "inevitable" and that without action it would happen again.

The grounding itself provides a useful case study into how vessel traffic can be managed near to an offshore hazard. In reaction to the grounding, several risk control measures were put in place to actively manage traffic flow in the area. By comparing the traffic flow before and after the incident, this paper aims to draw conclusions on how the grounding and subsequent risk controls have changed the behavior of vessels and therefore how successful the risk controls have been preventing a reoccurrence. The results of this assessment could be used to better inform and justify mitigation strategies taken to protect vessel traffic from grounding on other hazardous reefs. Furthermore, the results highlight the uncertainty and limitations inherent in risk models that use vessel traffic analysis and historical incident data as their basis.

1.1 *Grounding of the MV Rena*

On the 5ᵗʰ October 2011 at 0214 local time the 37,209 GT container vessel *MV Rena* ran aground at 17 knots on the Astrolabe Reef in the Bay of Plenty, New Zealand (Figure 1).

The vessel had been delayed and left Napier behind schedule and was under threat of missing the tidal access window to enter Tauranga. During the transit, the bridge team made several alterations to the passage plan to pass closer to the shore than planned to save time. Further, a north-westerly force 5 wind was setting the vessel to the south of its intended track. Insufficient monitoring of the vessels track by the bridge team failed to identify their grounding course which the vessel had held for more than an hour (Transport Accident Investigation Commission, 2014).

The grounding resulted in New Zealand's largest ever marine oil spill response, with the vessel carrying 1,733 tons of heavy fuel oil and 1,368 containers (Maritime New Zealand, 2015). The vessel broke up in the swell and the eventual cost of salvage exceeded US$500 million.

The New Zealand government put in place several controls to prevent future groundings on the reef. This included a 2 nm exclusion zone around the reef, which was monitored and enforced. A guard vessel was put on standby to patrol this exclusion zone. Virtual Aids to Navigation (AtoNs) were also activated in 2015 to provide additional marking of the hazard, in lieu of the significant cost required to physically mark the reef.

Following the grounding, the investigators recommended that Maritime New Zealand review the need to place Aids to Navigation on offshore reefs in New Zealand and to conduct a review of shipping routes to determine whether there was a need for offshore routeing measures around the New Zealand coast (Transport Accident Investigation Commission, 2014).

1.2 *Study Area*

This paper considers the Bay of Plenty region of the north island of New Zealand (Figure 1). Vessels typically approach the port either from the north, having come from Auckland or having rounded the tip of North Island, or from the east around East Cape, typically from other North Island ports such as Napier and Wellington.

Vessels approaching from the east are advised to keep at least 5 nm north of Volkner Rocks (White Island) and thence 3 nm to the north of Astrolabe Reef before turning for the pilot station (Port of Tauranga, 2011). There are four obstacles north of Motiti Island, namely; Astrolabe Reef, Okaparu Reef, Pudney Rock and Penguin Shoal to the north. Only Astrolabe Reef dries at low water.

Whilst this area lies within the harbor waters of the Bay of Plenty Regional Authority, the Port of Tauranga Vessel Traffic Services (VTS) do not actively monitor vessels on approach to the port.

Vessels laden with oil or other harmful liquid substances in bulk are to keep at least 5 nm off the land, any charted dangers or any outlying islands until reaching the position where alteration is required to make port (Port of Tauranga, 2011).

1.3 *Grounding Risk*

Groundings are a significant hazard to commercial shipping and therefore much attention has been directed by researchers to understanding and modelling their causes. The European Maritime Safety Agency's (EMSA) annual casualty report found that 17% of all casualty events between 2011 and 2014 involved a grounding (2015). Several authors have developed risk models to better understand the risk of grounding in a waterway (Pedersen, 1995; Kite-Powell et al., 1999; Fowler & Sørgård, 2000; Kristiansen, 2005). A key theme of this work is understanding the statistical relationship between several causal factors and the incident record as a tool for predicting the likelihood of grounding. Often this is achieved by analyzing vessel traffic and a historical record of groundings to identify these causal factors and determine some measure of their importance.

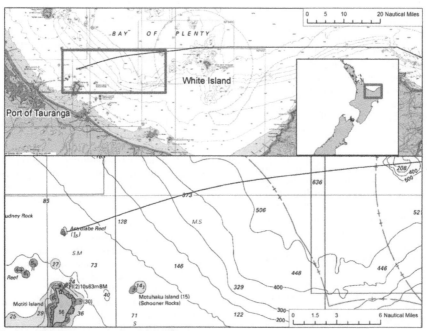

Figure 1. Study area and track of MV Rena (Data Source: Marico Marine's AIS Network).

Mazaheri et al. (2015a) provide an extensive list of the wide variety of factors that could influence the likelihood of a vessel grounding including; meteorological conditions, human factors, the traffic profile, waterway conditions and the presence of risk control measures such as VTS. Further work by these authors (Mazaheri et al. 2015b) also demonstrated the value which detailed analysis of grounding accident and incident reports could provide.

Several authors have considered the widely-accepted premise that there is a statistical relationship between traffic volume and frequency of grounding accidents, finding that the relationship is more nuanced than might be expected. Work in the Gulf of Finland published in Mazaheri et al. (2013; 2015a) found no significant correlation between traffic volume and frequency of grounding accidents, albeit a correlation exists for traffic distribution and grounding likelihood.

Unlike much of the published literature, this paper provides a retrospective analysis of one particular grounding in New Zealand. Through comparing the traffic before and after the incident, and considering the measures put in place following the accident, a better understanding can be achieved of how vessel traffic flow around a hazard.

2 METHODOLOGY

2.1 Data

The analysis conducted in this paper has utilized data from the Automatic Identification System (AIS) collected from a coastal surveillance network in New Zealand operated by the author.

Data was collected continuously from 1st January 2011 to 31st October 2016 and collated into a database. To ensure the accuracy of ship type and size information, a separate vessel information database was used to complete the ship attribute information where no records existed in the recorded dataset. Analysis was limited to cargo and tanker vessels inbound or outbound from Tauranga. All vessels engaged in salvage works, protection or any other activity at the *MV Rena* wreck were removed from the dataset.

2.2 Analysis

The analytical process used in this paper followed three phases. Firstly, vessel tracks were plotted in a Geographical Information System (GIS) and connected sequentially with lines to form vessel tracks. From this a density map of transits could be generated by counting the number of transits through grid cells. This would therefore indicate the routes of vessels through the study area.

Secondly, the passing distance of vessels from Astrolabe Reef was compared by measuring the closest point of approach (CPA) of different vessel types at different times during the data set. A linear transect perpendicular to the traffic flow (a gate) was created approximately due north of Motiti Island, through the Astrolabe Reef, and for a 5 nm. Each transit through this gate was recorded and by analyzing the distance of the transit from the reef, a passing distance could be calculated. By comparing the change in passing distance before and after the grounding, conclusions on the impacts on traffic flow could be drawn.

Finally, the effectiveness of risk controls put in place following the grounding was evaluated by using a simple geometric risk model to measure the change in likelihood of a grounding of vessels before and after the 4th October 2011. Equation 1 and Figure 1 gives a simple schematic of this model. Only the relative difference between grounding risk per year was required and therefore only a simple model was used, constituted by the number of vessel transits, geometric proportion overlapped with the reef and a causation probability, in this case given as the default grounding probability of the IALA risk modelling tool IWRAP.

$$Collision\ \mathrm{Pr}obability = N * Q_i * Pc \qquad (1)$$

where:
- N is the number of ships on a shipping route (kept as a constant at 1000 per year);
- Q_i is the proportion of a vessel type at risk. This is calculated using the proportion of the distribution of traffic which is obstructed by the reef (Figure 2); and
- Pc is the causation probability of a vessel not altering course to avoid a grounding – given as 1.6 x 10^{-4} (Friis-Hansen, 2008).

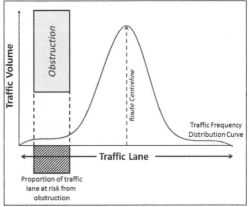

Figure 2. Schematic of grounding model.

3 RESULTS

3.1 Traffic Routeing

Figure 3 compares the density of commercial vessel transits in 2011 and 2012. In 2011, the most frequent of vessels approaching Tauranga from East Cape was to pass 6 nm to the north of White Island and pass between Astrolabe Reef and Pudney Rock, generally keeping 2 nm from Astrolabe Reef (marked with a 2 nm buffer). Since 2011, vessels have been more frequently using two additional northern routes to approach Tauranga. Firstly, vessels would pass north of Pudney Rock but south of the Penguin Shoal (depth of 10 meters chart datum) which keeps them well clear of the Astrolabe Reef (route B). Secondly, more vessels are seen keeping further offshore and passing north of Penguin Shoal before turning to port to approach the pilot station (route C).

Route A necessitates passing between two obstacles for which the width is less than 2 nm. As restrictions were placed on passing Astrolabe Reef, it appears that more bridge teams are planning their passages further offshore. Route B and C provide greater clearances from these obstacles. Whilst there is some additional steaming time between White Island and the pilot station, and therefore cost, they are not significant:

1 Route A – 49nm;
2 Route B – 50nm (+2%); and
3 Route C – 53 nm (+8%).

3.2 Passing Distance

Figure 5 shows the difference in transit distribution between 2011 and 2012. Whilst a minority of vessels are recorded transiting inside the 2 nm exclusion zone in 2012, the greater proportion have been offset to the north. Furthermore, there is almost a complete absence of transits to the south of the reef.

Figure 4 shows the average passing distance of commercial shipping from Motiti Island. The grounding of the Rena and subsequent action taken increased the passing distance of vessels by on average 2 nm. There has, however, been a downward trend in passing distance since 2011/2012 yet the passing distance remains at least 1 nm further offshore than in 2011.

If taken by year the difference in traffic flow of these vessels is more clear. Figure 6 shows the frequency distribution of transits by year between 2011 and 2016. The shift in peak traffic flow from approximately 10.5 km to 13.5 km between the pre and post grounding datasets can be seen. It is also notable that post-2011 datasets contain two clear peaks; firstly, the 13.5 km offset centerline of the traffic flow, but also a smaller cluster of transits at 12 km. This suggests a number of vessels are planning their passage at the limits of the exclusion zone and would likely have transited closer to the reef if the exclusion was set at 1 nm.

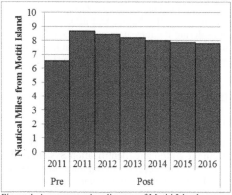

Figure 4. Average passing distance of Motiti Island.

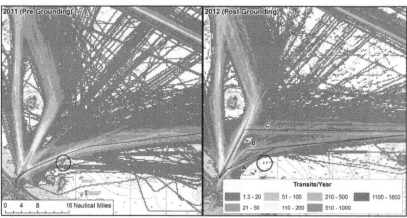

Figure 3. Comparison of commercial vessel routes in 2011 and 2012 (east-west passage marked with dotted line).

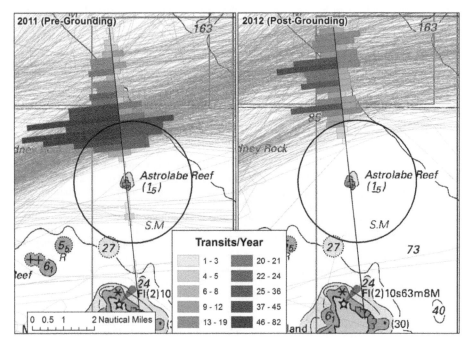

Figure 5. Vessel transit distribution from Astrolabe Reef (2011 and 2012).

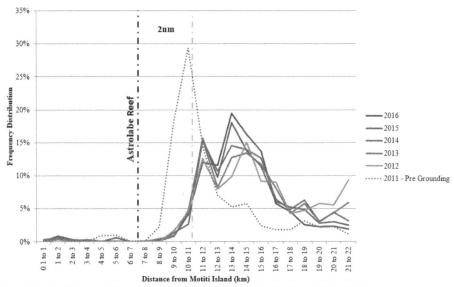

Figure 6. Passing Distance of Commercial Vessels from Astrolabe Reef by Year (2011 to 2016).

Finally, the number of commercial shipping transits which passed within the 2 nm exclusion zone is compared to the number transiting within 12 nm of Motiti Island (Table 1). The results show that on average there has been a twenty fold reduction in vessels entering the exclusion zone following the grounding. Both the New Zealand Almanac and Port

of Tauranga's Information for Masters advise shipping to pass 3nm to the north of Astrolabe Reef. The *MV Rena* herself set a waypoint 1 nm north of the reef.

Table 1. Incursions of 2 nm exclusion zone by cargo and tanker vessels.

Year	Transits	Incursions (2 nm)	%
		Pre-Grounding	
2011	712	296	41.6%
		Post-Grounding	
2011	80	2	2.5%
2012	548	21	3.8%
2013	660	24	3.6%
2014	533	13	2.4%
2015	832	21	2.5%
2016	811	16	2.0%

3.3 Risk Model and Effectiveness of Risk Controls

Table 2 shows the results of this analysis based on the configuration of passing traffic in each year's dataset. Between 2011 and 2016 the proportion of vessels at risk has decreased significantly and by extension so has the likelihood of another incident.

A combination of an enforced 2 nm exclusion zone, guard vessels and a greater awareness of passing vessels of the reef has therefore reduced the likelihood of a vessel grounding on the Astrolabe Reef by 74%, four times less likely than before. Since 2012, this likelihood has reduced further with a significant reduction between 2015 and 2016 of a further 46%, the result of a greater concentration of the vessel routes resulting in a tighter distribution and lower grounding likelihood. This reduction coincides with the 2015 implementation of virtual aids to navigation on seven hazardous reefs in the Bay of Plenty. This may well have had the effect of providing much clearer markings to vessels in routeing through this region, similar to the effect of traffic routeing measures such as traffic lanes.

Table 2. Modelling results for grounding risk at Astrolabe Reef.

Year	Geometric Probability	Likelihood / Year	Return rate (years)
2011	1.71%	2.74×10^{-03}	365
2012	0.44%	7.04×10^{-04}	1,421
2013	0.38%	6.11×10^{-04}	1,637
2014	0.38%	6.07×10^{-04}	1,649
2015	0.33%	5.25×10^{-04}	1,903
2016	0.18%	2.83×10^{-04}	3,528

4 DISCUSSION

The analysis conducted in this paper has considered how vessel traffic flow has been altered following the grounding of the MV Rena. The results show a clear and significant alteration in both vessel routeing and passing distances from the Astrolabe Reef immediately following the grounding.

Prior to the grounding of the vessel, the principal vessel route to and from Tauranga was centered 1.8nm from the reef. This was in clear contradiction to the Bay of Plenty pilotage directions which advised all vessels to keep at least 3 nm from the reef. This advice indicates that the reef had been identified as a hazard and that by advising vessels to keep a safe distance, the risk could be controlled. Whilst the reef was within the waters of the Bay of Plenty regional authority and the port of Tauranga had both pilotage and a VTS, it is clear that this was neither being actively monitored or enforced. Therefore, without active management this had a limited effectiveness as a risk control and was routinely ignored by vessels. It is also likely that decision makers were unaware that vessels were transiting close to the reef routinely prior to the grounding. As simple analysis as plotting vessel track density using AIS data would have immediately highlighted the difference between routeing advice and actual vessel movements and could have served as the impetus for additional steps to be taken. By comparison, following the grounding, the creation of a 2 nm exclusion zone was so well monitored that the approximately 20 vessels per year which entered this zone were both contacted by the harbor authority and often reported in the media.

The results also indicate that the measures taken following the grounding have significantly reduced the likelihood of another incident on the same reef. Whilst the principal cause of the incident was human error, decision makers took the approach that if they could increase the passing distance of vessels from the reef then the likelihood of a vessel straying off course to the degree to which it could impact the reef is reduced. This was achieved through a managed and enforced 2 nm exclusion zone, although other measures such as implementing traffic routeing schemes were also considered. The analysis of vessel traffic and a simple geometric risk model indicate that this had the desired effect with at least a 74% effectiveness resulting. Furthermore, the 2015-2016 reduction by a further 46% may be the result of the implementation of virtual aids to navigation on not only the Astrolabe Reef, but the other reefs in the region. With the main obstacles marked by these AIS beacons, they serve as clear boundaries to the established traffic route which has concentrated the traffic on a safe course. Other effects such as the greater awareness of bridge crews as they transit this region, a result of publicity of the MV Rena's fate, would also have contributed to a greater vigilance so it is not possible to provide definitive measures of the effectiveness of these particular controls.

A further point of interest is the implication of these results on the use of historical accident statistics in risk models. This paper has clearly demonstrated how the vessel traffic profile has changed significantly (more than 74%) following this incident. Given the low numbers of groundings within a study area, much research into grounding risk uses long duration historical incident databases and analyses recent vessel traffic data such as AIS to develop their model. If an incident alters vessel traffic movements, through active management or a greater awareness of the hazard by navigators, then more recent vessel traffic data is not representative of the factors which resulted in that grounding. In summary, it would not be possible to develop a risk model using recent data calibrated with historical incidents. This uncertainty in risk modelling was highlighted in Mazaeri et al. (2013; 2015a) who found no relationship between traffic volume and grounding risk but cautioned that their research used AIS data from 2010 but accident data from 1989-2010. During this time they recognized that traffic routeing had changed and a VTS as well as other risk controls had been implemented. It may therefore not be surprising that an incident in 1989 is not directly predictable from data in 2010.

This limitation is not easily resolved; reducing the accident data to only recent records reduces the sample size and increasing the traffic data increases computational requirements and is limited beyond the adoption of AIS. Therefore, this uncertainty should be considered by researchers in the methodological development of risk models, and where possible steps should be taken to overcome it.

5 CONCLUSIONS

This paper has demonstrated how vessel traffic flow passed the Astrolabe Reef in New Zealand has been altered following the grounding of the *MV Rena*. Vessels have transited further offshore and the distribution of transits has decreased resulting in a decrease in the likelihood of another grounding. Further work is required to repeat this analysis at other locations where a grounding has taken place to identify whether similar changes in traffic flow are evident. Similarly, an evaluation of the risk controls put in place after the grounding should also be considered to identify whether they have been more or less effective at controlling the traffic and

reducing the likelihood of a reoccurrence. Understanding the effectiveness of these controls is essential for decision makers to justify the needs and associated costs of implementing them preventively at hazardous locations.

Whilst this study has considered only a single case study at one particular location, the results hold lessons to researchers in developing grounding risk models which could be applied more widespread.

REFERENCES

European Maritime Safety Agency 2015. *Annual Overview of Marine Casualties and Incidents 2015.*

Fowler, T.G. and Sørgård, E. 2000. Modelling Ship Transportation Risk. *Risk Analysis* 20: 225–244.

Friis-Hansen, P. 2008. *IWRAP: Basic Modelling Principles for Prediction of Collisions and Groundings Frequencies.*

Kite-Powell, H.L., Jin, D., Jebsen, J., Papakonstantinou, V. and Patrikalakis, N. 1999. Investigation of Potential Risk Factors for Groundings of Commercial Vessels in U.S. Ports. *International Journal of Offshore and Polar Engineering* 9: 16–21.

Kristiansen, S. 2005. *Maritime Transportation: Safety Management and Risk Analysis.* Oxford: Routledge.

Maritime New Zealand 2015. New Zealand Coastal Navigation Safety Review.

Mazaheri, A. Montewka, J. & Kujala, P. 2013. Correlation between the Ship Grounding Accident and the Ship Traffic – A Case Study Based on the Statistics of the Gulf of Finland. *TransNav, the International Journal of Marine Navigation and Safety of Sea Transportation* 7(1): 119-124.

Mazaheri, A. Montewka, J. Kotilainen, P. Sormunen, O. & Kyjala, P. 2015a. Assessing Grounding Frequency using Ship Traffic and Waterway Complexity. *Journal of Navigation* 68(1): 89-106.

Mazaheri, A. Montewka, J. Nisula, J. & Kyjala, P. 2015b. Usability of accident and incident reports for evidence-based risk modeling – a case study on ship grounding reports. *Safety Science* 76: 202-214.

Port of Tauranga 2011. Port Information for Ships Masters.

Transport Accident Investigation Commission 2014. Final Report: Inquiry 11-2014 – Container ship MV Rena grounding on Astrolabe Reef, 5 October 2011.

Pedersen, P.T. 1995. Collision and Grounding Mechanics. *Proceedings of WEMT '95'. Copenhagen, Denmark*, The Danish Society of Naval Architecture and Marine Engineering.

Dynamic Positioning

Innovation Methodology for Safety of Dynamic Positioning under Man-machine System Control

R. Gabruk & M. Tsymbal
National University «Odesa Maritime Academy», Odesa, Ukraine

ABSTRACT: This paper presents an innovation methodology for assessment the safety of polyergatic system, which controls high precision navigation during dynamic positioning in locally confined area of technological work under the flow of disturbances. The probability of reliable interaction between group of dynamic positioning operators and machine was adopted as quantitative characteristic of dynamic positioning safety. The developed methodology enables the support in making decisions concerning safety of dynamic positioning and can be used to verify the validity of previous made decisions for commercial operation of mobile water transport object equipped with dynamic positioning systems of any class.

1 INTRODUCTION

The further intensive development of the continental shelf resources requires the use of mobile water transport objects (MWTO) equipped with dynamic positioning systems (DPS).

A modern MWTO, which performs dynamic positioning (DP) in locally confined area of technological work (this area could be considered as 500 m around offshore installation or MWTO itself) is equipped with complex of systems adjusted to receive and process a large flows of information and are important for a safe navigation and technological work performance. Navigation bridge represents workplaces, which are integrated with all MWTO's systems and forms polyergatic system.

The term "ergatic" could be used when one dynamic positioning operator (DPO) performs high precision navigation control. Realistically, group of DPOs interacting with machine system and term "polyergatic" system could be used in this case. The integrated bridge equipment complex (which includes DP workstations) enables the DPO to perform safely navigational tasks in the flow of disturbing events. From the point of view of the human-machine interaction system, the inputs of DPO into the laws of automatic motion control have guidance function.

The MWTO's Master is responsible for ensuring adequate procedures are in place to guarantee compliance with relevant national or flag and operating company's requirements with regard to DP and technological work operations.

Presently safety of DP operations assessment is based on Company Safety Management System (SMS) requirements and Failure Modes & Effects Analyses (FMEA) procedures.

Analyzing numbers of incidents with MWTO during DP operations, it is possible to conclude that new additional measures should be implemented to ensure safety. The new innovation assessment should be able to provide the information concerning safety of DP operations in real time mode, which will be great and reliable tool for DPO to support his final decision. To achieve this target it is necessary to consider DP process, human and technical component on polyergatic system, which performs control of MWTO high precision navigation process, taking into account factors of disturbance.

Following new methodology allows assessing safety of polyergatic system to ensure safety of MWTO during DP operations in locally confined area of technological work.

2 DYNAMIC POSITIONING

DP process in locally confined area of technological work represents different MWTO's subsystems interaction, which are targeting to ensure reliable positioning under disturbances from different sources.

MWTO hull during DP is affected by disturbances of environmental factors, which produce nonlinear external forces and try to move MWTO outside of position or track. Environmental factors (waves, current, wind, ice) represent hazard to DP safety and associated technological work performance.

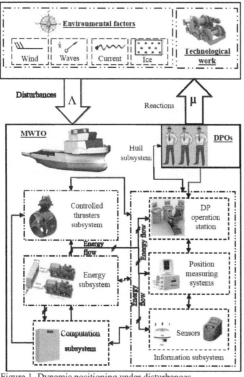

Figure 1. Dynamic positioning under disturbances.

Character of technological work affects greatly DP process. Relevant technical features and their interactions with MWTO could produce additional forces. This situation is typical for anchor handling, drilling, cable or pipe lay operations.

Technological work, environmental forces, internal interactions of MWTO's components, locally confined area characteristics form flow of disturbing events Λ, which are directed to group of professional and qualified DPOs.

The only way for the provision of the DP safety is the assessment by DPO the particular situation during specified space-time interval, as well as the adopting of adequate solution and its timely execution. In the analysis of factors and events the DPO takes into account the uniqueness of each event, dynamics in the situation development, etc. Thereby group of DPOs produces reaction flow μ.

Machine system represents itself combination of interacting subsystems, which automatically maintain the MWTO position and heading by means of active thrust. Hull – central subsystem, which contains all other subsystems and is the subject of nonlinear influence.

Controlled by DPS thrusters generate control forces to compensate disturbances. Thrusters are located accordingly to provide optimum capability and movement, with minimum interference with other thrusters and positioning sensors.

Computation subsystem on the basis of processed data computes signals that should compensate forces of external disturbances. Computation subsystem provides DP control options such as minimum power consumption, fine position control, taking into account barred zones for azimuth thrusters to protect equipment. The power for efficient function of all MWTO subsystems provided by energy subsystem.

Information subsystem consist of redundant computer workstations, which provide graphical user interface, operation information input and output. As well, it distributes data flows during the information interaction of DPS functional elements, such as position measuring systems and various sensors.

One of the most important technical characteristic of the system is the redundancy. Redundancy enables the MWTO safely terminate a DP operation after losing a critical component, system or subsystem. This concept is referred as "Single Point Failure" mode. Most MWTO have propulsion configurations, computers (for DP-2 and DP-3 class) and workstation configurations beyond the minimum.

Failures of the subsystem's normal functioning can be renewed by technical resumption of operation of components by replacing faulty blocks, which is typical for active components, or by changing the technical facility.

3 HUMAN COMPONENT

It is necessary to consider the DPO as a part of polyergatic system which is involved in the flow of events and which is defined as a sequence of similar events occurring one after the other at random times. The considered flow of events is called stationary, as its probabilistic characteristics don't depend on the choice of the starting point or, more specifically, the hit probability of coming of any number events at any time interval depends only on the length of this time interval and does not depend where exactly it is located at the time line.

The events flow in which the MWTO polyergatic system is situated during performance of technological works associated with DP operations is considered the ordinary one. The ordinary flow of events can be considered a random process.

The considered flow of events is in fact a flow without aftereffect. The condition for occurrence of the flow without aftereffects arises when the number of events falling at any time interval is independent of the number of events, which hit any other time interval that is not intersecting it. Actually, the absence of the aftereffect in the flow of events means that the events constituting the flow appear in one or more times independently from each other. That is, the considered flow of events is stationary, ordinary and has no aftereffects.

The ordinary flow of events without aftereffect is called Poisson flow. Thus, the considered flow of events means the stationary Poisson flow. The stationarity property of the considered flow of events determines its same intensity throughout the considered period.

The time interval between two successive events of a stationary Poisson flow has an exponential distribution:

$$\tilde{f}(t_\Lambda) = \Lambda e^{-\Lambda t_\Lambda},$$

where e – the Euler's number;
Λ- the events flow rate, 1/s.
t_Λ– the interval time between two consecutive events of a stationary Poisson flow, s.

The flow of reactions will remain stationary, ordinary and will has no aftereffect. It forms a stationary Poisson flow of events as well.

We shall consider the actual process of guarantee the safety and efficiency of MWTO DP, as well as carrying out of technological works from the position of a scenario description of polyergatic system states. The most important aspect during performance of technological works in locally confined area of DP is to ensure MWTO safety – i.e. the development of potentially dangerous scenarios prevention. The way of avoiding accidental outcome of potentially dangerous scenarios is the treatment of flow of events by DPO. During primary processing of events in the general flow of events, the DPO identifies important events that cause the need for his further action.

Important events include those events, which may be the beginning of a chain of cause-related events leading to an accident or can reduce the effectiveness of the process itself as a high precision navigation and technological work.

The random process occurring in any physical system has been called Markov process (or a process without aftereffect) provided that it has the property that in the future for each point in time the probability of any state of the system depends in the future only on its state at present and is not dependent on when and how the system has reached this state. That means that when the present is stabile the future will not be dependent on the past.

To generate an adequate model of polyergatic system it is necessary to conduct the full description of the characteristics of its components, as well as of all possible states of the said system. The status of DP polyergatic operation system and of the technological work in general are described as follows:

$$\Upsilon_0, \Upsilon_1, \Upsilon_2, \Upsilon_3, \Upsilon_4 ... \Upsilon_i ... \Upsilon_n,$$

where Υ_i - the state of polyergatic system with i-number;

n - the number of states of polyergatic system.

A prerequisite is that the number of these states is to be countable. The condition of a countable number of states of the system during the flow of Markov processes is important for the hierarchical division of polyergatic system into appropriate subsystems. The number of dedicated subsystems may vary depending on the purposes of the study. The bijection of states of the polyergatic system in general is conveniently written as follows:

$$n \leftrightarrow \Phi,$$

where Φ - all natural numbers.

To find the probability P(t) we shall draw up and solve the Kolmogorov-Chapman equation for the one-parameter family of continuous linear operators:

$$\frac{dP_i(t)}{dt} = \sum_{i=1}^{n} \Lambda_{ij} P_j(t) - P_i(t) \sum_{j=1}^{n} \Lambda_{ij}.$$

For the convenience it is possible to omit the t argument in P variables. Then the equation can be re-written as follows:

$$\frac{dP_i}{dt} = \sum_{i=1}^{n} \Lambda_{ij} P_j - P_i \sum_{j=1}^{n} \Lambda_{ij}.$$

The equations can also be formulated by using the marked graph of the system states and in accordance with the following mnemonic rule. The derivative of the probability of each state is equal to the sum of all flows of probabilities coming from other states into this one, minus the sum of all flows probabilities coming from this state to the other ones.

To find the final probability of states of the polyergatic system appearing during a random ergodic Markov process it is required to create an adequate system of equations consisting of differential Kolmogorov-Chapman equations, as well as by appearance of a normalization condition:

$$\sum_{i=1}^{n} P_i = 1$$

The DP process and potential emergency situations must be considered from the standpoint of the scenario description of polyergatic system conditions formed by human-operators and technical control systems (the DPS, technological work management systems) during performance of technological works in locally confined area.

During processing of flows, the events are divided into categories. The events of the importance category I are considered the beginning of the chain of cause-related events that could lead to an accident. These events include nonlinear disturbing effects of environmental factors, failures and malfunctions of technical equipment, mistakes made by DPO and other destabilizing events that may lead to an accident. The reaction to these events should be a priority. Since they exactly cause the development of dangerous scenarios.

The events of the importance category II are not considered the beginning of a chain of cause-related events that could lead to an accident. Such events may not require a primary DPO's reaction. These events primarily include the answers to requests for information concerning the carrying out of technological work, the coordination of execution of technological work and other events that require the action of DPO but are not able to cause an accident. This classification is most appropriate for the safe tasks execution in implementation of the technologycal works performed by MWTO during DP operations.

The evaluation of events takes place in such manner that the event of importance category I has priority in processing before the events included to the importance category II. This means that if the event of the importance category I takes place while all operators are busy and at least one of them is busy processing the event of the category II, he will immediately start the processing of the events of the category I.

The states of a safe and efficient operation of the polyergatic system during performance of the technological works associated with a dynamic positioning can be described as follows:
– Υ_{00} - important events have not occurred;
– Υ_{10} - there was one event of the importance category I and no one event of the importance category II;
– Υ_{11} - there was one event of the importance category I and one event of the importance category II;
– Υ_{01} - there were no events of the importance category I, but there was one event of the importance category II;
– Υ_{20} - there were two events of the importance category I and no one event of the importance category II;
– Υ_{02} - there were no events of importance category I, but there were two events of importance category II;
– Υ_{03} - there were no events of the importance category I, but there were three events of the importance category II;

– Υ_{12} - there was one event of the importance category I and two events of the importance category II;
– Υ_{21} - there was two events of the importance category I and one event of the importance category II;
– Υ_{30} - there were three events of the importance category I and no single event of the importance category II.

The corresponding marked state graphs of polyergatic control system of the dynamic positioning and technological work carried out consisting of one, two and three DPO implemented in a Markov chain are shown below.

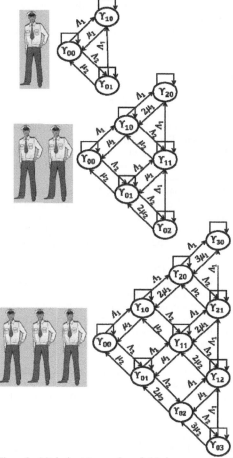

Figure 2. Marked state graphs of Markov processes for polyergatic system consisting of one, two and three dynamic positioning system operators.

During constructing the marked graphs we shall place the arrow of transition from one condition to another one on the left side. In this case a clear "route" picture transitions from one state to another is created. The stationary Poisson flows of events with importance categories I and II have respectively intensity Λ_1 and Λ_2. The stationary Poisson flows of reactions have respectively the intensity μ_1 and μ_2 for events with importance categories I and II

Accordingly to the marked graph of events we shall set up the system of differential Kolmogorov-Chapman equations for one DPO:

$$\begin{cases} \dfrac{dP_{00}}{dt} = \mu_1 P_{10} + \mu_2 P_{01} - (\Lambda_1 + \Lambda_2) P_{00}, \\[2mm] \dfrac{dP_{10}}{dt} = \Lambda_1 P_{00} + \Lambda_1 P_{01} - \mu_1 P_{10}, \\[2mm] \dfrac{dP_{01}}{dt} = \Lambda_2 P_{00} - (\Lambda_1 + \mu_2) P_{01}, \\[2mm] P_{00} + P_{10} + P_{01} = 1. \end{cases}$$

Based on polyergatic system conditions description, the corresponding probabilities of its stay in dangerously and safe states with respect to possible events of categories I or II shall have the following formula:

$$\begin{cases} P_{SI} = P_{00} + P_{01}, \\ P_{SII} = P_{00}, \\ P_{UI} = P_{10}, \\ P_{UII} = P_{10} + P_{01}, \end{cases}$$

where P_{SI}, P_{SII} - means that the polyergatic system is probably kept in a safety with respect to the events with importance categories I or II respectively;

P_{UI}, P_{UII} - means that the polyergatic system is probably in a potentially dangerous condition with respect to the events with importance categories I or II respectively.

For polyergatic system, which consists of two DPO:

$$\begin{cases} \dfrac{dP_{00}}{dt} = \mu_1 P_{10} + \mu_2 P_{01} - (\Lambda_1 + \Lambda_2) P_{00}, \\[2mm] \dfrac{dP_{10}}{dt} = \Lambda_1 P_{00} + 2\mu_1 P_{20} + \mu_2 P_{11} - (\Lambda_1 + \Lambda_2 + \mu_1) P_{10}, \\[2mm] \dfrac{dP_{20}}{dt} = \Lambda_1 P_{10} + \Lambda_1 P_{11} - 2\mu_1 P_{20}, \\[2mm] \dfrac{dP_{01}}{dt} = \Lambda_2 P_{00} + \mu_1 P_{11} + 2\mu_2 P_{02} - (\Lambda_1 + \Lambda_2 + \mu_2) P_{01}, \\[2mm] \dfrac{dP_{11}}{dt} = \Lambda_2 P_{10} + \Lambda_1 P_{01} + \Lambda_1 P_{02} - (\Lambda_1 + \mu_1 + \mu_2) P_{11}; \\[2mm] \dfrac{dP_{02}}{dt} = \Lambda_2 P_{01} - (\Lambda_1 + 2\mu_2) P_{02}, \\[2mm] P_{00} + P_{10} + P_{20} + P_{01} + P_{02} + P_{11} = 1. \end{cases}$$

Following system describes condition for two DPO:

$$\begin{cases} P_{SI} = P_{00} + P_{10} + P_{01} + P_{11} + P_{02}, \\ P_{SII} = P_{00} + P_{10} + P_{01}, \\ P_{UI} = P_{20}, \\ P_{UII} = P_{20} + P_{11} + P_{02}. \end{cases}$$

For polyergatic system, which consists of three DPO:

$$\begin{cases} \dfrac{dP_{00}}{dt} = \mu_1 P_{10} + \mu_2 P_{01} - (\Lambda_1 + \Lambda_2) P_{00}, \\[2mm] \dfrac{dP_{10}}{dt} = \Lambda_1 P_{00} + 2\mu_1 P_{20} + \mu_2 P_{11} - (\Lambda_1 + \Lambda_2 + \mu_1) P_{10}, \\[2mm] \dfrac{dP_{20}}{dt} = \Lambda_1 P_{10} + 3\mu_1 P_{30} + \mu_2 P_{21} - (\Lambda_1 + \Lambda_2 + 2\mu_1) P_{20}, \\[2mm] \dfrac{dP_{01}}{dt} = \Lambda_2 P_{00} + \mu_2 P_{11} + 2\mu_2 P_{02} - (\Lambda_1 + \Lambda_2 + \mu_2) P_{01}, \\[2mm] \dfrac{dP_{11}}{dt} = \Lambda_1 P_{01} + \Lambda_2 P_{10} + 2\mu_1 P_{21} + 2\mu_2 P_{12} - \\[1mm] \quad -(\Lambda_1 + \Lambda_2 + \mu_1 + \mu_2) P_{11}, \\[2mm] \dfrac{dP_{21}}{dt} = \Lambda_1 (P_{11} + P_{12}) + \Lambda_2 P_{20} - (\Lambda_1 + \mu_2 + 2\mu_1) P_{21}, \\[2mm] \dfrac{dP_{02}}{dt} = \Lambda_2 P_{01} + \mu_1 P_{12} + 3\mu_2 P_{03} - (\Lambda_1 + \Lambda_2 + 2\mu_2) P_{02}, \\[2mm] \dfrac{dP_{12}}{dt} = \Lambda_1 (P_{02} + P_{03}) + \Lambda_2 P_{11} - (\Lambda_1 + \mu_1 + 2\mu_2) P_{12}, \\[2mm] \dfrac{dP_{03}}{dt} = \Lambda_2 P_{02} - (\Lambda_1 + 3\mu_2) P_{03}, \\[2mm] P_{00} + P_{10} + P_{20} + P_{30} + P_{01} + P_{11} + P_{21} + P_{02} + P_{12} + P_{03} = 1. \end{cases}$$

Condition for system with three DPO:

$$\begin{cases} P_{SI} = P_{00} + P_{10} + P_{01} + P_{11} + P_{02} + P_{20} + \\ \quad + P_{12} + P_{21} + P_{03}, \\ P_{SII} = P_{00} + P_{10} + P_{01} + P_{11} + P_{02} + P_{20}, \\ P_{UI} = P_{30}, \\ P_{UII} = P_{30} + P_{21} + P_{12} + P_{03}. \end{cases}$$

Final probability P_S of polyergatic system safe functioning equals:

$$P_S = \frac{\Lambda_1}{\Lambda_1 + \Lambda_2} P_{SI} + \frac{\Lambda_2}{\Lambda_1 + \Lambda_2} P_{SII}.$$

4 TECHNICAL COMPONENT

To ensure the safety of navigation and quality during technological work performance it is necessary to ensure reliable operation of technical subsystems DPS representing the complex technical components of MWTO system. With a term "reliability of the technical subsystem" we shall mean its property to keep in time and within the fixed values of all the parameters that determine its ability to perform its required function in the given conditions, as well as exploitation conditions.

In order to ensure the safety during the practical MWTO operation while the DP is being active the classical approach shall not be effective, as it cannot provide visual information to make safety decisions in relation to DP at a particular time interval during specific MWTO operation. The traditional approach is not sufficiently flexible and, in particular, it does not consider the possibilities of MWTO DP operations in different conditions within a locally confined area. Both physical parameters of a locally confined area navigation of MWTO and emergency factors may impose obvious restrictions on the use of MWTO position measuring systems.

The proposed approach is to predict the possibility of reliable operation of the polyergatic system technical component taking into account the reservations of elements and subsystems in accordance with the class of the DPS.

The reliability of a component is marked as p_{RC}. It enables a component property to maintain a usable state and to perform the necessary functions to a limiting condition until the boundary state during the implementation of quality maintenance and repair system is reached. The reliability of operation varies with a time, which is a measure of a technical resource or service life – that is the total operating time of the component that in case of expiration of its service life is to be terminated regardless of its condition. Its lifetime is specified in certified documents. Mathematically the component's reliability can be written as follows:

$$p_{RC}(t) = e^{-\lambda t},$$

where λ- the failure rate.

For easier presentation of information on the subsystems and their components reliability, we shall develop a method of presentation in the form of reliability block-diagrams with following designations. On reliability block-diagrams, components without which the system cannot operate shall be represented as units connected in series. Duplicate components will reflect units connected in parallel. Let's assume that the reliability of all components is the same n, i.e.:

$$p_{RC1} = p_{RC2} = p_{RC3} = ... p_{RCi} = ... p_{RCn}.$$

The system failure happens when both of its components fail. In another case, the failure of system occurs when any of these components fails. Schematically it can be represented as follows.

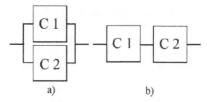

Figure 3. Reliability block-diagram. a) – redundant components; b) – components without redundancy.

To guarantee reliable operation it is necessary to provide adequate reservation of elements of the DPS. The minimum reservation is carried out according to the requirements of the Classification Society defined for the DPS of certain class and additionally may be carried out based on the features of the technological work process implementation and defined water area. The reservation is a technical measure to ensure the safety of MWTO's operation.

Using the algebra of events, subsystem reliability p_{RS} could be defined as follows:

$$p_{RS} = 1 - (1 - p_{RC})^2.$$

In case of redundancy by n-1 similar components, probabilistic reliability is determined by the following formula:

$$p_{RS} = 1 - (1 - p_{RC})^n.$$

If a system failure occurs when any failure of n components, the safety system without redundancy operation will be determined as follows:

$$p_{RS} = p_{RC}^n.$$

Final polyergatic system technical component reliability ps equals:

$$p_S = p_{RS1} \cdot p_{RS2} \cdot p_{RS3} \cdot ... p_{RSn}.$$

Practical computation process of polyergatic system technical reliability is based on the algebraic expressions that characterize the reliability of the systems, subsystems and components. Drawing algebraic expressions should be done by system approach methods and analysis of complex systems.

5 METHODOLOGY

During technological work related to the DP in a locally confined area, the control process over terminal control of high-precision navigation is carried out directly by polyergatic system of the specified MWTO. The proposed approach for assessment the safety of the polyergatic system, which includes the DPS and the group of DPO,

218

within the space-time interval of MWTO operation is carried out in two directions.

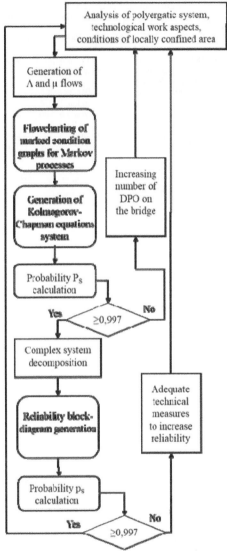

Figure 4. Information and logical model of methodology.

The first direction means the assessment of the DPO's possibility to carry out safe and effective management over the DPS and to perform particular technological work. Due to the considerable number of the loss and the devastating consequences of errors made in polyergatic system, the DPO plays a major role.

Generation of Λ and μ flows is based on the in-depth analysis of polyergatic system, technological work aspects, conditions of locally confined area.

After, as per described above, necessary to conduct flowcharting of marked condition graphs for Markov processes in ployergatic system. On basis of the graph, it is necessary to generate and solve Kolmogorov-Chapman equations system.

The probability P_S of a reliable processing of the flow of events, which disturb the system, as well as the approval and implementation of decisions carried out by group of DPOs within continuous uniform time interval is a numerical measure of the human safety component in polyergatic management system.

The normal distribution is a limiting law to which other distribution laws are aspiring under common similar conditions. Therefore, a value of 0.997 may be considered the criterion of reliability for human and technical components of the polyergatic system. In case of default of inequality it is necessary to call an additional DPO. The presence of additional DPOs, who are not involved in processing of events is a safety resource.

It should be noted that in the method presented the probabilistic reliability of DPO group evaluation begins with one operator. According to the International Guidelines for the Safe Operation of Dynamically Positioned Offshore Supply Vessels it should be at least two operators on the bridge. The assessment of one DPO can be carried out to investigate the possibility of one person to manage and control the processes of high-precision navigation.

The second direction means the assessment of the MWTO polyergatic system technical component reliability. For this purpose, it should be carried out the decomposition of all MWTO systems, subsystems and components. It is necessary to compose presented in this paper reliability block-diagrams, which will define interrelations between elements and their redundancy level.

Further calculations on probability of the respective components, subsystems and systems functioning reliability should be carried out. During this step the probability ps is calculated.

By analogy with the assessment of the polyergatic system human component the same test should be carried out with an inequality. If the obtained probability shows ps ≥ 0,997, then the system is in a safe side. If not – then it is necessary to take a complex of technical measures to improve the reliability of the technical component.

6 CONCLUSIONS

The paper draws the attention to the safety of dynamic positioning, where the notion machine – group of dynamic positioning operators (or vice versa) is labelled by the term „polyergatic". Controlled by polyergatic system movement during

DP operations in close proximity to offshore structures under flow of disturbances with high risk of collision forms potentially danger to navigation. One of the ways to control risks during DP operations is implementation of proposed new methodology for polyergatic system reliability assessment.

Described algebraic dependences were applied by using of "Polyergatic Systems Inspector" computer program with a Copyright Registration Certificate No 64517 issued by the State Intellectual Property Service of Ukraine. The application of the program confirmed its complete adequacy to real conditions and it can be used as a tool to support the adoption of solutions required for the safety of dynamic positioning.

Proposed methodology has an important role to ensure safety of MWTO during DP operations among many other renowned methods.

REFERENCES

Fossen, T. I. 2002. *Marine Control Systems.* Trondheim: Norwegian University of Science and Technology.

IMCA M 178. 2005. *FMEA Management Guide.* London: IMCA.

IMCA M 117. 2006. *The Training and experience of Key DP Personnel.* London: IMCA.

IMCA M 182. 2006. *International Guidelines for the Safe Operation of Dynamically Positioned Offshore Supply Vessels.* London: IMCA.

IMCA M 103. 2007. *Guidelines for the Design and Operation of Dynamically Positioned Vessels.* London: IMCA.

IMO MSC Circular 645. 1994. *Guidelines for vessels with dynamic positioning systems.* London: IMO.

Morgan, Dr. M. J. 1978. *Dynamic Positioning of Offshore Vessels.* The Petroleum Publishing Company. Tulsa, Oklahoma.

UK Offshore Operators Association/Chamber of Shipping. 2002. *Safe Management and Operation of Offshore Support Vessel.* London.

The Visual System in a DP Simulator at Maritime University of Szczecin

P. Zalewski, R. Gralak, B. Muczyński & M. Bilewski
Maritime University of Szczecin, Szczecin, Poland

ABSTRACT: The paper presents the visual system designed in Maritime University of Szczecin in conformance to the Nautical Institute's requirements of Dynamic Positioning (DP) simulator class B. The specific functionalities and capabilities of this visual system connected to a DP advanced trainer of Kongsberg K-Pos type has been described.

1 INTRODUCTION

Dynamic positioning operators training approved by the Nautical Institute (see Zalewski, 2012) must be conducted on the equipment meeting specifications currently set in (NI, 2016). To meet these specifications the DP Bridge simulator in Maritime University in Szczecin had to be modernized, especially in scope of its visual capabilities. The staff of the Centre of Marine Traffic Engineering in MUS performed this task by utilization of freely available graphics and rendering engine and by designing the vision's control application.

2 NI DP SIMULATOR SPECIFICATIONS

The main change in standards of DP Simulator Training has begun in 2012, finally leading to much higher simulator requirements, especially with regard to the visual system used for outside view. Since 2016 these requirements has included the following (NI, 2016):
- The simulator shall provide a realistic visual scenario by day, dusk or by night, including variable meteorological visibility, changing in time. It shall be possible to create a range of visual conditions, from dense fog to clear.
- For Class B, a visual system is required to increase realism and learning outcome. A visual system for Class B shall have a horizontal field of view of a single visual channel. Horizontally, the visual system for Class B shall be able to be panned 360 degrees.
- Simulated sea state visualization shall align with any changes in simulated weather. This need not be automatic. Manual entry of sea state parameters, by the instructor, is sufficient to meet this requirement. (Required by January 2020).

To meet these criteria the existing blind vision simulator must have been connected to the visual system with at least one channel visualisation of outside view.

3 VISION APPLICATION

The assumptions of the vision application are as follows.

Main software application must fulfil following requirements and elements:
- provide a satisfactory graphic representation of the simulated area, objects and the ship (rendering engine),
- adjust the visualization based on data from the simulator, including time of day, fog and wave height (networking and scene manager),
- allow to change the point of visualization (location of virtual camera),
- provide a realistic yaw, pitch and roll (physics and animation).

The rendering engine is the core part of any visualization system. It usually uses Direct3D or OpenGL application programming interface which provide a software abstraction of the graphics processing unit (Eberly 2006). Such engine can be developed to meet the fixed specification and be optimized for a specific hardware platform. This is the approach used in contemporary simulators like Kongsberg Polaris or Transas NTPRO.

For the purpose of MUS vision system it was decided to use an existing and free solution, namely Unity game engine which includes its own rendering engine and offers real-time global illumination, physically-based rendering, reflection probes, curve and gradient-driven particle system, full-screen post processing effects, low-level rendering access, deferred rendering and occlusion culling in terms of graphics. Additionally, it supports high level of detail for 3D models and a profiler that can analyse the performance of the GPU, CPU, memory, rendering, and audio. Unity engine also provides a fast and simple way for a rapid prototyping thanks to visual editor which lets the user to build a 3D scene by dragging and dropping objects into the scene window and running it without the need to build the application (Fig. 1). Each object in the scene can be modified using attached scriptable components. Behaviour and properties of any object can be controlled by custom scripts.

Unity supports programming in both C# and JavaScript languages and, at the same time, it supports Microsoft .NET sockets. This translates to efficient and simple communication with any other Windows application based on .NET framework. This was essential since a separate piece of custom software is responsible for catching and parsing NMEA sentences from the network and sending processed data to the visualization system.

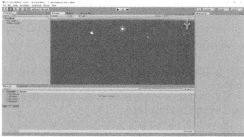

Figure 1. Unity engine editor window

As a complete game engine Unity handles not only rendering of each frame but also manages every sub-process in the application, including initialization, physics calculations, input events and game logic (Unity documentation 2017).

Figure 2. Unity editor, scene window - setting up DP vessel model

Water is a critical element in any ship simulator which can significantly improve the level of realism in a scene. It is very difficult to render properly due to high visual complexity present in the motion of water surfaces, as well as in the way light interacts with water (Kryachko, 2005). For this reason, it was decided to utilize Unity extension capabilities and use PlayWay Water - a complete water system available through Unity Asset Store, where independent developers and game studios can put their own extensions (assets) for use by other developers. This system provided high quality ocean surface with wave height dynamically set during simulation, accurate reflections and interaction system based on wave height. Similarly, we used UniStorm extension for a complete weather system that allows for a dynamic change of time of day and setting weather conditions, including rain, snow, fog and clouds.

Unity input manager and system of virtual cameras made it fairly straightforward to handle joystick input for looking around and changing the point of view.

Figure 3. Weather system: fog and change of time

The proprietary configuration of the Kongsberg DP Advanced Trainer made it impossible to directly access roll, pitch and yaw data through the network connection. To achieve approximate behaviour of the model in the visualization system it was decided to use the physics interaction provided with the water system for which a ship's hull volume had to be created and implemented in the form of so called Collider. Rotation of the collider was calculated for a 2D cross-section that has the same Y coordinate as the ocean plane. Height of the wave's mesh vertices were retrieved for the fixed number of evenly spread points on this section and the rotation of a whole Collider is calculated and set in each frame. This solution however resulted in roll angles much higher than those calculated by DP simulator. In order to bring roll values to acceptable empirical values the Collider's width had to be scaled up by a factor of 2 (Figure 4).

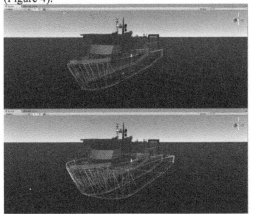

Figure 4. Ship model with attached colliders (green lines): collider based on ship geometry (upper image) and scaled collider (lower image)

Figure 5. Screenshot from the custom visualization system for DP simulator

4 CONTROL OF THE VISION IN DP SIMULATOR

Visualization must be related to the manoeuvring situation in DP simulator. This is accomplished by

"DP – vision configuration" C# program, shown in Figures 6-8. This program was designed to transfer information between simulator and vision program. The control application collects data about ship's heading and position from the simulator network. The data is parsed from NMEA0183 standard of HDT and GGA message type. In the next step the ship's position is recalculated to UTM. Procedures for these calculations are described in (Snyder, 1987). Every second message with all data is sent to vision program. Data format is as follows: *"Working,{0:f3},{1:f3},{2:f1},{3},{4},{5:f0},{6:f3},{7:f1},{8}"* where:

– {0} is position Northing with start offset [m],
– {1} is position Easting with start offset [m],
– {2} is heading [°],
– {3} is working status (1-on, 0-off),
– {4} is fog status (1-on, 0-off),
– {5} is fog range [m],
– {6} is wind speed [m/s],
– {7} is wind direction [°],
– {8} is time from midnight [s].

The heading is predicted by simplified Kalman filter. Otherwise, from the moment program loses heading data from DP Trainer, the screen will flicker and shake. The program interface has the functionality to change parameters in a fixed time, as presented in the Figure 8.

Figure 6. "DP – vision configuration" interface after start-up

Figure 7. "DP – vision configuration" interface after connection to DP simulator and visualization application

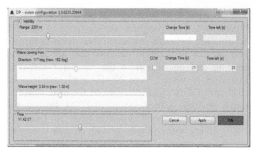

Figure 8. "DP – vision configuration" interface with change of wave height and direction indicated

5 CONCLUSIONS

The authors designed the visual system capable of integration with Kongsberg type DP Advanced Trainer leading to the DP Bridge Simulator of Class B.

The visual system presented has been well received during the reaccreditation visit of NI representatives. It can be easily operated by the simulator instructor and tuned to the parameters of external environment set in DP simulator while receiving online data of the DP ship in 3 DoF. This way the simulated sea state, ownship and targets visualization aligns with any changes in simulated weather.

REFERENCES

[1] Eberly, D.H.: 3D Game Engine Design: A Practical Approach to Real-Time Computer Graphics, CRC Press, 2006

[2] Kryachko Y.: Using Vertex Texture Displacement for Realistic Water Rendering, GPU Gems 2, Nvidia Corporation, 2005

[3] NI: The Nautical Institute Dynamic Positioning Accreditation and Certification Scheme Standard, January 2017, London, 2016

[4] NIMA Geodesy and Geophysics Department: World Geodetic System 1984, Its Definition and Relationships with Local Geodetic Systems, NIMA TR8350.2, 3rd Ed., Amendment 1, USA, 2000.

[5] Snyder J. P. Map projections: A working manual, Professional Paper 1395, U.S. Government Printing Office, Washington D.C., 1987

[6] Unity documentation: Manual and API Scripting, https://docs.unity3d.com accessed: 25.01.2017

[7] Zalewski P.: DP Simulator Course Manual, Wydawnictwo Naukowe AM w Szczecinie, Szczecin, 2012

Visualization of Data

Proceedings of 12th International Conference on Marine Navigation and Safety of Sea Transportation, TransNav 2017
21-23 June 2017, Gdynia, Poland

The System of the Supervision and the Visualization of Multimedia Data for BG

M. Blok, B. Czaplewski, S. Kaczmarek, J. Litka, M. Narloch & M. Sac
Gdańsk University of Technology, Gdańsk, Poland

ABSTRACT: Monitoring of country maritime border is an important task of the Border Guard. This task can be facilitated with the use of the technology enabling gathering information from distributed sources and its supervision and visualization. The system presented in the paper is an extension and enhancement of the previously developed distributed system map data exchange system. The added functionalities allow supplementation of map data with multimedia (telephone and radio calls, video (cameras), photos, files, SMS/SDS) and presentation of current and archival situation on a multi-display screen in the Events Visualization Post. The system architectures, functionalities and main system elements are briefly described and supported with preliminary analysis and test results.

1 INTRODUCTION

The Border Guard (BG) has to be equipped with a technology enabling communication, acquisition, exchange and visualization of data in different operational situations. Currently the Polish Republic maritime border is monitoring and supported by the Automatic National System of Radar Control for Maritime Areas of Poland (*Zautomatyzowany System Radarowego Nadzoru Polskich Obszarów Morskich - ZSRN*) which is an integrated security system (Gałęziowski A. 2005, Fiorini M. & Maciejewski S. 2013). The main limitation of this system is that BG mobile units are not connected with the central system, which have been addressed in our former project (Kaczmarek S. 2013-2015). However, even the extended system collects only the basic tactical data about observed objects, like position, speed and name, and provides communication means. At the same time, the assessment of the ongoing or archival situations requires integration of not only tactical map data but also of telephone and radio calls, videos (cameras), photos, files, SMS/SDS, AIS and radar data. The volume and types of information gathered by the BG require a specially designated system architecture, particularly when multiple, potentially multicomponent (multimedia), operational tasks have to be presented on a multi-display. The requirement for visualizing archival events besides the ongoing ones increases the system design difficulty. The paper presents a realization of such a system designed for the BG with tactical information collected from mobile units (vehicles, airplanes and vessels), Observation Points (OP) and Web Service. An important feature of this system is that it is distributed with collected information composed of telephone and radio calls, video (cameras), photos, files, SMS/SDS, AIS and radar data. Visualization of the current or the archival situation can be composed of an arbitrary combination of such components and must be managed by the personnel of the Events Visualization Post (EVP).

The crucial elements in the system architecture are Archive Servers (AS) based on the approach utilized in the cloud technology and the multi-display visualization post (EVP) for visualizing multimedia events. Visualization conditioning is completely managed by the visualization post (EVP) personnel. This paper presents the functionality, the concept and the realization of the system, hardware and software implementation, and the initial tests results. The project is co-financed by the NCBiR (National Centre for Research and Development) and implemented for the Border Guard (BG). The system presented in this paper is the extension and enhancement of the previous project (Kaczmarek S. 2013-2015) in which the Border Guard distributed map data exchange system have been developed (Blok et al. 2016a).

The rest of the text is organized as follows. The proposed system architecture and functionality is described in Section 2. The realization of functional elements of this system are presented in Section 3. The first and main test results and its analysis are described in Section 4. Summary and future work regarding our system are described in Section 5.

2 SYSTEM ARCHITECTURE AND FUNCTIONALITY

Distributed characteristic of the proposed system architecture (Fig. 1) is caused by requirements of BG (mainly strong support for mobile units including bidirectional data exchange and voice, video communication), geographically distributed sources of data (GPS, AIS, ARPA including also data provided from external WebServices and surveillance cameras) and a general rule for building complex IT solutions as interconnected set of sophisticated components performing specialized task for data acquisition, processing, storage and visualization with the aid of advanced technology.

Another distinguished feature of the developed system is its "all IP paradigm" which results in unified form of communication through TCP/IP network (including mobile IP radio links) accordingly for data, video and even interactive voice through application of Voice over IP (VoIP) technology for real time communication between BG personnel.

System architecture consists of stationary and mobile parts. The main functionalities of stationary part are gathered in the Central Supervision Centre (CSC) where the major elements are located. One of the most important element in the CSC is the CENTRE server with the custom MapServer application developed for storing in the SQL database and processing data used in constant visualization of digital maps on stationary consoles and on demand on consoles in mobile units. Data gathered in the CENTRE for map visualization originate from all system sensors (including far and mobile) and can be accessed by any mobile MapServer and its associate mobile console on Mobile Units (MU). The CENTRE server with the MapServer application is described in details in Section 3.1. Another key element in the CSC is the Events Visualization Post (EVP) which is a dedicated PC server with a custom application and a specialized hardware allowing (with the aid of the CENTRE and Archive Servers) presentation of a complex tactical situation with various data of several types (including digital maps, still images, audio/video, SMS/SDS) on multiple displays joined into large logical screen. The EVP is described in

details in Section 3.4. A vital role in the designed system plays a specialized entity named Archive Servers which are sophisticated solution based on cloud computing concept with a NoSQL database, a distributed file system, and other "state-of-the-art" technologies for "Big Data" storage and processing among cluster of application and archive serves available by WebService (http/JSON request) and Message Oriented Middleware communication channels. Archive Servers are used as an advanced "data base" for storage and retrieval of files, still images, audio, video and SMS/SDS data for the purpose of presentation in the EVP. Archive Servers are presented in detail in Section 3.5. The CSC is equipped with multiple Stationary Consoles which are used for presentation of tactical data on digital maps. Moreover, Stationary Consoles are also telecommunication terminals for interactive, real time voice communication among BG personnel accomplished with the aid of VoIP technology. That form of communication was a fundamental design demand for the proposed architecture as an advanced form of a modern dispatcher system. Thus, in the proposed system, an infrastructure for VoIP communication is provided including Telephone, Radio Servers and GSM gate for communication with a public mobile telecommunication network including possibility of sending and receiving SMS messages. It is worth noting that for the sake of investigation and inspection purposes every call and message in VoIP communication of the proposed system is recorded in the Recorder and stored in Archive Servers for future retrieval accordingly to demands. That feature was imposed by another important feature of the proposed system, which is possibility of simultaneous presentation of not only current, but also of all archival data from multiple sources including multimedia content (voice, video, images) and tactical data from maritime sensors and external WebServices both public and confidential.

All elements in the CSC are attached to a high speed LAN and to the dedicated IP based BG Wide Area Network (WAN). The BG WAN is used to communicate with multiple, geographically distributed autonomous OPs (Observation Points). The OPs belong to fixed part of the system architecture and are used as remote posts for mounting system sensor (AIS, radar, surveillance cameras) and radio antennas for communication with Mobile Units. Each OP is equipped with specially customized version of the MapServer application (described in details in Section 3.2) providing data from the set of sensors which cannot directly communicate through TCP/IP network.

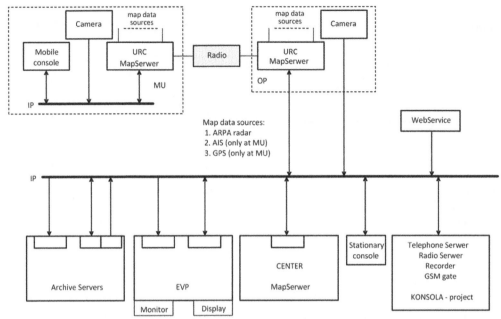

Figure 1. Diagram of the proposed multimedia distributed system (STRADAR).

In the system Mobile Units represent BG patrol units which can be land vehicles, airplanes and sea vessels. The system architecture is designed to provide advanced support for mobile units including bidirectional communication over broadband IP based (packet) radio link which is developed under another project (Stefański J. 2014-2017). Mobile Units consist of a specially designed server URC enclosed in a rugged case, appropriate set of sensors, a radio modem for broadband IP communication and a military class laptop (mobile console) connected to a high speed LAN. From the system architectural point of view consoles on mobile units are offered the same set of services (including VoIP) as stationary consoles with small limitations imposed by IP radio link communication. Operators of mobile consoles focus their main attention on situation around patrol units and information gathered from local sensors (GPS, AIS, ARPA, camera) so the URC is equipped with the special version of the MapServer supporting them in their duties.

Usage of an universal IP network in the proposed system allows application of respective protocols from TCP/IP family for communication between particular elements. Regarding data communication between consoles, the EVP and Mapservers a concept of communication based on Message Oriented Middleware is concerned. Particularly an ISO/ECMA standard protocol named Advance Message Queuing Protocol (AMQP) is applied. Moreover the AMQP and HTTP are used for communication with Archive Servers. Regarding interactive VoIP communication the SIP is used as signaling (control) protocol between terminal and telecommunication servers. In the proposed system media (voice and video) are carried by the Real-time Transport Protocol/Real-time Transport Control Protocol (RTP/RTCP). This concept allows application of Real Time Streaming Protocol (RTSP) for control of multimedia sessions, particularly video from surveillance cameras and video recordings stored in Archive Servers. The elements of the whole system are synchronized with the aid of the Network Time Protocol (NTP) and a central time server. Application of GPS in time synchronization is possible as a backup in the case of isolation of particular, mainly mobile, units.

3 REALIZATION OF FUNTIONAL ELEMENTS

In this section main elements of the system described in the previous section are presented in more details.

3.1 *CENTER*

The main functionalities provided by the CENTER are (a) map data gathering and providing, (b) tasks notifications processing and (c) Events Visualization Post (EVP) support. All these tasks are implemented in the software called the MapServer (Fig. 2).

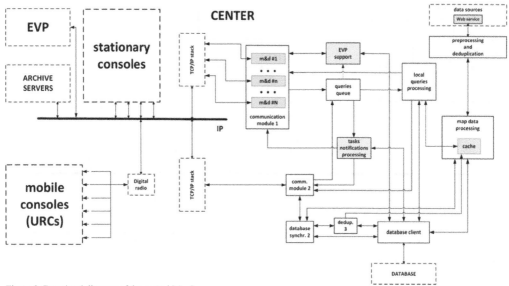

Figure 2. Functional diagram of the central MapServer.

The map data gathered in the CENTER come from locally attached sources (WebService), which are processed by the map data processing module, and from sources attached to mobile MapServers operating on mobile units (MU) (ARPA radar, AIS, GPS). The data collected by the MUs are sent to the CENTER during database synchronization sessions through the radio channel served by the communication module 2 and are processed by the database synchronization module. Map data duplicates, which are inevitable in case of data gathered from different sources operating in the same area, are removed by deduplication modules. After the deduplication, map data are inserted into the central database from where they can be accessed on demand. Requests for map data processed in the local queries processing module can be received from the EVP and stationary consoles through fixed network (communication module 1) or from mobile consoles through the radio channel and fixed network (communication module 2).

The other two main functionalities of the MapServer (tasks notifications processing and the EVP support) relate to system features enabling presentation of map and multimedia data in the EVP. The tasks notification processing module receives notifications containing information about tasks consisting of map and multimedia elements which are proposed for presentation in the EVP. These propositions can be created by the operators of stationary or mobile consoles and the EVP operator. Notifications received in the CENTER are stored in the central database and the updated list of tasks is sent to the EVP where the operator can either accept

for presentation or discard each task notifying the CENTER about their decision.

The EVP support module processes the EVP requests concerning visualization of task elements (map data, files/photos, SMS/SDS, video from cameras, audio – recorded calls). For each visualization start request the module retrieves the details of the requested task element from the central database, which have been stored there by the tasks notifications processing module. When the task element refers to map data, then the request is processed by the MapServer. For file/photo, SMS/SDS, video or audio task elements, a request is sent to the multimedia archive servers (AS), which respond with metadata (for visualization of archive data) or create a subscription (for visualization of current data). During a subscription the metadata associated with the requested type of current information are provided by the multimedia archive servers. The EVP support module gathers all metadata received from the multimedia archive servers and sends them to the EVP, which allows the EVP to fetch the demanded information directly from the multimedia archive servers in order to visualize current or archive events. When the EVP finishes visualization of a specific task, a visualization end request is sent to the EVP support module which updates the task data stored in the central database accordingly. Apart from requests regarding visualization of task elements, the EVP support module also handles requests for adding notes to multimedia data stored in the AS, which can be generated either by the EVP or consoles (both mobile and stationary) operators. All received add note requests are processed in the EVP support module and forwarded to the AS. The result of the performed operation is retrieved from the AS and passed to the originator of the add note request.

3.2 *Observation Point*

The central MapServer gathers map data from local sources, MUs as well as OPs and stores them in the central database. The MapServer in MU (Blok et al. 2016a) collects data from sources available at this unit and stores them in the local database at the same time synchronizing it with the central database using the best effort approach. Additionally, the mobile MapServer provides mobile consoles with map data stored in the local database or retrieved from the CENTER through radio channel.

In the current project map data are also collected in OPs, however, since there are no mobile consoles in OPs, the MapServer in OPs can be simplified with its functionality limited only to map data gathering and synchronization (Fig. 3).

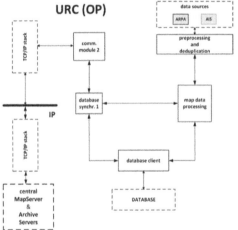

Figure 3. Functional diagram of the MapServer in the OP.

3.3 *Mobile unit*

Functionality for mobile units derives from the KONSOLA project (Blok et al. 2016, Czaplewski et al. 2016), which was to complement existing systems, currently used by the Border Guards, with the transmission of data on supervised objects, i.e. marine vessels, land vehicles, and aircrafts, from the mobile unit of the border guard to the stationary network, as well as, from the network to the mobile unit. In the current STRADAR project, the functionality for mobile units was extended by new tools for reporting events for visualization, i.e. for sending tasks for visualization of events in the Events Visualization Post (EVP).

A mobile unit, regardless of its type, is equipped with a Universal Radio Controller (URC), a mobile console, a radio link, an ARPA radar or an AIS receiver, and a GPS receiver. A mobile consoles are military class laptops, which are connected to URCs *via* local IP network. URCs are equipped with a local database, which has the information about supervised objects present within the sensor range of a mobile unit, the local MapServer, which is responsible for providing appropriate map data, processing incoming tasks for the EVP from mobile consoles, and providing communication with the CENTER *via* radio link. All the URCs are automatically uploading their data to the central server in order to synchronize databases. Each URC is able to download the data from the central server, which collects data about the naval situation from the network of AIS receivers and radars of the Maritime Office provided by the web service and other URCs. In this way, mobile consoles can visualize the data on supervised objects located beyond the reach of the sensors of the BG mobile unit. The structure of the equipment of a mobile unit is presented in Fig. 4.

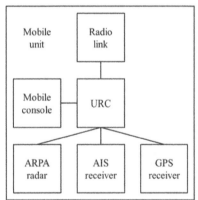

Figure 4. The structure of the equipment of a mobile unit.

Mobile consoles are equipped with the dispatcher application, which is C# .NET application for 32-bit or 64-bit MS Windows 7, 8.1 or 10. The dispatcher application for the mobile units allows the console operator to use the following functionality: data visualization on maps, radiocommunications, telephony, recording conversations, cross-network conference, intercom, SMS, file transfer, and reporting events (task generation) for EVP. In the context of data visualization on maps and event reporting, the dispatcher application consists of three tools: the MapControl, the BrowserControl and the TaskGenerationControl. All three controls cooperate with a local map server in the URC. Details about a structure and functionality of the MapControl and the BrowserContol can be found in (Czaplewski et. al. 2016). Details about a structure and functionality of the MapServer can be found in (Blok et al. 2016).

The TaskGeneratorControl is a tool for reporting events for visualization in the EVP. These reports are called tasks, and each task can include several (at least one) elements. In the project, there are 6 types of elements, which corresponds to the types of data that can be visualized in the EVP: map (visualization on the map), browser (of map objects), file/image, SMS/SDS, video, and audio.

During the generation of the task, the operator is able to define general task parameters:
– the operator's login,
– priority of the task,
– text note about the task,
and then, the operator is able to define parameters specific to the desired type of element. The presented below sets are search criteria for the Archive Servers or the MapServer. For each element, operator has to define at least one parameter. For "map" element, the set of parameters is as follows:
– time of situation,
– geographic coordinates,
– indexes of objects for detailed presentation,
– indexes of objects for presentation of trails,
– filtration of symbols on the map,
– filtration of labels on the map.
The set of parameters for "file/image" element is:
– type of data (file or image),
– identifier of file/image,
– login of the operator who created the file/image,
– source of the file/image,
– indexes of objects corresponding the file/image,
– geographic coordinates,
– time of creation of the file/image,
– text note about the file/image.
The set of parameters for "SMS/SDS" element is:
– sender number,
– receiver number,
– identifier of the SMS/SDS,
– text of the SMS/SDS,
– login of the operator who created the SMS/SDS,
– source of the SMS/SDS,
– indexes of objects corresponding the SMS/SDS,
– geographic coordinates,
– time of creation of the SMS/SDS,
– text note about the SMS/SDS.
The set of parameters for "video" element is:
– identifier of the video,
– login of the operator who created the video,
– source of the video,
– indexes of objects corresponding the video,
– geographic coordinates,
– time of creation of the video,
– text note about the video.
The set of parameters for "audio" element is:
– sender number or IP in case of VoIP,
– receiver number or IP in case of VoIP,
– forwarded number or IP in case of VoIP,
– identifier of the audio,

– login of the operator who created the audio,
– source of the audio,
– indexes of objects corresponding the audio,
– geographic coordinates,
– time of creation of the audio,
– text note about the audio.
Besides the task generation, the TaskGeneratorControl can store macros of defined elements and a history of generated tasks. Operators can save, load, edit, and delete macros containing predefined elements and their parameters. Moreover, operators can browse the history of sent tasks to quickly import any task, modify it, and send again at any time.

3.4 Events visualization post

The Events Visualization Post operates on a PC with high resolution multi-screen display and software designed for simultaneous visualization of data of different types and synchronized in time. The PC is configured for the purpose of allowing an operator to visualize tasks coming from the consoles and to set new tasks for visualization. The PC has been built on an Intel 64-bit processor (in this case Intel Core i7 4790K 4 GHz), 32 GB of RAM and two graphic cards, one being a processor integrated Intel HD Graphics 4600 and second being a dedicated graphic card with AMD Radeon Chipset (Club3D Radeon 7850 2 GB Eyefinity 6). Windows 8.1 Pro (64-bit version) has been chosen as EVP's operating system.

The basic concept for the EVP is to allow an operator to visualize elements of given tasks. For that purpose, a solution of using two screens has been implemented. We have proposed to connect the EVP to a Management Screen (which purpose is to allow the operator to manage the visualization process and allow them to receive and generate tasks) and a Multidisplay (which is tasked with displaying multimedia elements of tasks the operator works on).

To allow a big space for visualization a single display is not enough, therefore a solution of configuring a number of display devices into a single logical display have been installed (hence the name "Multidisplay"). According to Windows operating system one can plug as many displays to the PC as the number of graphic outputs available in the machine. To achieve a unified multidisplay a solution for creation of a logical screen has been decided on. AMD's technology Eyefinity 6 is a feature of novel graphic cards from the Radeon line that allows to create a single logical display from up to six physical displays connected to a single Radeon graphic card via miniDisplayPort interface. For the EVP six monitors have been connected to a single Eyefinity 6 compatible graphic card to create a logical display.

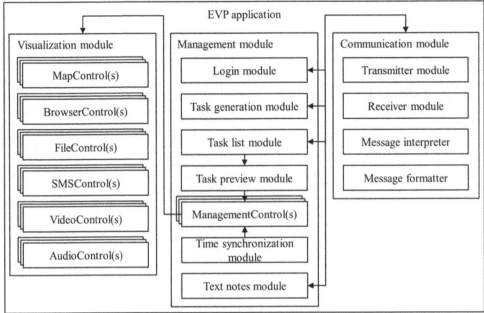

Figure 5. The general structure of software of the Events Visualization Post (EVP) application.

Since for the Management Screen a single physical monitor suffices, a second logical screen from the operating systems point of view have been attained by connecting a display to a DVI interface on the motherboard.

The Events Visualization Post (EVP) application is programmed in C# .NET for 32-bit or 64-bit MS Windows 7, 8.1, or 10. The EVP application allows for receiving event reports (tasks) from mobile consoles, generating new event reports (tasks) in the EVP, managing the list of tasks, visualizing the data of various types (data on the map, browser of naval objects, files, images, SMS/SDS, video, and audio), synchronizing the data in time, adding text notes to the data, and much more. The general structure of software is presented in Fig. 5.

In general, three main modules can be distinguished in the EVP application, namely:
- *Management module*, which is for the management of data acquisition, data visualization, communication, and the flow of the application,
- *Visualization module*, which is responsible for the presentation of various types of data on the multi-screen display,
- *Communication module*, which is responsible for the communication with the central server and the proper formatting of messages.

Management module consists of:
- **Login module**, which is used for logging the operator in the system and sets all the starting parameters,

- **Task generation module**, which provides task generation, task history, element macros, analogous to the TaskGeneratorControl functionality,
- **Task list module**, which is responsible for managing the list of tasks, including acceptance, preview, and rejection of tasks and elements,
- **Task preview module**, which is used for more detailed preview of elements, including browsing the metadata found in the Archive Servers,
- **ManagementControls**, which are used for controlling the presentation of different types of data in the visualization module,
- **Time synchronization module,** which is responsible for synchronizing ManagementControls and their corresponding visualization controls,
- **Text notes module**, which is used for creating custom text notes to the existing documents in the Archive Servers.

Communication module consists of:
- **Transmitter module**, which is responsible for sending outgoing messages to the CENTER,
- **Receiver module**, which is responsible for receiving incoming messages from the CENTER,
- **Message interpreter**, which is responsible for interpreting messages in JSON format and creating corresponding C# objects,
- **Message formatter**, which is responsible for formatting messages in JSON format on the basis of appropriate C# objects.

Visualization module consists of:

- **MapControls**, which are designed for visualization of marine objects on digital maps (Czaplewski et al. 2016),
- **BrowserControls**, which are designed for visualization of marine objects in the browser (Czaplewski et al. 2016),
- **FileControls**, which are designed for visualization of images in raster formats and files in the system file browser,
- **SMSControls**, which are designed for visualization of SMS and SDS messages from multiple sources in the form of chat,
- **VideoControls**, which are designed for visualization of current and archival video streams received from the Archival Servers,
- **AudioControls**, which are designed for visualization of current and archival audio streams received from the Archive Servers.

3.5 *Archive servers*

The Archive Servers (AS) receive, interpret and collect the data coming from external systems, and provide these data on demand. External systems, which are the CENTER, URCs, OPs, the EVP, the Recorder, telephone and radio servers, and stationary consoles, need fast access to current and archive data. In order to meet these requirements we have selected an architecture in which the AS form an independent system based on BigData technology. In our case this system is composed of data gathering, processing, storing, searching, and sharing modules operating in distributed infrastructure called a cloud. In contexts of the AS a cloud is an environment providing safe storage and fast access to huge amount of data. It guarantees organization of data coming from different sources and delivers means for analysis, characterization and enrichment of these data and their descriptions, as well as tools for discovering correlations within gathered data.

The architecture of AS is characterized with the following features:
- distributed technology ensuring continuity of operations,
- uniform utilization of resources of all devices forming the system,
- copies of data stored in several places,
- data storage using a format facilitating access to data,
- access control for all gathered data.

The AS functions are aggregated into modules based on data types the given module is predestined. Each module is an independent program which has implemented the following actions:
- message exchange – means for connecting to the outside world, accepting requests and distribution, based on cloud state knowledge, of tasks to places that promise the fastest execution,

- data processing and aggregation – linking incoming data with already present data as well as business logic, which is data processing based on rules applying to particular data type and knowledge of how the data could be used in the future,
- persistence – data in the form of formatted documents are stored in a NoSQL database. These data instead of being stored in a single shared central place are distributed among local storage media of the servers. For non-text large data (>>1MB) a decentralized and distributed file system is used.

The system demonstrator consists of four hardware servers, gigabit network switch, and 19" Rack mount case. Three of the servers are dedicated to archive management, while the remaining server is dedicated to visualization management, resources and subsystems monitoring as well as a disk resources server.

Within each of the three archive management servers virtual environments have been set up, which are used for creating functional clusters. The virtualization utilizes method called LXC (Linux Containers). The LXC is not an independent operating system running on the machine, but is a separate space within the operating system. It allows allocation of CPU, memory, disk space and network interfaces. For programs running in LXC containers processes, network access, user privileges and access to the files are separated. From the point of resource commitment the LXC virtualizes only the application and not the entire operating system, therefore it does not need a separate virtual machine with its own operating system. This means that the LXC generates little overhead because all the applications use standard libraries for system calls, I/O and networking.

The AS architecture does not impose restrictions on the number of processed messages, the size of the retained data or the number and complexity of the external requests. Their performance is limited only by the available processing power, memory and disk space. Horizontal scaling of the AS is recommended, as the NoSQL systems are designed and built to allow almost linear increase in performance by adding more machines to the cluster.

4 ANALYSIS AND TEST RESULTS

Models of main elements of the discussed system have been presented in the paper (Blok et al. 2016b). The models of the following elements were analyzed: CENTER, MUs, OPs, EVP and AS. The obtained partial results were used to analyze the overall system performance under several scenarios selected based on system functionality requirements. In sec. 4.1 conclusions from the abovementioned analysis and in sec. 4.2 functional tests results of the system under development are presented and briefly discussed.

4.1 *Results of performance analysis*

The performance analysis has been focused on the following parameters: average system reaction time, bitrate on the interfaces and load of processing units (processors) since these three values are critical to the operation of the analyzed system.

The results of the analysis show several critical points of the system. In the CENTER the main problem is high processing units load close to 90% resulting from large volume of collected map data, which are stored in and retrieved from the central database. This problem additionally results in increased delay introduced by the CENTER. Similarly, the processing unit load in the EVP is also close to 90%, which results from simultaneous processing of several multi-element multimedia tasks. As a result, the maximum reaction time for control messages is 3.208 seconds. For media maximum reaction time is equal to 3.479 seconds when media are uploaded to AS, while for media presentation in the EVP it is equal to 1.15 second for presentation of ongoing situation and 3.727 seconds for presentation of archival event.

The aforementioned problems can be solved with the use of more efficient multicore processors which can be readily adapted since the developed models and software assume a multithread implementation with processing threads designated to specific functionalities subsets. As a result, we can adjust the load of different functional components of the system.

Another critical point are radio modems, because of their limited bandwidth allowing only a single video stream, and Archive Servers because of large volume of uploaded and downloaded multimedia information. The problem related to Archive Servers can be simply solved by adding additional servers to the setup with software automatically distributing tasks between particular machines. The only problem indicated here, which cannot be solved within current project, is related to the maximum radio modem bitrate since it has been imposed by the specification of the concurrent project in which this modem is being developed (Stefański J. 2014-2017). However, because of modular structure of the developed system this radio modem can be readily upgraded to one with increased bitrate or video coder can be changed for one offering a smaller output bitrate.

4.2 *Functional tests*

At the current stage of system development the MapServer functionality tests are based on analysis of system response to locally generated requests. In the debugging mode the EVP or console requests addressed to the task processing or the EVP support module are injected into the communication module 1. Since all modules log all crucial actions along with precise times into separate files, it is easy to analyze how the given request is handled. Additionally, since many requests result in update of the data stored in the local database, the correctness of requests processing can be also assessed based on the analysis of the change of the database content.

In case of the EVP support module all the expected requests, with the exception of add note requests, come directly from the EVP. Differently, the requests in the task processing module also come from mobile consoles in which case they are handled first by the mobile MapServer which, if it is necessary forwards the request to the central MapServer and returns the obtained response to the console. The functionality of such requests is tested with mobile and central MapServers connected using IP network and the test request generated locally in the mobile MapServer. This requires analysis of logs and databases in both the mobile URC and the CENTER. The radio link failure in the connection between mobile and central MapServer can be simply simulated by switching the CENTER off which results in mobile requests being processed in the mobile MapServer without communication with the CENTER.

The tests of the task processing module covered:
- new task requests from the EVP processed only at the CENTER,
- new task requests from stationary consoles which communicate directly with the CENTER; in this case the updated list of tasks is sent to the EVP,
- new task request from mobile consoles which are passed by the mobile MapServer to the central MapServer; the updated list of tasks is sent to the EVP,
- operator id requests: local from the EVP and stationary consoles and remote from mobile consoles; if it is possible the global operator id is retrieved from the CENTER but in the case of radio connection failure the mobile console operator receives the temporary local id; additionally, when the EVP operator requests id, the current list of tasks stored in the central database is sent to the EVP.

The performed tests of the EVP support module included the following scenarios:
- visualization start requests for map data generated by the EVP (the address of the central MapServer is returned in the response),
- visualization start requests for archival files/photos, SMS/SDS, video, and audio generated by the EVP (the EVP support module retrieves the details of the requested task element from the central database, sends an HTTP/JSON

request to the AS, processes the response with metadata from the AS and forwards it to the EVP; if multi-page results are retrieved from the AS, subsequent HTTP/JSON requests are generated to the AS in order to get all pages and pass them to the EVP),

- visualization end requests for all types of task elements generated by the EVP (the EVP support module updates task list in the central database and sends a response to the EVP),
- add note requests generated by the EVP or consoles (the EVP support module forwards the request as a HTTP/JSON message to the AS and returns the result of the operation).

It is worth mentioning that in order to perform functional tests of the EVP support module a simulator of the AS was additionally implemented based on the Linux xinetd daemon and a set of bash shell scripts. This allowed us to verify the implementation of communication procedures between the EVP support module and the AS, which are based on HTTP protocol messages with JSON content.

5 SUMMARY

The main goal of the system presented in the paper is to supplement the basic map data gathered by the Border Guard with multimedia information (telephone and radio calls, video (cameras), photos, files, SMS/SDS) in order to be able to provide complete presentation of ongoing or reconstruction of archival events. To achieve this goal beside updating the previously developed system elements, two new crucial elements have been added. These are Archive Servers (AS) based on the approach utilized in the cloud technology and the multi-display visualization post (EVP) for visualizing multimedia events.

The presented system architecture have been analyzed based on models of its elements, which helped to determine potential weak points. Currently the all system elements are at the advanced stage of implementation with preliminary test performed and await full system integration stage.

This work has been co-financed by NCBiR (National Center for Research and Development), projects DOB-BIO6/10/62/2014.

REFERENCES

Gałęziowski, A., 2005, Automatic National System of Radar Control for Maritime Areas of Poland (in Polish), Przegląd Morski, 5, pp. 50-70.

Fiorini, M., Maciejewski, S., 2013, Lesson Learned During the Realization of the Automated Radar Control System for Polish Sea-waters (ZSRN), in: A. Weintrit (ed.) Advances in Marine Navigation, Marine Navigation and Safety of Sea Transportation, CRC Press/Balkema, London, pp. 217-221.

Kaczmarek S. (project manager) 2013-2015. Koncepcja oraz implementacja integracji informacji w rozproszonych elementach systemu wymiany danych Straży Granicznej (Concept and implementation of information integration in distributed elements of the Border Guard data exchange system), research project, National Centre for Research and Development (NCBiR), DOBR/0022/R/ID1/2013/03.

Stefański J. (project manager) 2014-2017. System szybkiej transmisji danych multimedialnych dla potrzeb ochrony morskiej granicy państwowej (System for rapid multimedia data transmission for the needs of protection of country maritime border), research project, National Centre for Research and Development (NCBiR), DOBR-BIO6/09/5/2014.

Blok M., Kaczmarek S., Młynarczuk M. and Narloch M. 2016a. MapServer – information flow management software for the Border Guard distributed data exchange system, Polish Maritime Research, 91(3), pp. 13-19.

Blok M., Czaplewski B., Kaczmarek S., Młynarczuk M., Narloch M. and Sac M. 2016b. Multimedia distributed system for visualization of ongoing and archival events for BG, The International Tech-Science Conference on „Naval Technologies for Defence and Security" NATCON 2016, pp. 61-76.

On Geoinformation Structures in Navigation According to ISO Series 19100 Standards

W. Pachelski

Military University of Technology, Warsaw, Poland

ABSTRACT: In front of a rich diversity of problems in land, sea and air navigation there is a need of proper and effective organization of the relevant information structures. This can be done with the use of general rules of geographic information modeling as formulated in International Standards of the series 19100, by means of the Unified Modeling Language. The paper gives a review of such rules for the topics of positioning, location based services, motion modeling of moving objects, as well as for models of transport networks of different kinds. There are presented only general views into the models, while all details, as attributes, data types, stereotypes and others can be get from the quoted sources.

1 INTRODUCTION

The subject of this study are those structures of geographic information, which are either directly related to certain issues of navigation, or indirectly associated with navigation system. The structures are defined in details in the ISO 19100 series of standards. Here they are discussed in an overall way to emphasize the general concepts of modeling navigation related information, along with explanations and comments necessary for the their interpretation. The structures are given with the use of the Unified Modeling Language (UML), which is commonly used for information modeling.

The above mentioned structures include:
- determination of the position of objects by means of the currently available technologies,
- location-based services, including: a reference model of the services, tracking, navigation and routing,
- description of the movement of moving objects,
- transport networks.

2 BASICS OF UML

As a formal tool for the description of these structures is used the UML (the Unified Modeling Language). Its main notational measures are presented in Figure 1.

Among many existing formal languages the most advanced in terms of rich measures is UML - the Unified Modeling Language. This language, officially defined in (ISO 19501), is widely used and recommended in ISO 19100 series standards and data specifications of INSPIRE to describe conceptual models built in all areas of geographic information.

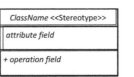

Figure 1. Class designation in UML.

The basic categories of the class diagrams in the UML include (ISO 19103 2015):
- Class: description of a group of objects that have the same meaning, attributes, operations, methods, and associations.
- Stereotype: an extension of the meaning of data types and classes.
- Attribute: a representation of features common to all objects of a particular class.
- Operation:-representation of the services.

Different types of relationships between the classes in UML shows the Figure 2, whereby:
- association: a semantic connection between two instances;
- navigation: directed association;
- aggregation: "components" are parts of "container";

- composition: strong aggregation – "components" do not exist without the "container";
- generalization: a relationship between an element and the subelements;
- dependency: the use of one element by another.

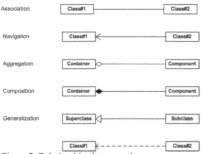

Figure 2. Relationships between classes.

UML is a language of general use in modeling information structures, that is in a conceptual modelling. There are two main roles the language plays in conceptual modeling of any real-world area, namely:
- the role of cognitive, i.e. a complete, unambiguous, coherent, etc., description of the real world in terms of geoinformation, showing just how this world is built on, and
- the role of technology, involving the creation of a universal and general pattern, independent of the diversity of environments, for implementing compatible applications in these environments.

In terms of this paper apply both of these roles.

3 POSITIONING SERVICES

Positioning services include a wide variety of measurement technologies, which provide coordinates within appropriate reference systems, together with related information. They are also used in a wide range of applications in many different fields, as shown in Table 1.

Table 1. Overview of technologies and applications of location-based services (Kresse & Fadaie 2004).

Positioning technologies	Areas using object positions
satellite (incl. GNSS)	geodesy
inertial INS (incl. Total Station)	cartography
line	navigation
integrated	intelligent transport systems
evolving	construction
	agriculture
	dispatch
	evolving

There is considerable diversity as to the forms (structures), contents, and accuracies of resulting information used by individual technologies, as well

as to the forms, contents and accuracies of information specific to individual fields. Hence the standard ISO 19116 defines such of general use information structures that are common for both various measurement technologies, as well as they generalize structures used in a variety of fields. The defined conceptual model is therefore a kind of interface for data communication on position between a wide range of positioning technologies and diversity of users.

The location service (PS_PositioningService) consists of obtaining and providing information about the position of a point or object. As indicated in Figure 3, this service consists of information contained in the three classes of the UML class diagram:
- PS_System - identifies the system and its capabilities,
- PS_Session - identifies the measurement session and performance of the measurement system,
- PS_ObservationMode – provides configuration and mode of operation of the measuring system and results of the measurements.

Additional classes shown in Figure 3, PS_InstrumentId, PS_Observation and PS_QualityMode, being the components of the classes PS_System and PS_ObservationMode, provide supplemental information on measuring instrumentation, the resulting observation value and the observation quality.

The summing - up of the overall structure of the information for location services is presented in Table 2.

Table 2. The overall structure of the information for location services

Class content and name	Attribute information
System information (PS_System)	System type
	System capabilities
	Referencing method
	Instrument identification
Session identification (PS_Session)	Session identification
	Dataset initiative
	Observer identification
Mode of operation (PS_ObservationMode) (PS_QualityElement)	Positioning reference system
	Link to reference system
	Coordinate transformations
	Quality element
Observation (PS_Observation) (PS_ObservationQuality)	Observation date time
	Result
	Object ID
	Offset
	Quality result
GNSS technology (PS_Observation)	Operating conditions
	Measurement conditions
	Performance indicators
	Solution vector
	Raw measurement

Figure 3. Major data classes of positioning services.

4 LOCATION-BASED SERVICES

Global navigation satellite systems created numerous applications that combine the obtained positions with other types of information, such as data on access to all kinds of services, depending on a current location, or data on optimal road links. The largest family of these applications is called Location Based Services (LBS). This is a general name and includes any technologies that provide current (online) location of the receiver position, but not only satellite navigation systems.

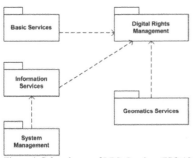

Figure 4. Subpackages of LBS_Services (ISO 19132:2007)

These services are computer applications, with a large variety of formats, data types, and data values, depending on the type of service and location of the participating objects, phenomena or events. In this situation, there is a need to establish certain conceptual framework to define set of general and universal notions and concepts applicable in this

area, as well as references to other standards in the field of geographic information.

This role plays the standard (ISO 19132:2007), which is comprised of the following three models:

- LBS_Participants: hierarchical structure of providers, brokers and users;
- LBS_Services, shown in Figure 4 and presented in more detail below;
- Message Data Model: types of the messages being sent between the participants.

In a similar way as subclasses of LBS_Resources of Digital Rights Management defined are Information Services and Geomatics Services. Their content services, as UML classes, is given in Table 3.

As indicated in Table 3, LBS include groups of services. In each of them listed are specific services as UML classes. Most of the services, apart from the System Management, depend upon the Digital Rights Management.

Digital Rights Management is intended to protect the rights of the owners of the data and services from misuse and unauthorised access. Each object or service which is assigned such a right, is referred to as a resource. This package contains:

- LBS_Resource: any resource of the system;
- LBS_License: digital recording of the rights;
- LBS_Right: a type of the rights to the resource;
- LBS_RightsCondition: restriction of the rights.

The dependence of the mentioned services means practically, i.a., that within the framework of Basic Services: LBS_ Routing, LBS_Navigation and LBS_Tracking are specializations of the class LBS_Resource in the Digital Rights Management, as shown in Figure 5.

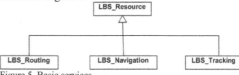

Figure 5. Basic services.

For further discussion of these services see Table 3.

Table 3. Characteristics of routing, navigation and tracking

Service class	Input/Output	Class Semantics
LBS_ Routing	Calculates optimum route. Input: a) the starting point, b) waypoints, c) endpoint and d) preferences. The returned response contains route description.	Identifies optimal route that meets the required criteria having regard to the specific nature of the type of request, the costs, and the types of restrictions and preferences.
LBS_Navigation	Navigation is based on the implementation of the route. The navigation service, given the route and tracking information, returns a stream of instructions showing the next or recovered items of this route. The input data is: a) route b) the identity of the client, and (c)) access to information. The resulting data are according to the same pattern.	"LBS_Navigation" is essential to navigation. Navigation is on the calculated route. Dynamic navigation is navigation, where the route can be changed on a regular basis due to changes of the conditions or of circumstances.
LBS_Tracking	Input: a) mobile client identity, b) access information, c) trigger information, d) disposal instructions.	Checks the permissions of the client and then returns the position of the indicated user.

5 MODELS FOR MOVING OBJECTS

ISO 19141 creates common and general information structure for applications that use information about moving objects, i.e. the ones that change their position over time. These structures relate to the geometry of the object moving as a rigid body, and include offsets and rotations, but do not include deformation of the object. So, the complex movement of the object has the following properties:

- The movement of the object is described using geometry models as described in ISO 19107, including reference systems using coordinates.
- The object moves along the planned route, allowing for deviations from the route.
- Movement can take place under the influence of physical forces, such as orbital, gravitational and inertial.
- The movement of the object can influence or be influenced by other factors, such as:
  - movement can take place according to a predefined route, as well as can be changed in known points (e.g., bus stops, points);
  - two or more objects can be "pulled" or "pushed" (e.g. train, barge);
  - two or more objects may remain temporarily in a fixed relationship (e.g. distance).

The movement model is based on a topological concept of one parameter set of geometries that can be interpreted as sets of "leaves" or trajectories, where each leaf represents the geometry of the object with the specified parameter values (such as the point position in time), and each trajectory is a curve that represents the way of the point motion as a function of the parameter (time).

The movement of the object is modeled as a combination of movements. Overall movement is expressed as a path or trajectory of a certain reference point of the object, such as its center of gravity. With a fixed trajectory the position of this center along the trajectory is determined using a linear reference system (as defined in ISO 19133) as the length along the curve.

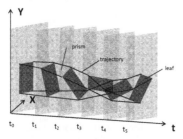

Figure 6. Object's movement and rotation

Figure 6 illustrates how the concepts of foliation, prism, trajectory, and leaf relate to one another. In this illustration, a 2D rectangle moves and rotates. Each representation of the rectangle at a given time is a leaf. The path traced by each corner point of the rectangle (and by each of its other points) is a trajectory. The set of points contained in all of the leaves, and in all of the trajectories, forms a prism. The set of leaves also forms a foliation.

These two representations of movement, the path and the position along this path, give the general location of a moving object. A separate component of the model is the rotation of the object around the selected reference point (center).

To describe this component a local reference system with the object's reference point as its origin is assumed. In this way, the "local" geometry of the object is described in the local reference system, while its location and orientation as the "external data" are given by mathematical representation of the local system in the global system. This situation is illustrated by Figure 7.

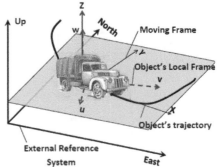

Figure 7. Global and local reference frames of the moving object.

This means that the prism of a moving object (defined as the set of all points in space, through which passes the property), can be thought of as a bundle of trajectories in local model of object's motion. Each leaf of the prism provides instantaneous and complete orientation of the object. Then the prism is called a foliation, meaning that there is a complete and separate representation of the geometry of the object for each specific time (called a "leaf").

Classes in the model of a moving object create an inheritance hierarchy of classes GM_Object and GM_Curve from the geometry models of ISO 19107. This is shown in Figure 8 and Figure 9.

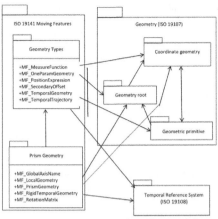

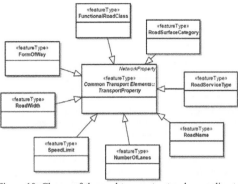

Figure 8. UML packages for moving objects modeling.

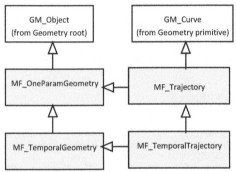

Figure 9. Specialization and inheritance of moving object classes

6 TRANSPORT NETWORKS

Among 34 themes covered by the directive INSPIRE (Directive 2007) there is the theme "Transport Networks", under which the detailed guidelines for the construction of information systems are included in the relevant data specification in the document (D2.8.I.7 2014).

In Poland technical regulations on databases of transport networks, mainly road and rail, are given in (Reg. 2011, 2013). For other type networks apply regulations cited in (Majczyna & Pachelski 2014), which, however, concern information structures only to a limited extent. In this context a comprehensive analysis and comparison of information models of transport network, as for their completeness and compliance with the INSPIRE regulations, is given in (Majczyna & Pachelski 2014).

In case of INSPIRE the basic and most general characteristics of the road transport are included in the abstract class TransportProperty of (D2.8.I.7 2014) shown in Figure 10.

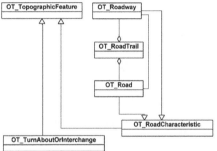

Figure 10. Classes of the road transport network according to INSPIRE (D2.8.I.7 2014).

In Poland the road transport network model is, according to (Reg. 2013), a part of a topographic database, as shown in Figure 11.

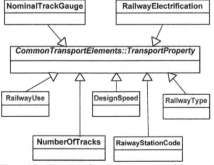

Figure 11. Classes of the road transport network according to Polish topographic database (class names are translated into English and attribute names are omitted).

In a similar way as for road network, basic classes of the INSPIRE model for railroad transport network are shown in Figure 12, and in Poland – as a part of a topographic database, as shown in Figure 13.

NominalTrackGauge   RailwayElectrification

CommonTransportElements::TransportProperty

RailwayUse   DesignSpeed   RailwayType

NumberOfTracks   RaiwayStationCode

Figure 12. Classes of rail transport model according to INSPIRE (D2.8.I.7 2014).

Figure 13. Classes of the railway transport network according to Polish topographic database (class names are translated into English and attribute names are omitted).

In (Majczyna & Pachelski 2014) there are also given author's suggested models for water transport network, as well as for air transport network. Both of them are consistent with the ones of INSPIRE and, in a simplified form omitting attributes, in order to avoid illegibility, are given in Figure 14 and Figure 15.

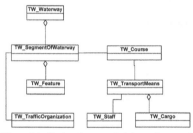

Figure 14. Basic classes of information model for water transport network

Figure 15. Basic classes of information model for air transport network.

Details of the above mentioned information models, as for attribute characteristics, stereotypes, data types and others, are given in the quoted paper. The paper gives also a review and recommendations for not discussed here cable lift transport.

7 CONCLUSIONS

Land, sea and air navigation covers a number of practical problems of information acquire, access, organization, processing. Its efficient use requires proper organization of relevant databases in terms of information structures. This has to be done in a general and universal manner, independent upon rich diversity of thematic scopes, individual user preferences and computer platforms. This goal can be achieved by means of methods of geoinformation modeling as formulated by International Standards of the series 19100 with the use of UML, as well as in some Data Specifications of INSPIRE. A review of such models relevant to navigation has been presented in this paper.

Such models give a deep insight into information structures pertaining navigation, that is in other words to make a precise, unique and complete picture of how this segment of the real world is built.

Besides of this goal, the models can be implemented as databases on different computer platforms in a consistent way thus providing interoperable means for efficient management of different navigation projects.

ACKNOWLEDGEMENT

This paper has been prepared within the project PBS No. 933 of the Faculty of Civil Engineering and Geodesy of the Military University of Technology. This support is herewith highly acknowledged.

REFERENCES

D2.8.I.7 2014. D2.8.I.7 Data Specification on Transport Networks – Technical Guidelines. http://inspire.ec.europa.eu/documents/Data_Specifications/INSPIRE_DataSpecification_TN_v3.2.pdf.

Directive 2007: Directive 2007/2/EC of the European Parliament and of the Council of 14 March 2007 establishing an Infrastructure for Spatial Information in the European Union (INSPIRE).

ISO 19116 2004. Geographic information – Positioning.

ISO 19132:2007. Geographic information – Location based services.

ISO 19133:2005. Geographic information — Location based services — Tracking and navigation

ISO 19141:2008. Geographic information - Schema for moving features.

ISO TS 19103:2005(E). Geographic information – Conceptual schema language.

ISO/DIS 19107. Geographic information – Spatial schema.

ISO/IEC 19501:2005. Unified Modeling Language Specification. Version 1.4.2.

Kresse W, & Fadaie K. 2004. ISO Standards for Geographic Information. Springer.

Majczyna A. & Pachelski W. 2014. Modele informacyjne sieci transportowych według dyrektywy INSPIRE i regulacji krajowych. *Logistyka* 6: /2014.

Reg. 2011. Rozporządzenie Ministra Spraw Wewnętrznych i Administracji z dnia 17 listopada 2011 r. w sprawie bazy danych obiektów topograficznych oraz bazy danych obiektów ogólnogeograficznych, a także standardowych opracowań kartograficznych, Dz.U. 2011 nr 279 poz. 1642.

Reg. 2013. Rozporządzenie Ministra Administracji i Cyfryzacji z dnia 12 lutego 2013 r. w sprawie bazy danych geodezyjnej ewidencji sieci uzbrojenia terenu, bazy danych obiektów topograficznych oraz mapy zasadniczej, Dz.U. 2013 poz. 383.

Hydrometeorological Aspects and Weather Routing

Proceedings of 12th International Conference on Marine Navigation and Safety of Sea Transportation, TransNav 2017
21-23 June 2017, Gdynia, Poland

Optimal Weather Routing Considering Seakeeping Performance Based on the Model Test

H.K. Yoon, V.M. Nguyen & T.T. Nguyen
Changwon National University, Changwon, Korea

ABSTRACT: Optimal weather route is dependent on the seakeeping performance of a ship, and its performance is highly related to hull form and operating conditions. Seakeeping performance can be modeled using Response Amplitude Operator (RAO) where the linear theory is valid. Seakeeping model test of 8600 TEU container ship was carried out in Changwon National University's seakeeping basin and its RAOs at various frequencies were used to predict the RMS motion values in irregular waves. Optimal weather route for a special candidate pilot was found using A* algorithm and the relationship between energy consumption and seakeeping performance.

1 INTRODUCTION

A ship sailing at sea will be influenced by environmental variations, such as wind, waves, current, and ice. These affect the ship's speed, fuel consumption, safety, and operating performance. The concept of ship weather routing has been practiced for a very long time. The optimal weather route of a ship is the process to determine the best route of oceangoing under different ship characteristics and weather circumstances. According to the "Safety and Shipping Review 2016", 'Foundered' is the main cause of loss, accounts for almost half of all losses and it is often driven by bad weather.

It is clear that the optimal weather route of a ship will save shipping operator's money by reducing the arrival time, and therefore also saving fuel. On the other hand, by avoiding the worst weather conditions, weather routing minimizes the risk of damage to the cargo or ship, as well as the risk of injury to crew or passengers.

In the past, many optimization algorithms have been developed to search an optimal route of a ship, e.g. the modified isochrones method, Dijkstra's algorithm, a new forward three-dimensional dynamic programming (3DDP), and the three-dimensional modified isochrones (3DMI). However, most of them apply only to single purpose routing problems, and the ship's safety is the only constraint value to avoid adverse weather conditions. For example, in 3DDP the only constraints on ship's

safety are the prevention of capsize and heavy roll (Wei et al., 2012), while in the 3DMI method, the ship's safety is only a function of significant roll response and significant wave height (Lin et al., 2013).

In the author's former works, the optimization of the ship route by only considering wind, wave conditions and safety values was modeled as a function of the minimum arrival time or minimum energy to avoid slamming and deck wetness (Nguyen et al., 2016). The relative motion and relative velocity of a ship that were used for prediction of the probability of slamming and deck wetness were taken from the ship's data. The highlight of this study is to develop an optimization algorithm that combines the available technologies in the area of weather forecast with the seakeeping performance of a ship, in conjunction with comprehensive ship operation cost

The present paper introduces the development of the algorithm to find an optimal weather route considering the seakeeping performance obtained from the model test result. The model test of the 8600 TEU container ship is carried out in Changwon National University(CWNU)'s seakeeping basin. The results of the optimal weather route of a ship can inform a captain about potential risk during navigation. In addition, the effects of temperature, wind and wave that change the total ship's resistance are considered.

2 TEST FACILITIES AND TEST CONDITION

2.1 *Towing tank test*

Figure 1 shows CWNU's square towing tank, in which the model test was conducted. A wave maker is installed at the end of the towing tank and wave absorber is at the opposite site. The wave maker can make waves with a height up to 20 cm, and a wavelength up to 3 m. Table 1 lists the principal dimensions of the CWNU square wave basin.

Figure 1. The CWNU square wave basin

Table 1. Principal particulars of CWNU's square basin

Item	Description
Length x Breadth x Depth	20 x 14 x 1.8 m
Water depth	< 1.65 m
Max. velocity of X, Y carriage	1.0 m/s
Max. wave height	0.3 m
Max wave length	3.0 m
Max. length of model	1.5 m

2.2 *Measuring equipment*

Figure 2 shows motion measurement device and wave probe. Heave, roll, and pitch are measured using each potentiometer. The weight of the equipment is compensated using a counterweight in the vertical direction. Table 2 lists the capacity of the motion measurement device

Table 2. Capacity of the motion measuring device

Item	Value
Maximum Heave motion	± 300 mm
Maximum Roll motion	$\pm 45^{o}$
Maximum Pitch motion	$\pm 30^{o}$

Figure 2. Motion measurement device and wave probe

Since the experiment is conducted under various wave directions and wave frequencies, it is necessary to evaluate the quality of the waves. Right photo in Figure 2 shows the wave probe that is used in this experiment. In addition, it is also necessary to

measure wave height during motion test in waves, in order to determine the phase between wave and motion. The phase information is estimated by measuring the incident waves from the wave probe mounted on the towing carriage.

2.3 *Definition of wave direction*

Figure 3 shows the definition of encountering wave direction that is used for this experiment. First, 'following sea' is defined by heading angles between 0° and 30° on the port side. Second, 'stern quartering' is defined by heading angles between 30° and 60° on the port side. Third, 'beam sea' is defined by heading angles between 60° and 120° on the port side. Fourth, 'bow quartering' is defined by heading angles between 60° and 120° on the port side. Fourth, 'bow quartering' is defined by heading angles between 120° and 150° on the port side. Finally, 'head sea' is defined with heading angles between 150° and 180° on the port side.

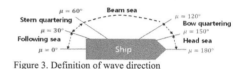

Figure 3. Definition of wave direction

Table 3. Test conditions in regular waves

λ/L	λ [m]	Wave frequency [rad/s]	Wave Period [s]	Wave height [cm]
0.5	0.56	10.45	0.60	1.4
0.6	0.68	9.54	0.66	1.4
0.7	0.79	8.83	0.71	1.4
0.8	0.90	8.26	0.76	1.4
0.9	1.02	7.79	0.81	1.4
1.0	1.13	7.39	0.85	1.4
1.2	1.35	6.74	0.93	1.4
1.3	1.47	6.48	0.97	1.4
1.4	1.58	6.24	1.01	1.4
1.5	1.69	6.03	1.04	1.4
1.6	1.81	5.84	1.08	1.4
1.7	1.92	5.67	1.11	1.4
1.8	2.03	5.51	1.14	1.4
1.9	2.15	5.36	1.17	1.4
2.0	2.26	5.22	1.20	1.4

2.4 *Test condition*

Most tests in regular waves are concerned with the experimental determination of the motion response amplitude operator (RAO). It is necessary to record the sinusoidal motions of the model, and to determine the motion amplitudes experienced for a variety of different wavelength or wave frequencies of which are related by so called dispersion relation. The experiment is carried out in regular waves in seven wave directions. The test conditions are 105 waves for 15 wave frequency conditions. Table 3 shows the set values for each test condition in regular waves.

3 MODEL TEST

3.1 *Model ship*

Table 4 shows the principal dimensions of the model ship of 8600 TEU container ship.

Table 4. Principal dimensions of model ship

Particulars	Unit	Real	Model
Scale ratio	-	1	285.630
L_{pp}	m	322.6	1.129
B	m	45.6	0.160
D	m	24.6	0.086
T(design)	m	13	0.046
C_B	-	0.5908	0.5908
∇	m3	112983	0.0048
Δ	kg	115909259.7	4.974
LCB	m	-6.625	-0.023
GM	m	1.5	0.0053

3.2 *Wave maker calibration*

Since seakeeping analysis is done for the assessment of the ship motion in waves, the wave characteristics, as well as the response of ship, are needed. The wave evaluation values are obtained through the amplifier and wave probe device in the model test. Therefore, it is necessary to establish the relationship between the measured wave height and the voltage form wave probe amplifier. Figure 4 shows the wave probe calibration process, and the typical relationship between wave height and output signal voltage of the amplifier.

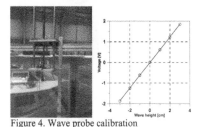

Figure 4. Wave probe calibration

Table 5. Evaluation input stroke for each wave frequency

Input Stroke [cm]	Wave period [s]	Measured Wave height [cm]	Target value [cm]	Error [%]
1.79	0.60	1.43	1.40	2.46
1.62	0.66	1.42	1.40	1.70
1.55	0.71	1.35	1.40	3.56
1.51	0.76	1.38	1.40	1.76
1.51	0.81	1.45	1.40	3.33
1.50	0.85	1.37	1.40	2.11
1.48	0.93	1.38	1.40	1.29
1.49	0.97	1.37	1.40	1.80
1.51	1.01	1.38	1.40	1.41
1.52	1.04	1.39	1.40	0.49
1.54	1.08	1.40	1.40	0.34
1.58	1.11	1.42	1.40	1.26
1.59	1.14	1.43	1.40	1.97
1.58	1.17	1.38	1.40	1.31
1.64	1.20	1.36	1.40	2.60

Wave maker calibration is also necessary for evaluation of the wave quality, and finding the input stroke of the wave maker system for each wave frequency. Wave maker calibration at different frequencies are performed at the center of the basin, in order to find the input stroke value in the wave maker system. Table 5 lists the results of the input stroke values of the wave maker system.

3.3 *Potentiometer calibration*

The motion measuring device is used to measure the heave, roll, and pitch components. In this system, a potentiometer, which is an instrument for measuring the variable resistance in a circuit, is used for measuring the motion. It is necessary to determine the relationship between motion displacement and digital voltage. Figure 5 shows the motions measuring the device and heave calibration result.

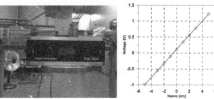

Figure 5. Potentiometer calibration

3.4 *Ballasting and inclining test*

In the seakeeping test, it is very important to exactly match the mass distribution of the model ship to the design waterline. In order to approximate the vertical and longitudinal mass distributions, the mass moments of inertia about longitudinal and transverse axis and metacentric height (GM) must be the same. Figure 6 shows the ballasted model ship.

Figure 6. Ballasting of model ship

The GM value is obtained through the inclining test. According to the target GM, Eq. 1 determines the angle of inclination at the time of application of the inclined moment by a small weight. Figure 7 shows moving the weight in order to change the GM during the test. Tables 6-7 show the results of the inclining test.

$$\phi = tan^{-1}\left(\frac{wl_y}{W \cdot GM}\right) \tag{1}$$

Table 6. Target value of metacentric height

Item	Model
Mass [kg]	4.974
GM [m]	0.0053
Moved weight [kg]	0.1
Lever l_y [m]	0.03
Heel angle ϕ (deg.)	6.49

Table 7. Measured heel angle

Item	Measured angle[deg.]	Real angle[deg.]	Error [%]
Start point	-0.3	-	-
Move weight to Port	-6.86	-6.56	1.04
Move weight to Stbd	6.04	6.34	2.35

Figure 7. Inclining test

3.5 Inertia test

According to ITTC(International Towing Tank Conference) recommendation, the radius of gyration in pitch motion is 0.25 of the ship length. In this study, inertia test was performed by inertia swing. Figure 8 shows the inertia swing, and the process of inertia swing calibration. Table 8 shows the process for inertia swing calibration with 4 kgf weight and different distance from the center of inertia of the inertia swing. The values A and B obtained through inertia swing calibration by using Eqs. 2–4 are -43.21 and 4.1045, respectively. Then, the measured mass moment of inertia value of the model ship is calculated by Eq. 5 and its value is 0.3925 kgm^2.

$$C = mlgT^2 - 4\pi^2 ml^2 \tag{2}$$

$$A = \frac{C_1 - C_2}{T_1^2 - T_2^2} \tag{3}$$

$$B = \frac{T_1^2 C_2 - T_2^2 C_1}{4\pi^2 (T_1^2 - T_2^2)} \tag{4}$$

$$I_s = -m_s l_s^2 - B + \frac{T_s^2}{4\pi^2} (m_s l_s g - A) \tag{5}$$

Table 8. Inertial swing calibration and inertia test

Item	Weight 1	Weight 2	Ship
m [kg]	4	4	4.974
l [m]	0.93	0.431	0.926
T [s]	1.936	1.784	1.978
C	0.153	24.502	-

Figure 8. Inertia swing calibration and inertia test

3.6 Installation of the ship model

The model is attached to the motion measuring device in the middle of the measuring frame of the towing carriage. The center of gravity of the model should match the thrust line as much as possible. The wave probe is installed at 1.703 m in front of the center of the model on the right edge of the yaw table, in order to avoid disturbance caused by the model. Figures 9-10 show the detailed set up of this experiment.

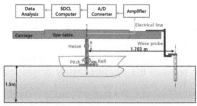

Figure 9. Model test setup

Figure 10. Real model test setup

3.7 Response Amplitude Operator (RAO)

The results of heave, roll, and pitch RAOs of the model for 7 wave directions in 15 wave frequencies were obtained as shown in Figures 11-13.

The relative vertical motion and relative velocity at bow area were calculated from the pitch and heave displacement with respect to the center of gravity based on the linear transformation. The relative displacement and relative velocity at the bow can be determined by Eqs. 6-7, respectively. In Eq. 6, ζ is the local wave depression, $\bar{\eta}_3$ and $\bar{\eta}_5$ are the heave and pitch response, respectively. Figures 14-15 show the relative motion and relative velocity at bow depending on wave direction for finding probability of slamming and deck wetness.

$$\xi_R = \bar{\eta}_3 - x_{B1} \bar{\eta}_5 - \zeta \tag{6}$$

$$\dot{\xi}_R = \omega \xi_R \tag{7}$$

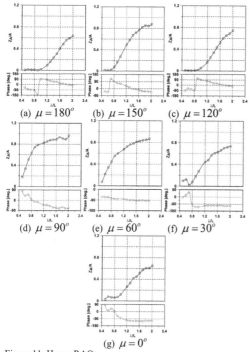

(a) $\mu = 180^o$　　(b) $\mu = 150^o$　　(c) $\mu = 120^o$

(d) $\mu = 90^o$　　(e) $\mu = 60^o$　　(f) $\mu = 30^o$

(g) $\mu = 0^o$

Figure 11. Heave RAO

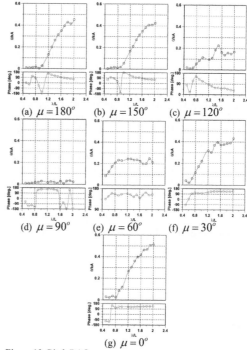

(a) $\mu = 180^o$　　(b) $\mu = 150^o$　　(c) $\mu = 120^o$

(d) $\mu = 90^o$　　(e) $\mu = 60^o$　　(f) $\mu = 30^o$

(g) $\mu = 0^o$

Figure 13. Pitch RAO

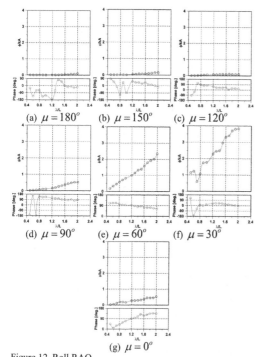

(a) $\mu = 180^o$　　(b) $\mu = 150^o$　　(c) $\mu = 120^o$

(d) $\mu = 90^o$　　(e) $\mu = 60^o$　　(f) $\mu = 30^o$

(g) $\mu = 0^o$

Figure 12. Roll RAO

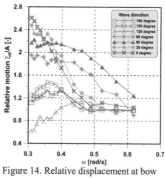

Figure 14. Relative displacement at bow

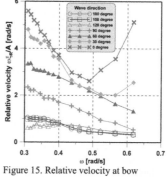

Figure 15. Relative velocity at bow

4 WEATHER ROUTING ALGORITHM

4.1 *Weather routing class*

A* algorithm was implemented in Optimal Weather Routing Setting Unit class (OWRSU). Figure 16 illustrates the implementation of OWRSU class in a flowchart. First, reading data of the ship and weather forecast data is initiated in the function ReadData, which data are kept in the Weather class and OwnShip class. These data involve ship information and weather forecast data, such as wind direction, wave direction, and wave height. After reading the data, the function Initialize is called, and this is used to initialize OwnShip, Weather, and Route classes. Next, FindRoute is executed to estimate RouteCost, which considers the speed reduction due to wind, wave, slamming, and deck wetness, in order to provide the cost of ship route segment. The final two stages are PrintData and Finalize.

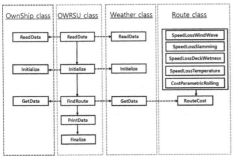

Figure 16. Flow chart for route finding

4.2 *Models of cost functions*

In order to calculate the cost of a path, A* algorithm employs an additive evaluation function $f(n) = g(n) + h(n)$, where $g(n)$ is the cost of the path being currently evaluated from start node s to the current node n, and $h(n)$ is the heuristic cost which includes geographic information and weather constraint. When the A* algorithm is implemented to search the optimal weather route of a ship, it is important to determine $f(n)$ of a node by combining weather forecast data with the characteristics of the ship. This optimization problem can thus be defined as follows:

$$f_{i+1} = g_{i+1} + h_{i+1} \qquad (8)$$

By optimizing the evaluation function f_{i+1} for the complete path, A* algorithm achieves the minimum time route or minimum fuel consumption route of a ship. At the same time, when considering f_{i+1}, the safety parameter of a ship such as avoiding the parametric rolling phenomenon at node i+1 is considered as the safety constraint. In order to avoid land and islands, this algorithm takes care of these hazards by simply prescribing a very large default value of evaluation function f_{i+1} to these grids.

4.2.1 *Model 1: Cost of minimum arrival time*

We assume that the engine of the selected ship is able to provide constant power output and speed reduction due to weather conditions. The ETA (Estimated Time Arrival) is used to establish the short time route from departure point to destination point by applying the available weather data from the middle ware so called SAS. Since there is an effect resulting from weather conditions at each node, the path cost function g_{i+1} can be estimated as sailing time from the start node to the current node $i+1$.

4.2.2 *Model 2: Cost of minimum energy*

We assume that the energy output of a ship can be changed in order to keep constant speed. It is noted that a ship sailing in heavy condition needs more engine power than in calm water. Therefore, the path cost function g_{i+1} can be estimated as the minimum energy to maintain constant ship speed.

4.3 *Various speed reduction parameters*

The method to avoid parametric rolling is recommended by IMO to avoid dangerous situations in adverse weather and sea conditions (IMO circular no. 1228). According to IMO Circ. 1228 to avoid parametric rolling, the course and the speed of the ship should be selected in a way to avoid conditions of encountering period T_E close to the ship roll period T_R ($T_E \approx T_R$) or one half of the ship roll period ($T_E \approx 0.5T_R$).

In ship navigation, the ship's motion can definitely influence on ship speed due to voluntary and involuntary speed reduction. Involuntary speed reduction is the result of added resistance of the ship due to waves and wind and changes in propeller efficiency due to waves. Involuntary speed reduction is obtained from empirical formulae suggested by many researchers in the past. we used Kwon's method(1981, 2005) to predict involuntary speed reduction under different weather condition. In addition, the effect of temperature is also considered in terms of involuntary speed reduction.

On the other hand, voluntary speed reduction means that the ship's captain reduces speed due to adverse weather, to avoid a certain excessive motion. However, this method depends not only on the subjective decision of the ship's captain, but also on his experience. It is necessary to consider involuntary speed reduction for finding the optimal route of a ship. This can improve seakeeping performance, such as reducing ship motion by avoiding slamming and deck wetness. So if slamming and deck wetness happen, the optimal route of a ship should consider involuntary speed reduction. The probabilities of slamming and deck wetness are often evaluated by relative motion and relative velocity at some particular point, such as on the bow area. In order to apply the result of the model test at various wave heading angles for the optimal ship route, it is necessary to calculate the relative vertical displacement and relative vertical velocity at the bow area. Many empirical formulae have been established for estimating the relative motion from the ship response. In this study, relative vertical motion at the bow area were calculated from the pitch and heave motion with respect to the center of gravity based on the RAOs taken from the model test in waves.

5 SIMULATION AND RESULTS

5.1 Simulation

The parametric rolling is the constraint value for both cases. The data set of weather forecasts is updated every 12 hours, and are obtained from SAS. The environmental data used to carry out the simulation in this study consist of forecasted data: swell direction, significant wave height, wind speed, and temperature. In addition, these data are used to estimate speed reduction under given weather conditions and parametric rolling, the probability of slamming, and the probability of deck wetness.

The sample weather data are given at each node with grid size of 1 degree in longitude and 1 degree in latitude. The weather forecast data at any given time and the ship's position in the simulation are obtained by linear interpolation of environment data around that position. In order to validate the capability of the A* algorithm in OWRSU class, and investigate how the weather conditions and ship geometry influence the optimal weather route, 8600 TEU container ship which is the same as the model ship was selected for simulation.

Table 9 shows the route of container ship for which the simulation was conducted. The simulation was carried out to evaluate the effectiveness of the optimization algorithm in two cases. In Case 1, we assumed that engine power is constant, so the optimization index is ETA, and the cost of each node depends on the speed reduction, slamming, deck wetness, and arrival time of the ship. On the other hand in Case 2, the optimization index is energy consumption, since the distance from departure to destination is different between paths using A* algorithm and great circle. The cost in each node only depends on the power increase and time.

Table 9. Simulation condition

Port	Location
Departure Tokyo, Japan	139° East, 35° North.
Destination: San Francisco, US	122° West, 38° North.

5.2 Results

Figure 17 shows the comparison of the ship speed estimated by both Great Circle (GC) and optimal weather route found by A* algorithm during the voyage of a ship. There is little difference between estimated ship speeds in case of GC and A*. When the ship follows the path suggested by A* algorithm, the reduced speed is significantly smaller than the one by GC. That is because the GC does not consider the weather forecast data through ship routing even though it is the shortest distance between departure point and destination point.

In addition, the graph shows ETA of the ship route suggested by algorithms A * is faster than one by GC. The arrival time of a ship can save 9.10 % if using A* algorithm as shown in Table 10. So, we can see that it is more efficient for finding ship route by using A* algorithm for ship moving from Tokyo, Japan to San Francisco, US.

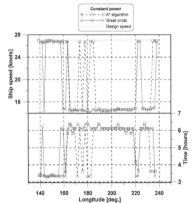

Figure 17. Comparison of speed and time in case of constant power condition

Table 10. Comparison of ETA

Algorithm	ETA [hours]	Time saving ratio[%]
GC	268.61	-
A*	244.17	9.10

In cases of constant speed condition, if ship follows the path suggested by A * algorithm, it will consume less fuel than route that passes through the GC to keep the design speed of ship as shown in Figure 18. Particularly, the energy consumption is saved 5.47 % as shown in Table 11. However, the ETA through the great circle is faster than the ETA of optimal path using A* algorithm because the great circle distance is the shortest path.

Table 11. Comparison of estimated energy consumption

Algorithm	ETA [hours]	Energy [KJ]	Energy saving ratio [%]
GC	166.39	3.4037E+10	-
A*	168.06	3.2175E+10	5.47

251

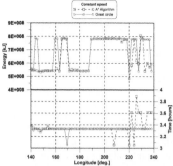

Figure 18. Comparison of energy and time in case of constant speed

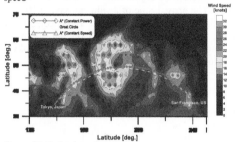

Figure 19. Optimal weather route

6 CONCLUSION

In this paper, the seakeeping model test of the 8600 TEU container ship was carried out in Changwon National University's square wave basin, and its RAOs at various frequencies were modeled for finding the optimal weather route minimizing the arrival time or minimum. A* algorithm for avoiding hazard situations has been proposed as an optimization method. The concluding remarks are as follows:

First, roll, pitch, and heave components have a clear effect on the relative vertical, relative velocity at the bow. The measured ship motions in various waves and heading conditions by performing the model test in the square wave basin. In particular, the experimental results of this study have been used to find the optimal route of a ship.

Second, A* algorithm was applied to 8600 TEU container ship. It is clear that A* algorithm is efficient on finding the optimal route based on the minimum arrival time, minimum energy, and safe operation of the ship. Based on simulation results with two routes, we found that OWRSU class can suggest the optimal route with different criteria, so that users can choose either the ETA, or the energy consumption of a ship. Simulation results indicate that the path suggested by A* algorithm is better

than the ones by GC for minimum arrival time and minimum energy.

Finally, the effects of environmental data, such as the effects of temperature, wind and wave, which change the total ship's resistance, have been considered. The advantage of this algorithm is that it not only provides the speed that changes with both time and ship's position, but also considers updated weather forecast data at current ship's position. This algorithm can provide quick results when looking for the optimum route. This study can be used to make a commercial contribution to the development of ship navigation.

ACKNOWLEDGEMENT

This research was supported by "Development of Solution for Safety and Optimal Navigation Path of a Ship Considering the Sea State" sponsored by Korean Evaluation Institute of Industrial Technology (KEIT).

REFERENCES

Hagiwara, H.: Weather routing of (sail-assisted) motor vessels (Ph.D. thesis), Delft University of Technology, Delft, 1989.

Takashima, K., Mezaoui, B., Shoji, R. 2009. On the Fuel Saving Operation for Coastal Merchant Ships using Weather Routing, *TransNav*, 3(4): 401-406.

Padhy, C.P., Sen, D., Bhaskaran, P.K. 2008. Application of wave model for weather routing of ships in the North Indian Ocean, *Nat Hazards*: 373-385.

Wei, S., Zhou, P. 2012. Development of a 3D Dynamic Programming method for Weather Routing, *TransNav*, 6(1): 79-85.

Lin, Y.H., Fang, M.C., Yeung, R.W. 2013. The optimization of ship weather-routing algorithm based on the composite influence of multi-dynamic elements, *Applied Ocean Research*, 43: 184-194.

ARJM, L. 1998. Seakeeping: ship behavior in rough weather, *Ellis Horwood Ltd.*

Allianz Global Corporate & Specialty, 2016. Safety and Shipping Review 2016

Nguyen, V.M., Jeon, M., Yoon, H.K. 2016. Study on the optimal weather routing of a ship considering parametric rolling, slamming, deck wetness, *PRADS' 2016*

Kwon, Y.J. 1981. The effect of weather, particularly short sea waves, on ship speed performance (Ph.D. thesis), *University of Newcastle upon Tyne*

Kwon, Y.J., Kim, D.Y. 2005. A Research on the Approximate Formulae for the Speed Loss at Sea, *Journal of Ocean Engineering and Technology*, 19(2): 90-93.

Spyrou, K.J. 2005. Design Criteria for Parametric Rolling, *Oceanic Engineering International*, 9(1): 11-27.

International Marine Organization, 2007. Revised guidance to the master for avoiding dangerous situations in adverse weather and sea conditions, *IMO circular no. 1228*.

Sen, D., Padhy, C.P. 2015. An approach for development of a ship routing algorithm application in the North Indian Ocean region, *Applied Ocean Research*, 50: 173-191.

Avoidance of the Tropical Cyclone in Ocean Navigation

M. Szymański & B. Wiśniewski
Maritime University in Szczecin, Szczecin, Poland

ABSTRACT: Article presents various methods of determining the tropical cyclone avoidance manoeuvre in ocean navigation. Determining of the manoeuvre was carried out with the use of the ORS (Onboard Routing System) systems Bon Voyage 7.0 and SPOS 7.0, the 1-2-3 rule, manual graphic anti-collision plot, CYKLON programme, and shore based weather routing by AWT (Applied Weather Technologies). True weather data from the voyage of the 9000TEU POSTPANAMAX container vessel from Yantian (China) to Vancouver (Canada) in August 2015 were used. During the voyage the vessel has encountered the typhoon MOLAVE.

1 INTRODUCTION

The key decision in tropical cyclone avoidance is the determining of the moment of the beginning of avoidance manoeuvre and the determining of the correct course and speed.

In this article the determination of the typhoon avoidance manoeuvre in ocean navigation by a big (LOA=336m, GT=97500, DWT=71274MT), powerful and fast (20knots), postpanamax container vessel on a voyage from Yantian (China, ETD= 07.08.2015 0300UTC) to Vancouver (Canada, ETA=19.08.2015 2100UTC) by various methods was analyzed.

Figure 1. Great circle route Yantian – Vancouver (Own study based on BVS 7.0)

The recommended route between the two above mentioned ports is the route shown in Fig. 1 (UK Hydrographic Office NP136 2014).However, in an initial, coastal stage of the voyage, typhoon SOUDELOR was encountered (Fig. 2). It prevented

the vessel from using the route shown in Fig. 1. Consequently, after a number of testing and consultation with AWT, the route through the Luzon Strait was chosen, clearing the typhoon SOUDELOR from the south. In an ocean stage of the voyage the typhoon MOLAVE was encountered. It was moving from the Philippine Sea towards Japan – see Fig. 3. The route takes into consideration navigational and legal restrictions – allowed routes in the Bering Sea and Aleutian Chain and the US and Canadian ECA zone (Szymański&Wiśniewski, 2016).

Figure 2. Track of the typhoon SOUDELOR and route through the Luzon Strait (own study based on BVS)

Figure 3. Track of MOLAVE forecasted for 5 days in JMA outlooks 9 and 10 August (JMA, 2015)

Symbols and abbreviations used in the article:
ETD – *Estimated Time of Departure*
Departure – Point of departure
ETA – *Estimated Time of Arrival*
Arrival – Pont of arrival
Troll – vessel's own roll period (transverse) in sec
nm – *nautical miles*, route distance
Hrs – *hours*, required steaming time
T_FO – *Total Fuel Oil*, Total fuel consumption en route
HSFO – *High Sulphur Fuel Oil*, consumption en route
LSFO – *Low Sulphur Fuel Oil*, consumption en route
MDO – *Marine Diesel Oil*, consumption en route
LSMDO – *Low Sulphur Marine Diesel Oil*, consumption en route
SC – *Calm Sea Speed*, ship's speed on calm seas for the optimized route
WxF – *Weather Factor*, influence of weather on ship's speed
CuF – *Current Factor*, influence of ocean surface current on ship's speed
SOG – *Speed Over Ground* for the optimized route
Fuel(USD) – Total fuel cost In USD for the optimized route

2 METHODOLOGY

The following tools were employed to determine the safest action to avoid the typhoon MOLAVE:
– ORS (onboard routing systems) BVS 7.0 – Bon Voyage System 7.0 (Applied Weather Technologies, 2014),
– SPOS Fleet Management 7.0.0.1 (Meteo Consult BV, 2009),
– 1-2-3 rule (Holweg, 2000),
– anti-collision plot,
– CYKLON programme (Wiśniewski&Kaczmarek 2012) (Wiśniewski 2012),
– shore based weather routing recommendations by AWT.
Weather data used were:
– the weather data file for ORS BVS containing weather analysis and prognosis up to 16 days in advance,

– the weather data file for ORS SPOS containing weather analysis and prognosis up to 9 days in advance,
– EGC forecasts, outlooks and advisories,
– surface pressure analysis and prognosis charts from JMA (Japanese Maritime Agency),
– typhoon prognosis charts from JMA (up to 120 hrs with 70% probability level),
– typhoon strong winds prognosis charts from JMA (up to 72 hrs).
Results obtained were compared. The best solution in terms of safety and possibility of route execution was chosen.

3 RESULTS

3.1 *Avoidance manoeuvre by the recommendation from shore based weather routing*

Shore based recommendations from AWT were received on 6th and 7th of August. They are shown in Fig 4 together with their navigational parameters.

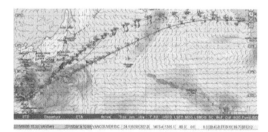

Figure 4. Routes recommended by AWT from August 6th and August 7th and their navigational parameters (own study based on BVS)

The recommended routes were moved south due to the typhoon MOLAVE developing south of Japan. They were programmed manually in AWT. Bold route on Fig 4 is from August 7th. Its corresponding navigational parameters in the table below the weather chart are highlighted blue. The route runs too close to the typhoon and generates weather alerts – exceeding the maximum wind velocity (34kts). Alerts are marked by violet circles visible along the route in Fig 4. Finally the AWT had accepted the captain's route passing the MOLAVE from the south and entering the Pacific via the Balintang Channel in the Luzon Strait.

3.2 *Avoidance manoeuvre with the use of the ORS BVS 7.0.*

Planning of the MOLAVE avoidance manoeuvre had begun on 9 August after the successful clearing of the typhoon SOUDELOR on the South China Sea. Optimization used was least fuel with fixed arrival time. Weather limitations were set for maximum 8m swell, seas and significant wave and wind maximum velocity for 34 knots.

Routes generated in the BVS until August 9th run along the coast of Japan from the Pacific side. They clear the MOLAVE from the polar side and place the ship between the typhoon and the land, in a narrow and restricted area (Fig. 5 and 6). The correct route clears the typhoon from the south. It is highlighted blue in Fig 5 (bold route). It was programmed manually onboard.

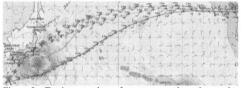

Figure 5. Testing results of routes passing the typhoon MOLAVE. (Own study based on BVS 7.0)

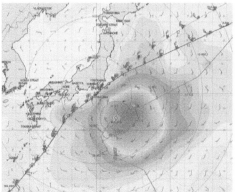

Figure 6. Result of testing in BVS from 9 August – position of the ship and typhoon for 11.08.2015, 0000UTC (Own study based on BVS 7.0)

3.3 Avoidance manoeuvre with the use of SPOS 7.0.0.1.

Ocean route testing began on 7 August. SPOS ORS enables to programme the minimum distance to the tropical cyclone and tropical depression (Szymański&Wiśniewski 2016).

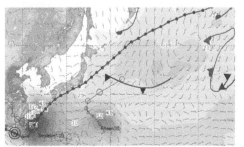

Figure 7. Avoidance manoeuvre in SPOS from 7, 8 and 9 of August (Own study based on SPOS)

Maximum wave heights and wind velocity were programmed as 8m and 34 knots. Minimum distance to typhoon was determined as 250Nm. Type of a chosen optimization was Optimum High&Wide, with speed of 19,5 knots, corresponding to the calm sea speed for the route optimized in BVS 7.0 ORS. Results of testing are presented in Fig. 7. Similar to the BVS system, routes generated in the SPOS system until August 9th run along the southern and eastern coasts of Japan and were clearing the typhoon form the polar side. They placed the ship between the typhoon and the land, in a too narrow and too restricted area. Due to that the routes were finally rejected by the captain.

3.4 Avoidance manoeuvre by the 1-2-3 rule

Blue colour in Fig 8. marks the position of the typhoon MOLAVE together with the 30knots wind zone according to the warning from 8 August, 1200UTC. Green colour circles mark the further position of the typhoon together with the 35 knots wind zones. The 24hrs, 48hrs and 72 hrs predicted zones' radii are increased by respectively 100Nm, 200Nm and 300Nm with regard to values from the warnings, in accordance with the rule methodology. For visualization of the 1-2-3 rule the CYKLON programme was used.

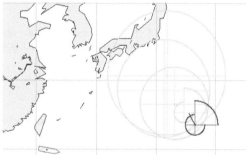

Figure 8. 30 knots wind zone of MOLAVE as calculated by 1-2-3 rule from 8 August for the next 3 days (Own study based on CYKLON)

3.5 Avoidance manoeuvre with the use of CYKLON programme and manual anti-collision plot

It is shown in Fig. 9 and 10. The software was fed with the data regarding the typhoon from 8 August 1200UTC, when the voyage began.

The safe course is 065°, the software calculated initially the dangerous sector 047° and 061° (Fig. 9), and after 12hrs, for updated positions and forecasts of MOLAVE's movement, the dangerous sector was between 054° and 064° (Fig. 10).

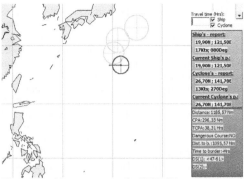

Figure 9. Avoidance manoeuvre by CYKLON programme on 8 August 1200UTC (Own study based on CYKLON)

Figure 10. Avoidance manoeuvre by CYKLON programme on 9 August 0000UTC (Own study based on CYKLON)

Results obtained by the manual plot are comparable with those obtained by the CYKLON programme.

4 CONCLUSIONS

1 Routes generated by both Bon Voyage and SPOS ORS do not meet the safety requirements for cyclone avoidance and passes the typhoon from the polar side. Calculations are based upon the user declared maximum allowed wave heights values and maximum allowed wind velocities and in SPOS additionally upon the minimum declared distance from the cyclone's eye.

2 1-2-3 rule is very ineffective and impractical in tropical cyclone avoidance. The 35 knots wind zone to be avoided, after 72hrs is almost 1000Nm in diameter. It is too play safe. It bears also a serious mistake in methodology: lack of differentiation of strong winds zone radii with regard to cyclone's quadrant.

3 Correct solutions to the problem were only obtained with the use of CYKLON programme and manual plot. CYKLON programme is more recommended than the manual anti-collision plot. Results obtained by both methods are identical, however accuracy in CYKLON programme, due

to methodology utilized (analytical-mathematical in CYKLON against manual-graphical in manual plot) is more correct and more accurate.

4 The best route was obtained by manual programming in the BVS according to results obtained in the CYKLON programme – see Fig 5, route highlighted blue. It is the best route obtained in all the testing and of all tools utilized. It is recommended to use the tool like CYKLON programme or similar as a complement to the ship's ORS in programming the part of the route passing in the vicinity of the tropical cyclone.

5 Shore based weather recommendations were generally correct. However, they run too close to the typhoon. Their quality is very much dependant on the knowledge and experience of the operator. Those are not known onboard and cannot be verified, thus caution is recommended.

ACKNOWLEDGEMENT

This research outcome has been achieved under the research project No 1/S/INM/16 financed from a subsidy of the Ministry of Science and Higher Education for statutory activities.

REFERENCES

[1] UK HYDROGRAPHIC OFFICE NP136 (2014) Ocean Passages for the World, 6th Edition 2014.
[2] JMA (Japanese Maritime Agency). (2015), weather charts. [Online] Available from: http://www.jma.go.jp/en/seawarn/ [Accessed: 8th, 9th, 10th August 2015]
[3] SZYMAŃSKI, M., WIŚNIEWSKI, B. (2016) 'Navigational and legislative constraints for optimization of ocean routes in the Northern Pacific Ocean', Scientific Journals of Maritime Academy in Szczecin, vol. 48, no 120, pp.96-105.
[4] APPLIED WEATHER TECHNOLOGIES. (2014) Bon Voyage System (BVS 7.0), Voyage Optimization Software, User Manual. 140 Kifer Court, Sunnyvale CA 94086.
[5] METEO CONSULT BV. (2009) SPOS Fleet Management 7.0.0.1. The Netherlands, 2009.
[6] HOLWEG, E.J. (2000) Mariner's Guide For Hurricane Awareness In The North Atlantic Basin, National Weather Service, National Oceanic and Atmospheric Administration.
[7] WIŚNIEWSKI B., KACZMAREK P., (2012) 'Ships' ocean route programming with adaptation to „Cyclone" program', Scientific Journals of Maritime Academy in Szczecin, vol. 29, no 101, pp. 174-181.
[8] WIŚNIEWSKI B., (2012) 'Ships' ocean route programming', Scientific Journals of Maritime Academy in Szczecin, vol. 29, no 101, pp.164-173.
[9] WIŚNIEWSKI B., SZYMAŃSKI M., (2016) 'Comparison of ship performance optimization systems and the bon voyage onboard routing system', Scientific Journals of Maritime Academy in Szczecin, vol. 47, no 119, pp.106-116.

Proceedings of 12th International Conference on Marine Navigation and Safety of Sea Transportation, TransNav 2017
21-23 June 2017, Gdynia, Poland

Prediction Method and Calculation Procedure of Resistance and Propulsion Performance for the Weather Routing System

E.C. Kim, K.J. Kang & H.J. Choi
Korea Research Institute of Ships and Ocean Engineering, Daejeon, Korea

ABSTRACT: In order to determine the ships' optimum route through a weather routing system, the accurate and quick prediction of resistance and propulsion performance is necessary. When a ship is going on a voyage in real sea, it faces with not only hull resistance in calm water but also added resistances due to external forces which come from wind, waves, current, etc. And it also faces with various added resistances due to drift and hull roughness condition. This paper studies the prediction method and calculation procedures for the resistance and propulsion performance of a ship operating in real sea environment. For the improvement of calculating process, an enhanced method has been developed, which utilizes pre-calculated database for the evaluation of the resistance and propulsion performance for a ship. This method has been applied to the weather routing module of KRISO Arctic Safe Voyage Planning (KASVP) system under development, which is supposed to be used for vessels operating in Arctic sea. This system shall be tested on the first research ice breaker of Korea, Araon, in the near future.

1 INTRODUCTION

It is a very important task not only to reduce the fuel cost but also to support the safe operation of the ship to take the optimal route according to the weather condition. The necessity of an weather routing system rapidly has increased due to the regulations of IMO.

The IMO Marine Environment Protection Committee (MEPC) at its sixty-second session in July 2011 adopted amendments to MARPOL Annex VI under resolution MEPC.203(62) which added a new chapter 4 to Annex VI on regulations on energy efficiency for ships for the requirements for technical and operational measures to improve the energy efficiency of international shipping. This regulation apply to all ships of 400 gross tonnage and above and entered into force on 1 January 2013 (IMO 2011, IMO 2012).

IMO adopted the International Code for Ships Operating in Polar Waters (Polar Code) and related amendments to make it mandatory under both the International Convention for the Safety of Life at Sea (SOLAS) and the International Convention for the Prevention of Pollution from Ships (MARPOL). This Polar Code entered into force on 1 January 2017. In order to establish procedures or operational limitation for polar water operational manual of the Polar Code, an assessment of the anticipated range of operating and environmental conditions shall be carried out (IMO 2014c).

The IMO Maritime Safety Committee (MSC) at its ninety-fourth session in November 2014 approved the e-navigation Strategy Implementation Plan (SIP) which contains a list of tasks required to be conducted in order to address prioritized e-navigation solutions. It is expected that these tasks should provide the industry with harmonized information in order to start designing products and services to meet the e-navigation solutions (IMO 2014a, IMO 2014b).

In order to improve the accuracy of weather routing system, it is important to predict not only the weather condition but also the resistance and propulsion performance in a given weather condition. To predict the resistance and propulsion performance of a ship, the estimation of resistance for the basic resistance in calm water and the added resistance due to external forces such as wind and waves must be considered.

The guidelines for verification of the Energy Efficiency Design Index (EEDI) described that the shipbuilder should calibrate the measured ship speed for the development of the power curves, if necessary, in accordance with ISO15016:2002 Standard or the equivalent (ISO 2002, IMO 2012).

Since then, ISO 15016:2002 Standard was revised in 2015 (ISO 2015), and the revisal had some changes in the effect of wave diffraction, however, the basic scheme of the revisal preserves ISO 15016:2002 Standard.

In order to find the optimum route of a ship, the various external forces on a target route and its surroundings must be calculated, which is the complicated and enormous work which requires a lot of calculation time (Hinnenthal 2008, Kim et al 2016). In this paper, the prediction method and calculation procedure of resistance and propulsion performance for the weather routing system were developed. In order to predict resistance and propulsion performance within a short time as possible while maintaining the accuracy of the results, resistance and propulsion performance on the various ship's draft conditions and weather conditions are calculated in advance and made into database.

2 PREDICTION METHOD OF RESISTANCE AND PROPULSION PERFORMANCE

The total resistance R_T is divided into basic hull resistance in calm water R_{hull} and added resistance due to various external force ΔR like formula (1).

$$R_T = R_{hull} + \Delta R \tag{1}$$

The basic hull resistance R_{hull} uses the resistance coefficient obtained from the model test. However, when there are no model test results, it is estimated from the model test results of a similar ship or through statistical analysis (Kim & Kang 2001).

The established methods to predict added resistance due to various external forces are like formula (2).

$$\Delta R = R_{AW} + R_{AA} + R_{\delta\delta} + R_{AR} \tag{2}$$

R_{AW} Resistance Increase due to Waves
R_{AA} Resistance Increase due to Wind
$R_{\delta\delta}$ Resistance Increase due to Drifting Angle
R_{AR} Resistance Increase due to Hull Roughness

The added resistance due to various external forces is calculated based on ISO 15016:2002 Standard, and proper methods are added to the content which ISO 15016:2002 Standard doesn't have and lacks (ISO 2002).

The added resistance due to waves is calculated by two stages according to ISO 15016;2002 Standard. First, the response function of the added resistance in regular wave is calculated in advance preparations. Using the result in main procedures, the added resistance in irregular wave is calculated on the specified wave condition. For the added resistance by ship motion due to waves, Maruo method is used, and for the added resistance by

wave diffraction in short waves, one of Faltinsen method, Kwon method or Fujii-Takahashi method can be selected. For the calculation of the added resistance due to waves, these waves are divided into wind wave and swell, both of which are calculated respectively (ISO 2002, Kim et al 2001).

In ISO 15016:2002 Standard, for the added resistance due to wind, one of the wind tunnel test results or JTTC standard empirical curves can be selected. JTTC standard empirical curves are made up of head wind resistance coefficient curve and directional coefficient curve using the data of a wind tunnel test of several ships. However, it is not enough to obtain the exact value as wind resistance must be calculated on the ships of all types with only these two curves. Thus, the Blendermann chart of wind resistance coefficients is added to the prediction method of wind resistance. Blendermann chart displays the results of each wind tunnel test on 22 ships (Blendermann 1990-1991). The wind resistance R_{AA} is calculated like formula (3). The ρ_A means air density, A_T front projected area above waterline, and V_R relative wind speed.

$$R_{AA} = C_X \frac{1}{2} \rho_A A_T V_R^2 \tag{3}$$

When a ship operates with a drifting angle, the added resistance due to drifting angle is calculated with SR208 method like formula (4) or KRISO method like formula (5) (Kim et al 2015). In formula (4), T_m means mean draft, β drifting angle of a ship. In formula (5), A_{drift} means projected area into drifting direction under the waterline, and A_{LH} projected area into front direction under the waterline.

$$R_{\delta\delta} = \frac{1}{4} \pi \rho T_m^2 V^2 \beta^2 \tag{4}$$

$$R_{\delta\delta} = 2 R_{hull} \frac{A_{drift}}{A_{LH}} \tag{5}$$

When a ship contacts seawater for a long time, as the influence on hull surface roughness cannot be ignored, the added resistance due to hull roughness must be calculated. In this paper, the added resistance due to hull surface roughness R_{AR} is calculated like formula (6) using 1977 ITTC powering performance prediction method (ITTC 1978). In formula (6), k_S means hull surface roughness, k_{S0} standard hull surface roughness where k_{S0} value is based on 0.15mm.

$$R_{AR} = \frac{1}{2} \rho S V^2 \frac{105}{L^3} \left(k_S^{\frac{1}{3}} - k_{S0}^{\frac{1}{3}} \right) 10^{-3} \tag{6}$$

The shallow water effect of a ship is calculated with Lackenby method like formula (7), which is a formula of indicating the reduction of a ship speed, not the increase of resistance (ISO, 2002). In formula (7), A_M refers to front projected area under the waterline, h water depth.

$$\frac{\Delta_V}{V} = 0.1242\left(\frac{A_M}{h^2} - 0.05\right) + 1 - \left(\tanh\frac{gh}{V^2}\right)^{\frac{1}{2}} \quad (7)$$

The shaft power P_T is calculated from total resistance R_T, quasi-propulsive efficiency η_D and shaft transmission efficiency η_T like formula (8).

$$P_T = \frac{P_E}{\eta_D \eta_T} = \frac{R_T V}{\eta_D \eta_T} \quad (8)$$

In formula (8), quasi-propulsion efficiency η_D is calculated like formula (9), when various propulsive coefficients such as wake fraction w , thrust deduction t , relative rotative efficiency η_R and propeller open water efficiency η_O are obtained through a model test or estimated through statistical analysis.

$$\eta_D = \frac{(1-t)}{(1-w)}\eta_R\eta_O \quad (9)$$

When the change in open water propeller efficiency η_O is considered, the influence on propeller surface roughness must be considered. Only its necessity is mentioned, however, there are no suggestions about the calculation method in ISO 15016:2002 Standard. In this paper, the change in efficiency according to propeller surface roughness is calculated using the 1978 ITTC powering performance prediction method (ITTC 1978).

3 CALCULATION PROCEDURE OF RESISTANCE AND PROPULSION DATABASE

The calculation procedure of resistance and propulsion performance has been developed. The calculation procedure is divided into two phases and calculated like Figure 1.

In Phase I, the resistance and propulsive coefficients in calm water and the added resistance due to various external forces are calculated in advance and made into database. In Phase II, the ship speed and shaft power of a ship at the specified weather condition are calculated using the database of resistance and propulsion performance.

Through a model test or statistical analysis, various resistance and propulsive coefficients in calm water for various draft conditions has been calculated out and included in database. The calculation of the added resistance due to various external forces consists of the effect of wind, waves, drifting angle and hull surface roughness. Table 1 shows the range of the variables which are calculated in advance in each calculation.

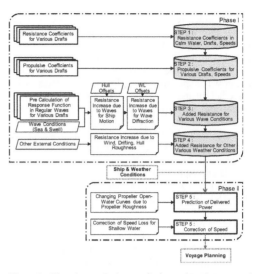

Figure 1. Flowchart of the prediction of resistance and propulsion performance for the weather routing system

Table 1. Database of resistance and propulsion performance for the weather routing system

	Resistance and Propulsive Coefficients		
Resistance Coefficient for Calm Water	Residual Resistance Coef. (or) Wave Making Res. Coef. & Form factor		
Propulsive Coefficient	Thrust Deduction Fraction Wake Fraction Relative Rotative Efficiency Qusi-Propulsive Efficiency		
Added Resistance	Added Resistance	1st Variable	2nd Variable	3rd Variable
Due to External Forces	Wave Motion	Incident Angle 0-180 deg. (19)	Significant Height 0-30.48 m (13)	Ship Speed
	Swell	Incident Angle 0-180 deg. (19)	Significant Height 0-30.48 m (13)	Ship Speed
	Wave Diffraction	Incident Angle 0-180 deg. (19)	Significant Height 0-30.48 m (13)	Ship Speed
	Swell	Incident Angle 0-180 deg. (19)	Significant Height 0-30.48 m (13)	Ship Speed
	Wind	Relative Direction 0-180 deg. (19)	Relative Velocity 0-30 m/s (11)	-
	Drifting	Drift Angle 0-180 deg. (19)	Ship Speed	-
	Hull Roughness	Hull Roughness 0-0.002 m (21)	Ship Speed	-

The added resistance due to waves is calculated using the wave height, wave period and wave incidence angle. However, as it takes a lot of calculation time and big data size when using those all of these three variables, Pierson-Moskowitz ocean wave spectrum is applied to the relation with wave height and wave period. Considering that wave height and wave period are related, and using the only wave height as a representative value, they can be calculated (DND Canada 2016). The waves are divided into wind wave and swell, and the resistance increase due to ship motion and wave diffraction are calculated respectively. The wave incidence angle is divided into 19 ones at intervals of 10 degrees from 0 to 180 degrees and wave height is done into 13 ones.

The wind direction for calculating the added resistance due to wind is divided into 19 ones at intervals of 10 degrees from 0 degree to 180 degrees, and the drifting angle for calculating the added resistance due to drifting angle is done into 19 ones at intervals of 10 degrees from 0 degree to 180 degrees.

The surface roughness for calculating the added resistance due to hull surface roughness is divided into 21 ones at intervals of 0.05mm from 0mm to 2mm.

In Phase II, using the database of resistance and propulsion performance like Table 1 which is calculated in Phase I, the total resistance and shaft power are calculated. In this Phase II, the open water propeller efficiency is calculated by the influence of propeller roughness, and the ship speed is corrected by the effect of shallow water. The calculation in Phase II is relatively simple, which is calculated directly in the weather routing system.

4 COMPUTER PROGRAM OF RESISTANCE AND PROPULSION DATABASE

This computer program to predict the resistance and propulsion performance has a function of predicting resistance and propulsive coefficients in calm water and added resistance due to various external forces, and calculating ship speed and shaft power. The Phase I of Figure 1 shows the flow chart of the computer program of manufacturing resistance and propulsion database. It consists of two individual programs, such as a program WS-WAVE for prediction of the added resistance due to waves, and a program WS-MAIN for manufacturing resistance and propulsion database in various weather conditions.

These computer programs have been applied to the weather routing module of KRISO Arctic Safe Voyage Planning (KASVP) system under development, which is supposed to be used for vessels operating in Arctic sea. For the validation

test, this system shall be tested on the first research ice breaker of Korea, Araon, in the near future.

Table 2 shows the front part of example output of resistance and propulsion database generated for constructing the weather routing system of the Aaron.

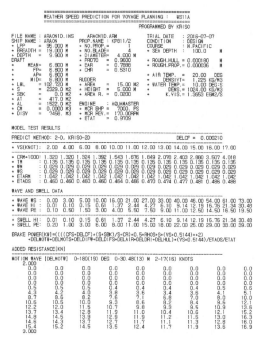

Table 2. An example of resistance and propulsion database for the Araon

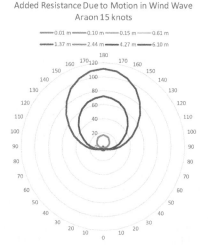

Figure 2. Resistance increment due to ship motion for the Araon

Figure 2 shows the resistance increment due to ship motion by the various incidence angles and wave height at 15 knot of design condition as an example.

Figure 3 shows the resistance increment due to wind by the various incidence angles and wind speed as an example.

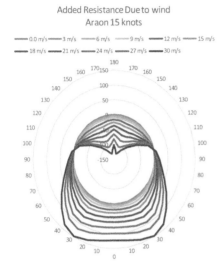

Figure 3. Resistance increment due to wind for the Araon

5 CONCLUSION

The prediction method and calculation procedure of resistance and propulsion performance for the weather routing system has been developed.

For the fast calculating process of the weather routing system an enhanced method has been adopted, which utilizes pre-calculated database for the resistance and propulsion performance to save the calculation time.

The database consists of resistance and propulsive coefficients and various added resistance by various draft conditions of a ship. The added resistance includes the influence of wind, waves, drifting angle and hull surface roughness condition. The added resistance due to waves consists of the effect of ship motion and wave diffraction, and wave is divided into wind wave and swell respectively.

The developed method and procedures were made into a series of computer programs, and these computer programs have been applied to the weather routing system. This system shall be tested on the first research ice breaker of Korea, Araon, in the near future.

ACKNOWLEDGMENT

This research is part of the project titled "Development of Safe Voyage Planning System for Vessels Operating in Northern Sea Route (PJT200642)" which has been supported by the Ministry of Oceans and Fisheries, Korea.

REFERENCES

Blendermann, W. 1990-1991. "Chapter 3.1 External Forces, Wind Forces of Manoeuvring Technical Manual", Schiff & Hafen 1990:2, 1990:3, 1991:4.

DND Canada. 2016. http://www.crs-csex.forces.gc.ca/boice/rp/hmcs-ncsm/rp/ann-eng.aspx, Home Page of Department of National Defence and the Canadian Armed Forces

Hinnenthal, J. 2008. Robust Pareto, Optimum Routing of Ships Utilizing Deterministic and Ensemble Weather Forecasts, Ph.D. Thesis, Technischen Universitat Berlin. pp.2-8, 26-57

IMO. 2011. Amendments to the Annex of the Protocol of 1997 to Amend the International Convention for the Prevention of Pollution from Ships, 1973, as modified by the Protocol of 1978 relating thereto, Resolution MEPC.204(62), pp. 1-17

IMO. 2012. Guidelines on survey and certification of the Energy Efficiency Design Index (EEDI), Resolution MEPC.214(63), pp. 7

IMO. 2014a. Draft e-Navigation Strategy Implementation Plan, Annex 7 of NCSR 1/28

IMO. 2014b. Report of the Marine Safety Committee on its Ninety-Fourth Session, MSC 94/21, pp. 36

IMO. 2014c. International Code for Ships Operating in Polar Waters, Resolution MSC.385(94)

ISO. 2002. Ships and marine technology - Guidelines for the Assessment of Speed and Power Performance by Analysis of Speed Trial Data, International Standard of ISO 15016:2002, pp.1-45

ISO. 2015. Ships and marine technology - Guidelines for the Assessment of Speed and Power Performance by Analysis of Speed Trial Data, International Standard of ISO 15016:2015, pp.1-85

ITTC. 1978. Report of Performance Committee, Proceedings 15th ITTC, pp.389-392

Kim, E.C. et al. 2001. Evaluation and Computer Program on the Speed Trial Analysis Method of the Ongoing Work in ISO/TC8, Proceedings of the Eighth International Symposium on Practical Design of Ships and Other Floating Structures, Shanghai, pp.525-532

Kim, E.C. & Kang K.J. 2001. Regression Analysis for the Resistance and Propulsive Coefficients based on KRISO Data Base, 2nd International Workshop on Ship Hydrodynamics, Uhan, China, pp.1-7

Kim, E.C., Choi H.J. & Lee, S.G. 2016. Analysis Method and the Related Computer Program to Predict the Towing Condition of a Towed Vessel Using Multiple Tugboats and Pushers, OCEANS 2016 MTS/IEEE Monterey

Kim, E.C., Kang K.J. & Lee, H.J. 2016. A Study on the Database Generation of Propulsion Performance for Ships Optimum Routing System, Journal of Navigation and Port Research Vol. 40, No. 3 (Korean), pp. 97-103

Proceedings of 12th International Conference on Marine Navigation and Safety of Sea Transportation, TransNav 2017
21-23 June 2017, Gdynia, Poland

A Mariners Guide to Numerical Weather Prediction in the age of 'Big Data'

H. Davies
Stratum Five Limited, Shoreham, United Kingdom

ABSTRACT: Weather affects safety and performance of vessels, cargo and seafarers. The application of powerful computers and high speed communications networks have led to the establishment of Numerical Weather Prediction which has transformed meteorology and underpins 'Big Data' services. Forecast accuracy has greatly improved and weather forecasts are now some 20% more accurate than they were only 5 years ago. However weather remains the biggest cause of total loss of ships and for the first time, many mariners are being exposed to computer generated weather forecasts which have been produced without vetting by meteorologists. This short paper is intended to provide an introduction to Numerical Weather Prediction for mariners, so that when they are being presented with NWP whether as part of a 'Big Data' service or on the bridge, that they will be prepared.

1 INTRODUCTION

Statistics from the International Union of Marine Insurance (IUMI) indicate that the leading cause of total loss of shipping between 1996 and 2015 was weather. It accounted for 25% of total losses in that period and 30% between 2001-2005. But it soared to an attention-grabbing 48% from 2011-2015. This is extraordinary because over this period, weather forecast accuracy had improved greatly. In fact a quiet revolution has taken place in weather forecasting, one which the marine industry has clearly not yet recognised and applied for its benefit. This revolution has been a result of the development of Numerical Weather Prediction or 'NWP' which has become all-pervasive.

The development of comprehensive interlinked databases and powerful computers which has enabled the production of NWP is now being applied more broadly to shipping and logistics to support 'Big Data' services. As weather conditions play a critical part in determining speed, consumption and passage time, NWP underpins many of these applications.

This paper is a summary of a book by the same name published by the Nautical Institute [1] and is intended to provide an introduction to the subject for mariners, so that when they are being presented with NWP, they will be able to identify the fact; and have the knowledge to ask some basic questions to establish its appropriateness, whether it is as a weather forecast product, or on their Electronic Chart Display and Information System (ECDIS) or back of bridge system.

2 WHAT IS NWP?

2.1 Background

NWP is in use throughout the world, such that all weather forecasts, including marine and shipping forecasts, are based on NWP. Although this paper focusses on atmospheric models the process is exactly the same for wave and current models.

The origins of NWP lie with the work of Wilhelm Bjerknes, who suggested in 1904 that the weather could be quantitatively predicted by applying the complete set of hydrodynamic and thermodynamic equations to carefully-analysed initial atmospheric states. The application of these ideas has been made possible through the development of a global meteorological observing system combining in-situ observations and remote sensing from space to provide the analysis, and of powerful computers to perform the calculations. The complete set of hydrodynamic and thermodynamic equations is applied to analyse the initial condition of the atmosphere and predict its future state.

2.2 Data Assimilation

The World Meteorological Organisation (WMO) is an organ of the United Nations and has created a mechanism for the free exchange of data and NWP products between accredited centres. The WMO estimates that over 100 million observations are made daily. This data is collated at major nodes and distributed on a dedicated communication. The observational data is very diverse and consists of data from terrestrial and space-based sensors and data are irregularly distributed in time and space. The first step in the NWP process is to assimilate these data and to interpolate them onto the computational grid.

Data is assimilated by breaking it down into key meteorological parameters and then giving these a weighting. This determines how far into the computational grid the influence of the observation is felt, both in the vertical and horizontal axes, always ensuring that the atmosphere is described in a physically consistent way in time as well as space.

Despite the many sources of data there are still considerable areas of the computational grid for which there is little data available. Some areas e.g. western Europe, are rich in data whilst others, particularly the oceans, are data-sparse. These are filled with values for each parameter for the appropriate time taken from forecasts generated by the previous computer run. The starting point for the analysis is therefore the previous forecast merged with all of the observational data received in the interim; this has the effect of creating continuity between successive forecast runs.

Although assimilated separately during the initial stage of the analysis, the different meteorological parameters are dependent variables and the assimilation scheme must ensure internal consistency within the whole volume. Even with modern Massively Parallel Processing these are non-trivial computations and data assimilation and the creations of the analysis takes between 1 and 2 hours. There are only a handful of centres in the world that can produce global NWP and all are state sponsored.

2.3 Forecast production

Once the analysis is complete the primitive equations are applied and the atmosphere allowed to evolve. At discrete intervals a snapshot of the conditions is output as a prognosis or forecast valid at that time. Thus, NWP results in the creation of databases of parameters such as pressure, temperature, wind speed and direction, on a 3 dimensional spatial grid with time as a 4th dimension.

Some models do not use grid points in the horizontal domain. Instead they describe the present and future states of the atmosphere by using sine waves. These are termed spectral models and the resolution is determined by the number of waves that are used to describe the atmosphere, this is often preceded by a 'T' for Triangular truncation. For example, the US Global Forecast System (GFS) model has T1534 spectral waves. This approximates to a horizontal resolution of 13 km. The use of spectral waves saves computational time and spectral models lend themselves more to longer-range forecasts than grid-point models.

The effective spacing between grid points determines the type of meteorological feature that can be resolved. The computer resources required increase with the resolution, as a rule of thumb halving the horizontal grid length requires a 16 fold increase in computer power. Therefore high resolution models are of a limited area and are known as mesoscale, regional or limited area models. Because of the global nature of shipping we will focus on global models.

3 ACCURACY

NWP forecasts are incredibly accurate. Anomaly correlation scores, which are a measure of accuracy, for the 5 day forecasts produced by the leading NWP models indicate that the best models are now better than 90% accurate at 5 days and still show skill in their 14 day forecasts.

For mariners this translates into reliable forecasts of key parameters such as surface wind speed. Verification of the European Centre for Medium Range Weather Forecasts (ECMWF) NWP forecast wind speed against the wind speed observed at manned observing stations in Europe (where the observations are reliable) indicates a forecast accuracy of +/- 1 knot at T+72 hours i.e. 3 days. Similarly the most skillful NWP wave analyses when compared to buoys are accurate to +/- 0.3 metre [2].

4 WHAT SYSTEMS USE NWP?

Currently, mariners primarily use NWP through weather forecasts and routeing services. These take many forms. Some companies send text and graphical forecasts by email, others host websites while others provide software applications and a data feed. Some routeing software applications such as 'Expedition LT' contain an NWP application which can download NWP data from a remote server, display the data and use it to provide routeing guidance. Some Electronic Chart Display and Information Systems (ECDIS) and chart plotters can also display NWP if the user subscribes to the weather data service. Fugawi Marine ENC, SeaPro,

MaxSea, RayTech, Expedition and Deckman are a few examples of software that can use NWP.

There are also services which use NWP behind the scenes and only provide the mariner a resulting product or recommendation, notably the ship routeing services provided by companies such as AWT, SPOS (Meteogroup) and WRI.

Other uses of NWP are more opaque and it may not be immediately obvious that NWP underpins the product, which will however only be as accurate as the NWP from which it is derived. For example the Charter Party Agreements in common usage contain definitions of 'Good Weather Periods' under which the performance and speed will be achieved. Usually these place upper limits on winds, waves and waves and adverse currents. If the conditions experienced exceed these, then the vessel does not have to perform as per the Charter Party. After the voyage the vessel performance is evaluated by reconstruction of the ship's track from the noon day reports and/or by AIS reports. The ship positions are then matched to the gridded output from NWP for the corresponding times and values for wind, waves and current extracted. The good weather periods are identified and speed and consumption calculated. If the performance achieved in good weather is less than that specified in the Charter Party, then the good weather speed and consumption is extrapolated to the entire voyage to calculate the additional bunkers consumed and the additional time taken for the voyage.

NWP is also present in tools such as OCTOPUS which are used by naval architects and marine engineers in the design of ships and offshore structures and by the oil and gas industry. Wind and currents from NWP are used to calculate drift for Search and Rescue operations and spreading following oil spillages.

As there is significant public interest in the weather, weather information lends itself very well to an 'internet' business model where income is generated from advertising on the website rather than charging users directly. Some sites provide a free to air product and offer a premium service for which there is usually a charge. This includes internet weather sites, mobile phone applications, weather toolbars for web browsers and notably an increasing number of chartplotters.

In combination with ship positional information and onboard sensors, NWP can provide powerful ship performance monitoring and analysis capabilities. 'Big Data' analysis will enable this service to be extended to incorporate self-learning algorithms to produce ship specific optimization. In the near future this will be an essential tool to reduce operating costs and to improve energy efficiency as required by the IMO Ship Energy Efficiency Management Plan (SEEMP) [3] and to minimize emissions as recorded by the IMO Energy Efficiency Operational Indicator (EEOI) [4] and the EC Monitoring Reporting and Verification (MRV) legislation [5].

5 POTENTIAL ISSUES

Although the NWP is accurate at source, once it is displayed on a third party system it is critical to know the source of the data and thus its timeliness, accuracy, and suitability for a particular application or spatial scale.

The importance of knowing the provenance of the data underpinning the bridge systems was evidenced during the U.S. Coast Guard investigation into the loss of the El Faro in Hurricane Joaquin with all 33 crew on 01 October 2015.

Information released by the U.S. Coast Guard reveals that El Faro was receiving weather support from Applied Weather Technology (AWT) to an onboard system. The data provided was sourced from the US National Centers for Environmental Prediction (NCEP), which is a leading NWP centre, and the National Hurricane Centre (NHC) by AWT and sent to the AWT onboard software 'Bon Voyage' for display. The U.S. Coast Guard Marine Board of Investigation heard that the NHC hurricane track sent by AWT to El Faro was provided was 21 hours old [6].

Even if the wind and wave data provided for display on a bridge system is the latest available, the data is usually not sent at full temporal or spatial resolution in order to reduce the file size and communications cost. This means that although the display on Bon Voyage and other software looks convincing, it can also be misleading.

Global NWP models have a temporal granularity of 3 hours and if the ship position report does not occur at 00/03/06 UTC etc then this means that interpolation in time is required between the forecast values or the ship position at the time of the forecast has to be estimated. Similarly the models output on a grid with values at discrete latitude and longitudes. The grids are currently around 0.25 degrees by 0.25 degrees and so if the ship position does not fall exactly onto the grid point then interpolation in space is required between the forecast values at adjacent grid points. Therefore voyage reconstructions and forecasts which use NWP to match weather conditions to ship positions should be used with caution as the temporal and spatial grid is unlikely to match the time and location of ship reports and a degree of interpolation will be required.

6 CONCLUSION

Mariners have always had to understand the weather and their knowledge and their requirements have evolved as marine and forecasting technology has developed. It is clear that NWP has become ubiquitous and will become embedded within 'Big Data' services. Mariners will undoubtedly make use of it either directly, through use of internet weather forecasts or indirectly through display of weather on ECDIS or use of vessel performance, routeing services etc. There is nothing inherently wrong with the use of NWP, as long as users are aware of its limitations. To operate safely, mariners must firstly recognise that the predicted weather is machine-generated without human meteorological input and make use of it in full knowledge of its limitations. Secondly, mariners should ensure that they know the source, timeliness, accuracy, of the NWP and make an informed assessment of its suitability for the intended purpose.

REFERENCES

[1] Davies, Huw, (2013) *Numerical Weather Prediction A practical guide for mariners*, Nautical Institute (ISBN 978 1 906915 40 7)
[2] http://www.ecmwf.int accessed 30 Jan 16
[3] MEPC.1/Circ.684 *Guidelines for voluntary use of the Ship Energy Efficiency Operational Indicator (EEOI)* dated 17 Aug 09
[4] MEPC.282(70) Addendum 1 *2016 Guidelines for the development of a ship energy efficiency management plan (SEEMP)* dated 28 Oct 16
[5] Regulation (EU) 2015/757 *on the monitoring, reporting and verification of carbon dioxide emissions from maritime transport, and amending Directive 2009/16/EC* dated 29 April 15
[6] http://jacksonville.com/news/metro/2016-05-18/story/storm-track-received-el-faro-day-sinking-was-21-hours-old

Proceedings of 12th International Conference on Marine Navigation and Safety of Sea Transportation, TransNav 2017
21-23 June 2017, Gdynia, Poland

Investigation of Ocean Currents in Navigational Straits of Spitsbergen

A. Marchenko
The University Centre in Svalbard, Longyearbyen, Norway

Z. Kowalik
University of Alaska Fairbanks, Fairbanks, USA

ABSTRACT: In the present paper we focus on narrow straits Akselsundet and Heleysundet, the parts of tourist and industrial routes. The Akselsundet Strait was used by coal boats on the way from the open sea to the coal quay Kapp Amsterdam located at the head of the Van Mijen Fjord near mining settlement Svea. The Heleysundet Strait is located between Barentsoya Island and Spitsbergen. In both straits the flow is driven by semidiurnal tide. The Norwegian Pilot (2016) reports the maximum current up to 7 knots in Akselsundet and 9 knots in Heleysundet. The current velocity in Akselsundet was measured with ADCP AWAC for the water column of 45 m depth in 2015/2016 according to the request of Port Captain in Svea. Surface velocities in the Heleysundet Strait were deduced from the ice trackers deployed on the drift ice. A simplified box model of tides in the fjord was developed for the estimates of slack water period. In the paper we compare numerical estimates of the slack water periods in Akselsundet with field data collected by the ADCP. In Heleysundet the slack water periods and current velocities are calculated and compared to the ice drift velocities.

1 INTRODUCTION

Navigation in narrow channels with currents driven by tides requires knowledge of maximal speed of currents, slack water time and duration. In the present paper we analyse these parameters in two straits in Spitsbergen: Akselsundet and Heleysundet (Fig. 1). The Akselsundet Strait connects the Bellsund Bay to the Van Mijen Fjord of the West Spitsbergen.

Coal boats up to 70.000 tons deadweight passed the strait on the way to and from the Kapp Amsterdam coal quay near coal mining settlement Svea at the head of the Van Mijen Fjord. Compulsory pilotage service is requested to pass the strait for vessels carrying dangerous cargo during slack water period (The Norwegion Pilot, 2016). Since 2016 the coal transport has been suspended and the future navigational activity in Akselsundet will be related to the tourism and research.

The Heleysundet Strait is located between East Spitsbergen coast and Barentsøya Island. The Norwegion Pilot (2016) reports that "several vessels recorded a tidal stream rate of 8-9 knots in the narrowest part. Even with high power engine vessels are at risk of becoming icebound as it presses together in the sounds' funnel-shaped entrance areas".

Kowalik et al (2015) developed numerical model of tides in West Spitsbergen Fjords and compared results to field measurements of tidal elevations. The model reproduces well tide measurements performed near the Akseløya Island. Marchenko et al (2015) considered box model of tides in the Van-Mijen Fjord using tidal constituents calculated by Kowalik et al (2015) to estimate the slack water period in Akselsundet.

Figure 1. Spitsbergen. Red rectangles show navigational straits Akselsundet (1) and Heleysundet (2).

According to the request of tag boats captain in Svea the Acoustic Doppler Current Profiler (ADCP) was deployed in Akselsundet to record sea current velocities over the entire water column. The deployment was performed from the tag boat in the October 2015. In the present paper we analyze the measured current and compare them to the numerical simulations.

In 2013 and 2016 Iridium ice trackers were deployed respectively on the ice in the Store Fjord and Olga Basin. The ice trackers provided GPS data of their locations each 10 minutes. They drifted in the Heleysundet Strait. In the present paper we describe the ice drift velocity calculated with the ice trackers data. We also use the box model to estimate the current velocity in the strait.

2 FIELDS MEASUREMENTS

To evaluate the tidal sea level change the measurements of water pressure at the bottom over several semidiurnal cycles were performed near Akseløya in 2010, 2011 and 2013 (Marchenko et al., 2015). The example of measurements in two locations (Fig. 2, left panel) is shown in Fig. 3. The phase shift between the tides in Birkelandodden and Hamnodden is smaller than 30 min depending on the moon phase. The Norwegion Pilot (2016) reports similar measurements in two locations at the Heleysundet Strait in 2008 (Fig. 2, right panel). The phase shift between tides in Henckeløyane and Kapp Bessels reaches half of the period of semidiurnal tide, i.e. more than 6 hours.

Figure 2. Locations of water level elevation measurements in Akselsundet (left panel) and Heleysundet (right panel). Point with coordinates marks the location of ADC deployment in October, 2015.

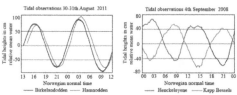

Figure 3. Recorded tidal heights versus the time: in Akselsundet (left pannel) and in Heleysundet (right panel).

On the November 4, 2015, the ADCP AWAC Nortek 400 kHz was deployed at sea bottom on 50 m depth in the Akselsundet Strait the location with coordinates shown in Fig. 2 (left panel). The ADCP was mounted upward looking inside the bottom

mount system Mooring Systems, Inc. The system includes Gimbaled Mount for the ADCP and lead weight of 60 kg. In addition it was fixed at the bottom with 200 m chain extended from the system to the anchor.

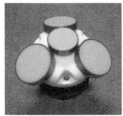

Figure 4. Mounting of ADCP AWAC 400 kHz in the bottom mount system Mooring Systems, Inc.

The ADCP measured vertical profile of the water velocity components in ENU system and water pressure with sampling interval 5 min from the November 4, 2015, to March 18, 2016, when the battery was drained. The measurements are shown in Fig. 5. The data give 3 components of the water velocity averaged over 1 m cell. The center of the first cell is located 2 m from the ADCP head. The data are collected from 45 cells.

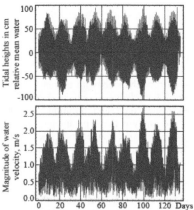

Figure 5. Tidal heights and water velocity magnitude in the last cell at the distance 46 m from the ADCP head. November 4, 2015 till March 18, 2016.

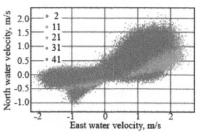

Figure 6. Horizontal components of the water velocity measured over the entire deployment period. Different depths

are shown by different colors. Distance from the ADCP head is pointed out in m.

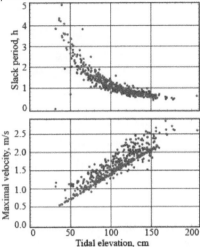

Figure 7. Slack water period (upper panel) and maximal velocity magnitudes (bottom panel) versus the tidal elevation over the semidiurnal cycle. Blue and pink points are respectively the data from the last cell and from numerical simulations.

Figure 6 shows almost the same water velocities over the entire water column with a reduction within the boundary layer near the seabed. Directions of the water velocities are very similar excluding slack periods. Maximal velocity magnitudes reach 2.5 m/s. Further we associate slack water period with velocity magnitudes smaller 0.5 m/s. Figure 7 (upper panel) shows that duration of the slack periods decreases with the increase of the tidal elevations. Minimal values of the slack periods about 30 min are observed at the spring tides. Mean value of the slack period is about 1.5 hours.

Figure 7 (bottom panel) shows the dependence of maximal velocity magnitudes (calculated over each semidiurnal cycle) on the tide elevation. Maximal values of the maximal velocity magnitudes above 2.5 m/s are observed during spring tides occurrence.

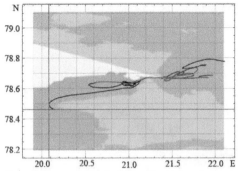

Figure 8. Fragments of trajectories of ice trackers deployed on the drift ice: Store Fjord in 2013 (black line) and Olga Strait in 2016 (blue line).

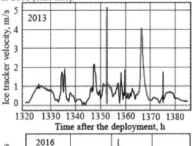

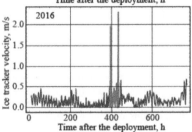

Figure 9. Velocity magnitudes of the ice trackers versus the time. Deployment year: upper panel 2015, lower panel 2016.

Sea current velocities in the Heleysundet Strait were calculated from GPS positions of the Iridium Ice Trackers (Oceanetic Measurements, Ltd) deployed on the drift ice in the Store Fjord (2013) and Olga Strait (2016). The ice tracker deployed in 2013 passed the Heleysundet Strait, the ice tracker deployed in 2016 approached to the strait and then turned back to the Olga Strait. Ice trackers provided the data of their GPS positions one time per 10minutes. Fragments of their trajectories are shown by black line (2013) and blue line (2016) in Fig. 8. Maximal drift speeds registered on June 28-29, 2013 corresponds to the brown colored trajectories. The red portions denote maximal drift speeds registered on May 18-19, 2016. The drift speeds are shown in Fig. 9 versus the time. Here the colors correspond to the colors of the trajectories in Fig. 8. Maximal drift velocities were registered outside the strait in the Store Fjord. Satellite image in Fig. 10 shows the ice gyre. In this location the drift velocity up to 5 m/s was recorded see, Fig. 9, upper panel.

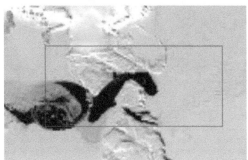

Figure 10. Satellite image of the Heleysundet strait from April 04, 2013 (Apollo Mapping).

3 NUMERICAL SIMULATIONS

To describe tidal current and water level elevation in the narrow passages a simplified box model was developed.

The model was applied to the strait connecting the Vallunden Lake and Svea Bay located at the head of the Van Mijen Fjord (Marchenko and Morozov, 2013). Afterwards this approach was used for the calculation of currents in the Akselsundet strait (Kowalik et al., 2015; Marchenko et al., 2015). In this approach equations of mass and momentum balance are integrated over the volume of the water jet penetrating in and out of the fjord through the strait. The speeds of the jet flow are expressed by the water level elevations at the ends of the strait.

$$S_1 u_1 = S_0 u_0 , \qquad (1)$$

$$\frac{u_1^2 - u_0^2}{2} + g(\eta_1 - \eta_0) = C \int_0^L R^{-1} u^2 dx \, \text{sign}(\eta_1 - \eta_0) , \qquad (2)$$

where $S_{0,1}$ are the areas of vertical cross-sections of the jet at the boundaries (ends of strait); $u_{0,1}$ are the water speeds averaged over areas $S_{0,1}$; η_0 and η_1 are the elevations of the water surface at the strait boundaries. Subscripts 0 and 1 are related to the sea side of the strait and to the fjord side of the strait, respectively. The hydraulic radius is calculated using formula $R=A/P$, where P is the wetted perimeter of the jet. $C=0.003$ is the drag coefficient, and L is the length of the jet.

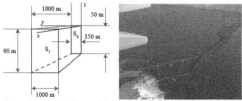

Figure 11. Dimensions of the jet flow in the Akselsundet Strait: left panel. Location of the jet flow on the aerial photo: right panel.

The shape of the jet is determined by seabed topography in the strait and specified in Fig. 11 (Marchenko et al., 2015). The integral in the right part of equation (2) is approximated by the formula

$$\int_0^L R^{-1} u^2 dx = \langle R \rangle^{-1} \langle u \rangle^2 L . \qquad (3)$$

The mean values of the hydraulic radius and the water speed are given by the following formulas

$$\langle R \rangle = \frac{R_0 + R_1}{2} , \quad \langle u \rangle = \frac{u_0 + u_1}{2} , \qquad (4)$$

where R_0 and R_1 are the hydraulic radii at the sea side of the jet and at the fjord side of the jet.

From equations (1)-(2) and formulas (3) and (4) it follows that

$$u_1 = -\text{sign}(\eta_1 - \eta_0) \left[\frac{2g|\eta_1 - \eta_0|}{S_1^2 S_0^{-2} + 0.5C \langle R \rangle^{-1} L (S_1 S_0^{-1} + 1)^2 - 1} \right]^{1/2} \qquad (5)$$

The law of mass balance in the fjord is written in the form

$$S_{VM} \frac{d\eta_1}{dt} = S_1 u_1 , \qquad (6)$$

where the area of water surface S_{VM} of the Van Mijen Fjord. It is equal to 500 km². Water level variation at the sea side of the strait (forcing) is determined by function

$$\eta_0 = A_{M2} \cos(\omega_{M2} t - G_{M2}) + A_{S2} \cos(\omega_{S2} t - G_{S2}) + \\ + A_{K1} \cos(\omega_{K1} t - G_{K1}) + A_{N2} \cos(\omega_{N2} t - G_{N2}) + \\ + A_{O1} \cos(\omega_{O1} t - G_{O1}) \qquad (7)$$

Above harmonic constants A, ω, and G have been specified in Marchenko et al. (2015). Water speeds $u_{0,1}$ and water surface elevation η_1 are calculated from equation (6) and formulas (1) and (5).

In the Heleysundet Strait water surface elevations $\eta_{0,1}$ at the both sides of the strait are determined by tides in the Store Fjord and Olga Strait (Fig. 1). In this case sea current speed in the strait is calculated directly by formula (5). Numerical simulations are performed with following characteristics of the strait: $R=H/2$, $H=65$ m, $S_1=S_0$, $L=3.6$ km, $C=0.003$, where H is the mean water depth in the strait. It is assumed that water surface elevation in the Store Fjord is described by formula (7). In the Olga Strait we use the same formula with phase shift of 6.21 hours.

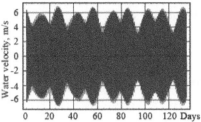

Figure 12. Simulated water velocities versus the time in the Heleysundet Strait.

Simulated slack water periods and maximal velocity magnitudes are shown in Fig. 7 by pink points versus the tidal elevation of the water level in the Akselsundet Strait. The results of numerical simulations compares well the measurements shown in Fig. 7 by blue points. Simulated water velocity in the Heleysundet Strait as a function of time is shown in

Dependence of the slack water period on the tidal elevation over the semidiurnal cycle is well illustrated in Fig. 13. Comparing this figure to Fig. 7 shows that the slack water periods in the Heleysundet Strait are much shorter than in the Akselsundet strait. This result fits in with the statement from the Norwegian pilot (2016) "The tidal streams in the sounds change quickly in the sounds as there are no slack periods worth mentioning".

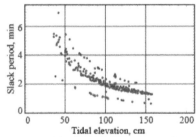

Figure 13. Simulated slack water periods versus the tidal elevation over the semidiurnal cycle in the Heleysundet Strait.

4 CONCLUSIONS

Sea currents were measured in the Akselsundet Strait over the entire water column of 45 m depth with ADCP AWAC 400 kHz with spatial resolution of 1 m and temporal resolution 5 min. We discovered that the water velocities are almost the same over the water column excluding slack water period. Reduction of the current velocities is observed at the distance 1-3 m from the bottom. Slack water period decreases with the increase of the

tidal amplitude and reach minimal values about 30 min at spring tide. Maximal magnitude of the water velocity increases with the increase of the tidal amplitude and exceeds 2.5 m/s at spring tide. Simplified box model was applied to reproduce the above observations. Computed slack water periods and maximal magnitude of the water velocity show dependence on the tidal amplitude well observed in the measurements. The model also demonstrates that water current in the strait is driven mainly by the water level slope over the strait.

Ice drift velocities up to 5 m/s were recorded at the western side of the Heleysundet Strait by ice trackers deployed in the Store Fjord in 2013. Satellite image confirms the existence of unfrozen ice gyre at this location. Ice drift velocity inside the strait was smaller probably due to the funnel shaped geometry of the passage which caused increased ice concentration. Numerical simulations performed with the box model show that maximal water velocity in the strait can exceed 6 m/s at spring tide. Slack water period in the Heleysundet Strait is estimated to be less than 10 minutes.

ACKNOWLEDGEMENTS

The authors wish to acknowledge the support of the Research Council of Norway through the SFI project SAMCoT, SIU foundation through the project HNP-2014/10008 SITRA, UNIS Logistic and Port Captain in Svea Lars Lundegaard.

REFERENCES

APOLLO MAPPING. Image hunters [Online]. Boulder, Colorado, USA. Available: https://apollomapping.com/, https://imagehunter.apollomapping.com/.

Kowalik, Z., Marchenko, A., Brazhnikov, D., Marchenko, N., 2015. Tidal currents in the western Svalbard Fjords. *Oceanologia*. 57, 318-327.

Marchenko, A.V., Morozov, E.G., 2013. Asymmetric tide in Lake Vallunden (Spitsbergen), *Nonlinear Processes in Geophysics*. 20, 935-944.

Marchenko,A., Kowalik, Z., Brazhnikov, D., Marchenko, N., Morozov, E., 2015. Characteristics of sea currents in navigational strait Akselsundet in Spitsbergen. *POAC15-00170*, Trondheim, Norway.

The Norwegian Pilot. Volume 7, Sailing Directions. Svalbard and Jan Mayen. Third Edition. 2016. The Norwegian Hydrographic Service and Norwegian Polar Institute.

APPENDIX

Analytical formula for calculation of tidal current in a strait

Let us assume that water level elevations due to the semidiurnal tide are described on different ends of a strait by the formulas

$$\eta_- = A\sin(\omega_{M2}t), \quad \eta_+ = \alpha A\sin(\omega_{M2}t + \theta),$$ (A1)

where $\alpha \in (0,1)$ and $\theta \in (0, \pi/2)$, and ω_{M2} is the frequency of the semidiurnal tide.

The slack water time is calculated from the equation

$$\sin(\omega_{M2}t) = \alpha \sin(\omega_{M2}t + \theta).$$ (A2)

The speed of water current in a channel with constant area of vertical cross-section is calculated by formula (5) as follows

$$u^2 = \frac{gH|\eta_+ - \eta_-|}{2cL},$$ (A3)

where c is the water drag coefficient at the bottom, H is the water depth in the channel and L is the length of the channel.

The slack water period is determined by the condition

$$u < u_{cr},$$ (A4)

where u_{cr} is a prescribed constant.

From formulas (A3) and (A4) it follows that the slack water period is determined by the condition

$$|\eta_+ - \eta_-| < \frac{2cL}{gH}u_{cr}^2.$$ (A5)

Conditions (4) and (5) are satisfied in a vicinity of the slack water time when

$$t = t_{sw} \pm \delta t,$$ (A6)

where $t=t_{sw}$ is the solution of equation (A2). Further we assume that $\omega_{M2}\delta t \ll 1$ and designate $\theta_{sw}=\omega_{M2} t_{sw}$. Then we have

$$|\eta_+ - \eta_-| = A|\cos\theta_{sw} - \alpha\cos(\theta_{sw} + \theta)|\omega_{M2}|\delta t| + \\ + O(\omega_{M2}^2\delta t^2)$$ (A7)

From condition (5) and formula (7) it follows that the slack water period is estimated with the formula

$$\delta t \approx \frac{2cL}{\omega_{M2}\Delta A}\frac{u_{cr}^2}{gH}, \quad \Delta = |\cos\theta_{sw} - \alpha\cos(\theta_{sw} + \theta)|.$$ (A8)

From equation (A2) it follows

$$\cos\theta_{sw} = \frac{1 - \alpha\cos\theta}{\sqrt{(1 - \alpha\cos\theta)^2 + \alpha^2\sin^2\theta}}.$$ (A9)

Simple estimates show that $\delta t \approx 42.6$ s when $u_{cr}=0.5$ m/s, $H=65$ m, $L=3.6$ km, $c=0.003$, $\alpha=1$ and $A=1$ m. The slack water period is $2\delta t \approx 1.42$ min.

Safety at Sea

Proceedings of 12th International Conference on Marine Navigation and Safety of Sea Transportation, TransNav 2017
21-23 June 2017, Gdynia, Poland

Navigation Safety and Risk Assessment Challenges in the High North

N.A. Marchenko
The University Centre in Svalbard, Longyearbyen, Norway

O.J. Borch & N. Andreassen
Bodø Graduate School of Business, University of Nordland, Bodø, Norway

S.Yu. Kuznetsova
Northern (Arctic) Federal University named after M.V. Lomonosov, Arkhangelsk, Russia

V. Ingimundarson
The University of Iceland, Reykjavik, Iceland

U. Jakobsen
The University of Copenhagen, Copenhagen, Denmark

ABSTRACT: The sea ice in the Arctic has shrunk significantly in the last decades. Partly as a result, the transport pattern has changed with more traffic in remote areas. This change may increase the risk of accidents. The critical factors are harsh weather, ice conditions, remoteness and vulnerability of nature. In this paper we look into the risk of accidents in Atlantic Arctic based on previous ship accidents and the changes in maritime activity. The risk has to be assessed to ensure a proper level of response in emergency situations. As accidents are rare, there are limited statistics available for Arctic marine accidents, therefore, in this study a mostly qualitative analysis and expert judgement has been the basis for the risk assessments. Implications for the emergency preparedness system of the region are discussed. The consequences of incidents depend on the incident type, scale and location.

1 INTRODUCTION

An understanding of risk factors, risk mitigating tools, and adequate rescue system capacities in different areas is necessary for sustainable development of the Arctic. Safe maritime operations in the Arctic are challenged by limited infrastructure, lack of rescue resources, long distances, a sparse population and harsh weather conditions. The paper is the part of MARPART (Maritime preparedness and International Collaboration in the High North) project, where researchers from all the countries of the "Atlantic sector" of the Arctic aim at finding ways to promote safety on the basis of activity and risk estimation and establishing the cross-institutional and cross-country partnership (Nord Universitet 2017).

In (Marchenko, Borch et al. 2015), we developed an assessment algorithm and presented a risk matrix for several sea areas of the High North region (Norway and Russia west of Novaya Zemlya). In (Marchenko, Borch et al. 2016), we considered available SAR resources and identified capacity gaps. Activity and probability of accidents differ in various parts of the Arctic for geographical, economic and historical reasons. In this paper, we elaborate on the western part that span the whole range of challenges related to remoteness, risk of ice and icing, and limited government resources and summarize assessment for the whole Atlantic Arctic. The experience and challenges of risk evaluation are discussed.

2 REGIONS OF THE ATLANTIC ARCTIC

The Atlantic Sector is divided it into five sea and land areas in accordance with the definition specified in the Artic Council's Search and Rescue (SAR) Agreement (Arctic Council 2011). We limit the Russian region by Novaya Zemlya and divide the Norwegian Sector into northern (Svalbard) and southern (Coastal) parts (see Figure 1) due to huge difference in the maritime activity and natural conditions.

These five regions are very unlike in terms of natural circumstances, ship traffic and preparedness system. Close collaboration is necessary in transboarding zones in the whole area where time is an important factor. Detailed characteristics of the eastern part (Numbers 3-5 in Fig. 1) are given in (Marchenko 2015, Marchenko, Borch et al. 2015, Marchenko, Borch et al. 2016). This discussion shows that in the Eastern part the capacities for emergency response is scarce when it comes to large scale incidents, where the larger cruise vessels represents a dominating threat. An important part of the risk assessment is thus the availability of both private and government emergency response resources that match the sea region activity.

Western part (1-2) is close to Svalbard by main features –relatively low but growing maritime activity, remoteness and harsh natural conditions.

Greenland (Region 1 on Fig.1) where Denmark has the responsibility of emergency preparedness in open waters, is a remote country with a huge territory, a small population, concentrated in the south-west part, and low industrial and transport activity. Only a few harbors, heliports and airports are located along the very extensive coastline (only 15% not covered by ice). Grounding and collision with ice are dominating risk factors. There are on average 80–90 SAR operations per year, performed by JRCC Greenland and police, and 200–300 person in distress every year (Joint Arctic Command 2016).

Iceland (Region 2 on Fig.1) is the largest and warmest SAR-region among our Arctic five regions, with a rather sparse population concentrated in coastal areas, developed harbor infrastructure in the south-western part and a rather intensive ship traffic. Bad weather, fire and human mistakes are the main reasons of ship distress. The Icelandic Coast Guard (ICG) is a key SAR actor in this region. It registers responds to many incidents per year. Fatal accidents on sea went down from 20 on average per year in the 1970s to two in the 21st century.

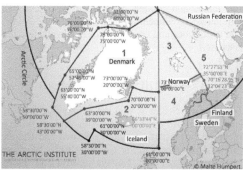

Figure 1. Considered Regions. Created on the base of Arctic Search and Rescue Agreement Map (Arctic Council 2011). 1-Greenland, 2-Iceland, 3-Svalbard, 4-Coastal Norway, 5-Russian sector of the Barents Sea

3 MARITIME ACTIVITY PATTERN

We assessed current (last five years) dynamics of ship traffic and maritime activity, using Arctic Havbase (Norwegian Coastal Administration 2017) – online resource, performing monthly AIS data since 2012. The number of port calls and passenger traffic over crossing lines are presented in Figure 2–5.

In all ports (except Hammerfest), the amount of port calls has increased in the last five years. The larger the port, the more significant was the increase in activity and the larger port the more significant was the growth in activity. The decrease of calls in Hammerfest can be y explained by lower price on hydrocarbon resulting in less visits by the offshore service vessel fleet.

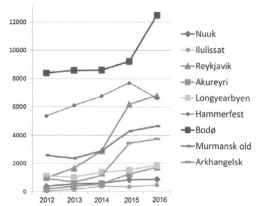

Figure 2. Number of port calls dynamics

The sizes of ports are clearly seen in the Fig.2 . Greenland and Svalbard ports are much smaller than Coastal Norway.

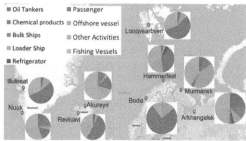

Figure 3. Type of vessels coming to Arctic Ports. Created on the base of (Norwegian Coastal Administration 2017)

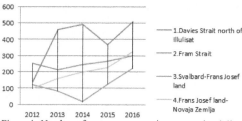

Figure 4. Number of passengers crossing conventional lines (data from (Norwegian Coastal Administration 2017)

The activity level in the far north remote waters can be estimated through the number of passengers crossing conventional lines (Figure 4) and through the type of vessels they used (Figure 5) – data available on (Norwegian Coastal Administration 2017). The activity is fluctuating (some years decreasing) and more than 20 times less that on traditional Maritime routes (see blue numbers in Fig.4.)

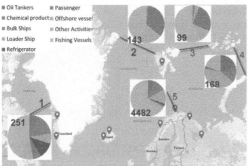

Figure 5. Types of crossing conventional lines vessels (data from the Arctic Havbase (Norwegian Coastal Administration 2017). Red numbers – line number (see fig. 4). Blue numbers-total numbers of passengers in one direction.

4 RISK ASSESSMENT THEORY

In this study, efforts are made to assess risk based on limited statistics. Risk has been defined as the amount of harm that can be expected to occur during a given time period due to a specific event. In this way, one can give indications of the level of risk. Risk is then the product of the probability that an accident happens multiplied by the severity of that harm. A typical risk matrix has rows representing increasing severity of consequences of a released hazard and columns representing increasing likelihood of these consequences (Trbojevic 2000). On a standard risk matrix, red cells indicate high risk, yellow ones modest, and green ones low (see risk matrix for Iceland as example -Table 2).

After identifying risk, measuring it, and estimating the consequences, a traditional risk management process encourages a response, which involves, among other things, a risk mitigation strategy (Crouhy, Galai et al. 2006) However, the assessment of risk in the Arctic sea regions is a challenging task, because the conditions are changing and there is a lack in incident statistics for calculating probabilities. There are possible variations of accidents, depending on ship and unwanted event type as shown in table 1. Grounding means that the ship hits land or an underwater rock. Damage due to collision includes both collision with other vessels/sea installations and sea ice. The category fire is about fire breaking out on board. The category violence means incidents of violent behavior towards persons and physical installations, from environmentalists stopping activity to terror

and piracy. The category other may include failure on the vessel such as construction and engine failure.

Table 1 Possible variations of accidents, depending on ship and event types

	Tourist	Cargo/tanker	Fishing
Grounding	T-G	C-G	F-G
Collision	T-C	C-C	F-C
Fire	T-F	C-F	F-F
Violent action	T-V	C-V	F-V
Other	T-O	C-O	F-O

The risk matrix approach has been widely used for initial discussions on preparedness improvement, because it provides a coarse-grained picture of risk levels as a basis for further assessments and for a discussion on priority needs both as to precautions and safety efforts, and allocation of preparedness resources.

The risk matrix approach, however, has its limitations (Cox Jr 2008). In general, risk matrices have limited ability to correctly reproduce the risk ratings because of difficulties of quantifying the two components of risk and their possible correlation. In most existing and available analyses, the risk level is difficult to assess because neither the probability nor the harm severity can be estimated with accuracy and precision (Cox Jr 2008). In particular, in Arctic waters, some accident types such as violent action and terror have not happened so no statistics exist for calculation of probability. We have limited understanding of possible consequences in this area. The risk assessment that is based on low incident occurrence, historically, may, in fact, be misleading. Such a traditional view suggests that one does not prepare for a certain crisis until it has already happened. Ian Mitroff (Mitroff 2004) claims that the "black swans", the crisis that an organization does not prepare for, may cause as great harm as the ones for which it is prepared for. He suggests that for management decisions, it is precisely those crises that have not occurred that need to be considered.

Therefore, the risk assessments in regions such as the Arctic have to be based on a combination of quantitative and qualitative information. Categorizing severity may require inherently subjective judgements about consequences and decisions on how to aggregate multiple small events and fewer severe events. Therefore, risk matrixes require subjective interpretation (Cox Jr 2008). Qualitative risk matrixes on emergency preparedness should be based on both the existing statistics and estimates from experts from professional and research emergency preparedness institutions.

For better reliability, the following factors should be taken into account in addition to incident statistics:

- the density of maritime traffic
- the increased capacity of fishing vessels
- the increased interest in cruise shipping in remote areas
- the increased size of the cruise ships entering Arctic waters
- the increased number of Arctic expedition cruise vessels contracted
- the number of oil and gas exploration licenses given in the High North, especially in Norway and Russia
- efforts from international organizations, governments and industries to increase safety in Arctic waters

As for categorization of consequences in case of a lack of statistics in the Arctic region, there is a need to learn from the largest SAR and oil spill response operations. Mitroff (Mitroff 2004) points out that such lessons from previous crises have too often been ignored, not learned. For this purpose, it is necessary to analyze past mishaps, on-scene drills and engage in exercises. Secondly, there is a need to distinguish the risk of consequences for the environment and for people. Consequences will always depend on different factors and preparedness and resource availability is one of the most important ones.

5 METHODOLOGY

In this study, the risk matrices estimate 1) the frequency level of different types of incidents with different types of vessels and 2) the severity of consequences for human health and the environment. A certain element of qualitative expert evaluations on specific risk areas or defined situations of hazard and accident (DSHA) serve as the basis for the matrix. The estimation of consequences is based on case studies of the effects of real incidents in different parts of the world illuminating accidents with different types of vessels. The analyses are also based on results from exercises showing the capabilities of mitigating the negative effects of accidents in Arctic waters. For our assessment matrices we use the moderate scenario of the accidents as a base for judgment on consequence.

Data for analyses include published reports on maritime activity in the Arctic, facts published by emergency preparedness institutions on relevant issues in Norway, Iceland, Russia and Greenland. In addition, risk assessments have been discussed with industry specialists, government officials, researchers, navigators, and representatives from SAR-related authorities, organizations and academic institutions from Norway, Iceland, Denmark, Russia and Greenland. The qualitative data was collected and discussed at the MARPART advisory board and

project group meetings 10 April 2015, in Murmansk, Russia; the MARPART advisory board and project group meeting and conference 25–26 February 2016 in Reykjavik, Iceland; and the MARPART project conference 17–18 October 2016 on the Hurtigruten from Bodo, Norway. The theoretical underpinnings for the method were discussed during the 23rd International Conference on Port and Ocean Engineering under Arctic Conditions (POAC) held in 14–18 June 2015 in Trondheim, Norway, and the 26th International Ocean and Polar Engineering (ISOPE) Conference held in June 26-July 1, 2016 in Rhodes, Greece.

6 MARITIME RISK IN THE ATLANTIC ARCTIC

Using developed algorithm (Marchenko, Borch et al. 2015) and available data on maritime accidents from authorities responsible for Search and Rescue operations (for example, Joint Rescue Coordination Centre) and summarizing reports (for example (DNV GL 2014), we have made a risk assessment and created the risk matrices for the all five regions. Here we perform unpublished yet matrices for Greenland and Iceland (Table 2-5).

Table 2. Risk matrix for people in Greenlandic waters. Risk level: red - high, yellow –moderate green – low; Possibility of accidents 1- Theoretically possible, 2 – Very rare,3 – Occurs, 4 – Relatively frequently, 5 – Frequently; Consequences: A - Insignificant, B – Minor. C – Moderate, D – Significant, E – Serious

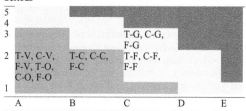

Table 3. Risk matrix for environment in Greenlandic waters. Legend and symbols see Table 2.

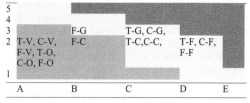

The risk assessment for the Search and Rescue Regions (SRR) of Greenland and Iceland evaluates the probability of different incidents and the level to which each incident can pose risk to people or the environment. The assessment is based on written reports of previous incidents and evaluation by

experts from Joint Arctic Command of Greenland and the Icelandic Coast Guard and the Environment Agency. It is important to note that the assessment is based on general estimation and does, therefore, not offer precise prediction of possible incidents in the Icelandic SRR. On the other hand, the risk assessment can serve as a basis for further analysis of dominating risk factors within the Greenlandic and Icelandic SRR. The assessment is done for the whole regions, as the risk factors are similar. However, given that vessel traffic is at a higher level in the south and south-western parts of Greenland and Iceland, accidents are more likely to take place in that area. The risk matrices (Table 2-5) evaluate the probability of the identified incidents and the possible consequences for people (passengers and crew) and the environment.

Table 4. Risk matrix for people in Icelandic waters. Legend and symbols see Table 2.

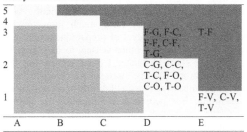

Table 5. Risk matrix for environment in Icelandic waters. Legend and symbols see Table 2.

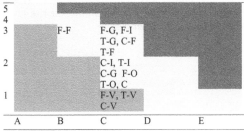

To assess the total risk and compare the regions, we estimated the share of events with different risk level, taking as 100% the total amount of chosen events. In our case, there are 15 different types of events: three defined types of ship (tourist, cargo, fishing) multiply on five defined types of accidents (grounding, collision, fire, violence and other (mostly technical failure)). Analyzing the risk matrixes for different regions, we counted the amount of events type on each risk level (Table 6). For example, for Russia, there are four types of high risk event for people life and health – collision of fishing ships, fire on cargo, fishing or tourist ship. For Svalbard, there are four other high risk types for people – all considered type of events with tourist

ship (collision, grounding, fire) as well as fire on cargo vessels. Large cruise ships are the main concern of SAR authorities on Svalbard. There are no other places in the world where cruise liners with 3000 tourists on board run up to 80°N. In the case of accident there are limited resources to save people in distress. The nearest ship to help may be several hours away, two Super Puma helicopters based in Longyearbyen are not enough for mass evacuation. As the Svalbard exercise (November 2015) has shown, acting very effectively these two helicopters can evacuate 80 persons during a 7 hours period on the distance from Longyearbyen 50 km (Svarstad 2015). One can compare this number with average cruise vessels with 2 thousand people on board in the Magdalena fjord (the main tourist attraction) on 180 km distance and average expedition ship with 150 persons on board in the Hinlopen Strait on the same distance, but with much more low probability to have other ship nearby.

Hypothermia is the main issue in case of large disasters in very remote place of the Arctic. Due to long distances, the help cannot arrive soon and most likely in emergency case people will need to wait for several hours. In an exercise testing survival in lifeboats and life rafts in ice-infested waters, even the youngest and best trained coast guard vessel crew faced problems after 24 hours in the life raft (Solberg, Gudmestad et al. 2016).

Among large accidents in the High North are the cruise liner *Maxim Gorkiy* (holed by ice at 60 NM west of Svalbard, 1989) (Kvamstad, Bekkadal et al. 2009, Hovden 2014) and the *Hanseatic* (grounded in Murchinsonfjorden, 1997), (Lorentsen 1997) when 575 and 145 passengers (respectively) and large part of the crews had been safely evacuated and the ships were recovered. The last accident occurred in the summer of 2016 – a Cruise ship *Ortelius* with 146 people (105 passengers) had to be towed for 2 days from the Hinlopen strait north of Svalbard back to Longyearbyen after engine failure (Sabbatini 2016).

In Coastal Norway we have 8 types of high risk level for people events – all possible (except violence) accidents with fishing ships due to a large amount and high activity; fire and collision for both cargo and tourist vessels. The fishing vessels represents the majority of vessels along the Norwegian coastline and is dominating the statistics of accidents at sea. This included the number of wounded and dead persons. For the first half of 2016 there were 123 persons wounded and 4 persons dead at sea. Three out of 4 deaths were at a fishing vessel.

The passenger vessels have few incidents. However, the consequences may be significant. One example is the grounding of the fast passenger vessel *MV Sleipner* in 1999 where 19 persons died. The larger passenger vessels/cruise ships are represented in the grounding statistics. One example is the grounding of the cruise ship *MV Marco Polo* in the Lofoten Islands in 2014 with 1096 persons onboard. There are, however, few incidents in total and no examples of severe accidents in the later years due to groundings and collisions. Fire and other problems like engine failure is very critical along the Norwegian coastline, and may lead to significant losses. The fishing fleet is facing challenges in this respect quite frequently.

Table 6. Type of events of different risk level (red - high, yellow – moderate, green – low) for regions under consideration

Greenland	Iceland	Svalbard	Norway	Russia
RISK FOR PEOPLE				
	6 F-G, F-C F-F, C-F T-G, T-F	4 T-C,T-G, T-F, C-F	8 F-O,F-G, C-F, F-F F-C,C-C, T-C, T-F	4 F-F, F-C T-F, C-F
6 T-G, C-G, F-G, T-F, C-F, F-F	9 C-G, C-C, T-C, F-O, C-O, T-O, F-V, C-V, T-V	10 F-G, F-C, T-O,C-O, C-C,C-G, F-F, F-V, C-V, T-V	5 C-G, C-O, T-G, T-O, T-V	11 F-G, F-O, T-G, C-C, C-G, T-C, T-O, C-O, C-V, F-V, T-V
9 T-V, C-V, F-V, T-O, C-O, F-O, T-C, C-C, F-C		1 F-O	2 F-V, C-V	
RISK FOR ENVIRONMENT				
			6 C-F, C-O, C-C, C-G, T-C, T-G	2 C-C, C-G
8 F-G, T-G, C-G, T-C, C-C, T-F, C-F, F-F	12 F-F, F-G, F-C, T-G, C-F, T-F, C-C, T-C, C-G, F-O, C-O, T-O	10 F-G, F-C, T-C, T-G, T-O, C-O, C-C, T-F, C-F, C-G,	7 F-G, F-F, F-O, F-C, T-F, T-O, T-V	11 F-G, F-O, F-F, T-G, F-C, T-O, C-V, C-F, T-F, C-O, T-C
7 T-V, C-V, F-V, T-O, C-O, F-O, F-C	3 F-V, T-V, C-V	5 F-O, F-F, F-V, C-V, T-V	2 F-V, C-V	2 F-V, T-V

Engine and fire problems occur in the passenger fleet also. The coastal steamer *Hurtigruten* traveling the coast with many vessels all year round occasionally faces such challenges. However, there are seldom severe accidents. One severe exception is the fire onboard Scandinavian Star in 1990 where 159 persons died out of 500 persons onboard. A more recent accident is the engine fire on board the Hurtigruten coastal steamer *Nordlys* in 2011, with 262 persons onboard. Two of the engine crew died and 16 persons were injured.

For Iceland, tourist vessels grounding, as for Svalbard, is estimated as a high risk events due to remoteness of touristic routes. The ships that have grounded or collided in the sea around Iceland have generally been smaller fishing vessels and older cargo ships that are not sailing according to a regular schedule. Large cruise vessels have not grounded around Iceland, but there have been incidents with smaller passenger boats. Should an incident involve a large vessel, it is obvious that the consequences could be very severe for the environment.

Till now we did not have violence act, except Greenpeace demonstrations, which are quite frequent, as they try to preserve fragile Arctic nature, but did not lead to large consequences so far. That's why all violence events are estimated and low and moderate risk. But in the case of such event, the estimation has to be revised significantly. This is one on the challenge of risk assessment in the region with lack of events and experience as a results.

Totally, regions with intensive traffic (coastal Norway, Western Russia) have higher risk level due to more frequent accidents, especially for environments. But they have much more rescue resources and higher level of preparedness and it is more possible to find other vessels to help ship in distress in short distance. Consequences for the environment are usually more severe closer to land, where pollution recovery is both difficult and time-consuming. For people it is opposite, the farther from the coast, the more complicated the rescue operations.

Estimation given in for modern level of activities and preparedness, with the forecast for small increasing, how it was in the last five years. But it the case of rapid growth of ship traffic, the risk level will change to the higher one.

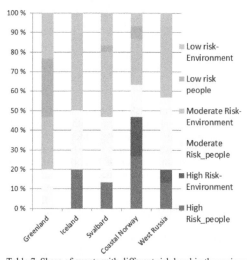

Table 7. Share of events with different risk level in the regions.

7 CONCLUSION

In this paper, we have shown that the increased traffic of oil and gas tankers, passenger ships and fishing vessels in the Atlantic Arctic may lead to negative incidents with such large consequence that mitigation efforts from a broad range of resources are needed. Efforts to reduce probability are imminent in the new regulations for ice infested waters, especially the Polar code. It is important that the operative standards following the Polar code such as navigation planning and polar water operation manuals are at a high enough safety level to reduce the probability of incidents, especially related to cruise vessels, Also, the risk assessments point to the need for emergency response plans, resource allocation and an organization of the preparedness system in an optimal way. This may also include strengthened cooperation across borders.

In this study we have shown that the validation of the risk assessment tools are important. Effective risk management decisions cannot be based exclusively on mapping ordered categorical ratings of frequency and severity, as optimal resource allocation may depend crucially on other quantitative and qualitative information. Therefore, distinguishing between the most urgent and least urgent risks in a setting with fast changing conditions and the lack of incident statistics, like the High North sea regions, is a challenging task, which needs subjective judgements and estimates. We have to take into consideration that non-expected accidents (black swans) may appear, and prepare for the consequences with a combination of accidents, such as fire, grounding, wounded and missing persons and oil pollution emerging in a series of incidents.

The risk matrices of this study also include damage to environment to illuminate the need for efforts in several emergency preparedness areas.

In the last decades, emergency preparedness resources in the Arctic, have been significantly strengthened through the addition of available vessels and helicopters. However, still the response time may be long and the capacity limited if major incidents occur. This calls for increased research efforts to learn more about how to reduce the probability of unwanted incidents. We also need to look closer on preparedness capacities both for the private actors in the region as well as on the government side. We also need to look into the competences of both the vessel crew and the emergency response resources to deal with the Arctic water challenges. This includes training and exercises on less likely large-scale incidents demanding efforts from a broad range of emergency response actors.

ACKNOWLEDGEMENTS

The authors wish to acknowledge the support from Norwegian Ministry of Foreign Affairs and the Nordland County Administration for their support via the MARPART project, and all MARPART partners for cooperation.

REFERENCES

Arctic Council (2011). Agreement on cooperation on aeronautical and maritime search and rescue in the Arctic. Nuuk, Greenland, Arctic Council.

Cox Jr, L. A. (2008). "What's Wrong with Risk Matrices?" Risk Analysis 28: 497-512.

Crouhy, M., D. Galai and R. Mark (2006). The essentials of risk management. USA, McGraw-Hill.

DNV GL (2014). Miljørisoko knyttet til potensiell akuttoljeforurensning fra shipstrafikk i havområdene omkring Svalbard og Jan Mayen: 58.

Hovden, S. T. (2014). 25 år siden Marsim Gorkij-ulikken. Svalbard Posted. Longyearbyen.

Joint Arctic Command (2016). Skibsfartens of luftfartens redningsråds årlige redegørelse for sø- og flyveredningdtjenesten i Grønland i 2015: 23.

Kvamstad, B., F. Bekkadal, K. E. Fjørtoft, B. Marchenko and A. V. Ervik (2009). A case study from an emergency operation in the Arctic Seas. In. A. Weintrit (ed.), Marine Navigation and Safety of Sea Transportation. Gdynia, Poland, CRC Press/Balkema, Taylor & Francis Group, London, UK: 455-461.

Lorentsen, N. (1997). Fire døgn på skjæret. Svalbardposten. 28: 6-7.

Marchenko, N. A. (2015). Ship traffic in the Svalbard area and safety issues. The 23rd Int. Conf. on Port and Ocean Eng. under Arctic Conditions (POAC 2015). Trondheim: 11.

Marchenko, N. A., O. J. Borch, S. V. Markov and N. Andreassen (2015). Maritime activity in the High North – the range of unwanted incidents and risk patterns. The 23rd Int. Conf. on Port and Ocean Eng. under Arctic Conditions (POAC 2015). Trondheim.

Marchenko, N. A., O. J. Borch, S. V. Markov and N. Andreassen (2016). Maritime safety in the High North – risk and preparedness. 26th International Ocean and Polar Engineers conference (ISOPE-2016). Rhodes, Greece: ISBN 978-971-880653-880688-880653; ISSN 881098-886189.

Mitroff, I. I. (2004). Crisis Leadership. Planning for the Unthinkable., John Wiley &Sons Inc.,. USA., John Wiley &Sons Inc.

Nord Universitet. (2017). "MARPART -Maritime preparedness and International Collaboration in the High North." from www.marpart.no.

Norwegian Coastal Administration. (2017). ""Havbase" (Sea base)."

Sabbatini, M. (2016). Ship out of luck: Governor tows vessel with 146 people back to Longyearbyen after engine failure. Ice people. Longyearbyen.

Solberg, K. E., O. T. Gudmestad and B. O. Kvamme (2016). SARex Spitzbergen April 2016-search and rescue exercise conducted off North Spitzbergen, University of Stavanger. 58.

Svarstad, S. M. (2015). Øvelse Svalbard. National helseøvelse 2014. Største i historien på Svalbard.

Trbojevic, V. M. C., B.J (2000). "Risk based methodology for safety improvements in ports." Journal of Hazardous Materials 71: 467–480.

Safety Management on the Bridge: Safety Cultural Factors For Crew Member's Safety Behaviour

X. Xiao & S. Nazir
University College of Southeast-Norway, Borre, Norway

ABSTRACT: This paper address navigational safety performance from safety culture perspective, by assessing attitudes and self-reported behaviour of deck officers. Building on Cooper´s safety culture model, the influence of organisational safety culture on its members' safety behavior was investigated. Multiple dimensions of safety culture which derived from MTOI (Man, Technology, Organisation and Information) safety management system framework were used to evaluate safety culture on the Bridge, their correlations and possible associations with safety behaviours were further assessed from empirical data. Multiple regression analysis determined eight significant safety culture dimensions, among which safety required resources and communication on safety both were found to be positively associated with all the safety behaviour dimensions. Plausible effect of ECDIS (Electronic Chart Display and Information System) transition, crew´s age and rank on safety behaviour were also identified. The results provide theoretical and practical guidance to understand the key factors associating with safety performance on the Bridge.

1 INTRODUCTION

Technology and automation were introduced to increase efficiency and safety in shipboard operations. Nowadays, working on the Bridge requires quality collaboration between operators and highly-integrated electronic navigation system to perform safety- critical operations. However, the human-automation interactions can lead to unexpected outcomes as the automation can create new error pathways and delay opportunities for error detection and correction (Ek et al. 2014, Nazir et al. 2014). According to Skuld (2014) and Standard (2015), there is an increasing trend of the reported incidents/ accidents which were found to be induced by Electronic Chart Display and Information System (ECDIS). Human error is widely believed to be the contributing factor in 90% of maritime accidents (Nazir et al. 2015). Human operator plays one of the important roles in the complex sociotechnical system. Their status, decision-making process and performance are inevitably affected by the environment they work in and the situations they encounter with (Nazir et al. 2013). The demands to maintain and improve navigation safety performance has raised the necessity to look into the emerging challenges within a wider context.

The current study was conducted in line with both the technological advancement and the safety challenges on modern Bridge. It proposes a systematic assessment approach that facilitates quantitative understanding (Ek et al. 2014) of the interrelationships between safety culture dimensions. The proposed approach can also be used to evaluate the effectiveness of safety management on the Bridge. By adopting the approach, the analysis result proved that core safety culture dimensions are validated cross safety-critical disciplines; it also emphasized that the determined safety culture factors should be maintained and reinforced as baseline in the formulation of safety management interventions; most importantly, it sends alarming message supported by empirical evidence that the industry calls for research attention and risk-free solutions which could accelerate the ECDIS transition.

1.1 Safety management and safety culture in Maritime domain

Human error was identified associating with majority of maritime accidents, which induced a common approach to control and monitor crew member´s behavior initially. This created a culture of punishment, where the sharp-end operators placed at the end of the error chain were often blamed (Veiga, 2002). Followed by a number of serious maritime accidents in 1980s, IMO set and mandated more stringent rules and regulations for the industry to follow in order to achieve higher standards, which led to the culture of compliance with external rules (ICS, 2013). Recent findings based on the Human Factor and Classification System (HFACS) framework and STAMP uncovered the organisational factors and latent conditions that could contribute to maritime accidents/ incidents (Chauvin et al. 2013, Kim et al. 2016).

With the adoption and mandate of ISM code by IMO, the culture of self-regulation was promoted in Maritime industry. Both vessel and onshore company are required to design and implement self-regulated Safety Management System (SMS). IMO also emphasized the importance of safety culture in maritime sector, as the application of ISM code support and foster the development of safety culture which largely determine the professionalism of seafarers (Håvold 2010, Ek et al. 2014). The reciprocal relationship is underpinned by Cooper (2000) safety culture model which is shown in Figure 1.

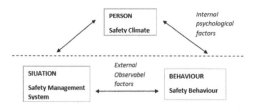

Figure 1. Reciprocal safety culture model (adopted from Cooper 2000).

Built on this model (Figure 1), the captured interactions between human factor (attitude, perception, beliefs etc.), organisational factors (safety management system, policy etc.) and member´s safety performance (safety behaviour) were adopted as the theoretical foundation in this study to identify the factors that impact safety performance on the Bridge.

1.2 *Safety culture dimensions*

Guided by Cooper´s (2000) safety culture model (see Figure 1), the studying variables should underpin the constructs of safety management system while vary themselves systematically. The Man, Technology, Organisation and Information (MTOI) framework proposed by Wahlström & Rollenhagen (2014), which consist of the four interactive sub-systems was adopted as the guidance for studying safety culture aspects that suit in the Bridge context. The MTOI model was found to be

appropriate for evaluating Bridge SMS because it entails holistic systematic approach to understand the macro safety behaviour found in socio-technical system (Wahlström & Rollenhagen, 2014), and it fits in the Bridge setting where the work is processed by the interactions among deck officers, information, navigation equipment and the wider organisation they belong to.

With extensive literature review in both Maritime safety culture and Bridge team management, 12 independent safety culture dimensions were chosen and classified under the MTOI framework:
Man system:
- COMP: competence (Håvold & Nesset 2009, Mearns et al. 2003)
- CSRP: shared responsibility (ABS 2014)
- CTSE: trust in the efficacy of safety system (Flin et al.2000, Grabowski et al. 2007, ABS 2014)

Organisation system:
- TMSC: onshore management safety commitment (Grabowski et al.2007, ABS 2014)
- LMSE: shipboard management safety engagement (Grabowski et al.2007, ABS 2014)
- MSJT: management safety justice (Reason 1997, Grabowski et al. 2007, Hollnagel et al. 2011)
- MFDB: management feedback (Grabowski et al. 2007, ABS 2014)
- RSLO: resource allocation (Håvold & Nesset 2009, Mearns et al. 2003, Grabowski et al. 2007, Glendon & Litherland 2001)

Information system:
- COMU: communication about safety (Mearns et al. 2003, Grabowski et al. 2007, Håvold & Nesset 2009, Parrott 2011, Glendon & Litherland 2001, ABS 2014)
- RPCT: reporting culture (Reason 1997, Håvold & Nesset 2009)
- LNCT: learning culture (Reason 1997, Hollnagel et al. 2011)

Technology system:
- WKEV: working Environment (Grech et al. 2008, Wahlström & Rollenhagen 2014).

1.3 *Aims*

The primary purpose of the study is to identify the determinants of safety culture on Bridge crew member´s safety behaviour. The secondary purpose is to look into the impact of organisational, occupational and demographic characters on their self-reported safety behaviour.

1.4 *Hypotheses*

The study aims to give answers to the following research hypotheses:

- H1: In line with previous research, the safety culture dimensions are positively correlated with safety behaviour dimensions.
- H2: Bridge crew members´ self-reported individual safety behaviour and Bridge Team group-level safety practice are associated with same safety culture dimensions, holding demographic and organizational factors constant.
- H3: Rank, age, ECDIS application, frequency of working on the same vessel and contract type are positively associate with Bridge crew´s safety behaviour.

2 METHODOLOGY

2.1 Dependent variables

Previous research validates the method to use self-reported safety as the proxy for shipboard safety, and it is still considered as a proactive measure of workplace safety (Lu & Yang 2011). Griffin & Neal (2000) and Neal et al. (2001) identified two types of safety behaviour: safety compliance and safety participation. In this study, the two components of safety behaviour were included and tested at individual level – the deck officer and group level-the Bridge team.

Therefore, the operational research model for this study can be framed as it is shown in Figure2.

Figure 2. The operational research framework in this study

2.2 Questionnaire development

Given the latent and forthcoming characteristics, perceptions and attitudes are most often assessed via survey questionnaires (Otedal & Engen, 2009). Based on the theoretical model presented in Figure 2, an analytical questionnaire was developed for this study. In the questionnaire, there are in total 86 questions asking for crew´s attitude/perception toward given statements in regard to safety management, besides that, 13 questions asking about the respondent´s background information, and another one open question leaving for respondent´s comment or suggestion for safety action on the Bridge.

Among the 86 questions, 22 self-developed questions are based on Bridge Team management and Bridge Resource Management component

(Parrott 2011, Adams 2006). The remaining questions are from review of published literatures and selected based on below criteria: 1) The theoretical foundation underpin the chosen question should be in line with studying variables in this study; 2) The ability to describe and measure underlying safety culture variables; 3) Be able to be measured at different organisational levels (Oltedal & Wadsworth, 2010); 4) Clear statement and easy to understand without ambiguity.

A pilot study (n=3) was carried out among three HSE managers with nautical background before conducting the survey. Incorporating the managers´ suggestion, wording adjustment has been made on some items for better understanding.

2.3 Participant, ethical consent and sampling

In this project, 70 vessels from three Norwegian shipping companies were included, 221 crew members who were working on the Bridge of the 64 vessels participated in the survey.

Top priority has been given to ethical considerations during the whole process of the study, much attention and care has been focused on protecting the anonymity and confidentiality of participants. Before sending the questionnaire to ship, all participating vessels and participants were well informed of the nature, purpose and scope of the survey, as well as how the collected data from them would be used. In addition, all three shipping companies confirmed to their vessels via written notice that they themselves would not get involved in the process or interfere with the analysis result at any level.

The survey was conducted in the March of 2016, 221 valid responses which from 64 vessels were collected. Table1 shows demographics profile of the respondents:

Table 1. Demographic profile of the respondents

Demographic profile of respondents					
		N			N
Age			**Job Position**		
< 31 years	36.2%	80	Captain	18.6%	41
31-40 years	32.6%	72	Officer	62.4%	138
41-50 years	22.2%	49	Mate	18.6%	41
51-60 years	8.1%	18	Missing	0.5%	1
Missing	0.9%	2	Total		221
Total		221			
Vessel type			**Nationality**		
Bulk	5.9%	13	Norwegian	3.2%	7
Combined	12.2%	27	Russian	2.7%	6
Oil Tanker	29.4%	65	Polish	23.5%	52
Car Carrier	30.8%	68	Filipino	43.9%	97
Offshore Vessel	8.6%	19	Croatian	3.6%	8
Ro Ro	4.1%	9	Latvian	0.9%	2
Cruise	8.6%	19	British	3.6%	8
Missing	0.5%	1	Indian	10.4%	23
Total		221	Ukrainian	1.4%	3
			Swedish	0.5%	1
			Romanian	0.9%	2
			Finish	0.5%	1
			Missing	4.9%	11
			Total		221

2.4 Statistical analysis

Factor analysis was applied to explore and confirm the latent underlying dimensions reflecting the concept of safety culture in the context of Bridge (Oltedal & Wadsworth 2010, Lu &Yang 2011). Based on the two sets of safety culture dimensions/facets, for hypothesis 1, Pearson correlation was performed to exam the direction and magnitude of the relation between safety culture dimensions as well as with safety behaviour variables; for hypothesis 2 and hypothesis 3, multiple regression was carried out to determine the possible association between safety behaviour with the dimensions of safety culture and other demographic / organisational factors (Mearns et al. 2001, Oltedal & Wadsworth 2010, Lu & Yang 2011); followed up by one way ANOVA, the impact of demographic and organisational attributes on safety behaviour were assessed (Oltedal & Wadsworth 2010, Lu &Yang 2011). All the analyses were performed by using SPSS V.24.

3 RESULT

3.1 Hypothesis 1

Original 12 safety culture scales and 2 safety behaviour scales (see Figure 2) reached a good reliability, Cronbach's α range (0.70-0.839). All the 12 safety culture scales were found to be positively correlate internally and with the two safety behaviour variables (r =0.40 to 0.70) at 0.001 significance level (2-tailed), indicating positive and strong relationship among the studying variables. The second sets of extracted factors also reported good internal consistency, Cronbach's α range (0.75-0.884), and the 13 safety factors were found to be more moderately correlated with each other (r = .30 to .49) at 0.001 significance level (2- tailed). Thus, the significant positive correlations between safety culture dimensions and safety behaviour expected in hypothesis 1 were supported by the Pearson correlation analysis results.

3.2 Hypothesis 2

Preliminary analyses were conducted to ensure no violation of the assumptions of normality, linearity, multicollinearity and homoscedasticity. As no specific pattern was found in the reported outliers after screening each scale, the located outliers were then replaced by a less extreme value (Pallant, 2016). A hierarchical multiple regression with forward stepwise method was applied in each analysis. The multiple regression results provided the basis that partially support this hypothesis, as the Bridge crew's self-reported safety behaviour and Bridge group-level safety practice were not associated with exactly the safety culture dimensions.

The final models which list in the last column of Table 2 and Table 3, both of the safety behaviour variables found significantly and positively associated with "Resource allocation" (Table 2: β =0.361, p < 0.001; Table 3: β =0.223, p < 0.001), and " Communication" (Table 2: β =0.299, p < 0.001; Table 3: β =0.160, p < 0.05). But the difference was that individual safety behaviour had a straightforward association with crew's competence (Table 2: β =0.189, p < 0.01), while at group level, the collective safety practice was most influenced by "Shared safety responsibility" (Table 3: β =0.364, p < 0.001) and further affected by "trust in the safety system efficacy" (Table 3: β =0.179, p < 0.05).

Table 2. Hierarchical regression analysis result (standardized β coefficients) on DV1: Crew self-reported individual safety behaviour.

	Model 1	Model 2	Model 3	Model 4
Control variables				
Rank	-0.18	0.066	0.118	0.106
ECDIS application	-0.172*	-0.056	-0.033	-0.02
Age	0.176*	0.038	-0.03	0.001
Contract type	0.125	0.08	0.051	0.043
Working on the same vessel frequency	-0.036	-0.008	0.037	0.03
Independent variables				
Resource allocaiton (RSLO)		0.623***	0.409***	0.361***
Communication (COMU)			0.400***	0.299***
Crew competence (COMP)				0.189**
F - value	2.983*	23.949***	30.182***	28.147***
R^2	0.072	0.429	0.527	0.544
ΔR^2		0.357***	0.097***	0.017**

Note: * p < 0.05, ** p < 0.01, *** p < 0.001.

Table 3. Hierarchical regression analysis result (standardized β coefficients) on DV2: Group level safety practice.

	Model 1	Model 2	Model 3	Model 4	Model 5
Control variables					
Rank	0.084	0.143*	0.177***	0.149**	0.174***
ECDIS application	-0.201**	-0.152**	-0.096*	-0.085	-0.078
Age	0.155	0.11	0.041	0.041	0.012
Contract type	0.055	0.046	0.022	0.034	0.019
Working on the same vessel frequency	-0.09	-0.038	-0.034	-0.018	-0.007
Independent variables					
Shared safety responsibility (CSRP)		0.661***	0.504***	0.398***	0.364***
Resource allocaiton (RSLO)			0.358***	0.260***	0.223***
Trust in efficacy of the safety system (CTSE)				0.230***	0.179*
Communication (COMU)					0.160*
F - value	3.375**	32.919***	41.042***	39.114***	36.413***
R^2	0.081	0.508	0.602	0.623	0.635
ΔR^2		0.428***	0.094***	0.022***	0.012*

Note: * p < 0.05, ** p < 0.01, *** p < 0.001.

By cross-validation, two facets of safety behaviour: safety compliance and safety participation were extracted by factor analysis. The significant findings on this two added DVs list in Table 4 and Table 5.

Table 4. Hierarchical regression analysis result (standardized β coefficients) on DV3: safety compliance.

	Model 1	Model 2	Model 3	Model 4	Model 5
Control variables					
Rank	0.113	0.178*	0.228**	0.288***	0.199**
ECDIS application	-0.163*	-0.08	-0.052	-0.032	-0.034
Age	0.157	0.06	-0.017	0.007	0.002
Contract type	0.068	0.045	0.021	-0.012	-0.02
Working on the same vessel frequency	-0.022	0.005	0.068	0.055	0.048
Independent variables					
Resource allocaiton2 (RSLO2)		0.476***	0.366***	0.307***	0.268***
Communication (COMU)			0.366***	0.291***	0.261***
Mutual trust (MUTS)				0.199**	0.173*
No-blame culture (NBLC)					0.135*
F - value	3.153**	12.872***	17.783***	17.070***	15.848***
R²	0.076	0.288	0.396	0.419	0.431
ΔR²		0.212***	0.108***	0.024**	0.012*

Note: * p < 0.05, ** p < 0.01, *** p < 0.001. (``RSLO2'' refers to software like training, manning, working time etc.)

Again, both of the safety behaviour variables found significantly and positively associated with ''communication'' (Table 4: β=0.261, p < 0.001; Table 5: β=0.211, p < 0.01) and ''Resource allocation'' (Table 4, RSLO2: β=0.268, p < 0.001; Table 5, RSLO1: β=0.272, p < 0.001).

Table 5. Hierarchical regression analysis result (standardized β coefficients) on DV4: safety participation.

	Model 1	Model 2	Model 3	Model 4
Control variables				
Rank	-0.124	-0.113	-0.043	-0.029
ECDIS application	-0.215**	-0.11	-0.093	-0.089
Age	0.176*	0.12	0.066	0.029
Contract type	-0.046	-0.099	-0.096	-0.097
Working on the same vessel frequency	-0.058	-0.046	-0.058	-0.015
Independent variables				
Management feedback & Interaction (MDBI)		0.448***	0.335***	0.262***
Resource allocaiton1 (RSLO1)			0.322***	0.272***
Communication (COMU)				0.211**
F - value	2.769*	10.700***	13.768***	13.668***
R²	0.067	0.252	0.337	0.366
ΔR²		0.184***	0.085***	0.030**

Note: * p < 0.05, ** p < 0.01, *** p < 0.001. (``RSLO1'' refers to hardware like safety tool, equipment, instructions etc.)

In addition, ''Safety compliance'' was also found to be significantly associated with ''mutual trust'' (Table 4: β=0.173, p < 0.05) and ''no-blame culture'' (Table 4: β=0.135, p < 0.05). And for ''Safety participation'', significant association was identified with ``management feedback & interaction`` (Table 5: β=0.262, p < 0.001), which reported the highest beta in this analysis.

3.3 *Hypothesis 3*

This hypothesis was partially supported, as the multiple regression results showed only the control variable ''age'' was positively associated with all the four safety behaviour variables, but the coefficients were proved significant only on ''individual safety behaviour''(Table 2: β =0.176, p < 0.05) and ''safety participation'' (Table 5: β =0.176, p < 0.05). In addition, the other control variable ''rank'' was found to have positive association with three safety behaviour variables except for ''safety participation'', and the coefficients on ''group safety practice''(Table 3: β= 0.174, p< 0.001) and ''safety compliance'' (Table 4: β =0.199, p < 0.01) were proved significant. The results suggest that the elder crew had higher self-reported individual safety behaviour and safety participation; while crew member with higher rank had higher group safety practice and safety compliance.

Contrast to the hypothesis, the two control variables' (ECDIS application, working on the same vessel frequency) effects on safety behaviour were found to be negative, but the negative influence of ''working on the same vessel frequency'' was not significant. Whilst, ECDIS application statistically proved negatively associated with all safety behaviour dimensions: individual safety behaviour (Table 2: β= -0.172, p< 0.05), group safety practice (Table 3: β= -.201, p< 0.01), safety compliance (Table 4: β = -0.163, p< 0.05), safety participation (Table 5: β= -.215, p< 0.01). The results indicate that crew working on the Bridge already adopted ECDIS as primary navigation method had reported lower safety behaviours.

No significant effect of ''contract type'' on safety behaviour was identified.

3.4 *One-way ANOVA*

The followed up one-way ANOVA found all four crew member´s safety behaviour dimensions were significantly differentiated by ECDIS application at the 5% significance level. Levene´s test was checked; the homogeneity of variance was not violated. The same results shown in Table 6 were obtained by t-test.

Table 6. Comparison of differences of safety behaviour dimensions based on ECDIS application.

Dimension	N	Mean	SD	SE	df	F	Sig.	Eta Squ.
Self-reported individual safety behaviour	217	35.91	3.146	0.241	1	6.228	0.13	0.03
Primary method	184	35.68	3.213	0.237				
In transition	33	37.15	2.425	0.422				
Group level safety practice	217	35.28	2.987	0.203	1	7.958	0.005	0.04
Primary method	184	35.04	3.007	0.222				
In transition	33	36.61	2.524	0.439				
Safety compliance	218	28.85	4.094	0.277	1	4.756	0.30	0.02
Primary method	185	28.60	4.139	0.304				
In transition	33	30.27	3.564	0.620				
Safety participation	220	23.70	1.860	0.125	1	8.820	0.003	0.04
Primary method	187	23.55	1.935	0.141				
In transition	33	24.58	1.001	0.174				

4 DISCUSSSION

The present study evaluated Bridge crew members´ perceptions and attitudes toward safety management measures on the Bridge, and tried to identify the predictor(s) of safety behaviour which in forms of individual safety behaviour, group-level safety practice, safety compliance and safety participation. Once the significant predictor(s) are identified and properly reinforced, it can be expected to generate more occurrence of the safety behaviour, thus can provide support to create a safer shipboard working environment.

The results indicate that ''safe navigation required resources allocation'' and ''communication on safety'' are of vital importance to influence Bridge crew member's safety behaviour, and must be taken into considerations by ship management.

Notably, ''resources'' in this study was constructed as an umbrella concept that include: safety document like policy, procedure and instructions, personally protective equipment (PPE), training and manning level (as on the Bridge, human consider as the most important resource).

''Competence'' was found to be associated with individual safety behaviour. This result lent support to Neal et al. (2000) and Griffin & Neal (2002), where they argued that safety knowledge and skill are directly responsible for individual differences in behaviour. On the other hand, group-level safety practices were found to be associated with two more abstract dimensions, the observed difference in line with Guldenmund (2007), in which the author stated that group level attitude tend to be congruent with organizational culture while individual level behaviour are more influenced by rule-based and skill-based processes. The statement further underpinned the predictive value of ''competence'' with regard to individual safety behaviour. The ''shared safety responsibility'' and ''trust in the efficacy of safety system'' both are believed to be the contributing factors to form a workgroup norms favouring safety (HSC 1993, Cooper 2000).

Harvey et al. (2002) suggested that the management feedback about employee's performance of safe behaviours on the job contributed significantly to the occurrence of safety behaviour. ''No-blame culture'' has been proved as a fundamental aspect contributing to organisational learning in Maritime safety research (Ek et al. 2014, Bhattachrya 2015). The workforce's perceptions of the general standard of worker's qualifications, skills and knowledge were also validated in the safety climate research by Flin et al. (2000). The empirical evidence obtained in this study as well as previous significant findings from related research all emphasized their importance to be included and reinforced in the safety management measures and daily work routines on the Bridge.

The findings regarding elder respondents and higher ranks were positively associated with safety behaviour and can be considered as reasonable, as elder crew members with sufficient technical know-how and work experience are more capable of handling various situations; and those who taking more responsibility with higher rank aboard are more likely to fulfil self-regulation in regard to company safety policy, instructions and objectives. However, the adoption of ECDIS seems, at least currently, hasn't fully exerted positive impact on the Bridge, as the reported safety behaviour from crew members who are working on vessel using ECDIS as primary navigation method were inferior to those still using conventional method. Due to lack of sufficient data, this particular area requires further investigation to draw more solid conclusion.

5 CONCLUSIONS

The study identified that ''safety resources'' and ''communication on safety'' were the two indispensable safety cultural mechanism that influenced Bridge crew member's safety behavior. The other affecting factors are competence, mutual trust, shared safety responsibility, trust in the safety system efficacy, management feedback and no-blame culture. Besides, Bridge crew members' safety behaviour was found to be associated with age, rank and ECDIS application. The findings provide useful guidance for practitioners to locate source of variation and formulate effective interventions to improve the safety performance on the Bridge.

Nevertheless, two major limitations should be highlighted. Using self-reported measures instead of objective audit result is the major limitation in this study. Thus the result generated by the empirical data can expected to exhibit certain level of bias.

Another limitation of the present study is the magnitude of generalization. The respondents were all from Norwegian controlled vessels, the survey sample was not adequate to be the representative of the research population, neither from the number of respondents nor the types of vessel covered.

The mentioned major limitations also highlight the directions for further study so as to address the research problems more properly.

ACKNOWLEDGEMENT

This work was supported by three esteemed Norwegian shipping companies. The authors would like to thank the ship managers and HSE auditors from these three companies who provided valuable information, generous support, and constructive comments on this research.

REFERENCES

American Bureau of Shipping (ABS) 2014. Guidance Notes on Safety Culture and Leading Indicators of Safety. Houston, Texas: American Bureau of Shipping.

Adams, M.R.2006. *Shipboard Bridge Resource Management*. Maine: Nor' Easter Press.

Bhattacharya, Y. 2015. Measuring safety culture on ships using safety climate: a study among Indian officers. *International journal of e-Navigation and Marine Economy, 3,* 51-70.

Chauvin, C., Lardjane S., Morel, G., Clostermann, J.P. & Langard, B.2013. Human and organisational factors in maritime accidents: Analysis of collisions at sea using the HFACS. *Accident Analysis and Prevention, 59,* 26-37.

Cooper, M.D. 2000. Toward a model of safety culture. *Safety Science, 36,* 111-136.

Ek, Å., Runefors, M. & Borell, J. 2014. Relationships between safety culture aspects – A work process to enable interpretation. *Marine Policy, 44,* 179-186.

Fenstad, J., Dahl, ø. & Kongsvik, T. 2016. Shipboard safety: exploring organizational and regulatory factors. *Maritime Policy& Management, 43*(5), 552-568.

Flin, R., Mearns, K., O´Connor, P. & Bryden, R. 2000. Measuring safety climate: identifying the common features. *Safety Science, 34,* 177-192.

Glendon, A. I. & Litherland, D.K.2001.Safety Climate Factors, Group Differences and Safety Behavior in Road Construction. *Journal of Safety Science, 39,*157-188.

Guldenmund, F.W.2007. The use of questionnaires in safety culture research – an evaluation. *Safety Science, 45*(6), 723-743.

Grinffin, M.A. & Neal,A.2000. Perceptions of safety at work: a framework for linking safety climate to safety performance, knowledge and motivation. *Journal of Occupational Health Psychology, 5,* 347-358.

Grabowski M., Ayyalasomayajula P., Merrick J., Harrald J. & Roberts K. 2007. Leading Indicators of Safety In Virtual Organisations. *Journal of Safety Science, 45,* 1013-1043.

Grech, M.R., Horberry, T.J. & Koester, T. 2008. *Human Factors In the Maritime Domain.* Boca Raton: Taylor& Francis Group.

Harvey, J., Erdos, G., Bolam, H., Cox, M.A.A., Kennedy, J.N.P., Gregory, D.T.(2002). Analysis of safety culture attitudes in a highly regulated environment. *Work& Stress, 16*(1), 18-36.

Hollnagel, E., Paries J., Woods D.D. & Wreathall J. 2011.*Resilience Engineering In Practice: A guidebook.* Surrey: Ashgate.

HSC, 1993. ACSNI Study Group on Human Factors. 3rd Report: Organising for Safety. Health and Safety Commission, HMSO, London.

Håvold, J.I. 2005. Safety-culture in a Norwegian shipping company. *Journal of Safety Research, 36,*441-458.

Håvold J.I.& Nesset E. 2009. From safety culture to safety orientation: Validation and simplification of a safety orientation scale using a simple of seafarers working for Norwegian ship owners. *Journal of Safety Science, 47,*305-326.

Håvold, J.I. 2010. Safety culture and safety management aboard tankers. *Journal of Reliability Engineering and System Safety 95,* 511-519.

International Chamber of Shipping (ICS). 2013. Implementing an effective safety culture. London: International Chamber of Shipping.

Kim, T. E., Nazir, S., & Øvergård, K. I. 2016. A STAMP-based causal analysis of the Korean Sewol ferry accident. Safety science, 83, 93-101.

Lu, C.S. & Yang, C.S. 2011. Safety climate and safety behaviour in the passenger ferry context. *Accident Analysis and Prevention, 43,* 329-341.

Mearns, K., Flin,R., Gordon, R. & Fleming,M. 2001. Human and organizational factors in offshore safety. *Work & Stress, 15(2),* 144-160.

Means, K., Whitaker, S.M. & Flin, R. 2003. Safety Climate, Safety Management practice and safety performance in offshore environment. *Journal of Safety Science, 41,* 641-680.

Nazir, S., Colombo, S. & Manca, D. 2013. Minimizing the risk in the process industry by using a plant simulator: a novel approach. *Chemical Engineering transactions, 32,* 109-114.

Nazir, S., Kluge, A. & Manca, D. 2014. Automation in Process Industry: Cure or Curse? How can Training Improve Operator's Performance. *Computer Aided Chemical Engineering,33,*889-894. doi:http://dx.doi.org/10.1016/B978-0-444-63456-6.50149-6

Nazir, S., øvergård, K. I. & Yang, Z. 2015. Towards effective training for process and maritime industries. *Procedia Manufacturing* 3,1519-1526.

Neal, A., Griffin, M.A. & Hart, P.M. 2000. The impact of organizational climate on safety climate and individual behaviour. *Safety Science 34,* 99-109.

Oltedal. H. A. & Engen, O.A. 2009. Local management and its impact on safety culture and safety within Norwegian shipping. In S. Martorell, C. Guedes Soares & J. Barnett (Eds.), Safety, Reliability and Risk Analysis: Theory, Methods and Applications (1423-1430). London: Taylor & Francis Group.

Oltedal, H. A. &Wadsworth, E. 2010. Risk perception in the Norwegian shipping industry and identification of influencing factors. *Maritime Policy & Management, 37*(6), 601-623.

Pallant, J. 2016. *SPSS Survival Manual (6th ed).* Sydney: Allen & Unwin.

Parrott, D. S. 2011. Bridge Resource Management for Small Ships: The Watchkeeper´s Manual for Limited-Tonnage Vessels. Blacklick, Ohio: McGraw-Hill

Reason, J. 1997. *Managing the Risks of Organizational Accidents.* Surrey: Ashgate.

Skuld, 2014. Electronically Guided Accidents. Shipping Regulations and Guidance, December 2014. Can be retrigged from https://www.skuld.com/Documents/Library/PI_Columns/PI_Issue2014Dec.pdf

Standard, 2015. Standard Safety Special Edition: ECDIS assisted grounding. Can be retrigged from http://www.standard-club.com/news-and-knowledge/news/2015/04/standard-safety-special-edition-ecdis-assisted-grounding,-april-2015/

Veiga, Jaime L. 2002. Safety Culture in Shipping. WMU Journal of Maritime Affairs, 2002, No.1, 17-31.

Wahlström, B. & Rollenhagen, C. 2014. Safety management – A multi-level control problem. *Safety Science 69,* 3-17.

Proceedings of 12th International Conference on Marine Navigation and Safety of Sea Transportation, TransNav 2017
21-23 June 2017, Gdynia, Poland

Naval Artificial Intelligence

E. Kulbiej & P. Wołejsza
Maritime University of Szczecin, Szczecin, Poland

ABSTRACT: The following paper focuses on an analysis of a comprehensive system of nautical vessel's artificial intelligence. Such a complex structure should compromise a number of parallel major and minor items of vessel's equipment in order to maintain correct management of the vessel itself. Stemming from individual interests of the authors, an attempt to generalize a theoretical basis for a naval artificial intelligence (NAI) is presented. The definition of NAI and its components is derived and discussed. Schematics of an algorithm is proposed; the NAI as a virtual managing party of a vessel's deck and engine department, maintaining safe navigation and the freightage of the cargo as well as harbour operations.

1 INTRODUCTION

Astral navigation was, for many years, one of the main methods of determining ship's position. The twentieth century brought a lot of invention on board the ships e.g. radar. Due to process of automation, the size of a vessel's crew has been continuously reduced, nowadays oscillating around the number of 15 crewmembers, for an average merchant ship. More and more systems providing assistance for the navigational bridge team are being developed and implemented, for instance Automatic Identification System (AIS), Automatic Radar Plotting Aid (ARPA) or even Electronic Navigation Charts system (ENC). Currently the youngest and most novel invention is the Navigational Decision Support System (NDSS), an example of which is Navdec, system created under the research team of professor Pietrzykowski from Maritime University of Szczecin (Wołejsza P., 2013; Wołejsza P., 2014).

The future upholds an increased need for technological advance, so that congestion of merchant transport is satisfied. In order to achieve that goal, brand new technologies might be necessary, such as an autonomous vessel. Such a unmanned ship is the ultimate stage of technological development of maritime merchant branch of economy. This paper aims to propose a system called Naval Artificial Intelligence (NAI), which should act as a virtual operator of a ship, providing full monitoring and conducting all sorts of services, commencing with ensuring safety of the object.

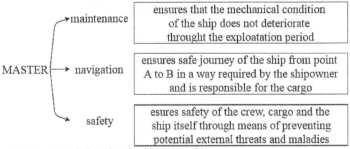

Figure 1. Master's duties onboard a ship (own study).

Figure 2. Naval artificial intelligence definition in form of a graph (own study).

2 SYSTEM OF NAVAL ARTIFICIAL INTELLIGENCE

The artificial intelligence is described as the capability of a computer-driven machine to imitate intelligent human behaviour. Its main objective would be to ensure navigation of the vessel, which is to be described as a safe voyage between the point of departure and the point of arrival whiles providing transfer of untouched cargo. This duty, main in the means of modern seaborne transport industry, is currently undertaken on average by a crew of fifteen people. Divided into two main departments: deck and engine, they control and maintain the vessel. The ship's master is the person directly entitled by the ship-owner to fulfil mentioned goal. He holds the immediate responsibility to provide several distinctive functions, as far as vessel's exploitation is concerned. They are shown in the figure 1.

It is clear that the master is the overseer of the vessel. Thus, module of artificial intelligence employed to run the ship should imitate an experienced captain's behaviour as it would directly inherit his major duties. But then again, master's duties limits to supervising and delegating most of the tasks, especially when it comes to physical activities. This phenomenon is not rare and rather typical of employees on the management level.

In order to achieve the premise of safe and effective voyage, vessel controlling system should execute more complex variety of tasks. It can be stated that Naval Artificial Intelligence is a system that undertakes navigational tasks, provides exemplar ship's maintenance, executes safe cargo handling, transportation and shore operations and establishes successful security of ship, while preventing from rendering herself dangerous to environment and other objects. This definition by a mean of a graph can be visually described as in figure 2.

Vessel's navigation sector will include mainly tasks connected to widely-known marine navigation but is not to be limited by it. Therefore NAI is to declare the most effective route from the point of departure to the point of arrival. The route itself is to be under constant scrutiny of the system's optimisation algorithms so it can be altered on a regular basis, dynamically answering to situation.

The voyage should be moreover continuously recorded and reported to the ship-owner and their remarks are to be taken into account in the navigational planning.

Own ship maintenance sector of the system ought to oversee ship's hardware, that is to say, provide correct use and ensure non-deteriorating performance of the main and auxiliary engines of the vessel. System of comprehensive detector would map the whole vehicle creating network of probes that sufficiently supervises hull, mooring equipment, electrical installation, ballast and other tanks' installations, fire-fighting installations and thus prevent physical maladies, which could render the ship inoperative due to one of aforementioned installation compromise.

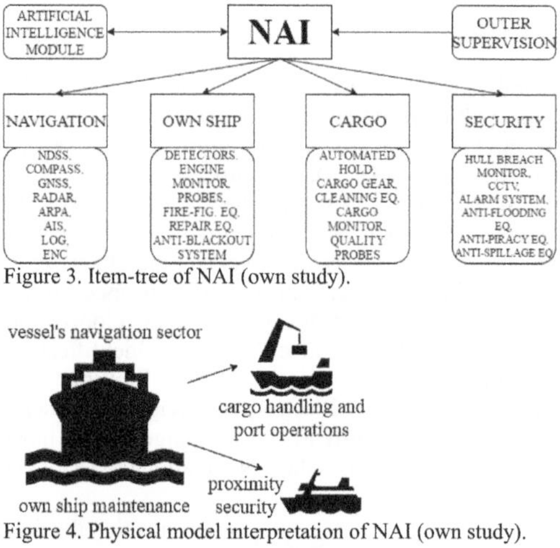

Figure 3. Item-tree of NAI (own study).

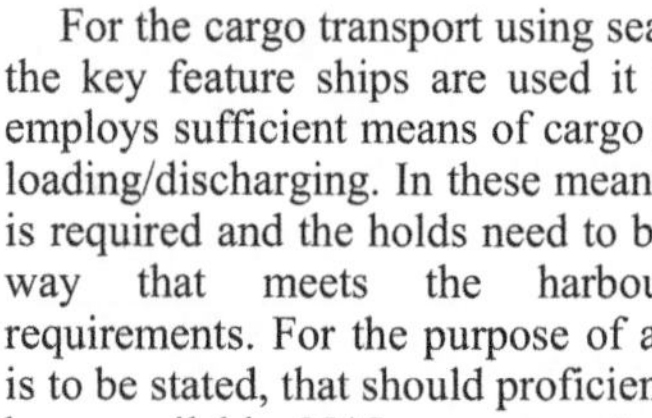

Figure 4. Physical model interpretation of NAI (own study).

For the cargo transport using seaborne methods is the key feature ships are used it is vital that NAI employs sufficient means of cargo protection and its loading/discharging. In these means mooring system is required and the holds need to be operational in a way that meets the harbours equipment's requirements. For the purpose of analysis of NAI it is to be stated, that should proficient mooring system be unavailable, NAI may use external help of shore facilities and workers. Then role of ship during the cargo transferring process can be either maintained by NAI or remotely dispatched to shore based office.

The last but not the least group of NAI functions is security segment. In contrast to safety of cargo, in regards to safety and security of the ship NAI should provide that route undertaken by the vessel is free of threatening encounters. In other words NAI has to, by successfully implementing Rules of the Road (COLREG), omit unnecessary closest points of approaches, undertake last moment manoeuvre is necessary and steer the vessel in respect to other watercrafts. In special situations NAI driven vessel should deploy itself for rescue mission and proceed to assistance for any object or person in need. The actions executed by NAI ought to be deprived of losses, but if these unfortunately are to occur, the damages to cargo, environment etc. are to be minimalised by all means. The situation when on-board a physical malady appears, for instance a fire or a leak, NAI should be provided with installed equipment or other means sufficient to cope with encountered problem.

Aforementioned particular and exemplar pieces of equipment and facilities that comprises for the whole item-tree of NAI are graphically presented in figure 3. On the other hand, physical model interpretation of NAI is shown in figure 4.

3 BEHAVIOUR MODEL OF NAI

In contrast to typical navigational equipment that is already available for maritime officers, NAI not only downloads data from different sources such as GPS and displays them after calculation but also applies own means of understanding. It is to say, that the data is processed and checked so all effects that may have a negative impact, for example relativistic effects (Kulbiej E., 2016), are taken into account. But then again, the facilities and gear installed on the vessel are only limbs, futile and otiose should be abandoned. It is NAI that functions as a brain of the vessel with presumably main and auxiliary engines in the role of heart and optical fibres or other ways of convening data as veins. To serve its purpose of problem solving, decision making and way leading NAI must be given within the ship a sort of authority, not to mention sufficient means or autonomy.

As a system of artificial intelligence, NAI is to solve encountered problems on the premise of several means of solution acquisition. Mainly and firstly the problem is to be compared to expertise based knowledge bank. If a suitable solution for the problem is found, it can be directly applied and the threat is compromised. Otherwise, one of preinstalled algorithms modules use is proposed. This is the autonomous step of decision making in contrast to previous automatic approach. If the problem is still unsolved, further means are

applicable. Specially designated shore personnel might be contacted, however immediate response is crucial. Provided that the NAI system declares evolutionary example of intelligence module it may execute a risk-minimalizing method based on dynamic and logical approach to the case. Important feature of NAI seems to be self-learning module and connection to a broader network of similar systems. Thanks to that, once encountered, the problem is then solved with automatic response should it reappear. The block schema of this problem solving algorithm is described in figure 5.

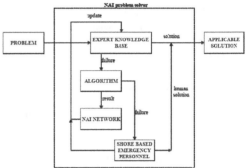

Figure 5. Problem solving algorithm (own study).

Provided no more mission threatening problems are encountered, NAI is to proceed to decision making component of own artificial intelligence module. In this step the situation need in-depth investigation in search for several particular (mainly numerical) traits. When the investigation is complete, the mission optimization process is commenced. The purpose of this action is purely of economical basis, as everything undertaken by NAI should lead to minimalizing costs of the voyage and cargo handling. For instance, the route is maximally shortened after taking into account untypical needs of weather navigation. A fully planned route with pre-set waypoints and an initial report is regarded as a fruit of this activity. The report is to be presented to the on-shore human crew, however the act of reporting should be executed by NAI on regular and continuous basis. One ought to never underestimate the fact that NAI is only artificial machine operating system.

The way leading activity of NAI is the continuous control over the vessel, namely deployment and use of provided equipment on-board accordingly to encountered problem or simply to execute tasks decided on in the decision making section. That means cooperation between different parts of machinery in order to fulfil the goal of cargo transportation.

The whole behaviour tree of NAI is proposed in a figure 6.

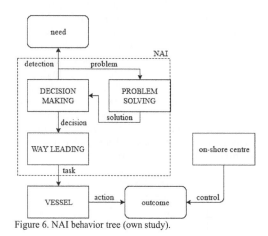

Figure 6. NAI behavior tree (own study).

4 COMPONENTS OF NAI

There are following components proposed in the NAI:

4.1 Autonomous vessel, which consists of three main elements: unmanned navigational bridge, unmanned engine room and communication protocol.

Nowadays most of existing ships have "autonomous" engine with A24 class, which means that all processes can be executed without crew. Alarm panels distributed throughout the ship inform engineers of any abnormal activities within engine room department. Similar panel can be installed in remote control centre located ashore. If any problem arises, it can be solved using procedure presented on Fig.5.

Navigational bridges are still required to be manned at least by an officer. Usually at night watch is doubled. Same situation is in restricted visibility, which from the point of view of common sense has no practical explanation. Visual observation carried out by watchman will not improve safety of ship. In such situation the officer relies on information received from radar, AIS and other electronic equipment (Pietrzykowski Z., et al., 2013). All these information can be transmitted to remote control centre located ashore. The decision how to continue voyage or how to avoid collision can be taken using procedure presented on Fig.4. If navigators have taken decisions basing on electronic sources in restricted visibility, they can also rely on these sources in good visibility. This is why, in authors opinion, such decisions can be taken by NAI, which is supervised remotely from control centre located ashore. Proposed solution has a lot advantages. First of all we can eliminate human factor, which is the

main reason of accidents at sea. For example number of collisions involving ships caused by human error is very high and is equal from 75 to 96% depends on time and a region of the world (Antão and Guedes Soares, 2008; Celik and Cebi, 2009; Harrald et al., 1998; Zhang et al., 2013a, 2013b). Algorithm does not become nervous in collision situation. It does not feel stress. It will generate anti-collision manoeuvre, which will be executed by autopilot and collision will be avoided. There many examples, when collision took place because of nervous action of navigators (Kulbiej E., Wołejsza P., 2016). In cited example, vessels would pass each other without any manoeuvre. Unfortunately, due to "coordinated" action of both navigators, an accident took place entailing many casualties and loos of m/v Baltic Ace.

Second of all reduces or eliminates salaries of crewmembers. One can say, that it can bring unemployment among seamen, who are over one million throughout the world (IMO, 2016)

Around half of them are officers both from deck or engine department. Estimates of the IMO (International Maritime Organization) and ILO (International Labour Organization) show that in 2020, will be missing about 40,000 officers of the merchant fleet. This is due on the one hand less popular profession (UK Government, 2011)

from the other with an increasing number of vessels that transport already about 90% of all goods in world trade. Lack of 40,000 officers, means that 5,000 ships have no staff. Those can be the pilot group of autonomous ships.

Third, but not least, algorithms can reduce fuel consumption. Medium size container ship (8,000TEU - Twenty feet Equivalent Unit) consumes about 260MT HFO (Heavy Fuel Oil) per day at full speed, which means that it burns 180l of fuel per minute. Thanks to the optimal anti-collision trajectory, shorten the time to complete this manoeuvre. In practice, the navigator assesses the size of turn basing on own experience. When the assessment is incorrect, improves own manoeuvre, but each rudder inclination is a measurable loss of fuel. Assuming that collision-avoidance manoeuvres is shorten only a minute a day, we can save 65 tons of fuel per year, which at a cost of about USD 320 per MT, gives savings of over 20.000USD, but in the case of low sulphur fuel savings are almost the double.

Above mentioned advantages will definitely reduce cost of sea transport. As most of the cargo are transported using this mode, total cost of goods will be reduced. To execute such a goal, an effective communication protocol should be implemented. Transmission between ship and shore can be executed via satellite. Speed which is presently available, enables video streaming and in coming years will be even faster.

In authors opinion, satellite communication will not be so effective for ship to ship information exchange. First of all it is not free of charge. It has relatively long delay, which can be important particularly during last moment manoeuvre. Moreover there are existing systems, which after some improvement, should meet requirements for autonomous vessels (Pietrzykowski Z., et al., 2014).

Among them AIS and DSC (Digital Selective Calling) are considered. Both systems enable to send short ship to ship messages.

Table 1. Available slots in AIS standard (IALA, 2009, 2011)

Message	Spare bits	Title
1,2,3	3	Position report
4	9	Base station report
5	1	Ship static and voyage related data
9	7	Standard SAR aircraft position report
14	968	Safety related broadcast message
18	8	Standard class B equipment position report
19	8	Extended class B equipment position report
21	1	Aids to navigation report

They can be used to convey intentions or agree manoeuver. Crucial issue is that such messages have to be sent autonomously. For example when anticollision system generates manoeuver, message with intended action have to be transmitted to other target within selected range e.g. 10 Nm. In other case, when avoiding collision is not possible by manoeuvre by one ship only, the coordinated manoeuvre has to be transmitted to dangerous target. It has to be, without delay, executed by rudder/autopilot.

What can be added to existing standard:

- Route to the next port of call, including all waypoints (WPs) or at least 5 next WPs – transmitted by each ship in Message 5: Ship static and voyage related data. Unfortunately there is only one spare bit in msg 5 in two slots.

Solution: to add third, fourth, fifth etc. slots to this msg to accommodate a/m information which are crucial to predict behaviour of other ships particularly in collision situation.

- Meteorological and navigational warning – transmitted by each ship in Message 14: Safety Related broadcast message. There are 968 spare bits in msg 14. It gives 161 available characters (6-bits ASCII) to send a/m information. Unfortunately this option is very seldom used by vessels to transmit a/m information. It's usually used to transmit test msg or "social" information.

Solution: awareness of seafarers about the opportunities.

- Anticollision manoeuvre – transmitted by each ship in Message 14: Safety Related broadcast message

Such information should be transferred to AIS an transmitted by AIS automatically, because the navigator in collision situation has no time to do so. This information is crucial for another vessel/vessels involved in collision situation.

Solution: direct and two way transmission between anticollsion system and AIS

The authors find also great potential to create local communication platform in NANET network (Nautical Ad-hoc Network), which can use vacated by analog TV radio frequencies. NANET uses Wi-Fi protocol to create ad hoc network. Due to 700 MHz frequencies, network has a range of at least 20 Nm, which is more than enough from the anticollision perspective.

Integral parts of the system being developed for automatic communication will be the language of communication, protocol and protocol interaction and negotiation strategies (Pietrzykowski Z., et al., 2016).

4.2 Berthing module

From the technical point of view, autonomous passage of the vessel from point A to B is relatively easy. Challenge arises, when other objects, which were not predicted in the sailing plan, obstruct ship's movement. Next challenge arises, when ship is approaching destination port. Basing on current technology, it is hard to imagine, that ship can berth without crew. Additionally, it is required by local authorities, that most of the ships have to be boarded by pilot, before coming alongside. Combining above mentioned facts, temporally solution for transitional period, can be introduced as follows: mooring team can board the ship together with the pilot on the roads and conduct the ship from the roads to the destination berth. Team for unberthing will board the ship alongside and sign off together with the pilot to pilot boat or helicopter.

When the ship is already alongside, the process of discharging/loading the cargo can be started.

4.3 Cargo module

There are already many terminals, where the cargo is handled by autonomous machines. Such container terminals exist in Hamburg or Rotterdam (Delta Terminal, 2017).

Also wet products are handled with limited presence of human. After connecting to manifolds, all the process are controlled by computers with man supervision. It is also easy to imagine, that cargoes in bulk can be transferred from/to shore by autonomous cranes and conveyors.

Challenges arise during sea passage. Some cargoes require monitoring (temperature, humidity etc.) and special care e.g. ventilation. We can image that such process is controlled by NAI, but what happens in case of motor failure in reefer container?

5 VESSEL'S MANAGEMENT

Low interest in the seafaring profession, shortages of highly qualified officers and still a high proportion of human error in the causes of marine accidents. All these factors result that shipowners think about replacing human by machine. There is currently no legal possibility to implement it. Shipowners shall continue to apply stringent Safe manning certificates, which precisely regulate the requirements for the number and quality of the crew. NAI technology follows current trends in transport. It enables to manage vessel from the level of single, virtual entity. Such management centre can be located in VTS, where all critical data will be delivered via satellite or other media. VTS becomes monitoring and supporting centre, while the basic crew is still on board. In the further future, when unmanned vessels become reality, it will be the total control and traffic management centre. The biggest challenge that arise that moment are security issues. Unmanned systems have to be well protected and security issues should be as much important as safety ones.

6 FUTURE OF NAUTICAL VESSELS

The autonomous vehicles is the strongest trend of the development of transport technology in the XXI century. Maritime transport is probably the most conservative transport sector, but even this sector will not avoid general trend and in near future autonomous vessels will dominate on the oceans. The process already starts, as usually first in the navy, which apart from small stand-alone objects, also operates unmanned destroyers like Sea Hunter (The Guardian, 2016). There are also few running or completed project, which aims to develop autonomous vessel e.g. commercial project led by Rolls-Royce or Saab, or research project MUNIN.

From technological point of view, we are ready to send an autonomous commercial vessel for example through the ocean. The biggest concern is avoiding other targets. For this reason anti-collision systems like NAVDEC were developed. IMO currently works on performance standards for such systems, which is the first step towards autonomous vessels (Wołejsza P., et al., 2013).

7 RESUME OR CONCLUSIONS

In an era of automation and the rapid development of autonomous systems, the shipping industry remains far behind, either on the side-lines of the mainstream. The main limitation is the international regulations, and generally tedious and long-lasting process of changing these provisions or to introduce new regulations. However, the steady increase in tonnage while reducing the number of people willing to work as a seaman, accelerate the process of change that from a social point of view, are highly desirable. The sailors will work on the land. Long lasting contracts away from home. This in turn should reduce the cost of transport, although the authors expect that in the initial period, the cost will be higher, as even on the mentioned Delta terminal, where the introduction of autonomous systems personnel costs increased by 20%. However, we believe that the effect on a global scale will ultimately lead to a reduction in the cost of crew, which at the same time reducing social costs, provides a solid base to conclude that such changes are desirable. The authors also do not have doubts that the money saved in this way, funds partly, if not fully, consumed in order to improve system reliability, increase bandwidth communications systems, and perhaps most importantly, their protection against interference from outside. Technologically, we are ready for these changes. It is the time for next steps. One of them are works on Performance Standards for Navigation Decision Support Systems (NDSS) for Collision Avoidance (CA) initiated by Poland at the 98 session of the committee MSC.

REFERENCES

Antão, P. & Guedes Soares, C. 2008. Causal factors in accidents of high-speed craft and conventional ocean-going vessels. Reliab. Eng. Syst. Saf. 93(9): 1292–1304.

Banas, P., Pietrzykowski, Z. & Wójcik, A. 2013. A Model of Inference Processes in the Automatic Maritime Communication System. Communications in Computer and Information Science 395: 7–14.

Celik, M. & Cebi,S. 2009. Analytical HFACS for investigating human errors in shipping accidents. Accid. Anal. Prev. 41(1): 66–75.

Harrald, J.R., & et al., 1998. Using system simulation to model the impact of human error in a maritime system. Saf. Sci. 30(1): 235–247.

http://www.imo.org/en/About/Events/WorldMaritimeDay/Documents/World%20Maritime%20Day%202016%20-%20Background%20paper%20(EN).pdf

https://www.theguardian.com/us-news/2016/apr/08/us-military-christens-self-driving-sea-hunter-warship

https://www.gov.uk/government/uploads/system/uploads/attachment_data/file/9015/seafarer-projections-2011.pdf

http://www.ect.nl/en/content/ect-delta-terminal

IALA, IALA Technical clarifications on recommendation ITU-R M. 1371-3, edition 2.4. 2009.

IALA, IALA Recommendation A-126 On The Use of the Automatic Identification System (AIS) in Marine Aids to Navigation Services Edition 1.5, 2011.

Kulbiej E. 2016. Relevance of the relativistic effects in satellite navigation. Scientific Journal of Maritime University of Szczecin 47: 85-90.

Kulbiej E. & Wołejsza P. 2016. An analysis of possibilities how the collision between m/v 'Baltic Ace' and m/v 'Corvus J' could have been avoided. Annual of Navigation 23: 121-135.

NAV 59/INF.2. 2013. Development of an e-navigation Strategy Implementation Plan. Report on research project in the field of e-navigation submitted by Poland: IMO.

NCSR 2/INF.10. 2015. e-navigation Strategy Implementation Plan. A study on ship operator centred collision prevention and alarm system submitted by the Republic of Korea: IMO.

Pietrzykowski Z., Hatłas P., Wójcik A. & Wołejsza P. 2016. Subontology of communication in the automation of negotiating processes in maritime navigation. Scientific Journal of Maritime University of Szczecin 46: 209-216.

Wójcik, A., Banas, P. & Pietrzykowski, Z. 2014. The schema of inference processes in a preliminary identification of navigational situation in maritime transport. Communications in Computer and Information Science 471: 304–312.

Wołejsza, P. 2013. Functionality of navigation decision supporting system – NAVDEC. Navigational Problems 1: 43–46.

Wołejsza, P. 2014. Navigation decision supporting system (NAVDEC) – testing in real condition. Annual of Navigation 21: 177–186.

Wołejsza, P., Magaj, J. & Gralak, R. 2013. Navigation Decision Supporting System (NAVDEC) - testing on full mission simulator. Annual of Navigation 20: 149–162.

Zhang, J.F., Yang, X. D., Zhang, D., Haugen, S. 2013b. Ship trajectory control optimization in anti-collision maneuvering. TransNav 7 (1): 89–93.

Software Updating Regime for Ships Necessity for Cyber Security and Safe Navigation

M. Bergmann
Comité International Radio-Maritime (CIRM), London, United Kingdom
BM Bergmann-Marine, Grosskrotzenburg, Germany

ABSTRACT: Navigating and operating a ship in modern world needs and utilizes a growing amount of Information and communications technology (ICT). This tools, which are used as decision support systems, need regular updating, same as every IT system. The current paradigms on ships not necessarily supports this need. As such CIRM and others are advocating for an updating regime. This is even more important as the risk of Cyber-attacks on ships and their systems are growing. The IMO is actively looking into Cyber Security, same as organizations like BIMCO or ICS.

The paper will look into the current ICT situation with focus on system updating. It will then try to describe an outlook over the next years. A second focus will be Cyber Security. It will provide a high-level overview on Cyber risks and the need for Cyber security on ships. While touching on aspects like IT security and security procedures, the main focus will be on the need for system updating. In a summary the paper will link the different aspects together and provide a recommendation on the way forward The paper will focus on the following topics:
- Current handling on ICT on ships
- Ship ICT and system updating
- Cyber Risk on ships ICT
- System updating as a necessity for Ship ICT and Cyber security

1 SHIPBOARD ICT

The use of Information and Communication Technology (ICT) is a base fact in today's world, both in private and in business. Figure 1 is an earlier 2007 projection of the internet usage.

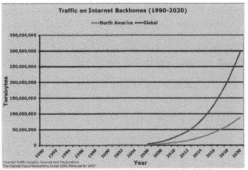

Figure 1. Traffic on Internet Backbones, CISCO, 2007

This development does not stop at maritime transport. While the maritime business traditionally is a very conservative business, the use of ICT on shore and on ship has become an integral part of the operations.

Looking on modern bridges, the use of electronic systems to support the decisions of the crew is of growing importance as the ship handling is becoming more and more complex. Besides the economic pressure and with that the need to improve efficiency of ship navigation, the need for advanced decision support systems is paramount. IMO has recognized that and started the e-Navigation initiative. IMO also agreed on a mandate for all SOLAS class ships to be equipped with ECDIS by end of the decade.

Connectivity of ships creates huge benefits as it enables the mariner to access real-time, semi-real-time and static data to increase his situational awareness. This connectivity, as it becomes an essential part of modern ship handling, requires ICT systems to transform the incoming data in meaningful data. Only then, the decision support systems can reach their goal to help safe and efficient navigation.

In addition to the navigational systems, like ECDIS, other ICT systems are becoming commonplace on modern ships. Engine room monitoring, motion sensors to support seakeeping, automated ship reporting to ship owners and operators, ICT supported propulsion and steering are only some examples of a growing use of those systems on board of ships.

2 DATA INTEGRATION ON BOARD

As the number of sensors on board of ships are growing, the collection of the useful data and combination of the different data streams, both from on-board sensors as well as from data received from shore, needs IT support. The systems will need to be interlinked so that data can be exchanged automatically for more efficient operations of the different components of the "ship organism". This basic concept of the "Internet of Things" (IoT) is starting to gain access to ships, same as it already has been introduced widely on shore.

However, as said the use of this data can only be beneficial, if it is analyzed correctly and presented to the mariner in a way that it increases his situational awareness for better decision making. The "Pyramid of Competence" (Figure 2), introduced in 2012, illustrates this need.

Pyramid of Competency

Figure 2. Pyramid of Competency, Michael Bergmann, 2012

Here the concept of "Big Data" is manifesting on ships. The concepts looks for a combination and analysis of data streams as they become available in the system network, in this case on board of a ship.

This analysis then can generate meaningful information for the mariners. Situational factors are defined by various data sensors, like speed, heading, engine performance or ship motion, define it. However, it also will be used to automatically initiate manage some system activities. Ship propulsion and steering may be automatically adjusted based on mariners parameters and sensor data analysis. Here "Big Data" is playing part of

"IoT" as the "Things" (i.e. computer subsystems) are connecting to each other and influencing their actions.

It has to be noted that the shipping industry has a high level of diversity. There are ships, for example modern tankers and cruise ships, which have a high level of ICT integration and are using most or all of the above mentioned ICT concepts. Other ships, like older ships or smaller fishing vessels, may have only a small subset of ICT systems with little interaction. So IoT and Big Data may not be utilized. Nevertheless, same as the IT usage overall is rapidly increasing as Figure 3 is illustrating, the migration of ships toward fully electronically enable is gaining speed.

1950-2005					
Number (world)	1950	1975	1985	1995	2005
Computers	60	650,000	50,000,000	200,000,000	822,150,000
Cell Phone Subscribers			700,000	89,000,000	2,065,000,000
People using the Internet			21,000	45,100,000	1,081,000,000
Number (U.S.)	1950	1975	1985	1995	2005
Computers	9	400,000	30,000,000	80,000,000	223,810,000
Cell Phone Subscribers			340,000	34,000,000	202,000,000
People using the Internet			19,000	28,100,000	210,000,000

Figure 3. ICT Development, David Houle, 2007

At the same time as ships are becoming more and more "e-enabled", the ship owners and ship operators are increasingly require up to date information from ships to improve the efficiency of their fleet as well as the shore side operations.

ICT has is no longer a potential future feature in maritime transport and on ships, but ICT is already established and is growing in importance

3 SHIP ICT AND SOFTWARE UPDATING

It is old news, that Software always requires updates. The "Draft Standard Software Maintenance of Shipboard Equipment" the CIRM (Comité International Radio-Maritime) and BIMCO (former: The Baltic and International Maritime Council) joined working group has developed, is clear in its assessment:

"As shipboard equipment becomes increasingly dependent on software, effective maintenance of this software is needed to ensure safe and efficient operation of the ship.

Effective maintenance of software depends on the identification, planning and execution of measures necessary to support maintenance activities throughout the full software lifecycle."
Draft Standard Software Maintenance of Shipboard Equipment, CIRM/BIMCO, 2016

4 CYBER RISK AND ICT UPDATING REGIME

With the development of ICT and the intrinsic connectivity of the IT systems the phenomena of "Cyber Risk" has evolved. The Institute of Risk Management (IRM) describes Cyber Risk as follows:

"'Cyber risk' means any risk of financial loss, disruption or damage to the reputation of an organisation from some sort of failure of its information technology systems."

Source: https://www.theirm.org/knowledge-and-resources/thought-leadership/cyber-risk/

This definition is very broad and includes two main aspects:
1 Cyber Risk as it relates to regular handling, and execution of ICT systems
2 Cyber Risk as it relates to external attacks on ICT systems

While in today's discussion the second aspect, mostly in conjunction with Cyber Security, is most prominent, the first aspect is worth looking into as well.

As we already know, ICT is an integral part of the operations and navigation of most ships. Quite a few of those systems run in the background and the mariner is only presented with the results of those systems. As the crew, in light of the "Pyramid of Competence" is relying on the preparation and presentation of underlying data by ICT systems as digestible information. Any malfunction of those systems or their components may result in false situational awareness and finally in wrong decisions. The risks posed there can reach from less optimal operational efficiency up to catastrophic loss of a ship through collision, sinking or grounding.

Various reasons can lead to wrong behaviors of ICT systems, specifically:
– Software bugs
– Input Sensor failures or malfunctions
– Connectivity issues of connected systems and subsystems

This kind of cyber risks are not unique to ships but common risks on any ICT system. Unfortunately, the awareness that shipboard systems are not different from those on shore is not always understood. Often the companies, which are not questioning at all the need for proper system maintenance and support at their offices, are expecting ship systems to run without issues and ignoring any requirement for updates or maintenance. In the maritime industry therefore, it could be better phrased as "This kind of cyber risk are not unique to shore based ICT systems, but also apply to ships ICT systems".

The additional complexity on board ships is that there is no "IT support department" available and the environment is incomparably more harsh than an office environment. The officers are often kept alone with software issues or system malfunctions. Given this specific difficulties the maintenance of shipboard systems is even more critical.

A key aspect of system maintenance, including those on ships, is a proper software-updating regime. Any software may have underlying software bugs, which may surface under certain circumstances. In addition as new functionalities or system components are added or existing once are exchanged, may require software updates. Only then, the first aspect of cyber risk can be mitigated.

Unfortunately, the IMO, while requiring certain IT systems on board of SOLAS ships, has not defined a software-updating regime. The earlier mention standard from CIRM and BIMCO, in the first half year of 2017 field-testing, tries to fill that gap. The standard defines how updating should be performed, which measures should be taken to ensure the update is successful and the systems in all complexity are working and collaborating as desired after finalization of the update.

The standard also defines early actions to mitigate cyber risks. It recommends preventive software maintenance by looking at three aspects:
– Periodic maintenance
– Conditional maintenance
– Predictive maintenance

A good maintenance regime for shipboard ICT systems, including a software maintenance regime is paramount to mitigate the system intrinsic Cyber Risk. This is a natural system requirement of any ICT and should not be questioned.

5 SHIP ICT AND CYBER SECURITY

The second, currently intensively discussion aspect of Cyber Risk relates to Cyber-attacks.

As demonstrated above the growing connectivity of ships open up higher risk of those attacks.

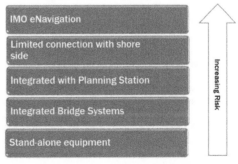

Figure 4. Cyber-attack risk, Michael Bergmann, 2006

Figure 4 illustrates this relationship. On a Stand-alone equipment, the risk of cyber-attacks are limited to external influence of sensors on those systems, e.g. through spoofing or jamming of satellite signals (GPS…). Any attack could only target the one system. Once you integrate different equipment, for example in Integrated Bridge Systems (IBS), a cyber-attack could identify the weakest link and once within the system attack other equipment. If the planning station is connected, and be it only by forwarding the route plan to the ECDIS via an USB stick, any encapsulated bridge system could be infected by a successful cyber-attack on the planning station.

If the ship systems are only connected to each other, the cyber risk is somewhat limited as it need to get on the system first, for example while an infected equipment is connected. An example could be a smart phone, which is connected to a system USB port to load the battery. Once systems are connected with the outside world, for example to download ENCs or for other reasons, the cyber-risk is growing largely. And once we move towards e-Navigation with constant or at least very frequent on-line times, Cyber Security, the mitigation of cyber-attacks, is essential to reduce this risk.

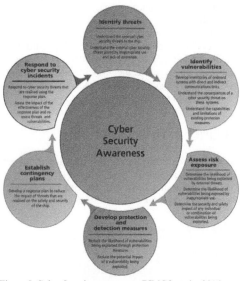

Figure 5. Cyber Security awareness, BIMCO et al., 2016

In its Cyber Security Guideline on board Ships, the consortium of key industry associations (BIMCO, CLIA, ICS, INTERCARGO, INTERNTANKO) not only tries to improve the cyber security awareness (see Figure 5), but also describes how to mitigate the risk and implement a cyber security system.

While the guideline, and not only it, is quite clear in that a major aspect of cyber security is the definition and execution of procedures, processes

and best practices associated with this topic, the technical aspects are of essence, too.

The recent publications on this topic, not only in maritime sector, have clearly identified options on mitigation. Building and maintaining firewalls as well as system encapsulations are necessary technical mitigation tools for cyber security. Figure 6 is illustrating a potential ship architecture realizing this technical mitigation tools. It looks into specific gateways separating a secured network from uncontrolled networks. Multiple those secured networks could be built on board of ships to isolate or encapsulate different ICT systems on board and with that further reduce cyber risk.

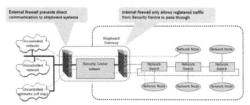

Figure 6. Ship Gateway Concept, Michael Bergmann, 2016

As systems are secured with firewalls, a non-debatably need, it has to be understood that those firewalls need very frequent updates to enable protection against new cyber-attack pattern and other cyber risks which have been encountered elsewhere.

This even more highlights the need for a solid software and system-updating regime.

6 CONCLUSION

The development in modern world in increased use of ICT systems does not leave the maritime sector out. Ships are receiving a growing amount of IT systems. Ships are connected more and more with shore to receive and send data. As those tools are finding their way in the daily operation, it increases the reliance of the crews on decision support systems with integrated data streams.

It is important to realize that those systems need maintenance throughout their lifecycle. The systems also need protection against cyber security, both through processes as well as through technical protections like firewalls and a cyber security topography and architecture. The maritime industry need to develop a sustainable lifecycle model and updating regime for shipboard ICT systems. In addition, the cyber security concepts need to move from guidelines to high-level standards, which are adhered to in all systems and across systems on board. Only then, the mariner can use the systems as true decision support systems.

The implementation of both aspects is essential to realize the positive effects of modern ICT systems in safety and efficiency of maritime transport.

REFERENCES

BIMCO statement on Cyber Security, 2016, https://www.bimco.org/about-us-and-our-members/bimco-statements/cyber-security

Guidelines for Cyber Risk Management (Islamic Rep. Iran), 2016, IMO MSC97/4

The Guidelines on cybersecurity on board ships (BIMCO et al.) , 2016, MSC96/4/1

Cybersecurity aspect of the ongoing work of the CIRM/BIMCO Joint Working Group on Software Maintenance (CIRM & BIMCO), 2016, MSC96/4/6

IMO, 2009, SOLAS (Consolidated Edition)

IMO, 2008, MSC 85/26 Annex 20

IMO Circular MSC Circ 1221

IMO NCSR 1/9 "DEVELOPMENT OF AN E-NAVIGATION STRATEGY IMPLEMENTATION PLAN"

IEC 60945, Fouth Edition 2002-08

IEC 61023, Edition 3.0 2007-06

IEC 61174 ed3.0 (2008-09)

IEC 61924-2 ed1.0 (2012-12)

IEC 62288 ed2.0 (2014-07) The Shift Age, 2007, David Houle,

Draft Standard Software Maintenance of Shipboard Equipment, CIRM/BIMCO, 2016

Cyber Risk: Resources for Practitioners, IRM, 2016

IEC 61162-460 Ed. 1.0, Maritime navigation and radiocommunication equipment and systems - Digital interfaces - Part 460: Multiple talker and multiple listeners - Ethernet interconnection - Safety and security

IMO International Ship and Port Facility Security (ISPS) Code, International Safety Management (ISM) Code, Resolution A.741(18) as amended by MSC.104(73), MSC.179(79), MSC.195(80) and MSC.273(85)

IMO MSC.1/Circ.1389, Guidance on Procedures for Updating Shipborne Navigation & Communication Equipment

IMO SN.1/Circ.266/Rev.1, Maintenance of Electronic Chart Display & Information System (ECDIS) Software

ISM Code, Chapter 5, Section 10, Maintenance of the Ship & Equipment

ISO 9001, Quality management systems – Requirements

ISO 17894, Ships and marine technology – Computer applications – General principles for the development and use of programmable electronic systems in marine applications

ISO/IEC 90003, Guidelines for the application of ISO 9001 to computer software

ISO/IEC 12207, Systems and software engineering – Software lifecycle processes

ISO/IEC 15288, Systems and software engineering – System life cycle processes

ISO/IEC 25010, Systems and software engineering -- Systems and software Quality Requirements and Evaluation (SQuaRE) -- System and software quality model

The Cyber Security Guideline on board Ships, BIMCO et al., 2016

Bergmann M. (2013). Integrated Data as backbone of e-Navigation in: Weintrit, A. (2013). Transnav Volumn 7, Number 3, September 2013, ISSN 2083-6473

Bergmann M. (2015) The Concept of "Apps" as a Tool to Improve Innovation in e-Navigation. TransNav, the International Journal on Marine Navigation and Safety of Sea Transportation, Vol. 9, No. 3, pp. 437-441.

Efficient and Extremely Fast Transport including Search and Rescue Units Using Ground Effect

K. Szafran & Z. Pągowski
Institute of Aviation, Warsaw, Poland

ABSTRACT: "Ground Effect" is known, not only theoretically, as an efficient and extremely fast means of transportation. Combining these types of aircraft with search and rescue (SAR) units and intervention over the surface of water or ice is a promising idea. Authors of the paper describe the new role of "ekranoplanes", also known as Wing-In-Ground (WIGs), in search, rescue and intervention operations at sea, including the Arctic region.

1 INTRODUCTION

The ground effect aircraft known as "ekranoplanes" have been used in different civil and military configurations, but practically never in search and rescue operations. Practice shows that the use of ekranoplanes in those operations is the most effective element of search and rescue operations at sea. After the collapse of the Soviet Union, the idea of new types of ekranoplanes was introduced after the catastrophe of the nuclear submarine "Komsomolets" in the Barents Sea in 1989. Comparative estimation of properties and characteristics of present ekranoplanes, airplanes, helicopters and vessels in these type of operations indicates that ekranoplanes are the natural choice. It has confirmed their speed of arrival of up to 550 km/h, range of operation of about 1500 km, including high search capacities at sea state up to number 5 etc. [1]

2 NEW WORLD RESCUE AND SECURITY TRANSPORT VISION

New world transport vision will promote coordination between operators and authorities responsible for rescue and security and enhancing new types of R&S units, devices and accompanying infrastructure, especially in regions with poorly developed transportation network such as the Arctic. Additionally, all modes of transportation have to be protected against terrorist and pirate attacks, which makes the requirements more complex.

The first version of ekranoplane was designed in 1991 by a Russian financial-industrial consortium named "High Speed Ships", which included "Alekseev Central Hydrofoil Design Bureau". To support search and rescue operations on the ocean and at sea within the range of up to 3,000 kilometers and with a cruising speed of up to 500 km/h, the military "killer of aircraft carriers" rocket ekranoplane "Lun" was modernized into ekranoplane "Spasatel", which included up to 80 seats and beds for refugees. The flight and sailor crews are 9+19. A passenger variant with capacity of up to 500 passengers can be delivered within 3 years. [2][3].

Figure 1. Ekranoplane "Lun"

Figure 2. R&S ekranoplane "Spasatel"

3 GLOBAL TRANSPORT VISION

Improving the global vision of transport, including WIG transportation, in times of economic competition, requires, among others, the cooperation of all transport players, regulators and industries by in the European and World markets. Ekranoplanes based on military requirements are not commercially efficient and lack connection to economic and civil global vision of transport (Spasatel was proposed in 2012 as a monument in the city park of Niznyj Novgorod, Russia on the bank of the Volga River). Global trends and targets are clear and they focus on the integration of all systems of transport, which must be safe, efficient and include new technologies. Many countries are interested in using the WIG (highlighted in yellow), or are ready for their production (highlighted in green), or conduct their own development and manufacturing (highlighted in red) acc. Fig 3 by [8]

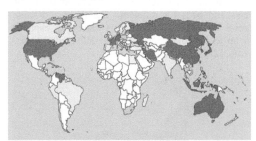
Figure 3. Vision of using the WIG by countries

In aviation it "could be integrated with all types of aircraft, and lead to rethinking aircraft architecture" also for operations in the Arctic [4][5][6][7][8].

In Europe this conception will be complementary to EPATS/SAT projects (http://www.epats.eu// http://www.epats.eu/SATRdmp/index.htm) on many points and will be fully integrated in the European Transport System, taking care of solutions needed by 2050. [9] A possible scenario for the use of small aircraft in 2035 is illustrated below (according to SAT Rdmp D2.3 Analysis of the impact of each

business case fig. 4, but without European seas (Baltic, North Sea, Mediterranean, Adriatic Sea, Black Sea). The "white spots" must be defined by new "ground effect" technologies thanks to investigation of various scenarios, analyzing gaps and problems, potential of the markets, integrated visions, new aircraft technologies, perspective recommendations and also conception of security and rescue ekranoplanes for this region, including the Arctic – possible area for oil & gas exploitation. Especially in the Arctic, the use of multipurpose ekranoplanes fits in with the economic vision.

In Russia currently some WIG designs are already in exploitation in the Arctic – e.g. EL 7 "Ivolga", EK 12, Orion 14, Orion 20, as well as demonstrators of technology such as "Tungus" and "Burieviestnik 24" [8]

Figure 4. EPATS vision

Figure 5. "Burieviestnik 24"

An interesting vision for the Arctic in Canada is the Nunavik Arctic Express – a concept by Charles Bombardier (grandson), prepared to explore the Artic region and to conduct search-and-rescue missions. It will have a composite cargo-friendly fuselage design with large side doors for containers and it will be able to fly at max speed of 200 mph, at 15-50 ft above the sea, ice or tundra. Propulsion will be provided by four 1,500 hp Pratt & Whitney PW100 turboprops

Figure 6. Nunavik Arctic Express

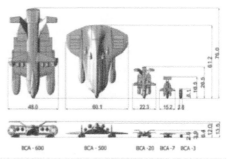

Figure 9. VSA Family of R7S units

Table 1. Main technical data of VSA family

Technical data	VSA -3	VSA -7	VSA -20	VSA -500	VSA -600
Velocity, km/h	120	250	350	500	500
Range, km	600	1500	3000	6000	6000
Starting mass, т	3	7	20	500	600
Max. mass., т	1	3	20	300	350
Person	8	20	90		

4 NEW CONCEPCTION OF THE SECURITY AND RESCUE EKRANOPLANES IN RUSSIA

Concepts of efficient and extremely fast transport including SAR units using ground effect must address a variety of challenges, including irregular migrants, operating conditions (day or night, visibility, wind conditions, icing, etc.), technology, social plus environmental issues, as well as challenges of the market. An interesting idea for the Arctic is a Russian vision presented by OOO "SKY+SEA" in "Gelendjik 2016" [9]fig. 7

Figure 7. Integrated R&S base for Arctic area

According to OOO "Sky +Sea", the use of aircraft or ekranoplanes and international co-operation in search and rescue operations will be the most effective form in future Arctic oil and gas fields, which has been confirmed in an existing situation in the area of Murmansk, where Norwegian aviation helps Russians agencies. After tests of demonstrator of technology ekranoplane VSA-7 a vision for family of R&S units has been prepared (fig 8 -9 + tab.1).

Figure 8 Tests of VSA-7 (summer 2016)

5 THE SECURITY AND RESCUE EKRANOPLANE

In 2015 acquiring new technologies acc. [11] was indicated by 12% of SAR organizations, but mainly unmanned aerial vehicles and new mobile phones were highlighted, alongside helicopters with fully-trained crews.

But why not take under consideration the family of economical ekranoplanes, with improving interoperability saving time, budgets and other resources?

An analysis of European capabilities and strategic plans for the main direction of SAR operations is needed, especially now when irregular migrants are a general problem for migratory routes from the Mediterranean sea regions to the EU [12] Tab. 2.

Table 2. Irregular migrants by Jan-Oct. 2016 [12]

Mediterranean route	Number of persons
Western African	509
Western	7888
Central	173055
Apulia & Calabria	5000 (?)
Eastern	180289

Where the numbers of migrants are small, Frontex has given strong indications that drug smuggling is the primary goal and suggests that the smugglers are using small planes and helicopters.

The current trends and challenges in the migratory routes are not clear, while there are barriers to improving SAR strategy, also concerning the identification of effective SAR units including ekranoplanes.

Frontex indicates the barriers for seven Mediterranean countries (Spain, France, Italy, Slovenia, Malta, Greece and Cyprus) plus Portugal, 50 separate authorities, under 30 ministries and 40 different agencies in the northern EU countries. This is why the European Patrols Network was started to resolve future challenges to the migration and come up with an adequate response to the current crisis in Europe, including application of new surveillance tools like satellites and UAVs. But it does not assist with the search of small targets based of existing intelligence and with fast response needed to help people on the sea [12]. The use of special ekranoplanes with UAVs on board situation is more practical for SAR operations. A vision of a new type of ekranoplane for the Baltic Sea (i.e for a specific length of sea waves) was created by Jacek Kończak – an independent consultant and former designer at the Institute of Aviation on Fig. 10 and 11. [14][15] [16]

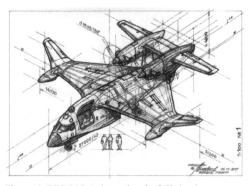

Figure 10. WIG/SAR-1 ekranoplane by J. Kończak

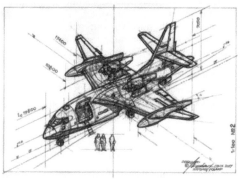

Figure 11. WIG/SAR-2 ekranoplane by J. Kończak

The source of this vision is connected with the transport visions realized from 1990-ies by Airfish 8 produced by a German enterprise Airfoil Development GmbH.[17] Bigger, 80-seat ekranoplanes for the Baltic Sea by Fischer Flugmechanik have been arranged by The German

Ministry of Economics. Airfish 8 - 001 has become the first ekranoplane to be registered in the Singapore Registry of Ships by the Maritime and Port Authority of Singapore. Hoverwing 50 known as WSH500 was produced in Singapore under license by Korean company Wingship Technology Corp. (WST Corp.) In 2013, production of 10 Hoverwing 20 ekranoplanes (a new design) was started by PT AGEC TECHNO (Indonesia). Wigetworks owns now the intellectual property and marketing rights for the AirFish family.

A luxury second generation ekranoplane is Hoverwing 80, announced by Pangiotis Zaglis in Athens, Greece in co-operation with Fisher Flugmechanik and AFB Developments GmbH from Germany.[17] Calculations indicate that due to different frequencies of sea waves it is necessary to develop big new generation ekranoplanes with new capabilities. [18] Some parameters for the new technologies are shown in Tab. 3 below.

Table 3. Technical data AF8 and Howerwing WIGs

Technical data	Airfish 8	HW-20	WSH 500	HW-80
Velocity, km/h	160	190	180	207
Range, km	200	900	300	800
Starting mass, т	no data	6,5	17,1	27,5
Max. mass., т	no data	10	22,5	39,5
Person/crew	8	23	41	80

China's activities in the South China Sea will probably need a new fleet of 100-seat A-050 modern ekranoplanes with max. speed up to 450km/h and a range of 5000 km, with maximum weight of 54 tons and a loading capacity of nine tons.

A050 ekranoplanes were offered by the well-known Alekseyev Central Hydrofoil Design Bureau based in Nizhny Novgorod in Russia.

Figure 12. AO50 ekranoplane

In China, Army Border Police has used XTW 5 ekranoplanes for years. During maritime SAR missions the hydroplane AG-600 is used, which can carry 50 people.

Figure 13. XTW-5 ekranoplane of Border Police in China

Figura 16. SAR station in the North of Russia

New concepts of large ekranoplanes that could carry up to 2000 passengers, for future long-range transport, with the range of 15 000 km, fueled by hydrogen, is being prepared by dr Levis and Serghides V.C. [22], fig. 14.

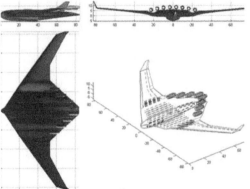

Figure 14. Vision of 2000-seat ekranoplane

Some of these types of visions using ekranoplanes, also for SAR missions, may appear futuristic, but some may become real in identified niches, such as the sea and the Arctic region [9],.[22], which is illustrated by fig.4,15,16

Figure 15. SAR missions in Canada 1998-2001 [23]

But the process of designing and buying new transport and SAR ekranoplanes is for many countries in its initial stage. Practice shows that the use of ekranoplanes in both kinds of operations will be the most effective with international co-operation during search and rescue operations, like the one existing in the above mentioned Murmansk. The main obstacle to the widespread introduction of ekranoplanes is the lack of international regulatory framework design, manufacture, technological and operational reliability, composite material of new generation, operational guidelines (i.e. in the Mediterranean or the Arctic), certifications etc. (The ekranoplane Airship 8 was certified in late 90's by Germanischer Lloyd (+100 A0 WIG – A , WH 0,5/1,5 EXP)) for the Baltic sea region). This problem involves the creation of new industries, transportation systems and SAR ekranoplanes at a competitive cost, to offer cheaper and efficient WIG transport between existing ports or with the port of the future for ekranoplanes, now without such governmental facilities even in Russia. It was, however, discussed during the recent conference in Gelenjik [10]

The Work Programme of EU [24] focuses on "Smart, green and integrated transport" and contains the great idea to develop new ekranoplane technologies, because it is oriented at cooperation of Europe with Japan, China and Canada and ongoing initiatives with United States, Russia, Brazil and Australia. It remains a question, however, who would take on the role of the co-ordinator - large, commercial cargo aircraft manufacturers like Boeing, Airbus or SME, or perhaps other international parties within a joint project of new generation of ekranoplanes and a new vision of transport ? The first steps have been made, but major business interest still has to develop. [25]

6 CONCLUSION

1 The European "Ground Effect" transport infrastructure needs formulation of:
  – general characteristics and a hierarchy list of seaports for intermodal transport, SAR operations and disaster areas

- general requirements for "Ground Effect" sea airports
- number of operations, flow of passengers in additional types of transport (GA, Business, etc. including irregular migrants)
- seaport distances and distribution from several points of view (tourist, business, special taxi system etc.)
- potential demand for ekranoplanes and their costs
- confirmed specifications and performance of currently produced "Ground Effect" aircraft
- overview of current R&D efforts conducted
- a future vision for the performance of "Ground Effect" aircraft including SAR operations
- new proposals for adequate research programs at HORIZON 2020, SME Instrument, MED Program in the area of aeronautics, aerodynamic design, composite materials, control systems, system of intermodal operations, technology implementation from GA and commercial aviation
- new programs of basic aerodynamic airfoil [26]
- classification of nomenclature and definitions at the EU and national levels, TRL classification of proposals submitted, overview through patent and research connected with WiG aircrafts, submitted in EU registers
- variants of cooperation with law enforcement agencies

REFERENCES

[1] Nevylov, A.V. & Wilson, P.A. 2002. Ekranoplanes – controlled flight close to the sea. Southampton, WIT Press.
[2] Search and rescue ekranoplan "Spasatel" http://www.hs-ships.ru/e_pages.phtm?f=2&p=29
[3] Yun, L., & Bliaut, A.& Doo J. 2010. WIG Craft andekranoplan –Ground Effect Craft Technology. New Yoortk etc.,Springer WIT Press.Clausen,
[4] U. Holloh, K-D., Kadow M., 2010Visions of the future transportation and logistics 2030, Fraunhofer IML | Daimler AG | DB Mobility Logistics AG
[5] EREA-vision for the future- Towards the future generation of Air Transport System,2010,Amsterdam, Association of EREA
[6] Warwick, G.,2014, Designer Airliners - Ice Road Flying,acc. http://aviationweek.com/blog/designer-airliners-ice-road-flying
[7] Górtowska, M.,2012, Available search effort of the WIG craft, Szczecin, Scientific Journals Zeszyty Naukowe Maritime University of Szczecin,
[8] Ganin, S. M. 2016, Technical, legal and environmental aspects of the WIG for use in Russia,PP " Gidroaviasalon 2016", Gelenjik
[9] Piwek K., Wisniowski W.,2016,Small air transport aircraft entry requirements evoked by FlightPath 2050, Aircraft Engineering and Aerospace Technology: An International Journal, Vol. 88 Iss: 2, pp.341 - 347
[10] OOO "Sky+Sea",Spasene po ldah, PP, Gelendjik 2016"
[11] Search and Rescue Servey Results 2015, 21 - 23 April, 2015 Action Stations, Portsmouth, United Kingdom
[12] http://frontex.europa.eu/trends-and-routes/migratory-routes-map/
[13] COMMUNICATION FROM THE COMMISSION TO THE EUROPEAN PARLIAMENT, THE COUNCIL, THE EUROPEAN ECONOMIC AND SOCIAL COMMITTEE AND THE COMMITTEE OF THE REGIONS Examining the creation of a European Border Surveillance System (EUROSUR) Brussels, 13.2.2008 COM(2008) 68 final
[14] Kończak J. – Independent consultant Fig.WIG/SAR_1 ekranoplane Fig. WIG/SAR_2 ekranoplane
[15] Pągowski Z., Szafran K., 2015,"Ground effect" Inter-Modal Fast Sea Transport, TransNav, the International Journal on Marine Navigation and Safety of Sea Transportation, Vol.8, No.2, pp.317-320
[16] Pągowski Z.T., Szafran K., Kończak J. "Ground effect", Transport on the Baltic Sea, In. A. Weintrit & T. Neumann (eds), Maritime Transport Shipping, Marine Navigation and Safety of Sea Transportation CRC Press/Balkema, Taylor & Francis Group, London 2013, pp. 221-234
[17] https://sites.google.com/site/hoverwingwigcraft/Home
[18] Nebylov A.V., Nebylov V.A.,2014, CONTROLLED WIG FLIGHT CONCEPT Preprints of the 19th World Congress, The International Federation of Automatic Control, Cape Town, South Africa. August 24-29,
[19] Baird, W., China in Talks to Acquire Russian Ekranoplane Wing-in-Ground Effect Vehicles http://thedragonstales.blogspot.com/2015/10/china-in-talks-to-acquire-russian.html
[20] China Army Aviation http://www.scramble.nl/orbats/china/army
[21] China seeks 100 passenger Wing in Ground Effect Ekranoplan and larger seaplanes for South China Sea,2015,http://www.nextbigfuture.com/2015/10/china-seeks-100-passenger-wing-in.html
[22] Levis,E.,Serghides,VC, 2014,The potential of Seaplanes as future large Airlines, Royal Aeronautical Society Bristol, http://hdl.handle.net/10044/1/18626
[23] Rescue Required: Canada's Search-And-Rescue Aircraft Program. SAR missions 1998-2001 http://www.defenseindustrydaily.com/rescue-required-canadas-searchandrescue-aircraft-program-03350/
[24] European Commision , Horizon 2020, Work Pr gramme 2016 – 2017 11. Smart, green and integrated transport, European Commission Decision C(2016)4 614 of 25 July 2016
[25] Pagowski Z. „GEIMSeaTS - Green Ground Effect Inter-Modal Sea Transport Study, http://www.kpk.gov.pl/wp-content/uploads/2015/10/Ground_Pagowski_green.pdf
[26] Szafran K., Shcherbonos O., Ejmocki D. 2014, Effect of duct shape on ducted propeller trust performance. Transaction of the Institute of Aviation. Nr 4(237). pp.85–91.

Belief Assignments in Nautical Science

W. Filipowicz
Gdynia Maritime University, Gdynia, Poland

ABSTRACT: Aleatory uncertainty is related to a measurement random and systematic errors. Stochastic distortions distribution remain estimated and imprecise in respect to their types and parameters. Uncertainty also embraces subjective assessment of each piece of available data. In traditional approach there is no room for modelling and processing all mentioned items. Moreover results of observations, final outcome quality evaluation, can be evaluated a priori before taking measurements. A posteriori analyses is impaired, it is not embedded into the scheme of traditional way of inaccurate data handling. To introduce the new idea one should explore alternative approaches towards doubtfulness modelling. Presented rational starts with basic interval uncertainty model.

1 INTRODUCTION

Mathematical Theory of Evidence (MTE for short), operates on belief assignments or functions. It exploits evidence sets and hypothesis frames, it is widely expected that evidence at hand supports each of considered hypothesis items, endorsement degrees varies on real scale (Dempster, 1968), (Shafer, 1976) Relations between involved universes/frames are encoded within belief assignments. Measures of included sup- port are belief and plausibility values. The theory also offers combination mechanisms in order to increase the informative context of the initial evidence. Combination scheme delivers results that include support regarding hypotheses space elements. The evidence is considered as a collection of facts and knowledge related to imprecision and uncertainty of measurements. In navigation, facts are results of observations such as taken bearings, distances or horizontal angles. A combination scheme is expected to enable obtaining best solution provided given observations/measurements and knowledge on their quality. The theory was already successfully implemented in many fields some of them are related to discussed area of application, see for example (Ayoun, 2001), (Filipowicz, 2014), (McBurney et al., 2002).

Results of observations can be affected by random and systematic errors. Deflections are governed by various distributions. Quality of measurements taken with different navigational aids differs. Prevailing are discrepancies in estimation of randomness distributions parameters. It is popular statement related to nautical observations that the mean error of a bearing taken with radar is interval valued within the range of [±1°; ±2:5°]. Random distortions are unavoidable although a seafarer knows much about it. The phenomenon is called aleatory uncertainty that cannot be eliminated but to certain extend might be evaluated. It is crucial that this sort of doubtfulness could be included into belief assignments. Mentioned evaluation of the mean error appears as a fuzzy figure and as such, fuzziness should be accepted and taken into account during computations (Yen, 1990). Most analyses exploit theoretical distributions but empirical ones can be used instead. The most important thing is the ability for objective evaluation of the obtained output along with measures indicating its accuracy meant as probabilities of alternative results (Filipowicz, 2014). There are publications devoted to implementation of MTE concept into nautical science (Filipowicz, 2009, 2010, 2011). Most of them discuss practical navigational aspects of the concept. This paper presents details regarding aleatory uncertainty modelling and processing.

2 INTERVAL UNCERTAINTY MODEL

Popular basic uncertainty model includes proposition and associated range of probabilities also called as belief interval (Lee et al. 1995). Given proposition z and range of real values [a, b] one can define the model in terms of the statement truth. It can take the form of equation 1.

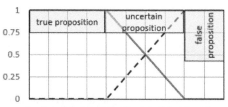

Figure 1. Interval uncertainty representation. Fuzzy probability sets membership functions

$$z : [a, b]; \quad a, b \in [0, 1] \, and \, a \leq b \quad (1)$$

where a = upper limit of probability that proposition z is true; b - a = range of uncertainty, possibility of the truth of z is defined by a descending function (see figure 1); and b = lower limit of probability that proposition z is false.

Proposition engaging z can be transferred into belief function (Denoeux, 2000) shown in Table 1. The assignment engages two elements hypothesis space Θ reflecting the truth (true or false) of considered proposition z. Thus one of the items is marked with z and another with $\neg z$ meaning negation of z. Within probability assignment all elements of power set of the considered frame might appear, consequently multiple items set of the frame: {z; $\neg z$} is present in the table. Last raw expresses uncertainty since it refers to the whole frame of discernment.

Table 1. Basic probability assignment for uncertainty model.

notation	probability value
m(z)	a
m(¬z)	1 - b
m(z, ¬z)	b - a

m(z)   probability mass that proposition z is true
m(¬z) probability mass that negation of proposition z is true
m(z, ¬z)   range within one can doubt on true or false of the proposition

Diagrams shown in figure 1 can be perceived as membership functions showing possibility of x belonging to two fuzzy probability sets. Within range [0; a) possibility of true proposition z is equal to one, then it descends and reaches zero at point b. Interval [a; b) can be seen as an amount of ignorance, doubtfulness in the truth of proposition z. Further on the statement cannot be true, contrary to $\neg z$ for which possibility of being true is one within the rightmost range. Value of a can be treated as

belief that z is true, and upper bound b is the plausibility that z is true (for proof see (Lee et al., 1995)).

3 UNCERTAINTY IN NAUTICAL SCIENCE

Aleatory uncertainty, which is related to random and systematic measurement deflections, is present in all nautical measurements. Observation made with a navigational aid is randomly affected, it can be treated as instance of random variable governed by some kind of distribution. It is often that Gaussian bell function is used although discrepancies in parameters of such distributions are frequently met. It is popular to state (Jurdziński, 2008) that the mean error of a distance taken with radar is a fuzzy valued figure representing the range of [±1%; ±2.5%] of the measured distance. Such evaluation of the mean error is a fuzzy figure thus fuzziness should be accepted and taken into account while constructing adequate model.

Example of aleatory uncertainty related to distance to a landmark observation is shown in figure 2. It is assumed that randomness is governed by Gaussian distribution. There are various dispersion parameters estimations available. Two of dispersions, one most optimistic with standard deviation σ_{min}, second assumed as pessimistic with deviation σ_{max}, are presented in figure 2. Measured distance is marked by point of y-axis and abscissa intersection. Given all mentioned above one can seek truth of proposition \is the true distance to the landmark represented by a point close to abscissa x_1". Measurement related proposition refers to easily established interval [$C \cdot p_{1min}$; $C \cdot p_{1max}$]. Where C depends on width of the considered neighborhood of abscissa x_1 since given diagram represents probability density function. All mentioned before mean that range valued uncertainty is transferable and applicable to nautical science. Membership functions showing possibility of various distances belonging to fuzzy probability set are the same as presented in figure 1. Thus, as stated above, possibility and probability can be used jointly in order to include uncertainty into defined mathematical model. The approach appears new but remains not of great value once single proposition is specified. One statement contributes to definition of a belief assignment. It is practical to have more observations (pieces of evidence) and find out the truth of a statement from multiple input perspective. Handful of measurements delivers crucial results once all items are combined. Association scheme is important from uncertainty propagation point of view. In nautical science scope of doubtfulness regarding each of observation should be transferred to results accuracy analyses. Statistical approach appears of limited value in this respect.

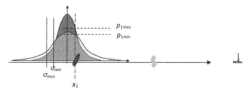

Figure 2. Aleatory uncertainty related to distance to a landmark observation

3.1 Combining evidence

Figure 3 shows two pieces of evidence related to two distances taken to a landmark with the same appliance. Note that both pairs of random dispersions are the same. For the new case one can seek truth of proposition "what is a support that the true distance to the landmark is represented by abscissa in the vicinity of x_1, taking into account two (all) available pieces of evidence".

Two belief assignments related to two distance observation made with the same navigational appliance are shown in second and third column of table 2. Uncertainty intervals for both measurements are respectively: $[C \cdot p_{1min}; C \cdot p_{1max}] = [0.32, 0.40]$ and $[C \cdot p_{2min}; C \cdot p_{2max}] = [0.48, 0.68]$.

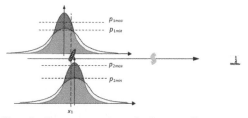

Figure 3. Aleatory uncertainty related to two distances to a landmark

Table 2. Basic probability assignment for uncertainty model.

	$m_1(..)$	$m_2(..)$	$m_C(..)$	$m_C^D(..)$	$m_C^Y(..)$
$\{z\}$	0.32	0.48	0.256	0.420	0.256
$\{\neg z\}$	0.60	0.32	0.338	0.554	0.338
$\{z, \neg z\}$	0.08	0.20	0.016	0.026	0.406
$\{\varnothing\}$			0.390		

$m_C(..)$ – non-normalized combined mass values
$m_C^D(..)$ - combined mass values normalized with Dempster method
$m_C^Y(..)$ - combined mass values normalized with Yager method

Results of conjunctive combination for above mentioned belief functions are presented in the right part of table 2. Results of association presented in column titled with $m_C(..)$ require conversion since there is some mass attributed to empty set (see last row of the table). Such assignment means inconsistency, conflicting situation that should be avoided. Once occurred, mainly during null generating association, must be eliminated. It is normalization that leads to belief structures,

assignments without unwanted inconsistency cases (Yager, 1996).

It should be noticed that support for proposition that the true measurement, distance is related to particular point vicinity is relatively small. Based on presented results one can be convinced that the true measurement is somewhere else with even higher degree of credibility. Mass attributed to uncertainty is the highest once Yager normalization is applied. Meaning of constant C remains unresolved. Therefore one can doubt on practicality of obtained results. It should be stressed that way of reasoning started from basic interval valued uncertainty scheme, then measurement aleatory doubtfulness was revoked in order to show coincidence in practical aspects of the approach. Considering the truth of this proposition, this feature seems impractical both in metrology and in nautical science. To make it practical question "does the true measurement involves given neighbourhood or any other one" should be substituted with "what are supports on representing the true measurement for each point out of a given set". Instead of discussing any other locations one should ask for particular ones those being elements of a hypothesis space. That was the main idea that lied behind proposition of universal model enabling calculating amount of cumulated support included in each piece of evidence for particular hypothesis (Filipowicz, 2010, 2011). The proposal gives a tool for alternative uncertainty modelling and further on processing.

4 ALTERNATIVE WAY OF NAUTICAL EVIDENCE ENCODING

Belief function can be perceived as pairs of values defined by formula 2. There is certain membership function μ and support for it embedded within given piece of evidence e_i, which is subject of subdivision to e_{ij} in discussed approach. In metrology and nautical science measurement is perceived as random variable instance, which is potentially systematically distorted. Last issues appears challenging for analysts once traditional approach is exploited.

Belief assignments embrace relations between evidence and hypothesis frames. It tells how a piece of evidence with its reference sets $\{e_{ij}\}$ supports each of the hypothesis points $\{x_k\}$ given their location μ_{ij} within each of the sets.

$$m(e_i) = \{(\mu_{i1}(x_k), m(\mu_{i1})), \quad ... \quad , (\mu_{in}(x_k), m(\mu_{in}))\}$$
$$m(\mu_{ij}) = f(e_{ij} \rightarrow \mu_{ij}(x_k)) \qquad (2)$$

where m = belief assignment, m = probability mass of support for x_k embedded within e_{ij}, which is a function of fuzzy location μ_{ij}, μ_{ij} = membership function related to j-th part of i-th piece of evidence, $x_k = k$-th hypothesis frame item.

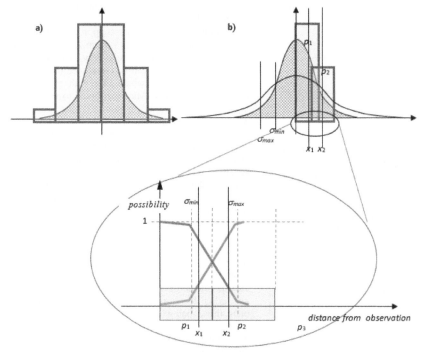

Figure 4. Adjacent confidence intervals with fuzzy limits and two respective poly-line membership functions

Given particular observation as well as knowledge regarding distribution one seeks support for the true measurement being located somewhere in the vicinity. Taking into account particular distribution and confidence intervals (C constant in above mentioned ranges) concept belief function can be readily upgraded. Assuming normal density distribution example set confidence intervals can be proposed. An option is taking cumulated probabilities calculated for intervals width of standard deviation.

As result one obtains stepwise function (see figure 4a)). Thus any point binary located $\mu_{ij} \in \{0, 1\}$ within each of the strips receives support for being the true measurement credibility equal to probability attributed to the range. The idea could be sufficient provided following are observed:

1 Histograms are exploited to define random deflections
2 Results accuracy confined to the fraction of the measurements standard deviation is satisfactory
3 Multiple solutions are meaningless
4 There are no discrepancies in observations standard deviation estimation
5 Exponential computational complexity is not of primary importance

Measures calculated on observations are called empirical. Empirical probability is used in terrestrial navigation. Thanks to various experiments one can estimate probabilities from series of test observations. Probability is approximated as the ratio of the number of those results that fall into a selected bin to the total number of observations provided it sufficient count. The empirical data enables estimating statistical probability. Histograms are widely used as graphical representation of empirical probabilities. It shows a diagram of the experimental data distribution. A histogram consists of bins that are adjacent rectangles, erected over non-overlapping intervals. The histogram is usually normalized and displays relative frequencies. It then shows the proportion of cases that fall into each of several bins, with the total equals to one. The bins are usually chosen to be of the same size for given range of available data. There is no universal rule to specify number of bins.

Empirical type of random variables distributions are usually converted to Gaussian ones although sometimes conversions are not theoretically justified. Thus empirical distributions inclusion into evidence representation seems natural and necessary. In this case, confidence intervals are substituted by histogram bins and final probabilities are replaced by relative frequencies of observations falling within the bin. Since available histograms differ, calculated frequencies are rather range valued than single figures. Empirical distributions of observational distortions with imprecise bin width and the same relative frequencies are analysed by (Filipowicz, 2011).

Figure 4a) presents six bins that are confidence intervals with crisp limits. Result stepwise function may cause multiple solutions of evidence combination. At contrary continuous and injective function guarantee uniqueness of the final result that is expected in nautical applications. Referring to point 4 of the above list it disables modelling of fuzzy valued accuracy estimation applicable in practice (for example [±X;_±Y] is popular notation in navigation).

Computational complexity of the combination algorithm proved to be exponential. A number of items, fragments of evidence, within considered belief assignments appears as a base of power complexity function. The function returns a maximum of steps required for association. In the worst case number of steps is limited by an exponential function $O(k^n)$, where k is a count of single belief assignments and n stands for the number of observations. Reduction of involved numbers improves computational efficiency. The reduction can be made through aggregation of evidence fragments or expansion of confidence interval.

4.1 Fuzziness and Nautical Evidence

As it is seen at figure 2 standard deviations of probability density dispersions differ. Thus instead of discussing confidence intervals with crisp limits one should try to introduce their fuzzy borders. As it is widely assumed standard deviations are interval valued usually separately defined for core and support of a fuzzy set. Two adjacent confidence intervals with fuzzy limits and two respective poly-line membership functions are shown in drawing 4b). Exploded insertion contains two membership functions showing degrees of inclusion into adjacent strips, that refers to two fuzzy probability sets: $\overline{0.673}$ and $\overline{0.278}$. Respective values are assigned to confined, single sided, with respect to measurement, possible location of the true measurement. In view of another piece of evidence the assumption is practical one. In order to avoid multiple solution example membership function descends starting from zero. At first inclination is small the diagram reaches 0.95 in _min. Then the function diagram is steeper it passes 0.5 in the middle of [σ_{min}, σ_{max}] range and reaches 0.05 at the rightmost border. Second presented inclusion function is complementary to the first one. Membership functions are those defining evidence and hypothesis frames relationship. It is fundamental that data and related knowledge at hand should support various correlated hypothesis. Thus one is expected to include values returned by discussed functions into belief assignments that depict mentioned

relationship. Given randomly and systematically distorted measurements and their aleatory uncertainty one seeks their support for particular location of the true measurement or an observer fixed position.

In metrology and nautical science it is required that evidence embedded support varies on real scale and possibly can be expressed by injective function in order to avoid multiple solutions. An injectivity is a feature of a mathematical function. It means that each element of its range arises from a single element of the domain. Due to common symmetry of the distribution function injectivity is required with respect to its single side. To calculate grades of belonging to j-th defined fuzzy sets one can use sigmoidal functions. Appropriate formulas of two complementary functions are specified by expression 3. For given fuzzy probability set descending function $\mu_j(x)$ specifies grade of x belonging to the set. At the same time function $\mu'_j(x)$ returns the same point membership within adjacent set.

$$\mu_j(x) = \frac{1}{1 + e^{a_j \cdot (x - j \cdot \sigma_m)}}$$

$$\mu'_j(x) = 1 - \frac{1}{1 + e^{a_j \cdot (x - j \cdot \sigma_m)}}$$

(3)

where j = number of considered fuzzy set, range established in the vicinity of the measurement isoline, σ_m = mean for standard deviations range [σ_{min}, σ_{max}], a_j = factor deciding on steepness of the membership function (to be calculated with algorithm I).

It should be stressed that equation 4 holds in order to obtain appropriate membership function. It specifies that taking into consideration the combined belonging fuzzy grade, one has to take into account the adjacent sets' related to the functions.

$$\mu_{j+1}(x) = \min(\mu'_j(x), \mu_{j+1}(x))$$

(4)

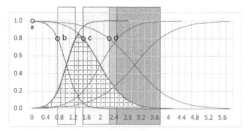

Figure 5. Three regions established at the right hand side of a measurement, their limits and sigmoidal membership functions

Ranges established at the right hand side of a measurement, assumed at the coordinate's intersection, are shown in figure 5. Diagrams were obtained for $\sigma_{max} - \sigma_{min} = 0.5 \cdot \sigma_m$, Intervals referring to fuzzy probability sets: $\overline{0.673}$, $\overline{0.278}$ and $\overline{0.047}$ related to single standard deviation width of confidence intervals, with their expanding limits are depicted. Respective steepness sigmoidal function parameter a_j was calculated with algorithm I. Crossed shape refers to combined membership function defining second fuzzy set, thus the set is subnormal one.

Algorithm I calculates membership function steepness factor. It enables to obtain coefficient aj guaranteeing smallest descending rate of the sigmoidal function regarding given standard deviations range (variable HR in algorithm I) and leftmost limit of an overlapping region (variable *bordervalue*), while crossing with y - axis is at close to one (function *inzero* controls this value). The algorithm calculates lowest location of points (see figure 5) *b*, *c* or *d* starting from their initial values equal to 0.999. They descends while maintaining fixed position of point *a*. The algorithm is valid for wide range of $\sigma_{min} < \sigma_m$. In case of $\_\sigma_{min} = \sigma_m$ membership become binary one, given point belong to the set or not. Partial inclusion is not allowed and reference to fuzziness is not valid. It should be stressed that equation 4 holds in order to obtain appropriate membership function. It specifies that combined belonging grade involves adjacent sets' related functions.

Algorithm I

1 calculate mean of standard deviations range, name it HR
2 assign: aj:=20; bordervalue:=0.999; step:=0.02
3 bordervalue:=0.999
4 repeat
5 aj:=-ln(1/aj - 1)/(1 - HR)
6 bordervalue:= bordervalue - step
7 until inzero(aj) $\leq$ 0.999

Combined membership function may define subnormal fuzzy set (for example see crossed shape in figure 4). In such case none from hypothesis points fully belong to defined probability set.

Belief assignment engaging single nautical observation and two points: x_1, x_2 as shown in figure 4b) is presented in table 3. The table contains columns designated to each of the hypothesis frame points and mass of evidence related to introduced confidence intervals. Rows count reflects number of specified ranges with the last entry devoted to uncertainty. From the table one can see that mentioned points are partially located within introduced probability fuzzy sets. Membership grades were calculated using sigmoidal functions, their parameters were obtained with algorithm I. Note that probabilities assigned to selected confidence intervals were reduced in order to accommodate uncertainty mass, which should be

treated as subjective assessment of the particular observation. From the table it is seen that point number one is located within the first and second range with degrees respectively 0.791 and 0.211. Final probability that this location represents the true measurement is 0.561, uncertainty is estimated at 0.05. Location of the true observation somewhere else is assigned probability of 0.389. This way of reasoning is formalized by formula 5. Using it one can evaluate probability for any point on its representation of the true measurement.

$$p(x_k) = \sum_{i=1}^{n} c_i \cdot \mu_{li}(x_k) \qquad (5)$$

where c_i = cumulated probability for established i-th confidence interval, $c_i \in C = [0.673, 0.278, 0.047]$, μ_{li} = membership of x_k within i-th fuzzy set with respect to l-th observation.

Injective function diagrams shown at figure 6 presents respective probabilities for single sided locations referring to taken measurement. The diagrams were obtained with expression (5) for two degrees of discrepancy in standard deviation estimation. It was assumed that for lower diagram: $\sigma_{max} - \sigma_{min} = 0.3 \cdot \sigma_m$ and for the upper one $\sigma_{max} - \sigma_{min} = \sigma_m$. In latest case range of more than zero probability extends compare to initial density function, which is very close to zero for relative distance equal to 3. General rule appears quite obvious "higher the discrepancy in evaluation of initial distribution characteristics wider the range of probability diagram". It appears as paradox less reliable observation is assigned higher probability. To reduce this phenomenon one has to exploit subjective assessment of the measurement (see table 3).

It is worth to stress that thanks to proposed transformation involving fuzzy sets discrepancies in estimating probability density distributions result in crisp amount of likelihood regarding considered statement truth. At the same time proposed approach delivers space for subjective assessment of the observation.

Table 3. Basic probability assignment for extended uncertainty model.

	$\mu_{lj}(x_1)$	$\mu_{lj}(x_2)$	m(..)
FS 1	0.791	0.243	$0.673 \cdot 0.95 = 0.639$
FS 2	0.211	0.739	$0.278 \cdot 0.95 = 0.264$
FS 3	0	0.05	$0.049 \cdot 0.95 = 0.047$
Un	1	1	0.05

$\mu_{lj}(x_i)$ - point x_i membership grades within fuzzy sets (FS) related to observation number one
FS i - i-th fuzzy set
Un - epistemological uncertainty, subjective assessment of the observation number one

It should be noted that when drawing final conclusions all pieces of evidence at hand are to be considered for a posteriori analyses. Informative context of each nautical observation is to be contextually perceived. For example scheme presented at figure 3 indicates that seeking for true measurement can be confined to range delimited by two depicted measurements. Handful of observations modify/increase values of the considered probabilities.

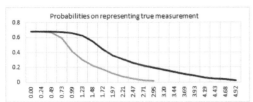

Figure 6. Single sided probabilities on representing true measurement for two cases regarding discrepancies in density probability estimations.

5 SUMMARY

Dealing with uncertain and imprecise evidence is a challenge in nautical science and practice. Formal descriptions of problems encountered in navigation involve models that accept imprecise, erroneous and therefore uncertain values. The concept should be followed regarding many practical problems, mention true measurement location, position fixing and systematic errors identification. Approach involving exploitation of belief assignments in nautical science was presented. At first basic uncertainty model was introduced. Then aleatory doubtfulness encountered in metrology is presented. Knowledge related to measurements random deflections fits into the model and therefore can be included into processing scheme. Uncertainty model is basically intended for fuzzy environment. Fuzziness can be meant in different ways although fuzzy probability sets are always involved. In metrology fuzzy membership functions are exploited to decide on dilemma regarding location of the true measurement given observations at hand.

Fuzzy sets may be associated with cumulated probability calculated for specified intervals once the bell function is considered. They can be related to bins when empirical distribution is involved. Fuzzy sets are defined by membership functions. Polyline and sigmoidal functions are used very often. Algorithm to calculate parameters of sigmoidal formulas was presented. Aleatory doubtfulness arises once one make nautical observations. They are systematically and randomly distorted. A seafarer knows much about it, his knowledge on diversity of probability density distributions is rather rich. To include the knowledge into computation scheme one has to invoke fuzzy concept. Thanks to the concept one can arrive at crisp probability enabling assessment of the truth

regarding statement: "is the true measurement represented by particular value". Upgraded model reduces aleatory uncertainty to a single crisp value. In proposed approach one can exploit Mathematical Theory of Evidence and use belief assignments in order to enrich upgraded model with epistemological uncertainty usually expressed by subjective assessment of each observation. Belief assignments are further subject to association in order to obtain a fixed position and its deep a posteriori credibility analysis.

Presented at figure 6 diagrams might be helpful once a sounder is implemented. The sounder is likely to generate probability that the ship position is at specified location given updating observations and nautical knowledge. Such logical device would operates on simple belief assignments rather than one presented in table 3. Simple structure engages two elements: a proposition and its contradiction.

REFFERENCES

Grove, A.T. 1980. Geomorphic evolution of the Sahara and the Nile. In M.A.J. Williams & H. Faure (eds), *The Sahara and the Nile*: 21-35. Rotterdam: Balkema.

Johnson, H.L. 1965. Artistic development in autistic children. *Child Development* 65(1): 13-16.

Ayoun, A., Smets, P. 2001. Data Association in Multi-Target Detection Using the Transferable Belief Model. *International Journal of Intelligent Systems* 16: 1167-1182.

Dempster, A. P. 1968. A generalization of Bayesian inference. *Journal of the Royal Statistical Society* B30, 205-247.

Denoeux, T. 2000. Modelling Vague Beliefs using Fuzzy Valued Belief Structures. *Fuzzy Sets and Systems* 116, 167-199.

Filipowicz, W. 2010. Belief Structures in Position Fixing. In Mikulski J. (ed.) *Transport Systems Telematics, Communications in Computer and Information Science* 104: 434-446. Berlin, Heidelberg: Springer-Verlag.

Filipowicz, W. 2011. Fuzzy Evidence in Terrestrial Navigation. In Weintrit A. (ed.) *Navigational Systems and Simulators, Marine Navigation and Safety of Sea Transportation*: CRC Press/Balkema, London, UK, 65-73.

Filipowicz, W. 2011. Evidence Representation and Reasoning in Selected Applications. In Jędrzejowicz P., Ngoc Thanh Nguyen, Kiem Hoang (eds), *Lecture Notes in Artificial Intelligence*: 251-260. Berlin Heidelberg: Springer Verlag.

Filipowicz, W.. 2014. Fuzzy evidence reasoning and navigational position fixing. In: Tweedale JW, Jane LC (eds.), *Recent advances in knowledge-based paradigms and applications, advances in intelligent systems and computing* 234: 87-102. Berlin Heidelberg: Springer Verlag.

Filipowicz, W. 2010. New Approach towards Position Fixing. *Annual of Navigation* 16. Gdynia: 41-54.

Filipowicz, W. 2015: On nautical observation errors evaluation. *TransNav* 9/4. Gdynia: 545-550.

Jurdziński, M. 2014. Principles of Marine Navigation. Gdynia: Akademia Morska (in Polish).

Lee, E. S., Zhu Q. 1995. Fuzzy and Evidence Reasoning. Heidelberg: Physica-Verlag.

McBurney, P., Parsons, S. 2002. Using Belief Functions to Forecast Demand for Mobile Satellite Services. In Srivastava R. P., Mock T. (eds), *Belief Functions in*

Business Decisions 281-315. Heidelberg: Physica-Verlag, Springer-Verlag Company.

Shafer, G. 1976. A Mathematical Theory of Evidence. Princeton: Princeton University Press.

Yager, R. 1996. On the Normalization of Fuzzy Belief Structures. *International Journal of Approximate Reasoning* 14: 127 – 153.

Yen, J. 1990. Generalizing the Dempster-Shafer Theory to Fuzzy Sets. *IEEE Transactions on Systems, Man and Cybernetics* 20/3: 559-570.

Inland Navigation

Proceedings of 12th International Conference on Marine Navigation and Safety of Sea Transportation, TransNav 2017
21-23 June 2017, Gdynia, Poland

The Analysis of the Possibility of Navigation the Sea-River Ships on the Odra River

W. Galor
Maritime University of Szczecin, Szczecin, Poland

ABSTRACT: The inland water transport is very useful method of cargo carriage. It permits on integration of transport in paneuropean corridors, in connection with sea shipping by sea-river ships. The main advantage of sea-river shipping is unique range of market. Sea –river ships can connect the area inside of land with oversea places without of indirect trans- shipment. It influents on decreasing of carriage cost and reduces the risk of damage, that caused by addition trans-shipment. The paper presents the analysis of conditions, which influenced on means on sea-river exploitation on lower part of Odra River.

1 INTRODUCTION

Sea-river ships can connect areas in the inland with overseas destinations without a need for intermediate handling. This affects lowering of transport costs and reduces the risk of damage caused by additional handling. Depending on the area of navigation, there are numerous physical and weather conditions limiting the process and safety of navigation. Parameters limiting the size of ships are: hydro-technical structures specifications, the availability of water bodies and their parameters such as bandwidth, the ability to maintain a constant water level, the hydro-meteorological conditions, the most common wind, waves and currents, fog, and freezing. The study of these issues is very important, especially in this period of intensive development of new transport links, based on inland waterways and the sea, especially between countries united in the European Union. New initiatives and modernization of waterways in Europe affects, and even makes it necessary to develop routes navigable in Poland, especially the connection to the network of European waters. For several years, distinct programs concerning renewal, modernization and development of the Odra river were developed. Most programs, however, were not carried out. Therefore, the questions debated in this work have not yet been fully answered. During the rapid development of countries - especially those belonging to the European Union - road, rail, air, sea and river transport plays an important role in national economy. Sea-river shipping involves sending ships to ports lying inland, several hundred kilometres from the sea. These can be different types of units (freight, passenger, recreational). In many European countries, inland sea-river shipping is an important element of the transport system. It can be noticed through an increasing development of new transport corridors to enable efficient transport of goods and provision of related services. The main advantage of sea-river shipping is that ships may operate inland and carry cargo to ports located in other countries without having to reload several times. This method significantly reduces the risk of damage to the cargo and transport costs. This field of transport is by far the cheapest transport subsystem, therefore its development should strive in Poland. Statistically, it can be observed that sea-river transport is approx. 300% cheaper than railways transport and approx. 800% cheaper than road transport. EU countries have continued to perform investments which are aimed at increasing the income of the discussed transport field and to improve security units involved in inland navigation. Shipping channels in Western Europe account for about 40% of the total length of waterways and a further increase in numbers can be observed. Poland, as a member of the European Union is obliged to conduct development of navigation on internal waters. It is also important that the Odra river comes in two channels: Canal Odra-Havel and Canal Odra-Spree. These channels connect the Odra river to a system of inland waterways, which allow for an access to the waterways of France, Luxembourg, Belgium, the Netherlands and Germany to the west; Switzerland

and Czech Republic to the south. On the north Odra is connected to the Baltic Sea. On the east, through Warta and Bydgoszcz Canal, Odra connects to the waterways of Vistula, while Vistula and the Bug river provide a connection to the waterways of Russia, Ukraine and Belarus. In the north, through the Vistula Lagoon, Kaliningrad port and waterways provide a connection to Lithuania and Belarus. Particularly noteworthy is the possibility of sailing ships on inland waters. In many European Union countries (the Netherlands, France, UK. Britain, Germany, Belgium) as well as in Russia, ships sail on inland waters up to several hundred kilometres inland. Sea-river shipping in Poland is possible primarily on the Odra river (the lower section), the lower section of Vistula river, whereas the Warta river and the Bydgoszcz Canal connect Odra and Vistula. Sea-river navigation is highly advanced on waters of the ports of Świnoujście, Police and Szczecin. Steps have already been taken to introduce ships on the lower section of Odra, south of the port of Szczecin (the port of Schwedt in Germany). In addition, undertaken studies have concluded that navigation on inland waterways is significantly different from navigation conducted on sea areas, coastal areas or ports.(Weintrit, 2010).

2 THE CHARACTERISTICS OF SEA-RIVER SHIPS

During the last few years there has been an intensifying trend of development of ships for different areas of navigation, ie. sea-river ships, which meet a number of requirements and parameters for navigating shallow waters. The dimensions of such ships have to be adapted to the conditions of river navigation. These are especially drought as it should be 2.50 - 5.50 m (average 3 - 4.5 m). Breadth of the ship is 11.4 - 11.5 m in accordance with V European class waterway (with a tendency to reach 12 m due to better conditions for loading containers). The total length of the unit is 70-90 m (typically 80 m), but can exceed 90 m. Deadweight is usually in the range of 2000-3000 t, sometimes up to 4000 t. The speed of sea-river ships is about 1-2 knots less than coastal ships. There are three types of inland shipping vessels (Żylicz, 1979).

1 Inland ships - permitted to continue to travel on specified sheltered marine waters in restricted weather conditions. This group is the most popular.

2 River-sea ships - designed to travel on inland routes and enclosed seas or open seas in restricted weather conditions. Ships of this type are the most popular in the basins of Danube and Rhine in Russia, in the UK and in third world countries, mainly in the Far East. Typical sea-river ships are marked by "restricted service" sign. This allows for navigation on open seas up to approx. 15 miles from the coast and under conditions up to approx. 6 ° Beaufort scale.

3 Sea-river ships - designed for sea travel without any weather restrictions and for inland travel on waterways of specified characteristics. They are actually either seagoing ships, limited to a specific climatic zone, or units of big cabotage (ie. limited to sailing on seas of north-western Europe, on the Mediterranean Sea etc.).

River-sea ships and sea-river ships should meet requirements for both inland and marine travel. The requirements for endurance, freeboard, construction security, stability, anchoring and emergency equipment, with minor exceptions, are the same as for sea ships. Ships meant for long voyages on inland waterways and short sea voyages should represent more features of inland vessels, ie .:

– high load capacity with small draft,
– sufficient seakeeping, speed and endurance,
– sufficient power to overcome decline and current of rivers in critical places and periods,
– good manoeuvrability in all situations,
– small air draft.

Table 1. presents the comparison of different ships in mixed, sea, inland and coastal shipping (proportions characteristic to transition zone ships), (Galor, 2015).

Table 1. The comparison the types of ships in sea-river shipping.

Ships types dead-weight [tons]	River ships	River-sea ships	Small Sea-river ships	Small sea ships
	800	700 - 2000	1500 - 4500	up to 1500
L/H	17 – 25	15 – 17	13,5 – 14,5	12,5 – 13,5
T/B	0,27 – 0,33	0,27 – 0,32	0,38 – 0,40	0,43 – 0,47
L/T	17 – 27	17 – 25	15,5 – 16	14,5 – 15
L*B*H	0,49 – 0,62	0,32 – 0,50	0,50 – 0,55	0,52 – 0,56

Where: L, B, T, H- length, breadth, drought and high of ship.

Due to the International Convention on Tonnage Measurement of Ships, they have a gross tonnage of 149, 199, 299, 499, 1599 RT (GRT). Such vessels achieve maximum tax reliefs for small vessels as of international conventions. Most meet the 499 RT value, which makes them the largest ships not subject to the SOLAS Convention. Ships of this size meet the requirements of sea-river navigation in Western Europe.

3 THE ODRA RIVER WATERWAY

The Odra River Waterway is located in the Transport Corridor of Odra Valley, which connects Scandinavia and Italy, as well as Czech Republic and Austria, through Polish territory. In Poland, this waterway is of utmost importance because it provides an opportunity of both navigation and cargo shipping. The waterway has a total length is 693.1 km. It consists of Gliwice Canal (41 km), Kędzierzyński Canal (6.1 km), the canalized section of Odra from Koźle to Brzeg Dolny (187 km), free flowing Odra from Brzeg Dolny to Szczecin including the Western Odra (Odra Zachodnia) (459 km). Like other Polish waterways, Odra River Waterway is very diverse in parameters and technical solutions. Better conditions for navigation on Odra exist on canalized parts (between Kędzierzyn Koźle and Brzeg Dolny) and the lower Odra River (from the mouth of Nysa river and especially Warta river, to Szczecin city), where shipping is possible for an average of nine months during the year (Figure 1).

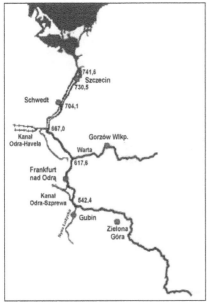

Figure 1. The Odra River waterway

In contrast, the most difficult navigation conditions exist between Brzeg Dolny city and the mouth of Nysa river. Periodically on this part of the waterway the depth drops to as low as 0.6-0.7 m. Meanwhile, the minimum depth condition allowing for cost-effective shipping in Poland is 1.3 m. Therefore, due to the current state of infrastructure, individual sections of the waterway can be operated by units of diverse specifications, as shown in table 2 (Galor & Galor, 2009). The most commonly utilized part of Odra River is the lower section (boundary section 542.4 - 704.1 km). Moreover, Eastern Odra and Regalica river are used for inland shipping. This part is the most commonly used section, serving as a connection to the Germany (through the Odra-Spree Canal and Odra-Havel Canal). In addition, this waterway has two connections which provide an access to Port of Szczecin and Port of Świnoujście.

4 THE NAVIGATION RESTRICTIONS ON THE LOWER PART OF ODRA RIVER

Safe operation of a unit requires avoidance of navigational obstacles and avoidance of collisions with other units. In conditions of limited visibility (fog, night) these actions are supported by the use of radar. This means to avoid the accident as unwonted event which can bring the losses.(Galor, 2010). These may include the following:
– loss of human life or health,
– damage or loss of cargo and vessel,
– environmental pollution,
– damage or destruction of hydraulic structures,
– the cost of rescue mission and others.

When navigating waters, a ship unit must meet two conditions of safe navigation:
– maintenance of safe supply of water under the keel,
– maintenance of a safe distance to navigational obstacles

The strongest effects of not adhering to the abovementioned conditions can be observed when a unit collides with a navigational obstacle (Galor, 2006). The obstacles can be either natural or artificial (hydrotechnical structures). The main factors limiting navigation are: the depth and width of the basin and navigational obstacles appearing on different parts of waterways - energetic power lines, hydrotechnical structures such as dolphin moorings and other obstacles mainly of natural origin, such as islands, shoals, boulders, stones, logs of trees, remains of hydrotechnical structures.

Bridges are often a significant navigation-limiting factor, as they provide a restriction to breadth and air draught of ships. Moreover, during limited visibility periods (especially at night) bridges are hard to navigate around. These constructions should be properly marked - radar corner are very important in this context.

Figure 2. The road bridge on Odra Zachodnia

323

Table 2. The technical parameters of Odra Border River (Odra gran.) section and Szczecin Water Area (Szczeciński Węzeł Wodny)

Name (km)	Length (km)	Waterway category	Deadweight [T]	Max.ships dimentions L- lenht, B- breadth	Min.height for High water level	Min. breadth.	Speed Restrictions
Odra river (542,4 -617,6)	75,2	II	301 -500	single ship up to: do 82 m L and 11, 45 m B, push and coupled convoy up to do 156 m L and 9,5 m B (upstream) and up to 125 m L and 11,45 m B.(down the river))	3,67 m railway bridge Kostrzyn	25 m bridges of railway and road in Kostrzyn, km 614,9 i 615,1	max - min. 4 km/h
Odra river (617,6 -697,0)	79,4	III	501-1000	single ship up to : do 82 m L and 11, 45 B, push and coupled convoy up to do 156 m L and 9,5 m B (upstream) and up to 125 m L and 11,45 m B.(down the river))	4, 14 m- railway bridge Siekierki km 653,9	50 m	max - min. 4 km/h
Odra river (697,0 -730,5)	33,5	V	1501-3000	single ship up to : do 82 m L and 11, 45 B, push and coupled convoy up to do 156 m L and 9,5 m B	5,27 m- road bridge Gryfino km 718,18	50 m	max - min. 4 km/h
Regalica river (730,5 -741,6)	11,1	V	1501-3000	single ship up to : do 82 m L and 11, 45 B, push and coupled convoy up to do 156 m L and 9,5 m B	3,06 m- railway brige (lift span) 6,30 m) Podjuchy, km 733,5	12,73 m (lift span of railway bridge Podjuchy, km 733,5	max - min. 4 km/h
Odra Zachodnia river (0,0-36,55)	36,55	V	1501-3000	single ship up to : do 82 m L and 11, 45 B, push and coupled convoy up to do 156 m L and 9,5 m B	3,68 m with width on 12,6m road bridge Długi, 35,95 km	11,91 m right span of railway bridge Szczecin, km 35,59	max. 16 km/h min. 4 km/h
Przekop Parnicki (0,0-1,3)	1,3	V	1501-3000	single ship up to : do 82 m L and 11, 45 B, push and coupled convoy up to do 156 m L and 9,5 m B			max. - min. 4 km/h
Przekop Klucz-Ustowo (0,0-2,7)	2,7	V	1001-3000	single ship up to : do 82 m L and 11, 45 B, push and coupled convoy up to do 156 m L and 9,5 m B			max. - min. 4km/h
Dąbie Lake (0,0-9,5)	9,5	V	1001 -1500	single ship up to : do 82 m L and 11, 45 B, push and coupled convoy up to do 156 m L and 9,5 m B			max. - min. 4 km/h

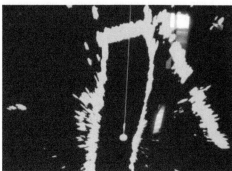

Figure 3. The presentation the bridge on radar

Figures 2 and 3 represent an example of bridge on Odra river with navigational marks (radars corner) and presentation on screen of radar (Adamczewski, 2010).

Table 3 below represents the list of bridges located on the lower section of Odra river, (Galor & Galor, 2005)

Table 3. The bridges on lower part of Odra river

Odra river, (border section km 542,4-704,1), Odra Wschodnia, Regalica

1 Road bridge in Świecko, km 580,0 Odra river, (management by German administration)
2 Railway bridge in Świecko, km 580,7 (management by German administration)
3 Road bridge in Słubice, km 584,0 (management by German administration)
4 Road bridge in Kostrzyń, km 614,9 (management by German)
5 Railway bridge in Kostrzyń, km 615,1(management by German administration)
6 Railway bridge in Siekierki, km 653,9 (management by Germany administration)
7 Road bridge in Osinów Dolny, km 662,3 (management by Germany administration)
8 Road bridge in Krajnik Dolny, km 690,5(management by Germany administration)
9 Road bridge in Gryfino, km 718,18 Odra Wschodnia river, (management by Polish administration)
10 Road bridge in Radziszewo, km 727,95 rz. Odra Wschodnia river, (management by Polish administration)
11 Railway bridge in Szczecin, km 733,7 . Regalica river , (management by Polish administration)
12 Railway- road bridge in Szczecin, km 734,6. (management by Polish administration)
13 Road bridge in Szczecin, km 733,35 (management by Polish administration)
14 Road bridge in Szczecin, km 737,6 (management by Polish administration)

All bridges are equipped with radar marks

Crosses of waterways are additional aspects which increase the difficulty navigation, because they introduce a higher risk of ship collisions. Table 4 represents the list of cross waterways on lower section of Odra river.

Table 4. The cross waterways on lower section of Odra river

Odra river, (border section km 542,4-704,1), Odra Wschodnia, Regalica

1 Eissenhuttenstadt, km 553,4 Odra river – the cross with Odra-Szprewa.Canal
2 Kostrzyn, km 617,6- the cross with Warta river.
3 Hohensaaten, km 667,2- the cross with Odra-Havela Canal
4 Bielinek, km 677,2 - the entrance to port in gravel mining.
5 Ognica, km 697, - the cross with Schwedt Canal
6 Szczecin, km 730,5 Odra Wschodnia river – the cross with Klucz-Ustowo canal
8 Szczecin, km 733,5. Regalica river – the entrance to Odyniec canal.
9 Szczecin, km 737,8 –the entrance to Dąbska Struga canal
10 Szczecin, km 739,6 - the cross with Parnica river (The border of inner sea waters) with Dąbski Nurt canal
11 Szczecin, km 741,6 . Regalica river - mouth to Dąbie lake
12 km 3,0 Odra Zachodnia river – access of Hohensaaten-Friedrichstahler – Wasserstrasse Canal
13 Szczecin, km 29,8 Odra zachodnia river- the cross with Klucz-Ustowo canal
14 Szczecin, km 33,5- the cross with Parnicki Canal
15 Szczecin, km 36,55 Odra Zachodnia river- the end of inland waterway (the border with inner sea watersa)

Energetic power lines crossing the waterways on lower part of Odra river:

– Widuchowa, km 705 Odra Wschodnia river
– Gryfino, 718,0- Odra Wschodnia river
– Radziszewo, 728,0 km- Odra Wschodnia river
– Area , 3,0 km -Odra Zachodnia river
– Szczecin-Pomorzany, km 31,5- Odra Zachodnia river.

5 SUMMARY

The European Union aims to shift transport streams from roads towards inland waterways. Despite minimal use of this type of transport in Poland so far, there is a need for renewal and development of this method. The lower section of Odra River, especially from the mouth of Warta River, is the prime area representing an opportunity for dynamic development. An important part of the development process should be an introduction of sea-river ships to the area, as this will allow to eliminate the need for transshipment from sea ships to inland units . It can be concluded that during the sailing season resulting from icing and periodic fluctuations in water levels, the main limiting factor for navigation is the specification of hydrotechnical structures. There are several road and railway bridges on the lower section of Odra.. The railway bridge in Podjuchy (Regalica river) limits the maximum width of ships to a value of 11.45 m. A major limitation is the minimum height of bridges, which is dependent on the state of waters and can be further reduced by lowered wheelhouses of ships.

ACKNOWLEDGMENTS

This research outcome has been achieved under the research project No. 1/S/IIRM/16 financed from a subsidy of the Ministry of Science and Higher Education for statutory activities of Maritime University of Szczecin.

REFERENCES

Adamczewski B., (2010):Ocena ograniczeń zastosowania radaru w żegludze śródlądowej. *Praca dyplomowa inżynierska*. Akademia Morska w Szczecinie
Galor A., Galor W., (2005): Nawigacja radarowa w Dolnym odcinku rzeki Odry. *Materiały Konferencji Naukowej „Inland shipping 2005"* .Szczecin..
Galor w., (2006). Wybrane zagadnienia określania pozycji jednostek w żegludze śródlądowej. *Zeszyty Naukowe Wyższej Szkoły Morskiej w Szczecinie, nr 3/2006..* .
Galor W., Galor A., (2009) Nawigacyjne ograniczenia eksploatacji portów w żegludze morsko-rzecznej, *Logistyka nr 3/09*.Poznań
Galor W., (2010), The model of risk determination in sea- river navigation. *Journal of KONBIN 2,3 (14,15)*.Warszawa.
Galor W. (2015): Zegluga morsko-rzeczna na polskich drogach wodnych. *Logistyka nr 4/15*, Poznań.
Weintrit A. (2010), Nawigacyjno-hydrograficzne aspekty żeglugi morsko-rzecznej w Polsce, (red). Akademia Morska w Gdyni, 2010.
Żylicz A., (1979) Statki śródlądowe. *Wydawnictwo Morskie*, Gdańsk

Proceedings of 12th International Conference on Marine Navigation and Safety of Sea Transportation, TransNav 2017
21-23 June 2017, Gdynia, Poland

Risk Assessment in Inland Navigation

E. Skupień & A. Tubis
Wrocław University of Science and Technology, Wrocław, Poland

ABSTRACT: On the study basis the authors found the presence of a gap in the research area of a risk management of transportation systems, especially: inland navigation. There is relatively small amount of publications on risk analysis for the transportation systems. Therefore, it seems important to carry out interdisciplinary research related to the identification of risk factors present in the systems of inland navigation.

The paper presents inland navigation risk analysis, conducted using the FMEA method, taking into account technical, economic and social aspects. The aim of the article is to present a procedure for the assessment of risk in inland waterways transport, and carry out risk analysis for transport companies.

1 INTRODUCTION

Nowadays, the importance of proper identification of risk associated with the business plays an increasingly important role. The need for control over the common risk has become an indispensable element in achieving goals, which is committed to the economic operator. The company, which is an organization focused on earnings is exposed to the effects occurring adverse events. Their occurrence may indeed have a significant impact on the degree of realization of the set goals. For this reason, many authors, including among others Aven & Zio (Aven&Zio 2014), develop risk analysis frameworks, focusing on issues such as how to understand and describe risk, and how to use risk analysis in decision making.

The process of transportation, due to its specifics is particularly vulnerable to numerous risks involved in its implementation. For this reason, the authors of many publications point out that extremely important is that companies involved in the transport identified as the greatest number of threats (both random and non-random), and also got to know the place in the transport chain the most vulnerable to the risk (Szczepański 2011). Only in this way in fact it is possible to attempt to reduce the impact or eliminate the factors generating the risk of occurring interference.

The aim of the article is to present a procedure for the assessment of risk in inland waterways transport, and carry out risk analysis for transport companies serving the freight on the Oder River. Assessing the risk would be the first step to prevent it. Therefore, in the first place, the most important definitions regarding the discussed research issues were presented. On the basis of literature review authors made a proposal for a procedure for risk assessment, taking into account the specificity of inland waterways transport. At the end it was presented the analysis of the risks for the handling of a freight on the Oder River.

2 LITERATURE REVIEW

There are many different views on what risk is and how to define it (Aven 2012; Hampel 2006), how to measure and describe it (Aven 2010, Kaplan 1997), and how to use risk analysis in decision making (Apostolakis 2004, Aven 2009). Overview of the basic definition of the concept of risk can be found, among others, in (Aven 2016). According to this review, the proposed definitions generally refer level of uncertainty to the probability of an adverse event and its consequences. Therefore, in the research on risk assessment in inland waterways transport described in this paper the authors also adopted this point of view.

The way one understands and describes risk strongly influences the way risk is analysed and hence it may have serious implications for risk management and decision-making (Aven 2016). The authors in their study relied on the guidelines of ISO 31000. According to it (PN-ISO 2010) the risk assessment is defined as a holistic process that involves three stages of the procedure: (1) risk identification, (2) risk analysis, and (3) evaluation of risk. These stages are included in the proposed model of risk assessment for the transport process. Risk analysis may use various quantitative and qualitative techniques, which are also described in the above standard. In the proposed model of risk assessment the FMEA technique was used. It is one of the risk analysis techniques recommended by international standards (Wang et al. 2012).

In the spotlight of the research conducted by the authors is primarily operational risk. It decides if the internal organizational processes are sufficiently effective, including immune to interference, that the organization is able to pursue their economic goals (Zawiła-Niedźwiecki 2012). Therefore since the early 2000s, there has been an increased focus on what has been defined as operational risk (Smallman 2000, King 2001, Ward 2001). Such risks relate to negative deviations of performance due to how the company is operated, rather than the way it finances its business (King 2001, Jorion 2006). It has been argued that there is a great need for improvement in the quality (as regards tools and formal processes to manage operational risk) and scope (such as identification of what risks to focus on) of Operational Risk Management. Companies frequently deal with operational risk issues as they occur, and often following a crisis or catastrophic event (King 2001). ORM is particularly important also for the organization involved in inland waterways transport processes.

Conducted by the authors analysis of publications in the EBSCO database from the years 2006-2016 dedicated to the risk management, indicates that for a water transportation, an extensive research are exclusively carried out in the field of maritime transport [including: (Bubbico et al. 2009, Brown et al. 2016, Yuebo & Xuefen 2014, Langard et al. 2015). Special mention in this case deserves an article (Goerlandt & Montewka 2015), in which the authors presented a detailed review of the literature devoted to the analysis of risk for maritime transport. Overview publications of the period 2011-2014 allowed the authors to define current problems undertaken in research on risk in maritime transport. These among the others are:
- Determine the ship collision probability and frequency in a sea area (Goerlandt & Kujala 2011, Jeong et al. 2012, Rasmussen et al. 2012, Suman et al. 2012, Weng et al. 2012).
- Determine the risk of oil spill and hazardous substances in a sea area (Montewka et al. 2011, Goerlandt et al. 2011).

- Quantify effect of risk reduction measures on accident risk in a waterway area (van Dorp & Merrick 2011).

However the issue does not correspond the problems of decision-making companies involved in the organization of inland waterways transport. For this reason it is necessary to define the area of analysis for risk assessment in inland waterways transport, taking into account the holistic approach proposed in ISO 31000.

3 INLAND NAVIGATION IN POLAND

Poland has favorable hydrological conditions for the development of inland waterway transport. The two main rivers Vistula and Oder link major economic areas of the country with seaports. These connections are not fully used. Vistula river is not navigable for almost its entire length. With a lack of possibility of transportation over the entire length of the waterway of the Vistula River, the waterway is not involved in handling of the Port of Gdańsk and the Northern Port. Inland navigation in Poland, for many decades, focused on the Oder.

The Oder has a dominant role of the inland navigation in Poland. This is due to the fact that it links the Upper Silesia with a Port of Szczecin and Swinoujscie. The second factor of that domination is the fact that the Oder is connected to the Western Europe waterways. In the structure of the country's transport, inland waterway transport has a negligible share. This share in recent years does not exceed 0.2%.

The transportation on the Oder River is dominated by bulk cargo such as coal, ore, aggregates and oversized constructions. On the Polish waterways the transport of containers, transport in RO-RO system, transport of liquid cargo and transport of dangerous goods is not conducted. This kind of loads are starting to dominate the waterways of Europe. All branches of transportation reduce the supply of bulk cargo. Increases the importance of the transport of containers, transport in the RO-RO system, transport of agricultural products, chemical industry, machinery and equipment. The transportation role of the Oder Waterway in recent years is further marginalized. Currently, transport on the Oder focuses on the Lower Odra, relations Szczecin - Western Europe. The regular transport in relation Gliwice - Szczecin disappeared, as well as on the canalized Oder. This is the result of further degradation of waterways in Poland.

The future development of inland navigation in Poland should focus on the role of the authorities in modeling hydrotechnical conditions but the risk connected to ship owners is also important.

4 PROCESS OF RISK ASSESSMENT IN INLAND WATERWAYS TRANSPORT

Requirements of the waterways of the international importance (class IV and V) meets only 5,9 % of length of waterways in Poland (214km). Other waterways can be classified as regional (class I, II and II) (GUS 2015). The greater part of the inland waterway fleet is decapitalized and requires restoration. Its age far exceeds the standard period of use and further exploitation is possible only thanks to the constant modernization. According to data from the Central Statistical Office (GUS 2015) majority of used pusher tugs (73%), almost half of the pushed barges (48.7%) and all self-propelled barge were produced in the period 1949-1979. The products carried in inland waterways transport on Oder Waterway are mostly coal, aggregates, metal ores and oversized goods.

Due to the fact, to collisions in inland navigation rarely ends with serious damages of a ship or health of people, they are almost never reported (to avoid a fine). For this reason authors decided not to take into account statistics devoted to reported inland navigation collisions. Authors defined the risks associated with transportation of cargo by inland waterways based on research on risk assessment in maritime transport and on the basis of cooperation with the Office of Inland Navigation in Wroclaw. Authors analyzed both external and internal factors that could disrupt the correct implementation of the process. Based on the conducted analysis, authors defined 7 basic risk groups:

- Ship collisions;
- Poor navigation conditions;
- Poor condition of infrastructure and loading equipment;
- Poor condition of the fleet;
- Insufficient financial of both: the ship owners and authorities;
- Shortage of qualified HR;
- Lack of interest in this branch of transport.

It should be noted that only 3 identified groups remain under the control of ship owners. Other risks stem from the environment in which transportation is implemented.

The occurrence of the event belong to the one of the groups mentioned above may cause disturbances of varying strength of impact on the realization of the objective. The objective defined for inland waterways transport process is to unproblematically accomplish carriage by planned cost, quality and logistics parameters. The aim of Risk Management is to prevent the possibilities of accruing undesirable events or limit the consequences of their occurrence. Due to the lack of the performed control by ship owners over the majority of the factors generating the risk, the main action taken by them will, however, reduce potential effects of the event.

Therefore, the proper identification of possible adverse events and assessment of accompanying risks is particularly important. The results of the proceedings constitute may then the basis for the planning scenario, allowing a flexible response to the disruption.

The risk assessment is carried out in three stages: (1) hazard identification; (2) an estimate of the likelihood and impact of hazards; (3) the identification of hazards, the level of risk is unacceptable by policy makers. Detailed course of the procedures is shown in Fig. 1.

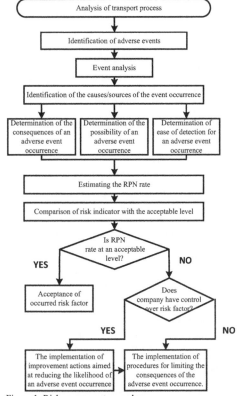

Figure 1. Risk assessment procedure

Due to the large diversity of the state of waterways in Poland, the presented risk assessment is illustrative.

5 ADVERSE EVENTS

On the basis of the 6 groups of risk, adverse events were defined. Table 1 shows those events.

Table 1. Adverse events in inland waterways transport

Ship collisions

SC1 ship damaged due to the collision with other ship or tourists bout;
SC2 ship damaged and people wounded or killed due to the collision with other ship or tourists bout;
SC3 ship damaged due to the collision with infrastructure (bridge, river bank, lock gates, river bed);
SC4 ship damaged and people wounded or killed due to the collision with infrastructure (bridge, river bank, lock gates, river bed);
SC5 ship sinking due to the collision;
SC6 infrastructure damaged due to the collision;
SC7 oil spills due to the collision;

Poor navigation conditions

NC1 closing of a navigation route due to too high depth of a waterway;
NC2 closing of a navigation route due to too low depth of a waterway;
NC3 closing of a navigation route due to ice cover on a waterway;
NC4 no possibility of full utilization of ships capacity due to too low depth of a waterway;
NC5 need of so-called wave support due to too low depth of a waterway;

Poor condition of infrastructure and loading equipment

IC1 longer time of lockage;
IC2 closing of a lock;
IC3 longer time of loading;
IC4 no possibility of loading in a given place due to a lack of loading equipment;
IC5 no possibility of loading in a given place due to a condition of a loading equipment.

Poor condition of the fleet

FC1 longer time of shipping;
FC2 no possibility of exploitation;
FC3 high costs service needed.

Insufficient financial of both: ship owners and authorities

IF1 no possibility of ship exploitation;
IF2 no possibility of ship equipping in modern technology;
IF3 no possibility of infrastructure exploitation.

Shortage of qualified HR

HR1 no possibility of sailing due to a lack of a crew;
HR2 greater probability of adverse events due to lack of experience of a crew members;
HR3 higher costs of crew.

Lack of interest in this branch of transport

LI1 lack of shipping orders;
LI2 more inconvenience of inland waterway transportation development.

The possibility of occurrence of the event was evaluated on a scale 1-5. This assessment was based on expert opinion. Effects of exposure for a company (a ship owner) were evaluated on a 1-4 scale, and ease of detection in a 1-3 scale. Detailed evaluation system is shown in Tables 2, 3 and 4.

Table 2. Possibility of occurrence level definition.

Possibility level	Estimated probability	Description of the probability level
1	Very high	The threat occurred in the last month
2	High	The threat occurred in the last 3 months
3	Medium	The threat occurred in the last 6 months
4	Low	The threat occurred in the last year
5	Very low	The threat occurred once per the last two years or more

Table 3. Effects of exposure level definition.

Exposure level	Recovery time	Description of the effects of exposure
1	High	High financial losses
2	Medium	Financial losses
3	Low	Small financial loss, loss of image
4	Slight	No financial loss, loss of confidence of clients

Table 4. Ease of detection level definition.

Detection level	Ease of detection	Description of the ease of detection
1	Low	Identification of a week or more
2	Medium	Identification within 1-3 days
3	High	Identification immediately after the occurrence

Table 5. Risk index evaluation.

Event	P_n	E_n	D_n	RPN_n
SC1	4	2	3	24
SC2	5	1	3	15
SC3	4	2	3	24
SC4	5	1	3	15
SC5	5	1	3	15
SC6	4	1	3	12
SC7	5	1	3	15
NC1	2	3	2	12
NC2	2	3	2	12
NC2	3	3	2	18
NC4	2	3	2	18
NC5	3	3	2	18
IC1	2	3	2	12
IC2	4	2	2	16
IC3	3	3	3	27
IC4	3	3	3	27
IC5	4	2	2	16
FC1	4	3	2	24
FC2	4	2	2	16
FC3	4	2	2	16
IF1	4	2	2	16
IF2	3	2	2	12
IF3	4	2	2	16
HR1	4	1	1	4
HR2	3	2	1	6
HR3	3	2	1	6
LI1	3	2	1	6
LI2	2	3	1	6

The risk index has been defined for all identified adverse events and expressed in accordance with the FMEA process, as a product described above 3 parameters.

$$RPN_n = P_n \cdot E_n \cdot D_n \qquad (1)$$

where:
RPN_n = risk index of appearance of n adverse event,
P_n = possibility of occurrence of n adverse event,
E_n = Effects of exposure on n adverse event,
D_n = Ease of detection of n adverse event.

Analysis of the results indicates that the ship owners particular attention should be focused on these events, which have the lowest RPN index. They are in fact events difficult to identify, which incidence is high, and at the same time are associated with significant financial consequences for the company. The acceptable level of risk determined on the basis of interviews with experts was set at RPN = 15. This is a product of the middle of the scale adopted for the estimated three-pointers (3 x 2.5 x 2). With such a specific scale, only half of the 28 identified adverse events is indicator of risk at an acceptable level. The remaining 14 events requires further analysis. However, they all cannot be treated in the same way.

One can certainly distinguish among these two groups of threats that the proceedings should be varied. A first group of events, the events resulting from the environment to which the owners have no effect. In these cases, one can only take measures to reduce the consequences of this adverse events. The second group are the events on which the ship owner has a direct impact. In these cases it is necessary to take immediate preventive action. These events should also be subject to constant monitoring by management.

The adverse events, with unacceptable RPN, resulting from factors not associated with ship owners, from groups connected to navigation conditions (NC) and infrastructure conditions (IC) should be looked after by the authorities responsible for the maintenance of waterways. Groups of human resource (HR) and lack of interest (LI) could be influenced by the government, starting with affecting the education of children.

The other group of adverse events, with unacceptable RPN is connected with ship owners. Ship collisions (SC) can be prevented by training of a crew and mitigated by hull construction and also training. Problems with modern technology onboard mentioned in insufficient financial (IF) can by partly solved by looking for founds in European Projects.

6 CONCLUSIONS

Increasing competition in the freight market, the increase in congestion on the roads makes inland waterways transport companies increasingly interested in the techniques of risk analysis and management. Due to the lack of experience and good practices in this sector, the ship owners are looking for solutions model, which will be defined not only the risk assessment techniques, but also areas that should be analyzed.

In this paper, the authors using a holistic approach to risk management, proposed by the ISO33000 standard, proposed a risk assessment procedure for inland waterways transport companies. The proposed solution takes into account the specific nature of inland navigation. In this procedure, the starting point for the analysis is carried out to determine appropriate process parameters to be tested. Identification of the conditions of the process, existing limitations and analysis of actions taken in the next stages of the process, a source of information about potential adverse events.

The researchers in the presented results focused on the events which in experts opinion affect the implementation process of the transport of goods with the use of inland waterways transport. The environment of inland navigation companies and changes in Polish legislation in recent times, requires further research focused also on the potential dangers.

REFERENCES

Apostolakis G.E. 2004. How useful is quantitative risk assessment? *Risk Analysis* 24: 515–520.
Aven T. 2009. Perspectives on risk in a decision-making context – review and discussion. *Safety Science* 47: 798–806
Aven T. 2010. On how to define, understand and describe risk. *Reliability Engineering&System Safety* 95: 623–631.
Aven T., 2012. The risk concept—historical and recent development trends. *Reliability Engineering&System Safety* 99: 33–44.
Aven T. & Zio E. 2014. Foundational issues in risk analysis. *Risk Analysis*, 34 (7): 1164–1172
Aven T. 2016. Risk assessment and risk management: review of recent advances on their foundation, *European Journal of Operational Research*, 253: 1-13.
Brown J., Hosseini A., Karcher M., Kauker F., Dowdall M., Schnur R., Strand P. 2016. Derivation of risk indices and analysis of variablility for the management of incidents involving the transport of nuclear materials in the Northern Seas; *Journal of Environmental Management* 171: 195-203
Bubbico R., Di Cave S., Mazzarotta B. 2009. Preliminary risk analysis for LNG tankers approaching a maritime terminal, *Journal of Loss Prevention in the Process Industries* 22: 634–638
Goerlandt F., Hanninen M., Stahlberg K., Montewka J., Kuja P. 2012 Simplified risk analysis of tanker collisions in the gulf of Finland. Trans Nav-Int J Mar Navig Saf Sea Transp 6(3): 381-387
Goerlandt F. Kujala P. 2011. Traffic simulation based ship collision probability modelling *Reliability Engineering and System Safety* 96(1): 91-107

Goerlandt F., Montewka J. 2015. Maritime transportation risk analysis: Review and analysis in light of some foundational issues. *Reliability Engineering and System Safety* 138: 115–134

GUS. 2015. *Transport wodny śródlądowy w Polsce w 2015*, www.stat.gov.pl

Hampel J. 2006. Different concepts of risk – a challenge for risk communication. *International Journal of Medical Microbiology* 296: 5–10.

Jeong J.S., Park G-K., Kim K.I. 2012 Risk assessment model of maritime traffic in time-variant CPA environments inwaterway. *Journal of Advanced Computational Intelligence and Intelligent Informatics* 16(7): 866-873

Jorion P. 2006. *Value at Risk*. New York. McGraw-Hill.

Kaplan S. 1997. The words of risk analysis. *Risk Analysis* 17: 407–417.

King J.L. 2001. *Operational Risk: Measurement and Modelling*. Chichester: John Wiley & Sons Ltd.

Langard B, Morel G., Chauvin C. 2015 Collision risk management in passenger transportation: A study of the conditions for success in a safe shipping company; *Psychologie française* 60: 111–127

Montewka J., Krata P., Goerlandt F., Mazaheri A., Kujala P. 2011. Marine traffic risk modelling – an innovative approach and a case study. *Proceedings of the Institution of Mechanical Engineers, Part O: Journal of Risk and Reliability* 225(3): 307-322

PN-ISO 31000:2010: Zarządzanie ryzykiem – techniki oceny ryzyka, PKN, Warszawa, 2010.

Rasmussen F.M., Glibbery K.A., Melchild K, Hansen M.G., Jensen T.K., Lehn- Schioler T., Randrup-Thomsen S. 2012. Quantitative assessment of risk to ship traffic in the Fehmarnbelt Fixed Link project. *Journal of Polish Safety and Reliability Association* 3(1-2): 1-12

Smallman C. 2000 What is operational risk and why is it important? *Risk Management: An International Journal* 2(3): 7-14.

Suman S., Nagarajan V., Sha O.P., Khanfir S., Kobayashi E., Malik A.M. 2012 Ship collision risk assessment using AIS data. *International Journal of Innovative Research & Development* 1 (10): 509-524

Szczepański M. 2011. *Ubezpieczenia w logistyce*, Wydawnictwo Politechniki Poznańskiej, Poznań

Yuebo J., Xuefen Z. 2014. Risk Management Discussion on Transport of Bulk Liquid Dangerous Goods. *Applied Mechanics and Materials* 638-640: 2019-2022

van Dorp R., Merrick J.R. 2011. On a risk management analysis of oil spill risk using maritime transportation system simulation. Annals of Operations Research 187: 249-277 [87]

Wang Y., Cheng G., Hu H., Wu W. 2012. Development of a risk-based maintenance strategy using FMEA for a continuous catalytic reforming plant. *Journal of Loss Prevention in the Process Industries* 25: 958-965

Ward S. 2001. Exploring the role of the corporate risk manager. *Risk Management: An International Journal* 3(1): 7-25

Weng J., Meng O., Qu X. 2012. Vessel Collision Frequency Estimation in the Singapore Strait. *The Journal of Navigation* 65(2): 207-221

Zawiła – Niedźwiecki J. 2012. *Analiza ryzyka operacyjnego z perspektywy teorii organizacji*. Zeszyty Naukowe Uniwersytetu Szczecińskiego 690: 179-188

Proceedings of 12th International Conference on Marine Navigation and Safety of Sea Transportation, TransNav 2017
21-23 June 2017, Gdynia, Poland

The Technology of Container Transportation on the Oder Waterway

J. Kulczyk & T. Tabaczek
Wrocław University of Technology, Wrocław, Poland

ABSTRACT: The paper contains the analysis of the possibility of container transportation on the Oder Waterway (the Gliwice Canal, the canalized stretch of the Oder River, and the regulated stretch of the Oder River), on the assumption that the waterway complies with conditions of class III European waterway. The analysis is based on the concept of modern motor barge, adjusted to hydraulic parameters of waterway. The vessel is designed for ballasting when passing under bridges. The amount of ballast water that enables transportation of two tiers of containers is given. The costs of waterborne transportation is compared to the costs of rail transportation of containers on selected shipping routes.

1 INTRODUCTION

The Oder Waterway (ODW) for many years will remain the most important waterway for inland waterborne transportation in Poland. Despite the considerable degradation it is still possible to transport cargoes from Szczecin to waterways of Western Europe, on the Gliwice Canal, and on the canalized stretch of the Oder River from Silesia (Koźle) to Wrocław. Construction and commissioning of the weir and lock in Malczyce will enable to restore the navigation along the entire Oder Waterway.

The Oder River connects five significant regions of Poland: Silesia, Opole Voivodeship, Lower Silesia, Lubusz Land and Western Pomerania. In year 2012 those regions provided almost 30% of gross domestic product (GDP). In year 2014 the share of those regions in the sold production of industry was 32% [6]. Besides the bulk cargo (coal, aggregates) that traditionally was transported on the ODW, also a number of cargoes that require special organizing of transportation gravitates towards the river. There is a considerable supply of dangerous cargoes in form of products from large chemical plants located by the Oder River and equipped with own loading berths that, at the moment, are not being utilised. There is a great number of factories of car industry, factories producing household appliances, equipment for energy production, chemical equipment, in the regions connected with the Oder River. Many products from above branches are highly demanding in terms of packing, packaging, composing of integrated freight units and safety in transportation. These products are perfectly fit for multimodal transportation in containers. Inland waterborne transportation is preferred for safety reason.

In economy of agglomerations located by the ODW the trade plays important role. It concerns in particular the agglomerations of Silesia, Wrocław and Szczecin. They are important centres of distribution, storage and logistics. There is a number of container terminals there, including the following [9]:

- Silesian Logistics Centre in Gliwice Port, handling capacity of 150,000 TEU per year;
- PCC Intermodal terminal in Brzeg Dolny, handling capacity of 110,000 TEU per year;
- PCC Intermodal terminal in Frankfurt (Oder), handling capacity of 100,000 TEU per year.

The agglomeration of Szczecin is connected to the Szczecin and Świnoujście Seaports where the amount of reloaded containers is continuously growing. In year 2015 there were 87,784 TEU reloaded only in Szczecin. It was about 55% more in comparison to year 2010. 5% growth was reported in first 7 months of year 2016 in comparison to corresponding period of year 2015 [10].

New container terminals are going to be built in Kędzierzyn-Koźle and in the region of Wrocław [1].

2 SUPPLY OF CONTAINERS GRAVITATING TOWARDS THE ODW

The estimates of cargo gravitating towards the ODW published before the year 2000 usually did not include the containerised cargo.

In this paper it is assumed that about 20 million tonnes of cargo will be shipped on the ODW per year. Such amount of cargo was shipped in year 1980. Then the bulk cargo was prevailing.

In the structure of transportation on the Rhine River in years 2013 - 2015 the amount of containerized cargo was 15.3 to 15.8 million of tonnes, that is 7.9% to 8.3% of total amount of shipped cargo [2]. At the average weight of cargo of 7.2 tonnes per TEU it yields about 2.2 million of TEU.

In European Union, in year 2015 the share of containerised cargo in total transport work (tonne-kilometres) was 10.3%. The estimates for container transportation in year 2016 suggested the growth by 10% in comparison to year 2015 [3].

For the purpose of the following considerations one may assume that the structure of transportation on the Oder River will not differ significantly from that in Western Europe. Assuming the 8% share of goods in containers, and the above estimated total supply of cargo of 20 million of tonnes, about 1.6 million of tonnes of cargo may be transported in containers on the ODW. At the average weight of cargo of 7.2 tonnes per TEU [2] there is an estimate of 220,000 TEU per year. This number of TEU exceeds the number of containers handled in the Szczecin and Świnoujście Seaports considerably [10]. However, besides the Szczecin and Świnoujście Seaports, the direction significant for container transportation on the ODW is also the western direction to the waterways in other EU countries, including the port in Hamburg. In year 2013 560,200 of TEU were reloaded in Hamburg in connection to shipment to and from Poland [7]. 20% of them were next moved to the south of Poland, the majority to the regions located by the Oder River. The estimated average weight of cargo in a single TEU was about 10 tonnes. Assuming that 10% of containers that are moved from Hamburg to the south of Poland may be taken by waterborne transportation, the next 11,000 TEU are estimated to gravitate to the ODW.

3 THE ODER WATERWAY - HYDROLOGIC AND HYDRAULIC CONDITIONS

Hydrologic and hydraulic conditions are the basic factors that determine the functional parameters of a waterway. They include:
- the degree of river regulation - canalized river, regulated river, or a freely running river,
- parameters of hydraulic structures - horizontal dimensions of locks,
- air clearances under bridges,
- width of fairway,
- radii of meanders,
- transit depths,
- minimum and maximum discharge in a river,
- icing periods,
- pauses due to high or low water level,
- working hours of hydraulic devices.

The above factors determine the class of waterway and, in consequence, the main particulars of ships.

In the case of canalized rivers and navigation canals the hydraulic conditions are constant. Ships are designed to have the maximum acceptable dimensions. Main dimensions depend on the depth of water in canal, the area of cross-section, and the dimensions of locks. Due authorities determine the maximum ship speed on waterway, particularly in canals. Speed limits are intended for protection of banks against the destruction due to wash waves generated by going vessel, and for protection of vessel against grounding. Usually the speed limit in a navigation canal is not higher than 10 km/h.

The ODW is, and will remain for very next years, the most important transport waterway in Poland. Is characterized by diverse hydraulic structures, various functional parameters and, generally, by the significant wear and tear of structures. Considering:
- the different functional parameters of waterway,
- the diverse hydrologic conditions,
- the functions of waterway,
- the need and potential for development of navigation conditions,

the ODW is divided into the following stretches:
1. the Gliwice Canal with 6 doubled locks of dimensions 72 m x 12 m;
2. the canalized Oder River from Koźle to Brzeg Dolny with 23 locks including 20 locks of dimensions 187 m x 9.6 m and 3 locks of dimensions 225 m x 12 m;
3. the regulated Oder River from Brzeg Dolny to the mouth of the Lusatian Neisse River (after building and commissioning of the weir and lock in Malczyce this stretch will range from Malczyce to the mouth of the Lusatian Neisse);
4. the regulated Oder River from the mouth of the Lusatian Neisse River to the mouth of the Warta River;
5. the regulated Oder River from the mouth of the Warta River to Szczecin.

The rules of classification and the classes of Polish waterways are given in the regulation of the Council of Ministers of the 7th of May 2002. Table 1 shows the classes of individual stretches of the ODW.

According to the classification of waterways binding in Poland the class III waterway shall allow

for the operation of vessels of the following main particulars:
- length of motor ship ≤ 70 m,
- length of pushed barge train ≤ 118 m - 132 m,
- beam ≤ 8.2 m - 9.0 m,
- draught ≤ 1.6 - 2.0 m.

Table 1. Classification of individual stretches of the ODW

5. The Oder River and the Gliwice Canal	Length and the kilometre of River	Class
a) the Gliwice Canal	41.2 km	III
b) from the lock in Kędzierzyn-Koźle to the lock in Brzeg Dolny (the canalized stretch of the Oder River)	183.5 km from km 98.1 to km 281.6	III
c) from the lock in Brzeg Dolny to the mouth of the Lusatian Neisse River	260.8 km from km 281.6 to km 542.4	II
d) from the mouth of the Lusatian Neisse River to the mouth of the Warta River	75,2 km from km 542.4 to km 617.6	II
e) from the mouth of the Warta River to Ognica (to the Szwedt Canal)	79.4 km from km 617.6 to km 697	III
f) from Ognica to the Klucz-Ustowo Cut and farther as the Regalica River to its estuary to the Lake Dąbie	44.6 km from km 697 to km 741.6	Vb
6. The Western Oder River from the weir in Widuchowa to the border of internal sea waters including the Klucz-Ustowo Cut	36,5 km from km 0 to km 36.5	Vb

Own elaboration based on the map "Inland waterways in Poland", as of year 2007, National Water Management Authority (Annex No. 2 to the regulation of the Council of Ministers of the 7th of May 2002)

The regulation of the Council of Ministers of the 7th of May 2002 determines also the width of the fairway: 40 m on a river and 35 m on a canal, the minimum radius of the bends (meanders): 500 m, and the minimum dimensions of lock chambers: length of 72 m, width of 9.6 m. The minimum recommended air clearance under bridges at the highest navigable water level is 4 m. This parameter is a significant confinement in transportation of containers or large-size cargo.

In the following assessment of potentials of container transportation on the ODW the present authors assume that ODW conforms to the requirements of class III waterway. This ensures the operation of ships at draught of at least 1.4 m for 90% of navigation season. Such conditions prevailed on the freely running Oder River from year 1980 to year 2008. In that period the transit depth of 1.8 m was reported for 27% of navigation season.

Air clearance under bridges is a significant confinement in transportation of containers. The limiting bridge on the stretch from Wrocław to Szczecin is the railway bridge in Krosno Odrzańskie (at 514.1 km of the river) where the air clearance at highest navigable water level (WWŻ) is 3.15 m. At the mean navigable water level (ŚWŻ), when ships can operate at draught up to 1.6 m, the clearance is higher by 0.8 to 0.9 m [5].

On the stretch of canalized Oder River the lowest air clearance of 3.37 m occurs under the road bridge in Ratowice, over the lock. On this stretch one may assume that air clearance under bridges is constant, independent from the level of water.

4 THE TECHNOLOGY OF CONTAINER TRANSPORTATION ON THE ODW

With the confinements in mind, assuming that a motor barge will be the basic unit in transportation on the ODW, the dimensions of a ship that best utilises conditions on the waterway are as follows:
- length overall L_{OA} = 70 m,
- breadth overall B_{OA} = 9.1 m,
- scantling draught T = 1.70 m,
- draught of light ship T_0 = 0.55 m,
- height H = 1.80 m,
- capacity (at T = 1.70 m) $P \approx 660$ tonnes,
- length of hold L_h = 51 m,
- width of hold b = 7.0 m.

The estimated operating volume of ballast tanks is 191 m^3.

The concept of such vessel was presented in report "Logistic conditions of combined transportation of coal in the transport corridor of the Oder Waterway" [4]. Report was written in the framework of research and development project No. 10-0003-04 financially supported by the National Centre for Research and Development. General arrangement plan of that vessel is shown in Fig.1. Transverse cross section of vessel, including dimensions of ballast tanks, is shown in Fig.2. Ballast tanks are intended for increasing draught in cases when air clearance under a bridge is insufficient and could cause a break in the voyage.

During the development of that concept it was assumed that vessel is a general-purpose ship for transportation of bulk cargo as well as containers. In the case of vessel dedicated for transportation of containers it is possible to increase the width of hold from 7.0 m to 7.512 m. That would increase the number of TEU in one tier from 20 TEU arranged transversely to 24 TEU arranged longitudinally in 3 rows.

In analysis of actual draught of vessel the present authors assumed that, in the range of draughts possible during operation, the draught is a linear function of loading condition. Based on the draught of light ship T_0 and the capacity at draught of 1.70 m one may estimate that load of 5.74 tonnes increases draught by 0.01 m.

Air clearances required when one or two tiers of containers are transported are presented in Table 2 and in Fig.3.

Transport of 1 tier of containers is possible regardless of the gross weight of one TEU. In the case of 2 tiers the required air clearance exceeds the limiting value of 4 m when gross weight of one TEU is equal or less than 14 tonnes.

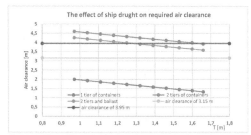

average weight of one TEU equals at least 16 tonnes, at ship draught of at least 1.66 m.

Ballasting enables operation of barge carrying 2 tiers of containers at high water levels. It increases the utilisation of fleet and the rationalization of transport costs.

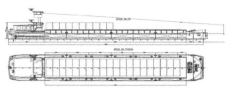

Figure 1. General arrangement plan of the motor barge BMN 700

Figure 3. The effect of ship draught on required air clearance

5 COSTS OF WATERBORNE AND RAIL TRANSPORTATION

The costs of transportation are determined for three example shipping routes. The distances are given in Table 3.

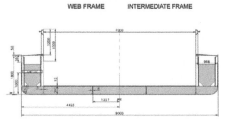

Figure 2. Transverse cross-section of the motor barge BMN 700

Table 3. Example shipping routes and distances

Shipping route	Distance by water [km]	Distance by rail [km]
Gliwice - Wrocław	200	149
Gliwice - Szczecin	680	512
Szczecin - Wrocław	480	363

The distance by rail was determined based on data given by PKP CARGO.

Table 2. The required air clearance under bridges

Motor barge BMN 700					
Draught [m]	Capacity [tonnes]	1 tier (20 TEU)		2 tiers (40 TEU)	
		Average gross weight of one TEU [tonnes]	Required air clearance [m]	Average gross weight of one TEU [tonnes]	Required air clearance [m]
0.97	244	12.2	2.00	6.1	4.60
1.04	280	14.0	1.93	7.0	4.53
1.11	320	16.0	1.86	8.0	4.46
1.18	360	18.0	1.79	9.0	4.39
1.25	400	20.0	1.72	10.0	4.33
1.32	440	22.0	1.65	11.0	4.25
1.39	480	-	-	12.0	4.18
1.46	520	-	-	13.0	4.12
1.53	560	-	-	14.0	4.05
1.60	600	-	-	15.0	3.99
1.66	640	-	-	16.0	3.92

Own elaboration

When ballasting the barge using ballast tanks the draught may be increased by 0.34 m. Then transport of 2 tiers is possible at average gross weight of 10 tonnes per TEU, at draught of about 1.3 m without ballast and 1.64 m with ballast. Without ballasting the transport of 2 tiers is possible only when the

The costs of rail transport were determined based on the PKP CARGO price list binding since the 1st of January 2016 [11]. Costs depend on distance, type of container, weight of container, and the state of container loading (loaded or empty). The basic rates are as follows: for route Gliwice - Wrocław 1,913 zł/TEU, Gliwice - Szczecin 5,035 zł/TEU, Szczecin - Wrocław 3,766 zł/TEU. Basic rates were multiplied by correction factors according to the following guidelines. For a 20' container not heavier than 22.0 tonnes the correction factor equals 0.75 and 1.00 for a 40' container. For empty containers - 0.5 and 0.8, respectively.

For rail transportation it was assumed that on a train of 40 platform carriages of kind S and type Sgs the half of total number of containers are empty. One platform carriage can carry three 20' containers or one 40' container. Transport costs of a single container on considered routes are given in Table 4.

The estimation of costs for waterborne transportation is based on cost analysis of coal transportation on the ODW [4]. The unit cost of one tonne-kilometre depends on distance and weight of cargo transported in specified time. For assessment of container transportation the mean costs were assumed depending on distance, with application of correction factor of 1.125. This factor corresponds to the ratio of gross domestic product in year 2015 to the gross domestic product in year 2011 [6]. After application of correction factor the cost of one tonne-kilometre (tkm) was in the range from 0.056 zł/tkm to 0.124 zł/tkm.

The external costs were determined based on data published in [8]. The exchange rate of 4.2 zł/euro was assumed for conversion. The external costs per one tonne-kilometre are: for rail transportation 0.047 zł/tkm, and for waterborne transportation 0.015 zł/tkm.

Table 4. Costs of container transportation on selected shipping routes

Shipping route	Costs [zł/TEU]				
	by rail			by water	
	20' cont.	40' cont.	External costs	20' cont.	External costs
Gliwice - Wrocław	398	432	53	310	38
Gliwice - Szczecin	1,049	1,133	180	476	128
Szczecin -- Wrocław	784	848	128	504	90

According to German data [8], assuming the above rate of exchange, the mean costs are estimated at 630 zł/TEU (150 €/TEU) for waterborne transportation, and at 903 zł/TEU (215 €/TEU) for rail transportation.

6 CONCLUSIONS

The presented analysis shows that, in the transport corridor of the Oder River, the costs of container transportation by water are lower than the costs of transportation by rail. The application of ballasting enables the transportation of 2 tiers of containers in operating conditions on the class III waterway. It is possible despite the low average weight of loaded containers, when the mean weight of cargo in a single 20' container is less than 10 tonnes. Hence, for most of the lifetime the capacity and the highest available draught (scantling draught) of a barge will not be utilised. It creates the opportunity for transportation of containers also at low water levels in the Oder Waterway.

REFERENCES

[1] Bogucki J., Lewandowski K., Kolanek Cz. et al.: Śródlądowy terminal kontenerowy. Uzasadnienie celowości budowy, Instytut Konstrukcji i Eksploatacji Maszyn Politechniki Wrocławskiej, Raport Nr SPR-94/2009

[2] Frühjahrssitzung 2016. Angenommene Beschlüsse (2016-I), ZKR Zentralkommission für Rheinschifffahrt, Strassburg, Juni 2016, www.ccr-zkr.org

[3] Market Insight, Europaische Binnenschifffahrt, Herbst 2016, Zentralkommission für Rheinschifffahrt, Europaische Kommission, Rotterdam 2016, www.ccr-zkr.org

[4] Kulczyk J., Lisiewicz T., Nowakowski T. et al.: Logistyczne uwarunkowania transportu łamanego węgla w korytarzu transportowym Odrzańskiej Drogi Wodnej, Instytut Konstrukcji i Eksploatacji Maszyn Politechniki Wrocławskiej, Raport Nr SPR-45/2011

[5] Kulczyk J., Skupień E.: Uwarunkowania transportu kontenerów na Odrze, Zeszyty Naukowe Transport, Z. 73, Politechnika Warszawska, 2010, pp. 61 - 77

[6] Rocznik Statystyczny Rzeczypospolitej Polskiej 2015, GUS, Warszawa, 2016

[7] Teuber M.O., Wedemeier J. et al.: Przewozy towarów między portem w Hamburgu i Polską – perspektywy rozwoju Unii Izb Łaby i Odry, Hamburgisches Weltwirtschaftsinstitut (HWWI), 2015

[8] Verkehrswirtschaftlicher und ökologischer Vergleich der Verkehrstrager Strasse, Bahn und Wasserstrasse, PLANCO Consulting GmbH, Essen, November 2007, http://www.planco.de

[9] www.pccintermodal.pl

[10] www.portszczecin.pl

[11] www.pkpcargo.com

Autonomous Water Transport

Safely Navigating the Oceans with Unmanned Ships

H. Stones

University of Southampton, Southampton, United Kingdom

ABSTRACT: Unmanned ships represent the biggest advancement in shipping in decades, but they pose some of the biggest regulatory challenges. This paper focuses on the legal requirements which have to be fulfilled as part of safe navigation. In relation to safe navigation the biggest legal challenges are posed by the International Regulations for Preventing Collisions at Sea, 1972 (COLREGS), the International Convention for the Safety of Life at Sea, 1974 (SOLAS), and the International Convention on Standards of Training, Certification and Watchkeeping for Seafarers, 1978 (STCW). The challenges posed by STCW can be avoided by not applying it, though it could be useful in developing regulations for remote control centres. By considering technological developments and remote controllers as fulfilling existing obligations in COLREGS and SOLAS, it becomes clear that unmanned ships can be considered, in law, as capable of navigating as safely as existing ships.

1 INTRODUCTION

Unmanned ships represent the biggest advancement in shipping in decades. They represent a culmination of technological advancements, including navigation and collision avoidance systems. They will transport goods more efficiently and with reduced human error. Therefore, this advancement is being welcomed to revolutionise the industry while improving safety. However, a little more consideration has to be given as to whether unmanned ships fulfil the legal obligations, which were designed for manned ships.

A lot of previous and current research in the industry has focused on the technical challenges primarily, although they have considered the law as well (e.g. Maritime Unmanned Navigation through Intelligence in Networks, and Advanced Autonomous Waterborne Applications Initiative). This paper's primary focus is the law, and what an analysis of the law can indicate as the fundamental areas that engineers can focus on when developing remote and autonomous systems for unmanned ships.

This paper discusses safe navigation in law, and thus will focus on Conventions and not the practice of navigation. The biggest challenges are posed by the International Regulations for Preventing Collisions at Sea (COLREGS), 1972, the International Convention for the Safety of Life at Sea (SOLAS), 1974, and the International Convention on Standards of Training, Certification and Watchkeeping for Seafarers (STCW), 1978. Each of these Conventions have improved maritime safety, thus unmanned ships must be able to fulfil their provisions, or potentially equivalents, in order to be considered safe enough to operate. This paper will show how selective application, purposive interpretations, remote control, and the use of systems, COLREGS, SOLAS, and STCW will not prevent unmanned ships becoming a reality while ensuring that they are safe. One of the key ways of holding legal obligations fulfilled is by considering the monitoring of the ship on shore through transmitted data from the ship to being equivalent to being on board.

By considering these technological developments as fulfilling existing obligations, it becomes clear that unmanned ships can be as safe existing ships (Rolls-Royce. 2016). Without holding unmanned ships to the same or equivalent obligations they represent a risk that the shipping industry (including the wider stakeholders of society, governance, and legal bodies) is unwilling to accept.

2 STANDARDS FOR SEAFARERS

STCW would appear to create an obvious problem for unmanned ships, as without a master and crew the requirements cannot be met. However, the preferred interpretation is that, since there is not a master and crew on board (Article 1(2)), STCW does not apply to unmanned ships (Veal et al. 2016). For instance, requirements for the master and chief mate in Regulation II/2 are for *"every master and chief mate on a seagoing ship…"* Since there will not be a master or chief mate 'on' the ship this Regulation does not apply.

2.1 *Challenges posed by STCW*

One of the challenges of interpretation avoided by this is Regulation VIII/2 2.1 which requires the physical presence of officers on the navigating bridge. The requirement for physical presence is problematic for an unmanned ship, as there is no officer physically present on the ship's bridge. Although this could be interpreted so that the remote control centre's virtual bridge fulfills this requirement, it is simpler to consider it as not applicable (especially for autonomous ships).

2.2 *Shore control centres*

Additionally, although STCW does not apply to unmanned ships, it may be worth using STCW for examples of some requirements that should apply to shore control centres (with some revisions). For instance, Regulation II/1 requires all officers on navigational watch to have a certificate of competency, as well as other requirements. It would be useful to also require this of remote controllers. At first it may also be useful to require seagoing experience (Regulation II/1 2.2), but as remote control becomes more common thus may not be necessary.

3 SAFETY OF LIFE AT SEA

However, SOLAS is more problematic than STCW. STCW regulates those on board, but SOLAS regulates the safety of life at sea (which extends beyond the master and crew) to create a safe environment for ships to operate.

3.1 *Application of SOLAS*

SOLAS applies *"only to ships engaged on international voyages"* (Chapter 1, Regulation 1(a)) unless it is stated otherwise. It applies widely to 'ships'. Therefore, regardless of the type or purpose of unmanned ships, SOLAS will apply to unmanned ships as it is widely accepted that they are ships (Van Hooydonk. 2014).

There are exemptions mentioned in Regulation 4 of Chapter I of SOLAS. However, the first, concerning ships that do not usually engage in international voyages, is not relevant to this paper. Regulation 4(b) provides on an exemption for a ship with novel features where the regulations would impede research.

Importantly this exemption does not apply to Chapter V: Safety of Navigation. Additionally, if this exemption were to apply to more of SOLAS it only applies to research, which would only be of use when developing unmanned ships, not once they are operating as part of the shipping industry.

3.2 *Application of Chapter V*

Chapter V is the most important chapter in relation to safe navigation. Chapter V applies to *"all ships on all voyages"* and *"all ships means any ship, vessel or craft irrespective of type and purpose"* (Regulations 1.1 and 2.3). Chapter V can also be applied to internal waters if the Administration so decides. Importantly it applies to all voyages, not just international voyages, and thus has wider application than other chapters in SOLAS (under Regulation 1(a) of Chapter 1).

Although the general exemption in Chapter I does not exempt Chapter V from application, there is another exemption in Chapter V, which states: *"The Administration may grant to individual ships exemptions or equivalents of partial or conditional nature, when any such ship is engaged on a voyage where the maximum distance of the ship from the shore, the length and nature of the voyage, the absence of general navigational hazards, and other conditions affecting safety are such as to render the full application of this chapter unreasonable or unnecessary, provided that the Administration has taken into account the effect such exemptions and equivalents may have upon the safety of all other ships. (Chapter V, Regulation 3.2)"*

This exemption would not provide a practical solution to a wide introduction of unmanned ships. Firstly, these exemptions will not be made lightly by Administrations as they have to consider the safety of other ships, so at first when there is not a lot of evidence from unmanned ships operating this will be of greater concern. As the fleet of unmanned ships increases it will also become potentially impractical for Administrations to have to grant the exemptions on an individual basis.

Additionally, as account is taken of proximity to shore, it may be easier to obtain an exemption when far from shore as there are fewer obstacles, or when near shore as closer to aid in an emergency. The factors are considered as to whether they make the regulations 'unreasonable or unnecessary', so any voyage with more navigational risks could be considered as posing too greater risk for the Chapter not to apply. Finally, depending on the nature of the voyage and the type of ship could mean the safety regulations are necessary.

Equivalents for certain regulations will be more appealing to Administrations. Although this will depend on what is considered to be a satisfactory equivalent and the ability of the technology to be a satisfactory equivalent.

3.3 Ships' routeing

An important ability for the navigation systems of the ship to have will be the ability to determine a safe route for the ship to take, and routeing aids have developed. However, as they are now, SOLAS provides: *"Ships' routeing systems contribute to safety of life at sea, safety and efficiency of navigation and/or protection of the marine environment. (Chapter V, Regulation 10.1)"*

This regulation considers such systems as an aid, as making a contribution, and thus not the only way of determining the route. Currently, the master and crew are on board to also determine the route. Systems are not currently considered able to replace humans to the extent that it is safer without humans, merely as enhancing the safety of navigation. Therefore, this Regulation does not facilitate the introduction of unmanned ships and would have to amended so that the systems can be relied upon more for remote and autonomous control.

3.4 Ship reporting system

The retention of data, and its transmission, is an important factor in developing autonomous and remote control technology as it allows for effective monitoring of what the ship has done, and is doing (Chapter V, Regulation 11.7). The systems are currently seen as contributing to the safety of lives, navigation, and the marine environment. However, such a system does not have to be technological, a log book can be recorded manually, and not in a digital format. The responsibility of reporting to authorities rests with the master of the ship. Thus, it must be asked without a master on the ship, and instead having a computer transmitting automatically to the authority, whether this regulation will be fulfilled. It is possible without a master then Regulation 11.7 does not apply to an unmanned ship. Also, if a person on shore is considered to be the master in a remote controlled system then they can fulfil this duty easily, as it does not state that they must be on board.

3.5 Manning and language

Requirements in relation to manning instantly appear to be problematic for unmanned ships by definition, and thus represent a greater challenge than some of the previous Regulations discussed. Regulation 14 provides: *"1 Contracting Governments undertake... all ships shall be sufficiently and efficiently manned. 2 For every ship to which chapter I applies, the Administration shall: .1 establish appropriate safe manning following a transparent procedure, taking into account the relevant guidance adopted by the Organization..."*

The first aspect to note is that it is for the State to determine manning. Therefore, arguably, this issue can be easily resolved if States want to facilitate the introduction of unmanned ships. States could decide that for the systems on board an unmanned ship that the ship is manned sufficiently and efficiently with zero (Veal et al. 2016). This could mean that States that do encourage unmanned ships will become popular flag States.

However, it could be argued that zero manning is not sufficient as there is no crew on board to intervene in a direct manner, nor efficient as personnel would have to be transported to the ship in order to perform maintenance. Although, this seems unlikely to prevent the introduction of unmanned shipping when it is a matter for the State, and a State that wants to encourage unmanned ships will use the former interpretation.

There are some Regulations relating to crewing that do not appear to be very problematic, and represent minor issues when compared to manning, but they also need to be resolved. Regulation 14.3 provides that: *"On all ships, to ensure effective crew performance in safety matters, a working language shall be established and recorded in the ship's logbook... Each seafarer shall be required to understand and, where appropriate, give orders and instructions and to report back in that language..."*

This Regulation does imply that the ship is manned through the term 'seafarer', and thus when interpreting the Regulation this implication should be ignored. A purposive interpretation allows it to be interpreted to be broader and apply to remote controllers as they are those controlling the ship.

In order to comply with SOLAS as much as possible, a working language should be established for the remote controllers within each remote control centre. However, working languages may be more complicated if remote control centres are location specific, and the ship is transferred between remote control centres throughout the voyage, as each centre may not use the same language. Therefore, it may be necessary to establish a working language for all remote control centres. Otherwise unnecessary delay could be caused during an incident resulting in loss.

3.6 *Bridge design*

In determining bridge design, placement of systems and equipment, and procedures, any decisions should aim to provide the navigators etc. with a *"full appraisal of the situation,"* and with the aim of navigating safely in all operating conditions (Chapter V, Regulation 15.1). The same should then apply to the remote control centre, especially those that use a virtual bridge. There should also be safe and effective resource management to ensure that the remote controllers use the bridge as well as a bridge at sea (Chapter V, Regulation 15.2).

The information that the pilot or bridge team use should be accessed conveniently, and continuously available (especially essential information), and be presented clearly and unambiguously in a standardised form (Chapter V, Regulation 15.3). The same standardised form can be used as on current manned ships, the greater challenge is the provision of essential information. Therefore, the transmission time from the ship to remote control centre must be adequate. Although the flow of information may be continuous it may not be effective to act on information that does not reflect the real-time situation of the ship, yet SOLAS requires the bridge to allow for *"expeditious, continuous and effective information processing and decision-making by the bridge team and pilot"* (Chapter V, Regulation 15.5). The best way to minimise this problem would be to improve communication systems and always transfer control to the remote control centre nearest the ship. It may also be necessary to set a requirement regarding how much a time delay there can be between ship and shore. Terminologically Regulation 15.3 could be avoided by arguing that the remote controllers are not the bridge team or pilot, but overall it is more logical to consider them as equivalent to a bridge team and pilot to encourage maximum compliance, so Regulation 15.3 needs to be fulfilled through the development of more 'expeditious, continuous and effective' systems.

If there is not a remote controller, but the ship processes information and makes decisions autonomously, this will need to be performed at the same or better speed and accuracy of the bridge team and pilot. Otherwise developing autonomous systems would be represent a regression in safety, instead of a development as the systems should represent.

The transferring of control from the ship's autonomous systems to shore may also have its own problems. The transfer would have to be instantaneous to avoid the chance of a potential incident developing and not acted upon quickly enough during the transfer.

There would also have to be a comprehensive summary of the situation to inform the new remote control centre quickly when transferring between centres. Firstly, this would have to include any specifics of the ships. Secondly, there would need to be voyage details. Thirdly, information about any incidents or damage that occurred, or the same for any that may be developing or in progress. Also, it is important that transfers do not happen automatically, as a remote control centre could be performing an important manoeuvre that should not be interrupted by the transfer process. There needs to be preparation and communication before, during, and after the transfer, which itself should be instantaneous.

Another aspect of bridge design that is regulated in SOLAS is the avoidance of unnecessary work or distractions that could have a negative impact on the bridge team or pilot (Chapter V, Regulation 15.6). The use of autonomous systems could aid in the minimisation of work to allow the bridge team on shore to be more effective when they do have to make decisions.

Semi-autonomous ships would be ideal for fulfilling the aim in Regulation 15.7, which reads: *"minimizing the risk of human error and detecting such error, if it occurs, through monitoring and alarm systems, in time for the bridge team and the pilot to take appropriate action."* The aim of autonomous systems is to remove a lot of the decisions from humans, so that they do not make an error, and when complemented through a checking and monitoring system on shore, any human error that may have occurred during the programming of the control system should be detected. When humans do need to make decisions on shore, having more than one person and a command structure will allow for the detection of errors in the remote control centre itself.

3.7 *Shipborne navigational equipment, maintenance, and voyage data*

The bridge as whole is not the only problem in relation to Regulations on the design of the ship, but also the equipment on the ship. Regulation 19.2 requires that *"all ships, irrespective of size, shall have"* certain equipment. It could be argued that if the equipment is in a remote control centre on shore then the ship does have equipment (as it has complete access, and is part of the operation of the ship). However, Regulation 19.2 is headed *"shipborne navigational equipment and systems."* The term 'shipborne' implies that it is physically on board the ship. Therefore, it must be considered that this equipment may still need to be on board. Some of the equipment would necessarily have to be on board, for instance a compass, but it is required to be independent of a power supply and this would not be of any use on an unmanned ship as the information could not be transmitted to shore or processed by the ship's autonomous control system (Chapter V, Regulation 19.2.1.2). Therefore, it may be more appropriate to have a different redundancy system for the compass that does involve a power supply.

Regulation 19.2.1.8 requires that when a bridge is enclosed that there is a sound reception system to allow sound signals to be perceived on the bridge, a similar requirement can be made so that sound reception system is required to transmit to a simulated bridge in the remote control centre. This will aid in providing multi-sensory perception, and fulfilling the need for a lookout (COLREGS, Rule 5).

The requirement for ships to have an Automatic Identification System (AIS) includes the requirement that it transmits to equipped shore stations, and ships, and receives information from other ships' AIS. This could function even better if more ships are operated from shore, and all ships are required to have AIS, and there will be more shore centres that receive the information, and then the control centre can directly act on that information.

Some bridge systems on ships are already integrated, as they would be on shore, so Regulation 19.6 requires that they are *"so arranged that the failure of one sub-system is brought to the attention of the officer in charge of the navigational watch by audible and visual alarms and does not cause failure to any other sub-system. In case of failure in one part of an integrated navigational system, it shall be possible to operate each other individual item of equipment or part of the system separately."*

All of the equipment required by Regulation 19 should be *"installed, tested and maintained as to minimize malfunction"* (Chapter V, Regulation 19.6). Maintenance presents one of the greatest challenges to unmanned ships, and the fulfilment of this provision will rely on the sensors and regular maintenance on shore instead of being able to react while at sea.

The provisions for ensuring that the ship is well maintained in Regulation of 16 of Chapter V may be more problematic as it requires that all reasonable steps are taken to ensure that equipment is in efficient working order. Again, there will need to be greater reliance on sensors and checks in port to ensure that the equipment is in working order otherwise it will risk more malfunctioning of the equipment. As on an unmanned ship malfunctioning at sea is more likely to be considered to not be in efficient working order.

The development of Voyage Data Recorders (VDRs) are an important aspect to making unmanned ships feasible, and will keep record of any pertinent information in relation to incidents especially (e.g. equipment malfunctioning, and maintenance required but not performed). In the absence of a crew on board to relay the circumstances and the events of an incident, the independent recording of data is necessary for investigations. VDRs will hopefully provide a more accurate and complete record of events, without the inconsistency of human testimony.

3.8 *Visibility*

In relation to the bridge, there are not just equipment Regulations, there are also provisions in relation visibility. For ships of 55 metres or more in length, there must be an unobscured view of the sea meeting some conditions (Chapter V, Regulation 22.1). Therefore, it is suggested that is replicated in a bridge simulator, to enable remote-controllers to have the same view as though they were on board the ship. This may require the amalgamation of many different camera angles, including those situated where the bridge would have been. The requirements for the view from the bridge windows based on a person's height, may also need to be replicated. Replicating these conditions may also aid seafarers in navigating from shore, as they will be accustomed to that view.

However, it must be noted, that the requirements for the dimensions and framing of the windows on the bridge are not necessary for an unmanned ship, whether controlled remotely or autonomously (Chapter V, Regulation 22.1.9). This can be resolved through the utilisation of the exception that allows the Administration to permit an alternative deign that allows for *"a level of visibility that is as near as practical to that prescribed in this regulation"* (Chapter V, Regulation 22.3). The cameras can provide for the same level of visibility for the form of control utilised without windows, which would not be practical to include when they do not in fact aid visibility in a remote control centre.

3.9 *Pilotage*

If a ship would currently utilise pilotage, then it is subject to Regulation 23 of Chapter V. However, it is thought that pilotage would operate differently as remote control develops, so that the ship is transferred to a local remote control centre, so that a pilot with local knowledge can navigate the ship for that part of the voyage (Maritime Unmanned Navigation through Intelligence in Networks. 2015).

3.10 *Track control systems*

There are many different systems that will contribute to the safe navigation of unmanned ship, which will be utilised by the equivalents of the master, crew, and pilots – one of which is the track control system. Regulation 24.1 provides that: *"In areas of high traffic density, in conditions of restricted visibility and in all other hazardous navigational situations where heading and/or track control systems are in use, it shall be possible to establish manual control of the ship's steering."* This Regulation may be problematic, given that 'manual control' implies a person on board the ship to physically take control. It is possible to interpret the Regulation, so that as long as remote control, with full control for each decision being made by the remote controller, is possible then manual (remote) control is possible. In order to fulfil this Regulation, it is also important that control is taken 'immediately': this does not mean instantaneously, for instance even if an officer on the bridge were to take control it would not be instant, as they would have to move, make decisions etc. Therefore, as long as the communication systems allow the remote controller to take control quickly, make decisions, and implement them through communications systems then this provision could arguably be met.

Regulation 24 goes on to provide that the officer on watch will have to be *"available without delay the services of a qualified helmsperson who shall be ready at all times to take over steering control."* Therefore, in addition to the measures to ensure compliance with subsection 1, the remote control centre will have to have a qualified helmsperson available.

If a ship is generally controlled by an autonomous system, and remote control is available as a redundancy measure and available to take control in such hazardous navigational circumstances, then this transfer of control shall be supervised by a responsible officer in the remote control centre (Chapter V, Regulation 24.3). In addition, the Regulation requires the testing of the manual steering after using heading and/or track control systems for a prolonged period, or prior to navigating in an area that is known to require caution (Chapter V, Regulation 24.4). This aids in ensuring the readiness of manual control at those times, and when a hazardous navigational situation arises unpredictably. Therefore, control would have to be taken on a regular basis to test the system that changes the method of control.

3.11 *Instructions*

In order to ensure that manual control is taken of steering gears and their power units, Regulation 26.3.1 requires that instructions are displayed on the bridge, and in the steering compartment. This Regulation may not be possible to entirely fulfil. The instructions could be posted in the remote control centre, as equivalent to the bridge. Although instructions technically could be in the steering compartment, the steering compartment would not have the ability to take manual control, so it would serve no practical benefit.

Although ships could still be constructed, so that they can be controlled manually on board, which

would then favour the posting of such instructions on the bridge, in the steering compartment, and additionally in the remote control centre; this writer does not favour this suggestion, as it would invite pirates to take physical control of the ship without a people on board to resist them. This would be an additional safety risk that should not be taken, especially when all ships, not just new unmanned ships, will need to have heightened cyber-security to prevent piracy through hacking.

As smarter ships develop, pirates will use smart technology, they may even venture into new areas of the oceans and target more ships. It is conceded that it may be possible to counter this through remote control overrides, but if the pirates sever the communication link they could take manual control as it would be favourable to allow autonomous control in general to override manual control (as it has been argued that it would be there for emergencies, and when autonomous control is not sufficient). Additionally, pirates may target the remote control centres to take remote control easily, or target multiple ships.

3.12 *Communicating with other ships, and distress*

When there is a danger or if it is voyage that is in an icy area the chance of danger is higher, there may be times when there is a danger to navigation, so the ship must be able to communicate this danger to all other ships in the area and the relevant authorities (Chapter V, Regulation 31.1). Therefore, they will need to be able to communicate with all types of unmanned ships, as well as manned ships.

It will be easiest to communicate with ships that utilise remote control centres, as soon as the information reaches the centre it can be disseminated between centres (and the information applied on the ships). However, communicating with manned ships will be more difficult, unless a remote control centre controls a means of communicating through sound and visual signals on the ship. Another solution would be to require a common method of communication on all ships that the ship can automatically relay information through to all types of ships.

It is well documented that the requirement to provide assistance to other ships in distress is another challenge for unmanned ships (Veal et al. 2016). Regulation 33.1 states: *"The master of a ship at sea which is in a position to be able to provide assistance on receiving information from any source that persons are in distress at sea, is bound to proceed with all speed to their assistance if possible informing them or the search and rescue service that the ship is doing so..."*

Most unmanned ships will not be in the 'position to be able to provide assistance', due to the lack of facilities and persons on board to aid those in distress, and will not be bound to provide assistance. Therefore, most unmanned ships will simply be required to record the information (having processed it by whatever means), and potentially forward it to other ships, and search and rescue.

3.13 *SOLAS Chapter V and beyond*

This section has focused on some of the regulatory challenges posed by Chapter V of SOLAS. However, there are many additional challenges in SOLAS and other sources of maritime law. For example, Regulation 5 of Chapter IX of SOLAS requires that *"the safety-management system shall be maintained in accordance with the provisions of the International Safety Management Code."* The challenges posed the International Safety Management Code (last amended in 2014), as worthy of considerations especially in relation to safety, but it is not within the scope of this paper.

4 COLLISION REGULATIONS

COLREGS are also well-known to be problematic for all unmanned ships, as Chapter V of SOLAS is, because it is so important for the shipping industry and safety for it to apply (Veal et al.). COLREGS applies to *"all vessels upon the high seas and in all waters connected therewith navigable by seagoing vessels"* (Rule 1(a)). And vessel is defined as *"every description of water craft, including non-displacement craft, WIG craft and seaplanes, used or capable of being used as a means of transportation on water"* (Rule 3(a)). This definition is sufficiently wide to encompass unmanned ships.

4.1 *Exemptions*

Since COLREGS applies, it is important to consider whether there is an exemption. The definitions of vessels that would be exempted in Rule 3(g)-(h) would not exclude unmanned ships by definition of being unmanned, but if their work or draught was such to constrain them.

Rule 3(f) refers to a *"vessel not under command,"* but an unmanned ship would not be classified as such (although the terminology would appear superficially be appropriate for an autonomous ship), as it is not due to an *"exceptional circumstance"* but the design of the ship.

There is an exception for vessels of a special construction or purpose in relation to lights, shapes, or sound signals, but the ship must comply as closely as possible with the Rules (Rule 1(e)). Close compliance with ordinary practice of seamen could be achieved by having seamen as the remote controllers.

Departure from the Rules is also allowed if *"necessary to avoid immediate danger"* (Rule 2(b)). However, this exception is allowed for danger, and special circumstances, including the limitations of the vessel. Therefore, although this exception is not applicable generally as the danger would be due to the design of the vessel it is allowed to be a contributing factor if there is danger by other means (Rule 2(b)).

4.2 *Perception of vessels, and keeping a lookout*

Some of the issues posed by COLREGS will now be discussed, since it has been established that there is no broad exemption for unmanned ships. Rule 3(k) deems vessels *"to be in sight of one another only when once can be observed visually from the other."* It must be considered whether the lack of a crew on any unmanned ship, means since there is no direct visual perception that such ships cannot be in sight. However, a better interpretation is to focus on the 'can' in the Rule, it does not require observation in fact, and 'observed visually' can be interpreted broadly to include the use of cameras relaying the information to shore or through the autonomous control system (and thus it becomes an issue of when the cameras and sensors observe other ships, and they must be able to observe as well as a human on the other ship to avoid confusion and collisions).

Rule 5 requiring a lookout by sight and hearing is considered to be a problematic requirement if it is interpreted narrowly, as 'sight and hearing' indicate a person as opposed to visual and audio. However, it is usually interpreted so that as long as there is an equivalent to sight and hearing then it is fulfilled (Veal et al.). Importantly, this is not just sight, and although a lot of discussion focuses on cameras and sensors, there must also be that audio element in the perceptive tools of an unmanned ship.

4.3 *Making decisions and determining risk*

COLREGS considers decision making, as well as perception in relation to preventing collisions. Rule 6 requires a judgement decision to be made in determining a safe speed, taking various factors into account. This process will be the same for the remote controller as it is for the master and crew. For autonomous unmanned ships it leads to the question of how an autonomous system will be programmed to make that decision. This does highlight the need for the ship not just to follow a pre-programmed list of responses, but collect data and make a decision based on that data, responses available, and wider information of the potential implications of those responses. This solution is also relevant for the section for ships in sight of one another (Part B, Section II).

In determining if there is a risk of collision *"all available means shall be used,"* and thus is wide enough not to require a crew (Rule 7). Rule 7 specifically refers to radar to ensure that it is used properly, and that assumptions should not be made from *"scanty information, especially scanty radar information."* This indicates a certain amount of scepticism regarding the reliability of radar, which will be more relied upon on unmanned ships as will other forms of technology.

One of the most important provisions in COLREGS for decision making concerns the duty of good seamanship. Rule 8(a) includes the duty to act with *"due regard to the observance of good seamanship."* This is especially problematic in relation to unmanned ships that are controlled autonomously as this duty cannot be programmed directly, and avoiding violations of the Rules in COLREGS may not be sufficient (Veal et al. 2016). The system can be programmed to act in accordance with the law, details on manoeuvres, and only to act differently when permitted by the law, and this may need to considered sufficient.

4.4 Restricted visibility

The lack of people on board, especially for autonomous ships can be considered as disabling the ship (at least as it is currently perceived). This is especially true when considering the duty of good seamanship above. This approach means Rule 19 for the conduct of vessels in restricted visibility could be relevant, as it could be argued that without people on board visibility is restricted, as the most established means of perception is not available. However, the changes in conduct that it would involve for conditions that are not present, and thus would lead to confusion and increase the risk of a collision.

4.5 Lights, Shapes, Sounds, and Signals

There are some issues with unmanned ships are purely practical, and although not in COLREGS would create a problem with compliance. For instance, if a bulb is blown this cannot be detected or remedied at sea, and thus the light signals will not comply with COLREGS' Rules on light and sound signals (in Parts C and D).

There have been recommendations to create new light signals for smaller unmanned craft to exhibit when not practical to exhibit ordinary lighting, which could be extended to exhibit not the inability comply with all provisions due to being unmanned and small, but simply as an unmanned ship (Norris 2013). Although unmanned ships should comply with COLREGS generally, this will make other users of the seas aware that the ship will be doing so differently and can encourage caution. Other writers have raised concerns that it will encourage other ships to violate the law as the unmanned ship will have to react, this writer suggests that this can be avoided through strong penalties for such behavior. Greater awareness will be especially useful when they are first introduced and there is more concern about how they will operate and interact.

5 CONCLUSIONS

This paper has considered three of the most important Conventions in shipping generally: STCW, SOLAS, and COLREGS.

STCW is the easiest of the three to apply to unmanned ships in the fact that it is not applicable as there is no one 'on board'. However, it has been noted that its provisions on qualifications and experience in particular may be useful when developing Regulations for remote controllers.

SOLAS, however, is more complicated to apply to unmanned ships. The requirement for manning can be resolved by the State determining that zero manning is sufficient and efficient. Other Regulations in SOLAS will depend more directly on the technology: e.g. transmitting data adequately, allowing for instant control changes etc. Thus, the analysis in this paper will hopefully encourage technical development

Finally, COLREGS poses its greatest challenge in the duty of good seamanship to autonomous ships. Other challenges can be resolved by considering the technology as equivalent to a person or a person onshore as equivalent to one at sea, which relies to the technological development (as SOLAS does). The suggestion for lights to indicate that the ship is unmanned will aid in ensuring safe navigation in relation to COLREGS, but also SOLAS.

REFERENCES

International Maritime Organization. 2003. COLREG Convention on the International Regulations for Preventing Collisions at Sea, 1972 Consolidated Edition 2003. London: International Maritime Organization.

International Maritime Organization. 2011. STCW Including 2010 Manila Amendments. London: International Maritime Organization.

International Maritime Organization. 2014. ISM Code International Safety Management Code with guidelines for its implementation 2014 Edition. London: International Maritime Organization.

International Maritime Organization. 2014. SOLAS Consolidated Edition 2014. London: International Maritime Organization.

Maritime Unmanned Navigation through Intelligence in Networks. 2015. D8.8: Final Report: Shore Control Centre.

Maritime Unmanned Navigation through Intelligence in Networks. 2016. Research in Maritime Autonomous Systems Project Results and Technology Potentials.

Norris, A. 2013. Legal Issues relating to Unmanned Maritime Systems Monograph.

Rolls-Royce. 2016. Remote and Autonomous Ships: The Next Steps.

Van Hooydonk, E. 2014. The Law of Unmanned Merchant Shipping – An Exploration. *Journal of International Maritime Law* 20:403.

Veal, R. & Tsimplis, M. & Serdy, A. & Ntovas, A. & Quinn, S. 2016. Liability for Operations in Unmanned Maritime Vehicles with Differing Levels of Autonomy.

The Concept of Autonomous Coastal Transport

A. Łebkowski
Gdynia Maritime University, Gdynia, Poland

ABSTRACT: The article presents a concept of an autonomous coastal transport system using autonomous, ecological vessels capable of transporting passengers or a passenger vehicle. The current solutions and development trends in the field of autonomous and unmanned vessels (ghost ship) are presented. The structure of autonomous vessels control system is discussed, along with a presentation of employed methods of artificial intelligence. The results of autonomous system simulated operation in a maritime navigation environment simulator are presented.

1 INTRODUCTION

The dynamic growth of transportation needs which is a consequence of expanding local communities and growing tourism in coastal regions and in-land waters aggregations is creating a demand for new, both in form and function, modes of transportation. New solutions are sought to shift the road transport to rail, water or airborne modes. One of the many possibilities of relieving the congested roads is to change the mode of transport from road to water. The water transport has the additional advantage when the landform causes the travel distance between two given points to be greater when travelling by land than by water. An example of such motions is the initiative of many local communities to introduce "water trams" between the most frequented places. The function of water tram can be fulfilled ad-hoc by existing manned vessels.

Figure 1. Concept ecological vessels E/V Orcelle, and Vindskip [1].

The alternative long-term solution with the goals of functionality, reduction of environmental impact, low operating costs while retaining the maximal possible safety standards is to deploy a fleet of fully autonomous vessels.

Figure 2. Wind Challenger Project [2,3].

Figure 3. LNG Fueled Container Ship by Kawasaki H.I. [4].

One of the design goals is to engineer a unit which is not a source of pollution for the environment. The designers employ various solutions to this problem, such as LNG/LPG propulsion (Figure 1÷3), sail propulsion, or even completely forgoing the conventional engine in lieu of electric propulsion powered by photovoltaic panels or hydrogen fuel cells [1-6].

Some ships augment the conventional propulsion with an aerodynamic Flettner Rotor (Figure 4,5) which harnesses the Magnus effect. [5,6].

Figure 4. Viking Grace with one Norse-power Rotor Sail [5].

Figure 5. E-Ship1 with four Rotor Sails [6].

In order to minimize the movement resistance losses, the design optimizes the aerodynamic and hydrodynamic drag by shaping the submerged and surface parts of the hull and the submerged propulsion system components.

The autonomous vessels offer a possibility of extending the traditional timetable based service to include travel to every possible harbour or landing in the area of operation. Additionally, the autonomous units can be used to collect flotsam and jetsam on the waters, and in the area of ports and marinas.

Presently, there are two fundamental concepts of unmanned vessel operation available. The first is basing on a remote control scheme, possible with the aid of data links and camera systems, while the second employs artificial intelligence methods to steer the vessel, without the need of operator supervision. Examples of the latter concept designs can be found in the proposed projects of Rolls-Royce Company aimed at the forwarders using seagoing vessels (Figure 6) to move the cargo at the long range between the ports [7].

Figure 6. Drone Cargo Ship by Rollce-Royce [7].

Another proposal was presented by a consortium consisting of Massachusetts Institute of Technology (MIT), Delft University of Technology (TUD) and Wageningen University and Research (WUR) [8]. It is a concept of an urban water transport (Figure 7) intended for cities with large available network of waterways, such as: Venice, Amsterdam, Rotterdam, Bruges, Stockholm, Bangkok, Suzhou, El-Gouna or Petersburg.

Figure 7. AMS ROBOAT [8].

Figure 8. Car / passenger ferry concept by Kongsberg [9].

Kongsberg, in turn, intends to deploy a fully autonomous ferry (see Figure 8) for transportation of passengers and cars on a permanent line in the area of Trondheim [9].

In this paper, the author presents a concept of coastal transport organization using small, autonomous and ecological vessels for ferrying passengers and cars.

2 STRUCTURE OF AUTONOMOUS COASTAL TRANSPORT

In the last few decades many solutions and algorithms, which employ artificial intelligence methods to ship steering and support the navigator in taking his decisions as a part of various DSS systems were presented in the literature [10-36]. A subset of presented methods can be applied to run the autonomous vessels after performing certain modifications.

The author proposes the application of an agent system for management of autonomous traffic in coastal and inland waters [23-25]. The first time an agent system was recommended for management of vessel traffic was in 2007 [23]. The agent system consists of an information technology agent platform on which the agents are placed, who perform particular tasks. The agent platforms are installed in computers controlling the autonomous vessels and on the shore stations.

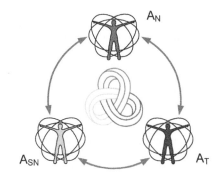

Figure 9. The structure of the agent system information platform.

The singular agent platform consists of three agents: A_T – trajectory agent, A_{SN} – navigation agent, A_N – negotiation agent (as pictured in the Figure 9). The goal of the trajectory agent A_T is to determine the autonomous ship route, basing on the data received from the shore station, and relayed by the navigation agent. Simultaneously, the trajectory agent is responsible for correcting the ship route in the event of a collision threat (from other vessels moving through the area and navigational limitations). The trajectory agent's function is also the precision control of the autonomous ship

including precision berthing and unberthing. The navigation agent A_{SN} gathers the data on actual navigational situation around the ship and performs its thorough analysis looking for possible collisions with dynamic objects as well as static objects, which consist mostly of technical equipment of autonomous vessel landing. The data describing the geometry of the navigation environment in which the autonomous ship is operating is supplied mainly from ECDiS system in cooperation with a system of LIDAR sensors, and optionally from vision cameras. The negotiation agent A_N conducts negotiations with agents installed in other platforms of autonomous coastal and inland waters transport system. The agent can interact with agents on other autonomous vessels as well as it can receive and send data to shore station agents. This system structure allows the relaying of data between the autonomous units operating on the given waters in order to avoid collisions as well as to share the data on occasionally occurring ship operation dangers.

The configuration and goals of agent platform of a shore station is somewhat different than autonomous vessel agent platform. The shore station agent platform also consists of three agents: A_C – communication agent, A_{SN} – navigation agent, and A_N – negotiation agent. The task of the communication agent is to process the requests from clients/users of the autonomous transport system. The communication agent analyses the requests filed by users, calculates and then passes to the navigation agent point of boarding the geographic coordinates along with the request type – transport of passengers only, or with a vehicle. The navigation agent not only relays the data on the navigational situation on the waters, but also the data on the state of autonomous units (fuel level, battery state of charge). The shore station negotiation agent has a higher decisive coefficient than negotiation agents on the autonomous vessels. It allows the station agent to organize the vessel traffic by issuing commands to the trajectory agents on the vessel agent platforms [24].

In order to use the autonomous transport vessel, the user/client firstly has to log himself in via the web interface. Next, when he wishes to use the transport, he has to send an SMS message from a phone number which he registered during the log-in procedure. The message should contain: the place of boarding, the starting date and time, desired form of transport (passengers only or vehicle with passengers; number of passengers), and desired destination. In reply, the user will receive information on confirmation of selected parameters of reservation, on necessity of making certain changes in the reservation (e.g. about lack of possibility of unloading a vehicle in a given destination), or about the closest possible time the service can be available. After arriving at the boarding place, the user should confirm his arrival by entering a code in the touchscreen on board of the autonomous vessel. In the next step, the user will be instructed via the interactive communication interface of the necessary actions that need to be performed in order to safely enjoy the voyage. This step is especially important during transport of the vehicle and the actions required to load the vehicle and secure it.

The steering of the autonomous ship is performed automatically according to the user set destination point or predefined route. The user during the autonomous transport ordering process has the possibility of choice from listed destination points in a given area, as well as touristic and sightseeing routes. The goal of the agent system is to direct the available autonomous vessels to the boarding places in such way as to minimize the customer waiting times.

The structure of the agent system for controlling the traffic of ecological autonomous vessels is presented in Figure 10.

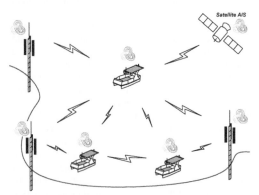

Figure 10. Autonomous, ecological vessel traffic agent management system structure.

The communication in the coastal water transport system using small, autonomous, ecological vessels can be accomplished with digital radio transmission in VHF band, GPRS system and AIS system.

The goal of the agent system is to autonomously and automatically conduct transport operations on given waters. In order to make it possible, it is required to design a way for communication between agents operating on the same platform as well as a way for data exchange between all the platforms operating in the system. The actions of agents in the platform are characterized as independent, performed without operator participation. The agents have a possibility to exchange data with the operator, but they can also try to negotiate their manoeuvres with the navigators of other vessels operating in the given area. The task of the agents is also to communicate with other

agents on the same platform, and with agents on other platforms in order to properly interpret the dynamic changes in the navigation environment, and react to them by making proper and optimal decisions.

3 AUTONOMOUS ECOLOGICAL VESSEL

In order for the coastal transport system using small autonomous ecological vessels for transportation of passengers to be efficient, it needs to use properly equipped crafts. The structure of equipment of such autonomous craft is presented in Figure 11.

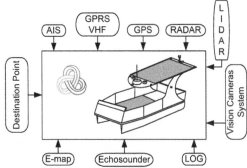

Figure 11. The Structure of the autonomous vessel equipment.

The operation of autonomous, ecological vessels is accomplished by processing of information provided by devices and systems installed on board of vessel, along with the data relayed wirelessly by radio, from the shore stations. The steering of the autonomous craft is performed mostly using the data from the positioning system (GPS/GLONASS/GALILEO), AIS, radar system, ARPA device, anemometer, log, echo sounder, and electronic map system. Additionally, especially during berthing operations, the signals from LIDAR sensor system and optional vision cameras system are used. After specifying the destination point, the system begins to follow the calculated route. The route is calculated based on actual geographic location, and the geographic location of destination point. The evolutionary algorithm [24] is used to calculate the route for the autonomous vessel. The trajectory agent A_T is responsible for the correct implementation of the route. Whenever the potential collision threat is detected, the actual navigational situation is analysed, and proper action is undertaken depending on the severity of the threat. The possible actions include entering negotiations with encountered vessels in order to define the optimal route. When the negotiation is not possible, the optimal route is calculated, or the anti-collision manoeuvre is defined, following the process of automatic steering along computed route. This process is performed by the trajectory agent, which is also responsible for precision steering of the vessel during berthing and unberthing operations, and manoeuvring in narrow passages. The listed actions are possible thanks to data from the positioning system, the radar system, and especially, the LIDAR sensor system, with sensors placed in four corners of the autonomous vessel or optionally, the vision cameras system. The data is processed before presenting it to the trajectory calculation algorithm and precision control algorithm in order to safely steer the vessel along the calculated route. The 3D objects detected and identified in the marine environment by the radar, AIS and LIDAR or cameras system are brought down to a 2D plane. This allows every object projected to the 2D plane to become a collision threat to a ship with installed autonomous agent platform. Depending on the motion parameters, the detected objects are classified as either static or dynamic. For such formulated navigation limits, the methods of artificial intelligence can be applied to determine the optimal ship route. In the described system of autonomous vessel traffic control and optimal ship routing in coastal and inland waters, the evolutionary algorithm technique has been used.

The autonomous, ecological vessel is fitted with dual electric drives with power of 5kW each, supplied from a bank of batteries (Figure 12). Additionally, for precision steering the ship uses two thrusters: a bow thruster and a stern thruster, with 1kW of power each. The energy stored in the battery bank allows the continuous operation for about 10 hours. In addition, the battery bank can be recharged by an on-board engine powered generator, which allows the vessel operation even with completely discharged battery bank. On the roof of the vessel, and in the forward part of the deck, there are photovoltaic panels, supplying the energy to the main propulsion motor inverter, or used to charge the battery. The ship is also fitted with a connector allowing automatic connection of a shore charging station, when moored at a landing.

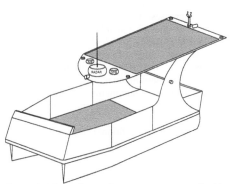

Figure 12. View of exemplary autonomous vessel with agent system.

4 VERIFICATION OF THE AGENT SYSTEM USING TO MODELLING THE AUTONOMOUS COASTAL TRANSPORT

In order to verify the operation of the autonomous coastal transport system using small, autonomous, ecological vessels, a maritime navigation environment simulator was employed. This simulator has the capability of conducting simulations of navigational scenarios with pre-set parameters describing the waters on which the simulation is performed.

The simulator consists of a central unit – a server, on which the navigational environment is simulated, and local client workstations emulating each autonomous unit. The tested navigational scenarios involved the verification of system operation in the event of collision situation between the autonomous vessels. The developed mathematical model of the autonomous vessel includes the dynamic properties of the hull, dual main propulsion motors with fixed pitch propellers, one in each hull, a blade rudder, and two transverse thrusters: a bow thruster and a stern thruster. The ship dynamics model includes the influence of hydro-meteorological disturbances including: wind, sea, currents, and changes in ship dynamics due to effect of shoal areas [25].

The verification of the agent system for management of autonomous traffic in coastal and inland waters was performed on the basis of a series of computer simulations using the maritime navigation environment simulator. Below, an example navigational situation is presented, in which the autonomous vessels were navigating through the Gulf of Gdansk waters (Figure 13).

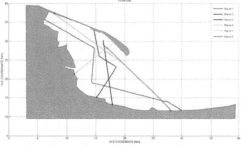

Figure 13. The navigational situation, with autonomous vehicles navigating the Gulf of Gdansk.

6 vessels took part in the navigational situation, 4 of them were fitted with an agent system. The motion parameters and weather conditions parameters are presented in the Figure 14.

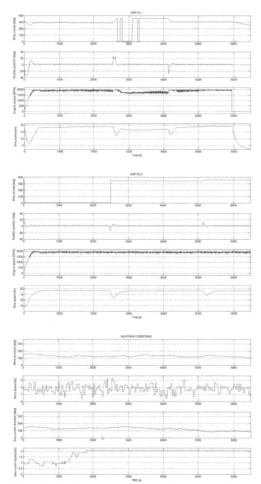

Figure 14. Motion parameters of selected vessels, and plots of weather conditions.

The routes of vessels 1 and 2 were planned in a way to transport the passengers from their boarding place to destination simultaneously with activation of touristic mode, which allows the vessel's route to be altered to pass, near or through, areas of interest or with scenic views. Basing on the simulation results it can be stated, that the system has correctly carried out the ship steering process, from the starting point to the destination point. In a situation where there was a risk of collision, in this particular case the Closest Point of Approach (CPA) distance was exceeded, the system has correctly altered the actual ship's route. With the application of the agent system for controlling the traffic of autonomous vessels it is possible to reduce the value of limiting CPA distance, due to the fact that the individual agent platforms have precise knowledge of other nearby ship routes, as they all cooperate among the agent system. Defining a smaller values of CPA

parameter has a direct effect of shrinking the areas around other vessels operating in the given waters, violation of which increases the danger to the navigation safety. The defined areas are taken into consideration by a specialized evolutionary algorithm [24], as one of the artificial intelligence tools which the trajectory agent A_T of the agent system uses to derive the optimal ship route for current navigational conditions. This approach allows determination of the shortest, and simultaneously safe, ship route. Unfortunately, this comfort cannot be guaranteed by the agent system in case when on the given waters there is a vessel, with which the contact cannot be established, as it prevents the negotiation agent A_N from accomplishing its task. In that case, the agent system falls back to the data available from the radar system, LIDAR and AIS, and applies the rules of the International Maritime Organisation's COLREGs regulations to determine the route.

5 CONCLUSIONS

- The performed simulations have shown that the application of the agent system for controlling the coastal traffic of ecological autonomous vessels brings positive effects.
- The research have demonstrated that even in the event one vessel loses contact with the shore station, its autonomous functioning in the navigational environment is still possible. The application of the agent system reduces the risk of collision with other objects, raising the overall safety level on the given waters.
- The main obstacles in deploying the autonomous vessels are the regulations and legal and financial issues in the event of eventual accidents.
- The practical implementation of the system increases the safety level of users of given waters, including the time throughout Search And Rescue operations, when autonomous vessels with vision cameras can take part in the operations.
- Deployment of fully automated, autonomous vessels increases the navigation safety and efficiency of coastal and inland water transport. The algorithms used to calculate the ship route fully respect the International Maritime Organisation's COLREGs regulations, which cannot be said about many navigators.
- The introduction of autonomous vessels allows new possibilities of ship design in the future, e.g. utilizing the spaces taken by crew accommodations for additional cargo space.

REFERENCES

[1] Shadbolt P., 'Vindskip' cargo ship uses its hull as a giant sail. CNN 11.2015, www.edition.cnn.com.

[2] Ouchi,K., Zhu,T., Hirata,J., Tanaka,Y., Kawagoe,Y., Takashina,J., Suzuki,K.: Development of Energy-saving Windshield for Large Container Ship, Conference proc., the Japan Society of Naval Architects and Ocean Engineers, 21, p.159-162, 2015.11 (in Japanese).

[3] Hane,F., Aoki,I., Ouchi,K.: Development of Auto-pilot for Sailing Ship, Conference proc., the Japan Society of Naval Architects and Ocean Engineers, 21, p.171-173, 2015.11 (in Japanese).

[4] www.global.kawasaki.com, (12.2016).

[5] Norsepower Rotor Sail Solution. www.norsepower.com, (01.2017).

[6] Daily News. E-Ship 1 wins wind propulsion innovation award. No. 4, 2016.

[7] Autonomous ships The next step. Ship Intelligence Marine. Rollce-Royce 2016. www.rollce-royce.com/marine.

[8] Amsterdam to Get World's First Fleet of Autonomous Boats. www.ams-institute.org/roboat-qa/, (09.2016).

[9] Automated Ships, Kongsberg to build fully-automated vessel for offshore operations. www.safety4sea.com, (11.2016).

[10] Escario J.B., Jimenez J.F., Giron-Sierra J.M., Optimisation of autonomous ship manoeuvres applying Ant Colony Optimisation metaheuristic. Expert Systems With Applications, Vol. 39, Issue 11, 2012, p.10120-10139.

[11] Escario J.B., Jimenez J.F., Giron-Sierra J.M., Optimization of Autonomous Ship Maneuvers applying Swarm Intelligence. IEEE International Conference on Systems Man and Cybernetics Conference Proceedings 2010.

[12] Ottesen A.E., Situation Awareness in Remote Operation of Autonomous Ships. Shore Control Center Guidelines Norway 2014.

[13] Gierusz W., Łebkowski A., The researching ship "Gdynia". Polish Maritime Research, Vol. 19, 2012, p.11-18.

[14] Kula K., Autopilot Using the Nonlinear Inverse Ship Model. In.: A. Weintrit (ed.), Activities in Navigation: Marine Navigation and Safety of Sea Transportation, CRC Press/Balkema, London, UK, 2015, p. 101-107.

[15] Kula K., On-line autotuning of PID controller for desired closed-loop response. 20th International Conference on Methods and Models in Automation and Robotics (MMAR) 2015, p. 707-711.

[16] Kula K., Model-based controller for ship track-keeping using Neural Network. IEEE 2ND International Conference on Cybernetics (CYBCONF 2015), 2015, p. 178-183.

[17] Levander O., Autonomous ships on the high seas. IEEE Journals & Magazines, Vol. 54, Issue 2, 2017, p. 26-31.

[18] Johansen T.A., Perez T., Cristofaro A., Ship Collision Avoidance and COLREGS Compliance Using Simulation-Based Control Behavior Selection With Predictive Hazard Assessment. IEEE Transactions on Intelligent Transportation Systems. Vol. 17, Issue 12, 2016, p. 3407-3422.

[19] Perera L.P., Oliveira P., Soares C.G., System Identification of Vessel Steering With Unstructured Uncertainties by Persistent Excitation Maneuvers. IEEE Journal of Oceanic Engineering, Vol. 41, Isuue 3, 2016, p. 515-528.

[20] Lisowski J., Computational intelligence methods of a safe ship control. Procedia Computer Science, Vol. 35, 2014, p. 634-643.

[21] Lisowski J., Optimization-supported decision-making in the marine game environment. Solid State Phenomena, Vol. 210, 2014, p. 215-222.

[22] Lisowski J., Game control methods in avoidance of ships collisions. Polish Maritime Research, Vol. 19, 2012, p. 3-10.

[23] Łebkowski A., Dziedzicki K., Agent System in Directing the Movement of the Ship. I[st] International Tech-Science Conference NATCON 2007, Scientific Journal of Polish Naval Academy, No. 170, 2007, p. 1-12.

[24] Łebkowski A., Evolutionary Methods in the Management of Vessel Traffic. Information, Communication and Environment 2015, p. 259-266.

[25] Łebkowski A., Control of ship movement by the agent system. Polish Journal of Environmental Studies Vol.17, No. 3C, 2008.

[26] Łebkowski A., Negotiations between the agent platforms. Scientific Publications Gdynia Maritime University, Gdynia 2013.

[27] Mohamed-Seghir M., Safe ship's control in a fuzzy environment using a genetic algorithm. Solid State Phenomena Vol. 180, 2012, p. 70-75.

[28] Mohamed-Seghir M., The branch-and-bound method, genetic algorithm, and dynamic programming to determine a safe ship trajectory in fuzzy environment. Procedia Computer Science, Vol. 35, 2014, p. 348-357.

[29] Mohamed-Seghir M., Computational Intelligence Method for Ship Trajectory Planning. 21ST International Conference On Methods And Models In Automation And Robotics (MMAR), 2016, p. 636-640.

[30] Rybczak M., Linear Matrix Inequalities in multivariable ship's steering. Polish Maritime Research, Vol. 19, 2012, p. 37-44.

[31] Gierusz W., Tomera M., Logic thrust allocation applied to multivariable control of the training ship. Control Engineering Practice, Vol. 14, Issue 5, 2006, p. 511-524.

[32] Gierusz W., Waszkiel A., Determination of suction forces and moment on parallel manoeuvring vessels for a future control system. Mechatronic Systems, Mechanics And Materials, Book Series: Solid State Phenomena, Vol. 180, 2012, p. 281-287.

[33] Tomera M., A multivariable low speed controller for a ship autopilot with experimental results. 20th International Conference on Methods and Models in Automation and Robotics (MMAR), 2015, p. 17-22.

[34] Miller A., Interaction Forces Between Two Ships During Underway Replenishment. Journal of Navigation, Vol. 69, Issue 6, 2016, p. 1197-1214.

[35] Weintrit A., Neumann T., Safety of Marine Transport Introduction. 2015, p. 9-10.

[36] Weintrit A., E-Navigation Revolution - Maritime Cloud Concept. Communications in Computer and Information Science, Vol. 471, 2014, p. 80-90.

Proceedings of 12th International Conference on Marine Navigation and Safety of Sea Transportation, TransNav 2017
21-23 June 2017, Gdynia, Poland

Platform for Development of the Autonomous Ship Technology

S. Ahvenjärvi
Satakunta University of Applied Sciences, Rauma, Finland

ABSTRACT: A model-scale platform for development and testing of the autonomous ship technology is being built at the Faculty of Logistics and Maritime Technology of Satakunta University of Applied Sciences (SAMK). The platform, called ELSA, consists of a model-scale ship equipped with electric motor driven propulsion, batteries and solar panels. The 8,4 meter ship, Kaisa, was originally a 1:15 scale towing model of a passenger cruise ship. Kaisa will be equipped with a comprehensive set of navigation equipment complemented with machine vision cameras and a lidar (Light Detection And Ranging) sensor. The real-time navigation data, the visual image and the lidar data is transmitted via a 4G communication link to the remote control centre, which is located in the main building of the Faculty of Logistics and Maritime Technology of SAMK. The remote control centre is equipped with displays for real-time presentation of the visual image from the onboard cameras and the control panels for remote manual steering of the vessel. The computers for target detection, decision making and automatic track control will be located in the remote control centre. Different ways of using this cost-effective model-scale platform for demonstration, development and testing purposes, as well as for the education of engineers and seafarers at SAMK, are discussed in the end of this paper.

1 INTRODUCTION

Unmanned and autonomous cargo ships have become one of the hot topics in the discussion of the future of sea transportation. However, different variations of automatic, though not unmanned, ships have existed already for decades (Manley, 2008). Fully automatic dynamically positioned (DP) vessels have been used by the offshore industry since the 1970's. The technology of on autonomous cargo ship was studied and demonstrated in Japan already in the 1980's. Fully autonomous Unmanned Surface Vessels (USV's) are already widely used in ocean research, coast guard and also in military applications.

The autonomous shipping is seen as a possibility for maritime transport to meet tomorrow's challenges. The key arguments are the improvement in safety and the reduction of costs, i.e. improvement in competitiveness. Also improved energy efficiency and protection of environment support the idea of using unmanned ships for transportation of goods and raw materials over longer distances.

Recent European research and development projects on this field are the MUNIN-project (MUNIN, 2016) financed by the EU, and the Norwegian ReVolt project by DNV GL supported by Transnova, Norway (DNV GL 2017). The third major European project on this subject is AAWA, financed by a group of Finnish companies and the state-owned Finnish Funding Agency for Innovation, TEKES.

One of the fundamental outcomes of the MUNIN-project was the finding that the unmanned vessels can indeed contribute to the aim of a more sustainable maritime transport industry and that the autonomous ship bears the potential to reduce operational expenses, reduce environmental impact and attract seagoing professionals. Also the fully autonomous, unmanned, battery powered and electrically driven concept ship ReVolt was estimated to have a considerable potential for cost savings compared to an ordinary diesel-run ship, over a million euros annually (Tvete 2014).

2 A MODEL-SCALE PLATFORM FOR DEMONSTRATION AND TESTING OF THE AUTONOMOUS SHIP TECHNOLOGY

An unmanned autonomous ship can be equipped with fully autonomous operation modes as well as with remote manual control modes. The ship's onboard control system can get assistance from the remote operator if the onboard decision making system has difficulties in solving the actual navigation problem at sea. The autonomous ship can thus operate fully independently or it can sail remotely operated. The operating status of an unmanned ship can vary, based on the operational situation, between fully autonomous execution, partially remotely assisted operation and direct remote control (Rødseth et al., 2014).

Development of unmanned and autonomous ship technology has gained a lot interest in Finland, due to the highly advanced ship building and information technology branches of the Finnish industry. Sea transportation in general is extremely important for the Finnish economy, since more than 80% the imported and exported goods to and from Finland are transported on ships. Introduction of autonomous shipping is seen as a major future trend in the sea transportation technology, and Finnish maritime industry has the potential and the desire to be at the front line of development of this new technology.

Although the idea is not new and the basic technology of an autonomous ship already exists, there is still a lot to be studied in this area. The development towards fully autonomous sea transportation systems is a time-consuming process. Introduction of unmanned cargo ships in the international sea transportation will not happen over one night, but gradually, step by step (Levander 2015). The most challenging problems may not be the technical ones. Many questions about international legislation, reliable and safe data transmission, cyber security, training of operators, interaction between manned and unmanned vessels, port operations, general safety management etc. need to be studied and answered before unmanned cargo ships can take any role in international sea transportation.

The Faculty of Logistics and Maritime Technology of Satakunta University of Applied Sciences is located close to the Port of Rauma and next to the Rauma Seaside Industry Park, an industrial area hosting major Finnish maritime technology companies including the RMC shipyard and the world-class azimuth thruster manufacturers Rolls-Royce and Steerprop. It is part of the strategy of SAMK to be active in the development of knowledge and solutions for the sea transportation of tomorrow.

SAMK decided to build a model-scale platform for demonstration and testing of the autonomous ship technology in co-operation with Rolls-Royce Ltd and WinNova Ltd, the institute for vocational education in Satakunta. The platform called ELSA utilizes the 8,4 meter miniature training ship *Kaisa*, which was built in 1994 for training of practical ship handling and harbor manoeuvers to sea captain students at Rauma Maritime College. *Kaisa* was originally a 1:15 scale towing model of a passenger cruise ship "Society Adventurer", later "Hanseatic" (Figure 1) which was built in Rauma, Finland in 1991. The towing model was upgraded in 1994 by equipping it with two hydraulic rudders, two AC-motor driven fixed-pitch main propellers and a bow tunnel thruster. The electric power for driving the propellers is produced by a 3 kW gasoline aggregate. The total power of the main propellers is 0,75 kW and the maximum cruising speed of the ship is five knots (Markkanen, 1994).

photo by By Genossegerd - Own work, CC0, Wikipedia
Figure 1. MS *Hanseatic*, originally MS *Society Adventurer*

Figure 2. The 1:15 model ship *Kaisa*

3 THE STRUCTURE OF THE PLATFORM

A block diagram of the overall structure, the instrumentation and the signal transmission of the system is shown in Figure 3. The electric power for the equipment is provided by a battery pack, which enables several hours' continuous operation of the propellers and the instrumentation of *Kaisa*. The ship is equipped with solar panels to load the batteries. Shore connection is also available for quick loading of the batteries.

Kaisa will be equipped with navigation equipment consisting of a GPS receiver, a radar, an AIS receiver, a wind sensor, a GPS compass with pitch & roll measurement capability and an echo sounder. The main propellers and the bow thruster are AC motor driven. Frequency converters are utilized in controlling the RPM of the propellers. Although the system is built for development of the technology of an unmanned ship, there will always be one person onboard *Kaisa* for safety purposes. The onboard operator has the possibility to take over the control of the propellers and the rudders at any time using the local control panel, should something unexpected happen. *Kaisa* will also be equipped with a lidar (Light Detection And Ranging) sensor and a set of machine vision cameras to provide the remote control centre with visual information of the surroundings and near-by objects.

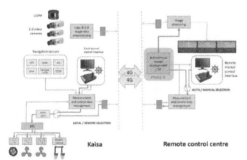

Figure 3. Block diagram of the system.

The remote control centre is located in the main building of the Faculty of Logistics and Maritime Technology of SAMK. The control centre is equipped with large LCD screens for displaying the image provided by the onboard cameras and a sensor data display and a HMI panel for remote manual steering of *Kaisa*. The lidar and machine vision data is preprocessed by an onboard computer and the data is analysed in the remote control centre. The track control / dynamic positioning (DP) computers for fully autonomous operation of *Kaisa* are also located in the remote control centre.

4 THE DATA LINK TO THE REMOTE CONTROL CENTRE

The data transmission link between the vessel and the remote control centre is a crucial part of the system. The data link has to fulfill seven basic criteria:

Firstly, it must have enough capacity for smooth real-time transmission of the data produced by the radar and other navigation sensors, the lidar and the machine vision cameras from the ship to the remote control centre. Secondly, the latency of the data transmission must be short enough for efficient monitoring of the traffic and for stable track control of the ship. Thirdly, the data transmission link has to be safe against interruptions and other kinds of failures. The data link should consist of two independent transmission channels to have redundancy against failures. Fourthly, the data transmission link should be safe against cyber threats, e.g. intentionally caused disturbances and hacking. Fifthly, the onboard equipment should be light and small enough for the 8.4 meter model ship. Sixthly, the coverage area of the data transmission link must enable testing of the system at sea, although not very far from the port of Rauma. The seventh requirement deals with operating expences to keep the costs of the data transmission on a reasonable level.

Alternative wireless data transmission techniques for autonomous ship operations were studied in the MUNIN project (Rødseth et al. 2014). According to those investigations, access to bandwidth on the order of 1 megabit per second would be needed if live video is used. However, since other requirements for the data transmission between the ship and the control centre were somewhat different in MUNIN than in the ELSA project, the conclusions about a suitable data transmission solution are not equal.

For the ELSA platform, the best data transmission solution seems to be the 4G network. The equipment is light-weight and relatively cheap, data transmission can be made redundant by using two sets of equipment and two independent 4G channels from two operators. The data transmission capacity of 4G network is sufficient even for transmission of real-time visual image and preprocessed lidar data. The latency requirement for remote control of the ship's movements is rather tight, since in the model-scale ship everything happens faster than in real life. The coverage of the 4G network is sufficient in the port of Rauma, which will be the main testing area for *Kaisa*.

Moreover, when a 4G network is used in communication between the ship and the remote control centre, the location of the control centre can be changed. The equipment can be installed on a van and moved to virtually any location with a decent 4G coverage.

5 THE USE OF THE "ELSA" PLATFORM

The main goal of the ELSA project is to create a useful and a cost-effective environment for testing, demonstration and development of the autonomous ship technology. Even though a model-scale environment is not 100 % realistic and it has some limitations, it has some advantages over full-size testing facilities and simulators. ELSA offers the possibility to run practical tests and demonstrations cost-effective in the real environment. The capital costs, the operation expenses and the safety risks are smaller, when the size, mass and the speed of the ship are scaled down. How about simulators? Couldn't testing of the autonomous ship technology be done in a simulator environment more flexible, with greater choice of scenarios and even more cost-effective? The answer is that simulator testing can never fully replace testing in real environment, since there are unknown facts that need to be studied in real environment, e.g. the operation of sensors and target identification algorithms in different weather conditions (also when the temperature is below zero, the sea is covered with ice and it is snowing!), the operation of the data transmission link, operation of the electric power generation and the batteries in low temperatures etc.

The platform can be used for research purposes in many different ways. Some interesting research areas would be, among others:

- the use of lidar and machine vision cameras for detection of objects
- algorithms for controlling the ship in abnormal traffic situations
- safety of data transmission between the ship and the remote control centre
- ergonomics and functions of the human-machine interface at the remote control centre
- training needs and good ways of building competence of the operators of the remote control centre
- utilization of solar energy and other forms of renewable energy onboard a battery powered ship
- necessary modifications in legislation and classification rules for autonomous ships
- autonomous operation in port, automatic berthing systems etc.
- interaction between an autonomous ship and an intelligent fairway

The ELSA project will also give nice opportunities for many engineering students of Satakunta University of Applied Sciences to increase their knowledge about the autonomous ship technology. Several Master's and Bachelor's degree theses will be published during the ELSA project on topics related to autonomous ship technology.

6 CONCLUSIONS

Within the coming decade we might witness a break-through of autonomous ship technology in sea transportation. It has been predicted in several occasions that unmanned cargo ships will gradually replace manned ships, beginning from short routes and special applications and expanding to international sea transportation. There are many reasons to draw this conclusion. Autonomous ship technology has the potential to bring many benefits to the ship owners and to the public, for example the following ones have been presented (Haugland, 2016):

1 Increased safety
2 Reduced operational cost
3 Reduced construction costs
4 Increased environmental sustainability
5 Increased social sustainability
6 Increased competitiveness
7 Reduced risk of piracy

Satakunta University of Applied Science takes actively part in the development of ships for the future by building a model-scale platform for demonstration and development of autonomous ship technology, together with Rolls-Royce Ltd and WinNova Ltd. The platform, called ELSA, consists of a model-scale ship equipped with a full set of navigation instruments, a lidar sensor and machine vision cameras. The real-time navigation data, the visual image and the lidar data is transmitted via a redundant 4G communication link to the remote control centre located in the main building of the Faculty of Logistics and Maritime Technology of SAMK. The visual image from the onboard machine vision cameras is displayed on the LCD displays of the remote control centre. The control centre is equipped with computers for target detection, intelligent decision making and automatic track control of the ship.

The ELSA platform will be used for research, development and training purposes in the coming years. Some areas of interest will be the detection of objects using the lidar and machine vision information, development of intelligence for management of unexpected traffic situations, issues related with the safety of 4G data transmission between the ship and the remote control centre, automated operations in port and interaction between ther autonomous ship and an intelligent fairway. One of the important topics for future research is the safety of autonomous shipping, as stated in the AAWA whitepaper (Jalonen et al. 2016). Different solutions for tackling the safety concerns listed by Jalonen et al. can be studied and demonstrated by utilizing the ELSA platform, e.g. ability of automation to navigate safely on coastal fairways, ability of automation to reliably detect small vessels and floating objects on route, problems caused by disturbances, malfunctions and vulnerabilities in data communication connections etc.

Designing and constructing the ELSA platform will offer nice opportunities for students of SAMK to get familiar with the technology and challenges of autonomous ships. Several Master's and Bachelor's degree theses will be written and published about autonomous ship technology during the ELSA project.

REFERENCES

Ahvenjärvi S. 2016. The Human Element and Autonomous Ships. *TransNav, the International Journal on Marine Navigation and Safety of Sea Transportation*, Vol. 10, No. 3, pp. 517-521, 2016

DNV GL 2017. The ReVolt – A new inspirational ship concept, at: https://www.dnvgl.com/technology-innovation/revolt/index.html [acc. 10.2.2017]

Haugland B. 2016. Towards Unmanned Shipping. at: https://www.linkedin.com/pulse/towards-unmanned-shipping-bj%C3%B8rn-kj%C3%A6rand-haugland [acc. 10.2.2017]

Jalonen R. & Tuominen R. & Wahlström M. 2016. Safety and security in autonomous shipping - challenges for research and development. In AAWA whitepaper *"Autonomous Ships The Next Steps"*, Rolls-Royce plc, at: http://www.rolls-royce.com/~/media/Files/R/Rolls-Royce/documents/customers/marine/ship-intel/aawa-whitepaper-210616.pdf [acc. 25.2.2017]

Jokioinen E. 2016. Advanced Autonomous Waterborne Applications (AAWA) Initiative, Rolls-Royce plc, at: http://www.maritime-rdi.eu/media/18556/Advanced-Autonomous-Waterborne-Applications-AAWA-initiative.pdf [acc. 10.2.2017]

Levander O. 2016. Unmanned ships. Presentation at the seminar *Älykäs meriteollisuus*, Rauma 17.11.2015, available at: *www.hel.fi/static/kanslia/elo/rollsroycemarine.pdf [acc. 10.2.2017]*

Manley J. 2008. Unmanned Surface Vehicles, 15 Years of Development, Proc. Oceans 2008 MTS/IEEE Quebec Conference and Exhibition (Ocean'08) pp. 1-4 2008.

Markkanen A. 1994. Pienoismallialus, Sea Captain's Thesis, Rauman merenkulkuoppilaitos, Rauma, Finland (in Finnish)

MUNIN PROJECT 2016. Reports of the MUNIN project at: http://www.unmanned-ship.org/munin/ [acc. 10.2.2017]

Rødseth O. & Tjora Å. 2014. A System Architecture for an Unmanned Ship. In Volker Bertram (ed.), *13th International Conference on Computer and IT Applications in the Maritime Industries, COMPIT'14*, Redworth, 12-14 May 2014, pp. 291-302, Technische Universität Hamburg-Harburg, 2014

Tvete H. 2014. The Next Revolt, *Maritime Impact 2/2014*, at: http://www.gcaptain.com/wp-content/uploads/2014/09/ReVolt-Details.pdf *[acc. 10.2.2017]*

Safety Qualification Process for an Autonomous Ship Prototype – a Goal-based Safety Case Approach

E. Heikkilä, R. Tuominen & R. Tiusanen
VTT Technical Research Centre of Finland Ltd, Tampere, Finland

J. Montewka & P. Kujala
Aalto University, Espoo, Finland

ABSTRACT: Autonomous and remote controlled technologies in shipping are becoming a reality with the rapid advances in communication technology, sensor systems, and navigational decision-making software. The emergence of the new autonomous technologies brings along modified and new risks and potential hazards, as well as needs for changes in predicting and controlling the risks to ensure safe operation. In this paper, we present a safety qualification process using a goal based safety case approach that we have specified and applied to facilitate qualification of an autonomous ship prototype for a series of proof-of-concept demonstration trials at sea. As a case example, the goal-based methodology is applied on the situational awareness system of a proposed autonomous vessel prototype.

1 INTRODUCTION

The development towards a higher automation level in transport systems is a clear global trend. Remote controlled and autonomous vehicles garner wide interest in the fields of road and rail traffic (Litman, 2015; Kim et al., 2015), as well as in aviation (Lee, et al., 2016). The sea surface, however, is somewhat different. The today's ships already include a great amount of automation, but fully autonomous commercial systems have not yet been implemented. While technical challenges exist especially related to sensor technology, decision-making intelligence, and system robustness, arguably the biggest problem is the integration of the autonomous ships into the existing maritime transport environments and their regulatory frameworks in a safe and secure way (Rødseth & Burmeister, 2012).

1.1 Scope and objectives

The work presented here has been conducted as a part of the Advanced Autonomous Waterborne Applications (AAWA) initiative. AAWA is a technology project that aims to enable a wider use of autonomous technologies and remote control in future commercial ships (AAWA, 2016). This includes development of an autonomous vessel prototype for testing and demonstration purposes.

The methods described in this paper have been developed to enable planned testing activities to demonstrate autonomous navigation capabilities as installed on an existing ship. Relevant approval from the competent safety authorities is required to perform such tests. Thus, the process introduced in this paper is designed to support the collection of safety evidence regarding the new technology applied, as well as to facilitate communication between the different stakeholders involved in the approval process.

1.2 Challenges in increasing automation level

The increasing level of autonomy in shipping can be seen as beneficial from economic, ecologic and social points of views. As with all new technologies introduced, however, many uncertainties are also present. The risks of safe operation in commercial unmanned shipping have been studied broadly in the MUNIN project (Rødseth & Burmeister, 2015), as well as in the early phases of the AAWA project (AAWA, 2016).

When testing new technologies for the first time in real marine environment and conditions one must be prepared for unexpected situations. Technical failures are likely to occur and software bugs in control algorithms may cause malfunctions. Safe prototype testing at sea must be ensured with special arrangements and specific safety instructions. Generally, it is clear that all current rules and regulations dealing with e.g. manning and watchkeeping will not be fully complied with the autonomous technologies.

The aim of the qualification procedure presented in this paper is to enable tests in areas within national waters, where the national marine safety authority can issue the required test permits. This requires that sufficient evidence is presented about the autonomous operation of the vessel, showing that in all stages of its development it will be at least as safe as the current corresponding vessel complying with the valid rules and standards.

1.3 Safety qualification of new technologies

To be able to attain permissions from the relevant authorities for field testing a new technology, sufficient evidence about safe operation of the technology in the intended use needs to be produced and presented in a clear manner. The process of gathering and structuring this information is called safety qualification. Qualification activities are not only needed when a novel technology is created, but a similar process needs to be carried out also when using proven technology in a new application area or environment

To be able to develop and qualify an acceptably safe system, the related relevant risks need to be recognized and controlled. In all technologies, absolute safety is unobtainable and a certain level of acceptable risk needs to be defined. Acceptability includes achieving the tolerable level of risk, but also the concept of lowering the risk to the lowest reasonably practicable (ALARP) level, regardless of the risk level in absolute terms.

General qualification methods and business specific frameworks for technology qualification have both been published. For marine applications, the available qualification frameworks are available mainly from the classification societies. These include guidelines for general technology qualification (e.g. DNV, 2011; Lloyd's Register, 2014; Bureau Veritas, 2010) as well as specific instructions for assessing cyber-enabled or autonomous vessels (Lloyd's Register, 2016).

Additionally, Formal Safety Assessment (FSA) by International Maritime Organization (IMO) is widely used as a guideline for evaluating the risk impact and efficiency of safety measures and regulations in the maritime domain (IMO, 2002).

2 METHODS

2.1 Safety qualification process for an autonomous ship prototype

As none of the existing safety qualification procedures were found directly applicable for the application at hand, a tailor-made safety

qualification procedure was developed. The structure of the documentation created during the qualification process was given a strong focus, as the primary function of the process is to support communication about the system's safety to acquire operating permissions from the approving authorities.

The process flow is mainly based on the DNV guidance note DNV-RP-A203 (DNV, 2011), with additional views from the corresponding guidance documents for technology qualification by Bureau Veritas and Lloyd's Register. The documentation is presented as a structured safety case, using Goal Structuring Notation (GSN) visualization language (Kelly & Weaver, 2004). Thus, the specification of safety goals is incorporated as a separate step. The process flow is presented in Figure 1 and it consists of the steps described in more detail below.

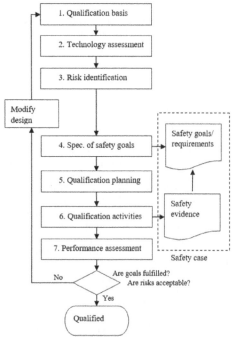

Figure 1. Overview of the safety qualification process applied in AAWA. Adopted from Heikkilä (2016).

1 Qualification Basis: Description of the technology, its operation and the operating context (i.e. environment and conditions) and identification of the relevant legislative requirements and acceptance criteria set by standards and authorities.
2 Technology Assessment: Functional assessment of the technology and degree of novelty assessment.
3 Risk Identification: Identification of risks relevant for the safe operation of the system and the planned risk control techniques.

366

4 Specification of Safety Goals: Formulation of safety goals based on the identified risks and planned risk control techniques in order to achieve the safety equivalence.

5 Qualification Planning: Development of a plan for the qualification activities.

6 Execution of the Plan: Collection of evidence as stated in the qualification plan.

7 Performance Assessment: Assessment of whether the evidence produced meets the authority requirements and the specified safety goals.

The process is of iterative nature, i.e. it shall be re-initiated for the different phases of the technology development process if major technological changes are introduced or the requirements for system performance are revised.

2.2 *Goal-based approach and the structured safety case*

The data creation and collection is only one part of the qualification process. The purpose of the final result of the qualification process is to communicate an argument about the safety of the system, and its compliance with the set objectives and requirements. Both the evidence and the argument are crucial for communicating safety: an argument without evidence is unfounded and unconvincing – and evidence without argument is unexplained, so that it is unclear if (and how) the safety objectives have been satisfied.

When used for demonstrative or persuasive purposes, e.g. for receiving authority approval for a new technology, the data compiled is said to constitute a *Safety Demonstration*. To increase the clarity of the demonstration, the data can be represented as a *Structured Safety Case*. This in turn can be used as a part of a broader *Assurance Case* in addition to economic and technological system assessments.

A powerful way to communicate the qualification results for demonstrating safety is through a goal-based approach. The aim of building a structured, goal-based safety case is to establish the link between the safety requirements and objectives (represented as goals) and the corresponding safety evidence. This can be done by functional decomposition of the system, i.e. representation of the safety-related system functionality as a set of operational goals (Fig. 2).

The safety case, when visualized with the GSN modeling language, consists of a tree structure of safety goals. These are complemented by the operating context, the different strategies (technologies) used for achieving the goals, as well as the evidence produced to show that all the requirements are fulfilled.

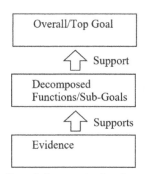

Figure 2. Components of a safety argument when using a goal-based approach. Adopted from DNV (2011).

The strength of a structured safety case in a novel technology development process is that it can be incrementally built as the technology advances or the knowledge on risks gets deepened. The safety case should be initiated at an early stage of development so that relevant hazards and the associated risks can be identified and dealt with at the earliest possible stage. Although the safety case is typically a set of documentation used for gaining permissions to commence operation of a system, the safety case should be maintained throughout the lifecycle of the system. This enables more efficient assessment of possible further modifications or additions to the system (DNV, 2011). By clarifying the role of the individual risk controls, it can also help avoiding dangerous degradation of the risk controls over the operational phase being unnoticed or ignored.

3 CASE APPLICATION: SITUATIONAL AWARENESS SYSTEM

The goal-based approach introduced above was applied in a case study to support the qualification of an autonomous vessel prototype for proof-of-concept tests in the actual sea environment. The basic elements required for autonomous and remote operation of the prototype vessel are presented in Figure 3.

The part of the case study presented in this paper as an example of the goal-based approach focuses on the situational awareness (SA) system. The purpose of the SA system is to produce a view of the surrounding world using various sensors, and to detect and classify relevant objects as input for performing productive and safe navigational decisions. This requires the physical sensor system, as well as the software to perform the object detection and classification. Activities related to docking and undocking, as well as matters concerning physical security of the SA system, are excluded from this example.

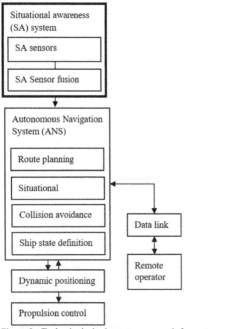

Figure 3. Technological elements proposed for autonomous and remote controlled operation, with the SA system highlighted. Adopted from AAWA (2016).

Various kinds of SA systems have been developed in different industries, e.g. for automotive use. The application area at sea, however, is completely novel. Thus, qualification of the entire SA system is clearly needed.

The risks related to operation of the autonomous vessel prototype were identified in workshops involving personnel with relevant experience of marine operations. Examples of risks identified as relevant regarding the SA system are presented in Table 1.

Table 1. Results of a risk identification including only risks relevant for the SA module.

Risk event	Probable causes	Contributing factors
Collision (with other vessel, floating or static obstacle)	Obstacle not detected in due time	Sensor resolution not sufficient Sensor system not fully operational (malfunction, view blocked, etc.) Environmental conditions (rain, fog, etc.)
	Not classified correctly (type & size)	Limitations in classification Uncharacteristic properties of obstacle

Based on the identified risks, the goals for safe performance were formulated. The overall goal is to show that the autonomous system operates as safely as a corresponding traditional vessel. Thus, the whole safety goal structure needs to argue how this

overall target is to be achieved. The requirements for the safe performance of the SA system are formulated accordingly: the system capability needs to match or exceed the capability of human lookout. The formulation of safety goals based on the identified risks is clarified in Table 2.

Table 2. Safety goals formulated based on the identified risk

Safety goal	Description
Sensor system capability matches or exceeds human watchkeeping capability	Risk of missed objects needs to be similar or smaller when compared to traditional lookout in all operating conditions.
Sensor status data is available	Other systems need to have the information about the sensor system status (broken sensors, etc.)
Sensor system is redundant for losing individual sensors	The system needs to be able to detect objects, even when individual sensors are disabled.
System reliably detects objects larger than 1 meter, from sufficient distance to react	All relevant sized objects need to be detected so that there is time to react if needed.
System classifies detected objects to pre-defined classes	To be able to perform according to COLREGS, the obstacle type needs to be known.

Based on the defined goals, a goal-based model was established. The overall goal for the SA system performance, with the related sub-goals, is presented in Figure 4. The elements are represented using GSN modeling language as follows: goal - rectangle, context - rounded rectangle, strategy - parallelogram, and evidence - circle.

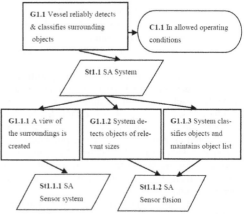

Figure 4. The overall goal G1.1 and the major sub-goals for the SA system, showing the roles of SA sensors and SA sensor fusion as strategies to achieve sufficient level of safety in allowed operating conditions (context C1.1).

Safety goal structures for the SA sensor system and the SA sensor fusion are presented in Figures 5 and 6. The required safety evidence is linked as the lowest level of the structure.

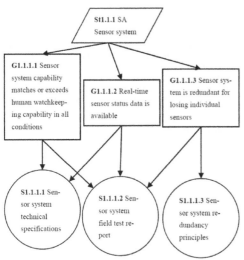

Figure 5. The sub-goals and required safety evidence (represented as circles) for the SA sensor system.

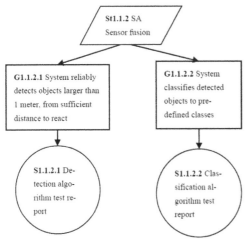

Figure 6. The sub-goals and required safety evidence for the SA sensor fusion.

The detailed contents of the required safety evidence and the methods to be used for collecting it need to be decided in the qualification planning phase. The methods can include (DNV, 2011):

– Analysis of documented earlier experience from similar applications
– Analytical methods (standards, etc.)
– Numerical methods (simulation testing, etc.)
– Experimental methods in simulated or actual field environments

The actual method selection and specification of the detailed evidence is excluded from the case example presented here.

4 DISCUSSION AND CONCLUSIONS

The structured safety case, represented using the GSN modeling language, has been used successfully for assuring safety of complex technologies in different domains, such as aerospace (Witulski et al., 2016) and automotive industries (Palin & Habli, 2010). The maritime sector, however, is considered as a new application area for such approach.

A structured argument regarding the system's safety is established by the goal-based safety case. The created safety case serves as the basis for collecting the system documentation and safety evidence throughout the technology development and approval process.

The major strength of the goal-based approach lies in the communicative power of the visual representation of system's safety requirements, making the link between the safety requirements and evidence easily comprehensible. Designed to be used as a communicative method, however, the goal-based approach does not directly offer tools for prioritizing the multitude of safety goals, i.e. the qualifier needs to have an understanding of the significance of the various goals when planning the qualification activities.

When developing a complex novel technology, the goal-based model can be used as a means of communication between the technology developer and the approving authorities to confirm that sufficient confidence for safe operation of the system has been achieved to grant the permissions for field testing and operation. It can also be used to provide guidelines for planning the testing activities by defining what additional evidence needs to be produced during the trials, and how the trials should be organized to answer these needs.

REFERENCES

AAWA. 2016. Technologies for marine situational awareness and autonomous navigation. In AAWA, *Remote and autonomous ships. The next steps.* London: Rolls-Royce plc. http://www.rolls-royce.com/~/media/Files/R/Rolls-Royce/documents/customers/marine/ship-intel/aawa-whitepaper-210616.pdf

Bureau Veritas. 2010. Guidance Note NI 525: Risk Based Qualification of New Technology – Methodological Guidelines. Neuilly sur Seine Cedex: Bureau Veritas.

DNV. 2011. Recommended Practice DNV-RP-A203: Qualification of New Technology. DNV.

Heikkilä, E., 2016. Safety Qualification Process for Autonomous Ship Concept Demonstration. M.Sc. Thesis. Espoo: Aalto University.

IMO. 2002. Guidelines for Formal Safety Assessment (FSA) for Use in the IMO Rule-Making Process (MSC/Circ.1023 - MEPC/Circ.392). London: IMO.

Kelly, T. & Weaver, R. 2004. The Goal Structuring Notation – A Safety Argument Notation. York: University if York, Department of Computer Science and Department of Management Studies.

Kim, J., et al. 2015. Automatic train control over LTE: design and performance evaluation. *IEEE Communications Magazine* 53(10): 102-109.

Lee, Y. S., et al. 2016. An Overview of Unmanned Aerial Vehicle: Cyber Security Perspective. *Asia-pacific Proceedings of Applied Science and Engineering for Better Human Life* 4(1): 128-131.

Litman, T. A. 2015. Autonomous Vehicle Implementation Predictions: Implications for Transport Planning. Victoria, Canada: Victoria Transport Policy Institute

Lloyd's Register. 2014. Guidance Notes for Technology Qualification. London: Lloyd's Register.

Lloyd's Register. 2016. Cyber-enabled ships – a Lloyd's Register Guidance Note

Palin, R. & Habli, I. Assurance of automotive safety – A safety case approach. *Computer Safety, Reliability, and Security. SAFECOMP 2010. Lecture Notes in Computer Science, vol 6351.* Berlin: Springer.

Rødseth, Ø. J. & Burmeister, H.-C. 2012. Developments toward the unmanned ship. *International Symposium "Information on Ships" (ISIS).* German Institute of Navigation.

Rødseth, Ø. J. & Burmeister, H.-C. 2015. Risk Assessment for an Unmanned Merchant Ship. *TransNav, the International Journal on Marine Navigation and Safety of Sea Transportation* 9(3): 357-364.

Witulski, A. et al. 2016. Goal Structuring Notation in a Radiation Hardening Assurance Case for COTS-Based Spacecraft. *Government Microcircuits Applications & Critical Technologies Conference.* NASA.

Optical Target Recognition for Drone Ships

M. Fiorini
Leonardo S.p.A., Rome, Italy

A. Pennisi
Katholic University Leuven, Leuven, Belgium

D.D. Bloisi
University of Verona, Verona, Italy

ABSTRACT: Remote controlled drone ships without crews on board are expected by the end of the decade. To achieve the goal of developing (semi-)autonomous boats, reliable vision-based methods for vessel detection, classification, and tracking are needed. In this paper, we present a machine learning approach for vessels detection from a moving and zooming camera. In particular, the proposed method is supervised and derives from a fast and robust people detection algorithm. Quantitative experimental results have been obtained on a publicly available data set, which contains images from real sites, demonstrating the effectiveness of the approach. Ground truth annotations and the code of the proposed algorithm are both released for the community.

1 INTRODUCTION

Drones are widely attracting attention in the media after the Internet activist Pirate Party has managed to interrupt Chancellor Angela Merkel and Defence Minister at a CDU campaign event in Dresden, Germany on September 16, 2013 making use of a miniature drone started circling above the audience.

UAVs (Unmanned Aerial Vehicles) technology is widely available to hobbyists and environmental scientists at affordable prices. Applications include land management, animal conservation, crop monitoring, and disaster mapping. The marine and maritime sector is not excluded from the possibility of using UAVs based solution. For example, underwater robots are able to gather environmental information, scour and sediment transport analysis, meanwhile drone ships will enable new business models in near future. According to Oscar Levander, head of the innovation marine unit at Rolls-Royce, remote controlled drone ships without crews on board may generate, by the end of the decade, a similar disruptive effect as the one provoked by Uber, Spotify and Airbnb on other industries. Totally autonomous ships are just a step ahead.

Optical tracking features are now present at different stages of development and integration in almost all surveillance applications, fixed or mobile, equipped with cameras. However, in order to allow those technologies to be used for autonomous vessels, targets recognition, i.e., classification, are needed. The target identification process, coupled with a decision support software module, allows to rise warning issues for potential collisions and to modulate speeds. Moreover, in the context of maritime boarder control and Search And Rescue (SAR), vessels patrolling are still a widely used procedure. These operations require considerable effort and resources, which could be considerably reduced by autonomous patrolling vessels.

The aim of this work is to give a general overview of existing optical based recognition solution in the maritime context and to present a machine-learning based approach for vessel classification and detection, which is a fundamental requirement to achieve autonomous navigation.

The remain of the paper is organized as follows. Section 2 provides an overview of existing techniques for vessels, humans, and floating objects detection. The proposed approach is presented in Section 3. Qualitative and quantitative experiments are shown in Section 4. Finally, conclusions are drawn in Section 5.

2 RELATED WORK

Maritime environment represents a challenging scenario for automatic object detection due to the complexity of the observed scene: High frequency background objects (e.g., waves on the water surface), boat wakes, and weather issues (e.g., heavy raindrops) contribute to generate a highly dynamic scenario (Bloisi et al., 2014).

2.1 *Vessel Detection and Tracking*

Vessel detection in the maritime scenario requires the monitoring of large areas at different resolution levels. Indeed, the objects of interest can have very different size, ranging from few to hundreds of meters in length. SeeCoast system (Rhodes et al., 2006) detects, classifies, and tracks vessels by fusing electro-optical (EO) and infrared (IR) video data with radar and AIS data. The detection is carried out by estimating the motion of the background and segmenting it into components. However, motion-based vessel detection can experience difficulties when a boat is moving directly toward the camera or is anchored off the coast due to the small amount of inter-frame changes.

Maximum Average Correlation Height (MACH) filters are employed for vessel classification by Sullivan & Shah (2008). Vessel detections are cross-referenced with ship pre-arrival notices in order to verify the access of vessels to the port. As reported by the authors, such an approach tends to misclassify small boats. Fefilatyev et al. (2010) propose a system exploiting a non-stationary camera installed on an untethered buoy. After detecting the horizon line, a color gradient filter is applied to obtain a grayscale image with intensities corresponding to the magnitude of color changes, then thresholding on the grayscale image is used to find objects of interest. The algorithm is limited by the assumption that all marine targets are located above the horizon line. ASV (Pires et al., 2010) is an automatic optical system for maritime safety using IR, GPS, and AIS. To detect relevant objects, the sea area is segmented and its statistical distribution is calculated. Any irregularities from this distribution are supposed to correspond to objects of interest. However, due to wakes, such an approach can produce false positives. An object detection system for finding ships in maritime video based on Histogram of Oriented Gradients (HOG) is described by Wijnhoven et al. (2010). Since the calculation of the detection features involves a significant amount of computational resources, real-time performance can be obtained only by means of hardware acceleration with programmable components such as FPGAs.

2.2 *Floating Objects Detection*

Floating objects detection plays an important role in USVs (Unmanned Surface Vehicles). Indeed, obstacles in the operational environment can be floating pieces of wood or other debris, which presents a significant challenge to continuous detection from images taken on-board (Kristan et al., 2016). Snyder et al. (2004) state that transitory obstacles, such as floating debris, are best detected and analyzed at navigation time with visual means. They propose a system for fully autonomous navigation in a river scenario. Obstacles and objects of interest are tracked across multiple cameras (by using feature clusters in aspect-elevation space) and then mapped onto the 3D world. Distant moving objects are detected and tracked by clustering feature points, while nearby movers are detected with motion blobs. Stereo rigs are used by Huntsberger et al. (2011) for obstacle and moving objects detection on a USV. However, since a large baseline is required for granting a large field of view coverage, this can create instability for small vessels. A method for detecting water regions in videos by clustering color and texture features is proposed by Santana et al. (2012). Fefilatyev et al. (2009) use the horizon line position to eliminate all edges not belonging to floating objects: it is assumed that within an image, all objects of interest lie above the horizon line. A similar idea is exploited by Wang et al. (2011): They first detect the horizon line and then search for a potential obstacle in the region below the horizon.

The main drawback of approaches based on the horizon line detection is that situations in coastal waters, close to the shoreline, cannot be easily handled, since the edge of water does not correspond to the horizon.

2.3 *Humans Detection*

In many USV applications, humans are considered just a special case of obstacles to be avoided (Almeida et al., 2009). However, distinguishing between human and non-human detections is important in order to devise different and more opportune navigation strategies, for example in case of casualty detection for search and rescue operations. Differently from a static floating objects, a person may be swimming or diving, possibly without being aware of an approaching USV, and this poses serious concerns for human safety. In recent years, the use of thermal images from Forward-Looking Infrared (FLIR) cameras for target detection and tracking has become popular in various application domains (Sanna & Lamberti, 2014).

Independent FLIR cameras have also been mounted on vessels and used for target detection and tracking at sea (Kim & Lee, 2014). Martins et al. (2013) uses a combination of thermal and colour images to detect casualties from a USV in the context of the FP7 project ICARUS. A simple horizon detection algorithm was applied to the thermal image to limit the search space to the water surface only, then targets detected by both (fixed and calibrated) cameras were tracked using Kalman filtering to deal with false positives and temporary false negatives. Although computationally efficient, the detection coverage of the proposed solution is limited by the narrow field of view and by the minimum focusing distance of the thermal camera, which made the system unsuitable for tracking humans in proximity of the USV. To increase the water surface area covered by the cameras, the latter can be mounted on a stabilized pan-tilt unit (PTU), as proposed by Bibby & Reid (2005). Underwater human detection can be achieved instead by using multibeam or mechanically-rotated sonars, similarly to what is already done for obstacle detection with autonomous surface vehicles (Heidarsson & Sukhatme, 2011).

3 PROPOSED APPROACH

To detect the boats, a method based on Aggregated Feature Channels (AFC) has been adopted. The method has been presented by Dollar et al. (2014) and is made of three main steps:
1 Feature extraction;
2 Pyramid computation;
3 Classification.
Each step is detailed below.

3.1 *Feature Extraction*

To represent a target, a set of feature channels is extracted from the images. Given an image I and a function Ω that represents the process of extracting features, we indicate a channel as $C = \Omega(I)$. Then, the information contained in each channel is aggregated over multiple pixel by summing (Σ) every block of pixels in C, and smoothing the resulting lower resolution channels. In such a way, a feature is a single pixel lookups in the aggregated channels. The process is shown in Figure 1.

The feature vector contains the following features: normalized gradient magnitude, histogram of oriented gradients (6 channels), LUV color channels, and integral image. The channels are divided into 4×4 blocks and the pixels in each block are summed. Finally, the channels are smoothed by applying a Gaussian filter.

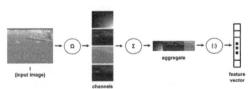

Figure 1. Functional architecture of the process for generating the feature vector. The input image is split in channels, then the channels are aggregated using a lower resolution. As in (Dollar et al., 2014), the feature vector is composed by single pixel lookups in the aggregated channels.

3.2 *Pyramid Computation*

A feature pyramid is a representation of an image at multiple scales. Usually, the number of scales s is between 4 and 12 in a log-space starting from $s = 1$. The common approach to build a pyramid is to compute a set of feature channels at each scale. However, this can be computationally expensive. Instead of computing for each scale a large amount of features, Dollar et al. (2014), propose a way to approximate the features within channels. Being I_s the scaled image of I, the feature channels C_s is computed by using an approximation of the information contained in $C = \Omega(I)$. In particular,

$$C_s \approx R(C,s)s^{\lambda\Omega} \tag{1}$$

where $R(C,s)$ represents the resampled features by s, while λ is a constant. Equation 1 is valid not only for the images, it is valid also for any corresponding window w_s and w in I_s and I, respectively.

C_s is computed at one scale per *octave*, where an octave is the interval between one scale and another with half or double its value. While, at the intermediate scales, it is computed as

$$C_s \approx R(C_{s'}, s/s') (s/s')^{-\lambda\Omega} \tag{2}$$

where $s' = \{1, \frac{1}{2}, \frac{1}{4}, \ldots\}$ is the nearest scale for which $C_{s'} = (\Omega I_{s'})$.

The above described approach represents a good compromise between speed and accuracy. Indeed, the cost of evaluating for the approximated scales is within 33% of computing $\Omega(I)$ at the original scale, and moreover, the channels do not need to be approximated more than half an octave.

1.1 *Classification*

A boosted tree classifier is used for detecting the boats. The classifier combines 2048 depth-two trees over all the candidate features (i.e., the channel pixel lookups) in each window. Since a vessel can be contained into a bounding box with height x and width $2x$, the size of the window is equal to 128×64 pixels. The detector has a step size of 4 pixels and 8 scales per octave.

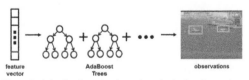

Figure 2. Adaptive Boosting is used to obtain the observations about the objects of interest. As in (Dollar et al., 2014), a set of decision trees is built over the feature vector provided by in order to distinguish object from background.

Figure 2 shows the process for obtaining the final observations, where multiple vessels can be detected in the same image. For training the classifier, a total of 288 boat samples from four different videos has been collected. The annotated images and the videos used for the training stage, as well as the video sequences used for the testing phase (including ground-truth data) are available at: https://github.com/apennisi/fast_vessel_detection

4 EXPERIMENTAL RESULTS

An experimental evaluation has been carried out on visual data coming from a real site to validate the proposed approach. In particular, we have decided to use a video from the publicly available Maritime Detection, Classification, and Tracking (MarDCT) data base (http://www.dis.uniroma1.it/~labrococo/MAR/). MarDCT (Bloisi et al., 2015) contains a collection of videos and images captured with different camera types (i.e., fixed, moving, and Pan-Tilt-Zoom cameras) and in different scenarios. The aim of MarDCT is to provide visual data that can be used to help in developing intelligent surveillance system for the maritime environment.

Figure 3. Boat detection using the proposed method on different videos from the MarDCT data base. The algorithm runs at multiple scales and can detect vessels of different size with varying lighting conditions.

Qualitative Evaluation. The proposed approach has been applied to several sequences. In particular, we tested 6 different videos from a moving and zooming camera. The experiments show (see Fig. 3) that our approach achieves good results in most of the videos, also considering that a small data set

(288 samples) has been used for training the classifier. In the top left image of the Figure 3, the big cruise boat is not detected since no samples of similar boats are contained in the training set.

Quantitative Evaluation. In order to evaluate the robustness of the detection module, we have carried out quantitative quality measurements by calculating the Precision, the Recall (or True Positive Rate - TPR), and the F1-score. The used metrics are defined as follows.

$$Precision = \frac{TP}{TP + FP} \qquad (3)$$

$$Recall = \frac{TP}{TP + FN} \qquad (4)$$

$$F1\text{-}score = 2\frac{Precision \cdot Recall}{Precision + Recall} \qquad (5)$$

where TP is the number of the true positive observations, FP is the number of false positives, and FN is the number of false negatives. F1-score gives a weighted average of the precision and recall.

The results of the experiments are shown in Table 1 and Table 2 and are totally reproducible, since the algorithm's code and the testing video are provided.

Table 1. The proposed approach has been tested extracting 116 frame samples from a video from a moving and zooming camera

Num. Frames	Num. Samples	TP	FP	FN
116	82	66	30	14

Table 2. Quantitative Results. Three different metrics has been used to measure the quality on the proposed approach on a real video.

Precision	Recall	F1-score
0,688	0,825	0,375

Runtime Performance. The detector has been implemented by using MATLAB for the training phase of the classifier and C++ for the testing stage. The training phase is offline and should be computed once, while the code of the testing stage has been realized to allow a real-time computation. In particular, for 640×480 images, the complete pipeline runs at about 30 fps (frames per second).

5 CONCLUSIONS

In this paper we have presented a fast and robust method for vessel detection from moving and zooming cameras. The proposed algorithm represents a valid baseline for building a Maritime Unmanned Navigation system. In particular, our algorithm is derived from a solid supervised method originally conceived for people detection, which runs at real-time speed (Dollar et al., 2014).

A quantitative experimental evaluation has been carried out on videos coming from the publicly available database MarDCT, containing data from different real sites captured with varying lighting conditions.

As future work, we intend to extend the method to run with images coming from omni-directional (360°) cameras and to add a module for the coastline detection. Indeed, tracking the coastline can provide useful information for the heading (yaw) of the drone ship, while pitch and roll values could be obtained by inertial sensors on board.

REFERENCES

Almeida, C. et al. 2009. Radar based collision detection developments on USV ROAZ II. In *Oceans - Europe*, 1-6.

Bibby, C. & Reid, I. 2005. Visual Tracking at Sea, In Proc. of *IEEE Int. Conf. on Robotics and Automation*, 1841-1846.

Bloisi, D. D. Pennisi, A. & Iocchi, L. 2014. Background modeling in the maritime domain. Machine Vision and Applications 25(5): 1257-1269.

Bloisi, D. D. Iocchi, L. Pennisi, A. & Tombolini, L. 2015. ARGOS-Venice Boat Classification. In *IEEE Int. Conf. on Advanced Video and Signal Based Surveillance*, 1-6.

Dollar, P. Appel, R. Belongie, S. & Perona, P. 2014. Fast feature pyramids for object detection, IEEE Trans. Pattern Anal. Mach. Intell. 36: 1532–1545.

Fefilatyev, S. Goldgof, D. B. & Lembke, C. 2009. Autonomous buoy platform for low-cost visual maritime surveillance: design and initial deployment. In Proc. SPIE 7317, Ocean Sensing and Monitoring, 73170A.

Fefilatyev, S. Goldgof, D. B. & Lembke, C. 2010. Tracking Ships from Fast Moving Camera through Image Registration. In *Proc. of the Int. Conf. on Pattern Recognition*, 3500-3503.

Heidarsson, H. & Sukhatme, G. 2011. Obstacle detection and avoidance for an Autonomous Surface Vehicle using a profiling sonar. In *IEEE Int. Conf. on Robotics and Automation*, 731-736.

Huntsberger, T. Aghazarian, H. Howard, A. & Trotz, D. C. 2011. Stereo vision based navigation for autonomous surface vessels. JFR 28(1): 3–18.

Kim, S. & Lee, J. 2014. Small Infrared Target Detection by Region-Adaptive Clutter Rejection for Sea-Based Infrared Search and Track. Sensors 14(7): 13210-13242.

Kristan, M. Sulić Kenk, V. Kovačič, S. & Perš, J. 2016. Fast Image-Based Obstacle Detection From Unmanned Surface Vehicles. IEEE Trans. on Cybernetics 46(3): 641-654.

Martins, A. et al. 2013. Field experiments for marine casualty detection with autonomous surface vehicles, In *Oceans - San Diego*, 1-5.

Pires, N. Guinet, J. & Dusch, E. 2010. ASV: An Innovative Automatic System for Maritime Surveillance. Navigation 58(232): 1-20.

Rhodes, B. J. Bomberger, N. A. Seibert, M. & Waxman, A. M. 2006. SeeCoast: Automated port scene understanding facilitated by normalcy learning. In *Proc. IEEE Military Communications Conference*, pp. 1–7.

Sanna, A. & Lamberti, F. 2014. Advances in Target Detection and Tracking in Forward-Looking InfraRed (FLIR) Imagery. Sensors 14(11): 20297-20303.

Santana, P. Mendica, R. & Barata, J. 2012. Water detection with segmentation guided dynamic texture recognition. In *IEEE Robotics and Biomimetics*.

Snyder, F. D. Morris, D. D. Haley, P. H. Collins, R. T. & Okerholm, A. M. 2004. Autonomous river navigation. In *Proc. SPIE 5609, Mobile Robots XVII*, 221.

Sullivan, M. D. R. & Shah, M. 2008. Visual surveillance in maritime port facilities. In *SPIE Optics and Photonics*, 697 811–697 811.

Wang, H. Wei, Z. Wang, S. Ow, C. Ho, K. & Feng, B. 2011. A vision based obstacle detection system for unmanned surface vehicle. In *Int. Conf. Robotics, Aut. Mechatronics*, 364–369.

Wijnhoven, R. G. J. van Rens, K. Jaspers, E. G. T. & de With P. H. N. 2010. Online Learning for Ship Detection in Maritime Surveillance. In *Proc. of the 31st Symposium on Information Theory in the Benelux*, 73-80.

Communications and Global Maritime Distress
and Safety System (GMDSS)

Voice Subtitle Transmission in the Marine VHF Radiotelephony

O. Shyshkin & V.M. Koshevyy
National University "Odessa Maritime Academy", Odessa, Ukraine

ABSTRACT: Imperceptible transmission of speech-to-text recognized characters by means of audio watermarking is addressed for application in the standard VHF radio telephone communication. Practicability of this design is based on advances in automatic speech recognition, especially taking into account utilization of IMO Standard Marine Communication Phrases and interactive recognition process. Watermarking performances are provided by applying power preserving algorithm in critical Bark bands; OFDM like data transmission in parallel through narrowband frequency channels and error correcting encoding. Watermarking system allows embedding up to 16 five-bit characters per sec in synchronous regime under influences of intersymbol interference and additive noise of signal-to-noise ratio 15 dB. Embodiment of the design is based on standard VHF communication installation and has to provide efficiency of voice messaging by means of text visualization.

1 INTRODUCTION

User needs are a priority criterion in the concept of e-navigation when introducing new technologies in navigation. The Nautical Institute conducted a questioning of seamen regarding their wishes on improvement of GMDSS. The navigators expressed opinion in relation of VHF communication that "the use of voice for traffic management and SAR operations was considered essential, however if in the future it would be possible for a text version of the voice message to augment the voice that would be very beneficial" [1].

The implementation of navigator's interests can be realized on the basis of modern technologies - speech recognition and digital steganography. An idea on text accompanying of phone message was expressed in [2].

Speech recognition (SR) is exclusively extensive direction of modern technologies. It has a wide range of applications in domains such as health care, military, telephony, dictation, robotics, and home automation. The SR is used in developing systems for speech-to-text conversion, language recognition, and audio search engine [3].

This article is not aimed to the development of speech recognition methods. Our goal is analysis of factors influencing the speech recognition process on the bridge of a ship. Another object is designing a system for steganographic (or watermarking) embedding the recognized text and subsequent extracting it at a receiving part.

2 SPEECH RECOGNITION OVERVIEW

Speech recognition is the ability to recognize spoken words only and not the individual voice characteristics such as emotions, timbre, loudness, male/female attribute, accent, so on. Also we consider SR only for audio signal, i.e. phone recognition without any visual perception.

SR technology from its origin in the middle of previous century achieved perfection today, passing the following events [4].

– 1950s and 1960s: speech recognition (SR) takes first steps. The first SR systems could understand only digits. In 1962 IBM demonstrated machine, which could understand 16 words spoken in English.

– 1970s: SR technology made major strides thanks to interest and funding from the U.S. Defense Advanced Research Projects Agency (DARPA) speech-understanding system. The Harpy SR system could understand one thousand words corresponding to average three-year-old child.

– 1980s: Over the next decade, thanks to new approaches to understanding what people say, SR vocabulary jumped from about a few hundred words to several thousand words, and had the potential to recognize an unlimited number of words. One major reason was a new statistical method known as the hidden Markov model (HMM). Rather than simply using templates for words and looking for sound patterns, HMM considered the probability of unknown sounds' being words. However, SR programs didn't work for continuous speech. The programs took discrete dictation, so you had ... to ... pause ... after ... each ... and ... every ... word.

– 1990s: Automatic SR comes to the masses. In the '90s, computers with faster processors finally arrived, and speech recognition software became viable for ordinary people. In 1990, Dragon launched the first consumer speech recognition product, Dragon Dictate, for price of $9000. Later Dragon Naturally Speaking application recognized continuous speech at about 100 words per minute. However it needed the training program for 45 minutes, and it was still expensive at $695.

– 2000s: SR technology flew up with the arrival of Google Voice Search app for the iPhone. The impact of Google's app is significant for two reasons. First, the tiny on-screen keyboard of mobile phones serves as an incentive to develop better, alternative input methods. Second, Google had the ability to offload the processing for its app to its cloud data centers, harnessing all that computing power to perform the large-scale data analysis necessary to make matches between the user's words and the enormous number of human-speech examples it gathered. This whole process lasts reportedly only seconds, depending of course of the speed of the network you are connected to.

– Present and the future: Microsoft has made a major breakthrough in SR (October 2016), creating a technology that recognizes the words in a conversation as well as a person does. The researchers reported a word error rate (WER) of 5.9 percent [5]. The explosion of SR apps indicates that speech recognition's time has come, and that we can expect plenty more apps in the future.

It is known substantial achievements for SR systems operating in hard environment, for example in war aviation. The problems of achieving high recognition accuracy under stress and noise pertain strongly to the helicopter environment as well as to the jet fighter environment [6].

The performance of speech recognition systems is usually evaluated in terms of accuracy and speed. Accuracy is usually rated with word error rate (WER), whereas speed is measured with the real time factor. Other measures of accuracy include

Single Word Error Rate (SWER) and Command Success Rate (CSR).

However, speech recognition (by a machine) is a very complex problem. Vocalizations vary in terms of accent, pronunciation, articulation, roughness, nasality, pitch, volume, and speed. Speech is distorted by a background noise and echoes, electrical characteristics. Accuracy of speech recognition varies with the following factors [3]:

1 Vocabulary size;
2 Speaker dependence vs. independence;
3 Isolated, discontinuous, or continuous speech;
4 Task and language constraints;
5 Read vs. spontaneous speech;
6 Adverse conditions.

Let us analyze these factors in relation to maritime communication.

1 Maritime communication should be based on the using of IMO Standard Marine Communication Phrases (SMCP). SMCP should be made as often as possible in preference to other wording of similar meaning. As a minimum requirement, mariners should adhere as closely as possible to them in relevant situations.

The SMCP includes phrases which have been developed to cover the most important safety-related fields of verbal shore-to-ship (and vice versa), ship-to-ship and on-board communications. The aim is to reduce the problem of language barriers at sea and avoid misunderstandings which can cause accidents.

The limited volume of SMPC vocabulary is very beneficial for SR quality.

2 Systems that do not use training are called "speaker independent" systems. Speaker independent system is more suitable for bridge communication because of periodical officer of the watch replacement on a bridge and crew exchange.

Speech for communication considerably differs from common place verbal dialog, which is characterized by continuous speech. The professional telephony looks rather discontinuous then continuous speech. The main guidelines to follow radiotelephony rules are:

– Use all prescribed words, codes and phrases;
– Speak clearly without emitting extraneous sounds;
– Establish, when there is time, a pitch that is most comprehensible, as well as the optimal distance of mouth from microphone, tempo and volume;
– Follow the established order of transmission where applicable.

3 Speaking and listening on the radio are distinct activities from normal speaking and listening, and this is particularly important to remember when speaking. Radio transmission clarity requires a shift from natural habits of speech into a more

self-conscious procedure emphasizing clarity, brevity and certainty. Following the directives of pronunciation is imperative for the operator. The operator must always remain conscious of precision in pronunciation, and refrain from inserting extra sounds into his speech - even such reflexive habitual sounds as 'ah', 'er', 'um', etc. At the initiation of any transmission the operator should be aware of the microphone placement, so that he is not to close or too far from the microphone. In urgent situations it is common for the speaker to 'mouth' the microphone, blurring the sounds so that they become incomprehensible.

These SMCP instructions work in favor of WER reduction.

4　Language constraints are of particular importance in the light of internationally trading vessels with crews speaking many different languages since language problems may complicate SR process. Variable situations and communication priorities may also cause errors in recognition. But again SMCP standard demands for navigational and safety communications from ship to shore, shore to ship, ship to ship must be precise, simple and unambiguous, so as to avoid confusion and error in understanding by another party and therefore in automatic voice recognition.

5　Reliable SR should be done both for readable text and the natural voice.

6　Acoustic environment in standard conditions is quite favorable, and may not have a destructive effect on the recognition quality.

The most significant factor affecting the quality of recognition is the realization of interactive voice translation into text. During voice to text translation recognized words should be displayed with minimum delay in the running line on the display. If an error occurs the operator repeats a word or a whole sentence, ensuring correct recognition. 100 percent conversion accuracy can be achieved in this mode.

3　STEGANOGRAPHIC TEXT EMBEDDING

First of all, estimate the necessary text symbol rate in the telephone channel.

This rate directly depends on the speech rate. The rate of speech indicates the average number of words a speaker says per minute. In personal conversations if a speaker wants to be sure the listener understands his points, he may speak slower than 100 words per minute (WPM) [7]. Obviously this assessment regarding SMCP rules may be accepted also for telephone communication. An average English word consists of 4 – 5 letters. Including space it comes to 5 – 6 symbols per word.

So estimation of 500 – 600 symbols per minute may be accepted as a base.

Taking 5 bit format per each symbol encoding, one can obtain the necessary text bit rate of 42 – 50 bps. Because of text redundancy, the averaged bit rate can be reduced even 2 - 3 times as compared with the above figures using effective variable length coding procedures (or entropy encoding), example Huffman encoding [8].

To solve the problem of supporting voice message by transmission the recognized text we use steganographic technique for bit stream embedding directly into the audio signal. Steganographic data embedding calls neither additional frequency or time channel resources nor replacement of existing radio installation. Employment of such system is fully compatible with commonly used equipment.

Steganography implies to concealing certain data information within another information product: computer file, audio, image, video, etc. Another similar technology for hidden transfer of information is digital watermarking. Both steganography and digital watermarking employ steganographic techniques to embed data covertly in the host signals. But whereas steganography aims for imperceptibility to human senses, digital watermarking is typically used to identify ownership of the copyright.

Audio watermarking (AW) approach was used for automatic identification of VHF radiotelephone transmission in maritime communication [9,10]. AW encoding algorithm is based on the key operations:

1　short fast Fourier transform FFT;

2　partitioning of audio frequency interval to a certain number of narrowband subchannels;

3　quantization of correlation coefficient $\tilde{x} = (x,u)/\|x\|\|u\|$ that leads to certain modification in amplitude of frequency harmonic x to s while minimizing the introduced distortions $\|x - s\| = \min$ and preserving the harmonic power $\|s\| = \|x\|$ within subchannel.

Above designations are:

$x = (x_1, x_2, ..., x_L)$ - origin (host) signal vector,
$s = (s_1, s_2, ..., s_L)$ - watermarked (modified) vector,
$u = (u_1, u_2, ..., u_L)$ - certain random vector.

Processing in the frequency domain (item 1) and partitioning the whole frequency band to narrowband channels (item 2) are motivated by insuring watermarks robustness against intersymbol interference.

Besides robustness WM should be imperceptible by hearing. Auditory WM insensitivity depends not on the objective mean-square deviation of the signal, but is determined by the subjective human perception. The main property of audio perception lies in the existing of the critical bands and masking phenomena. A critical band can be related to a bandpass filter whose frequency response corresponds roughly to the tuning curves of auditory neurons. A critical band defines a frequency range in psychoacoustic experiments for which perception abruptly changes as a narrowband sound stimulus is modified to have frequency components beyond the band [11].

If a noise signal bandwidth is gradually expanded while keeping its total energy, the auditory perception will be constant until the critical bandwidth is attained. As soon as the signal bandwidth is spread out over a critical band, the sound will be perceived abruptly louder.

This phenomenon is explained by anatomy and physiology of inner ear. In the inner ear a snail-shaped formation cochlea is placed. In the cochlea there is basilar membrane which makes significant signal processing in converting sound signal into neural stimuli. Different local areas along basilar membrane resonate to own sound frequency. This leads to a tonotopic organization of the sensitivity to frequency ranges along the membrane, which can be modeled as being an array of overlapping band-pass auditory filters. The auditory filters are associated with points along the basilar membrane and determine the frequency selectivity of the cochlea, and therefore the listener's discrimination between different sounds. The bandwidth of the auditory filter is just coincides with appropriate critical band. The auditory filter responds more likely on total sound energy in critical band than on spectrum details.

Critical bands notion explains frequency masking effect [11]. If a signal and masking sound are presented simultaneously then only the masker frequencies falling within the critical bandwidth contribute to masking of the signal. The larger the critical bandwidth the lower the signal-to-noise ratio and the more the signal is masked.

We applied hearing mechanism based on critical bands notion to auditory perception of embedded watermarks. The designed algorithm is based on partition audio frequency band to sub bands that coincide with critical bands. So every subchannel contains variable number of frequency coefficient in ascending mode from low to high frequency (see Table 1).

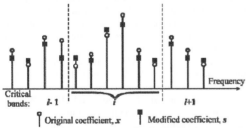

Figure 1. Harmonic modification in the critical bands

Table 1. Coefficient distribution in critical Bark bands

Bark band number	Cut-off frequency, Hz	Bandwidth, Hz	Number of coefficients
1	100	80	6
2	200	100	6
3	300	100	6
4	400	100	7
5	510	110	7
6	630	120	8
7	770	140	9
8	920	150	9
9	1080	160	11
10	1270	190	12
11	1480	210	13
12	1720	240	16
13	2000	280	18
14	2320	320	20
15	2700	380	24

Standard telephone channel covers the first 15 (from 24) critical or Bark bands. Distribution of Fourier coefficient is shown in the 4th column Table 1 under the following assumption: sampling frequency $F_s = 8$ kHz, FFT dimension $N = 512$.

Data embedding process in our method is as follows. Fourier coefficients from 1st to 174th for every signal time frame of duration $T_f = N / F_s = 64$ ms are allocated through subchannels with accordance Table 1. Zero coefficient (direct current) isn't used. Every subchannel carries one data bit. Due to bit insertion harmonic amplitudes are slightly modified so that the correlation coefficient $\tilde{x}$ has taken position on even or odd quantization level depending on the bit. Original and modified coefficients for i subchannel are shown in Fig. 1 with circles and black squares, respectively. Due to the variable subchannel bandwidth the length L of signal vectors x, s, u vary from 6 to 24.

It is essential that the power within every subchannel remains unchangeable. Power preserving improves auditory insensitivity of performed modifications because of critical bands effect.

The algorithm [9] also deals with the subdivision of sound spectrum into narrowband frequency subchannels. However, this subdivision has only purpose to combat intersymbol interference (ISI) like orthogonal frequency division multiplexing (OFDM) technology. Possible imposition of a sub-channel to the border of adjacent critical bands leads to a redistribution of the signal power in these bands, and as a consequence - to the audible artifacts. The proposed method eliminates this drawback while keeping OFDM technology against MSI.

Detection reliability of embedded text after channel passing is increased by means of error correcting code [12]. Rather simple BCH block code (15,5,3) was used. It has the next parameters: block length $n = 15$, number of information bit per block $k = 5$, number of corrected bits per block $t = 3$. The probability of block error decoding can be obtained from the formula [12]:

$$P_B \approx \frac{1}{n} \sum_{i=t+1}^{n} C_n^i \, p^i \left(1-p\right)^{n-i} \tag{1}$$

where

$$C_n^i = \frac{n!}{i!(n-i)!} - \text{binomial coefficients,}$$

p - channel bit error probability (BER).

The value of BER depends on watermark robustness and external channel interferences. Trade-off robustness - fidelity is achieved by selecting the correlation coefficient quantization step in the embedding algorithm.

In the experiments the test phrase "obviously navigation is the prime application of GPS devices" was used. It contains 50 symbols including spaces. Voice track is about 5 sec. This phonogram was used as a carrier signal for character sequence at a rate of 16 char per sec. Each character is represented randomly by five bits. So that the useful bitrate was 80 bps under sampling frequency $F_s = 8$ kHz and FFT dimension $N = 512$.

Experiment was carried out only in the part of watermarking and not included voice recognition stage. Considering that the symbol rate after recognition process changes substantially, and the watermarking operation is carried out synchronously at a constant rate, special symbols are used to fill the spaces and maintain symbol synchronization.

The total number of symbols in the phonogram including special was 75. Numbers of correctly decoded symbols (including special symbols) are shown in the Table 2, depending on the Signal-to-Noise Ratio (SNR) in the channel and quantization step.

Fig. 2 illustrates the waveform of the test voice message after watermarking and additive noise influence of SNR=15 dB and dispositions of correctly decoded symbols (dark marks). Subjective estimations of watermarked signal artifacts may be interpreted as "practically inaudible" ($\Delta = 0.5$) and "slightly noticeable" ($\Delta = 1.0$).

Table 2. Numbers of correctly decoded symbols

SNR, dB	Quantization step	
	$\Delta = 0.5$	$\Delta = 1.0$
30	72	75
25	71	73
20	62	68
15	52	59

4 CONCLUSION

Progress in speech recognitions technology voice-to-text creates conditions for improving maritime radiotelephony in the direction of formation a supplementary channel for text transmission. In parallel with the voice text displaying is one of the mariner's wishes for a better contact between navigators during VHF communication. The solution of this problem within the framework of compatibility with existing VHF GMDSS radio installation is possible by using the digital audio watermarking technology of voice messages.

Simulation has shown the ability of text transmission at a rate of 16 symbols per sec within audio watermarked voice signal. WM imperceptibility is achieved by applying power preserving embedding algorithm within critical Bark bands.

Resistance to channel MSI and additive noise is provided by OFDM like data transmission in narrowband channels and error correcting encoding. Further researches should be taken in optimal choosing of WM parameters L, Δ and BCH code parameters n, k, t.

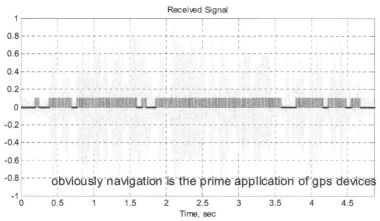

Figure 2. Watermarked audio signal after noise effect, disposition of correctly decoded symbols (dark), appropriate text (below).

REFERENCE

[1] Scoping exercise to establish the need for a review of the elements and procedures of the GMDSS. Shipboard User Needs. Submitted by the Nautical Institute / COMSAR 15/INF.3, 31 December 2010.

[2] Afonin I.L., Slozkin V.G. "Increasing of authenticity of information being transferable in GMDSS communication systems" (in Russian) / Visnyk SevNTU, Vyp. 149/2014, pp. 91 - 95.

[3] Sreenivasa Rao, Manjunath K.E. Speech Recognition Using Articulatory and Excitation Source Features. Springer, 2017.

[4] http://www.pcworld.com/article/243060/speech_recognitio n_through_the_decades_how_we_ended_up_with_siri.html

[5] https://thenextweb.com/microsoft/2016/10/18/microsofts-speech-recognition-is-now-just-as-accurate-as-humans/

[6] https://www.eurofighter.com/the-aircraft#cockpit

[7] Wong, Linda (2014). Essential Study Skills. Cengage Learning. ISBN 1285965620.

[8] Tokunbo Ogunfunmi Madihally Narasimha Principles of Speech Coding . CRC Press Taylor & Francis Group 2010

[9] Shyshkin O., Koshevyy V. Hidden Communication in the Terrestrial and Satellite Radiotelephone Channels of Maritime Mobile Services. In.: A. Weintrit & T. Neumann (eds): Information, Communication and Environment. Marine Navigation and Safety of Sea Transportation. A Balkema Book, CRC Press, Taylor & Francis Group, London, UK, 2015, ISBN: 978-1-138-02857-9. pp. 13–19.

[10] Shishkin A.V., Koshevoy V.M. Stealthy Information Transmission in the Terrestrial GMDSS Radiotelephone Communication. TransNav, the International Journal on Marine Navigation and Safety of Sea Transportation, Vol. 7, No. 4, pp. 541-548, 2013

[11] O'Shaughnessy D. Speech Communication. Human and Mashine. IEEE Press, 2000

[12] Sklar B. Digital Communications. Fundamentals and Applications, 2nd Ed., Prentice Hall PTR, 2001

VHF/DSC – ECDIS/AIS Communication on the Base of Lightweight Ethernet

V.M. Koshevyy & O. Shyshkin
National University "Odessa Maritime Academy", Odessa, Ukraine

ABSTRACT: Utilization of Digital Selective Calling (DSC), which is one of the core procedures in the GMDSS is still practically ignored on a ships because of improper manual control interface. Mariners appeal to creation of standardized and user friendly organized interface. The proposed VHF/DSC – ECDIS/AIS integration removes handmade DSC operations while keeping the necessary protocol of VHF radiotelephony under GMDSS requirements. Recently elaborated Lightweight Ethernet (LWE) standard allows connection of all marine electronic tools, including VHF DSC installation, to shipboard network. Integration of DSC device into shipboard network must be foreseen in any case in integrated navigation system and in communication architecture of unmanned vessels.

1 INTRODUCTION

The current state of maritime terrestrial radio communication procedures in the GMDSS using digital selective calling (DSC) is rather far away from notions "user needs", "integration", "presentation of information", etc. declared by the documents concerning development of e-navigation strategy [1]. The introduction of the DSC as one of the key GMDSS subsystem pursued the goal of improving terrestrial radiocommunication by allocation a calling channel (channel 70 on VHF), the use of digital codes, a certain processing automation of received messages and other innovations.

However, experience has shown that obligatory compliance with the radio procedures using DSC and especially manual operations with the DSC equipment proved to be a stumbling block for the average navigator. This disadvantage of procedure is critical in urgent situations [2].

A radical solution to the problem by our opinion is integration of Electronic Chart Display Information System (ECDIS), Automatic Identification System (AIS) and VHF DSC radiocommunication system for joint processing and presentation of information from these sources [3, 4, 5]. It gives smart addressed VHF communication on the base of conventional ships' installation and in strictly compliance with obligatory DSC procedures. So VHF/DSC communication becomes indeed a real-time system with respect to the current navigational tasks.

The problem of integration of radio DSC data in the Integrated Navigation System (INS) is obviously relevant in connection with the rapid development of unmanned navigation [6, 7]. It is clear that an unmanned vessel shall automatically maintain functions of radiocommunication, including remote monitoring and control from the side of Shore Control Center [8].

In support of the relevance of the problem being addressed should be noted that a number of international manufacturers of marine equipment are already producing integrated VHF/DSC/AIS stations [9] (for example, IC-M506, GX2100), intended primarily for yachts and pleasure boats. Such a radio combines three devices in one unit: VHF radiotelephone, DSC controller and AIS transponder. In addition to its common functions of radiotelephone and DSC on channel 70 the station allows making DSC calls using AIS marks due to integrated AIS transponder without manual entering of Maritime Mobile Service Identification (MMSI) of the called vessel.

The article addresses essential challenge of improving VHF communication in relation to user needs and navigation trends. The possibility of application of LWE network is discussed to solve the problem.

2 ECDIS – AIS – DSC INTEGRATION

DSC is a key subsystem of line-of-sight (LOS) GMDSS communication. DSC in fact replaced the first step of the radiotelephony – voice call – by digital call on channel 70 with a subsequent transition to the working channel. Manual keyboard procedures for establishing addressed radio telephone communication are tedious compared to the convenience of the use of common mobile phones on a shore.

Therefore an imperative utilization of DSC as a part of maritime VHF communication is hardly introduced in a practical application. In particular existing VHF/DSC radios require numerous manual operations to compose correct call. Thus entering only MMSI digits needs nine elementary manual actions at DSC controller key board. In the whole the time period needed for digital call – acknowledgement procedures, that are executed in the ideal condition and by skilled navigator is comparable to responding time in urgent circumstances. But the worse is that the navigator just neglects DSC procedures and immediately takes up the receiver on channel 16 relying on desired ship-to-ship communication. In this situation the question "who is who?" cannot be settled identically and at once. Instead, omitting the DSC, the navigator doesn't get secure, clear and addressed VHF communication.

In the present embodiment the VHF/DSC communication is not a real-time system, especially in urgent situation, in relation to external navigation processes. That's why navigators in practice ignore DSC procedures.

Proposal to improve VHF/DSC communication in the frames of conventional installations and procedures has been proposed by means of ECDIS – AIS – DSC integration [3, 4, 5]. Such integration provided the appropriate ECDIS software updates could give the ability to:

1 providing DSC communication automatically directly from ECDIS;
2 displaying the calling vessel by blinking AIS mark on called vessel's ECDIS (and red blinking mark in the case of distress call) and thus to make immediately the process of attachment of calling vessel to current navigation situation.

When a necessity for establishment VHF communication appears in extraordinary situation, the navigator needs to be able to account on quick access to clear voice communication, without wasting any time on fulfilling unnecessary operations for this, and should be able to concentrate on the main task, in particular, connected with safety navigation (with proper observance of the International Regulations for Preventing Collisions at Sea (COLREG), of course).

This suggestion is fully in compliance with e-navigation strategy which includes further development of means of communication and navigation and the implementation of modern digital information technologies in navigation.

The proposed technology implements one of the basic principles in execution of communication and navigation tasks: transition from handling with raw data to clear understandable information at ECDIS display. In other words, human-machine interaction is translated into a higher level of data processing, leaving the lower one for automatic execution routine operations.

COLREG requires from officer of the watch a Correct and timely action in an emergency situation, in particular for the collision prevention. Rule 8 "Action to avoid collision" says: (a). Any action to avoid collision shall be taken in accordance with the Rules of this Part and shall, if the circumstances of the case admit, be positive, made in ample time and with due regard to the observance of good seamanship.

Coordinated actions are required to perform a safe maneuver in a complicated situation. This doesn't assume collusion of mutual violation of the Rules, but attention attraction, clarifying of intentions, coordinated actions to perform maneuvers safely, taking into account all factors. The most effective way in such situation is using VHF/DSC radio. Avoiding collision and other incidents directly depend on its quickness, reliability and mutual clear understanding of communicating parties.

Implementation of this integration requires a corresponding interface on the part of the DSC equipment. The interface specification NMEA 0183, Ver. 3.01 provides the required sentence for connecting the VHF/DSC controller: DSC* – Digital Selective Calling Information [10] (see Figure. 1). This message is marked as "Designated by IEC for use with IMO maritime electronic devices as required by IMO in the SOLAS convention (1974 as amended)".

DSC sentence provides bidirectional information transfers. This sentence is used to receive a call from, or provide data to, a radiotelephone using Digital Selective Calling in accordance with Recommendation ITU-R M.493 [10]. System configuration (wiring) and the Talker ID are used to confirm if the sentence is transmitted or received.

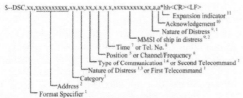

Figure 1. Format of DSC NMEA-0183 sentence [10]

The problem is that the current standards for DSC equipment do not provide an interface for automatic/remote control and monitoring. New standards IEC 61097-3 GMDSS DSC and IEC 62940 Integrated Communication Systems are currently in the final stages of development. It is expected that after the adoption of appropriate standards (if they contain specified requirements), hardware manufacturers get the right stimulus.

3 AMENDMENTS TO RESOLUTIONS NEED

From the other hand for the implementation of ECDIS – AIS – DSC integration need to amend performance standards for Integrated Navigation Systems (INS), resolution MSC.252(83) and ECDIS, resolution MSC.232(82).

Such amendments would correspond to the principles that underpin IMO's e-navigation concept (COMSAR 14/WP.6, 2010, paragraph 50.14). In particular, one core objective related to the e-navigation concept states: "integrate and present information on board and ashore through a humane machine interface which maximizes navigational safety benefits and minimizes any risks of confusion or misinterpretation on the part of the user".

Shipboard user needs relating to GMDSS have been analyzed and submitted by the Nautical Institute at COMSAR 14 [11]. Particularly it was noted that the lack of standardization makes onboard familiarization difficulties with GMDSS consoles. Mariners almost universally suggested that there should be far more standardization of interface, and/or that GMDSS functions should be integrated into general bridge communications systems to improve familiarization through everyday use.

It was also noted that many communication and navigation tasks are directly associated with regards to collision avoidance, port operations and SAR operations. Thus integrating navigation and communication tasks could aid the effective planning, execution and reporting functions for a ship's voyage.

The Correspondence Group at NAV 59, 2013 among nine main categories and practical e-navigation solutions prioritized four potential e-navigation solutions [1]:

– S1: improved, harmonized and user-friendly bridge design;
– S2: means for standardized and automated reporting.
– S4: integration and presentation of available information in graphical displays received via communication equipment;
– S9: improved communication of VTS service portfolio.

The above-mentioned priorities also relate to the interaction with other GMDSS systems for receiving maritime safety information (MSI) such as NAVTEX, Inmarsat-C, high frequency narrow-band direct printing (HF NBDP) telegraphy and Electronic Navigation Chart (ENC) updating. Therefore amendments to current ECDIS performance standards should be done from a wider perspective, considering connection different types of communication means to ECDIS.

4 UNMANNED VESSEL PERSPECTIVE

Today design and research of unmanned ships is developing at an unprecedented rate [6, 7]. Within the MUNIN project (Maritime Unmanned Navigation through Intelligence in Networks) the concept of deep-sea transportation of non-dangerous goods, with autonomous sailing showed its effectiveness and feasibility.

MUNIN developed a technical concept for the deep-sea navigation of the unmanned ship as illustrated in Figure 2. The on-board structure scheme includes integrated bridge system (IBS), autonomous ship controller (ASC), Dedicated LOS communication system, Communication Controller, and some other modules. It's important in this context that all modules interact by means of bidirectional interfaces. Namely, on-board LOS communication system (including AIS and GMDSS) is linked via bidirectional interface with IBS.

The MUNIN concept relies on a manned Shore Control Centre (SCC), responsible for the operation of the ship. Some operations, including LOS communication, the unmanned vessel should perform autonomously, especially when satellite communication fails or is limited. But in general LOS communication should be supervised from SCC. Therefore LOS communication including VHF DSC should be remotely monitored and controlled from the shore.

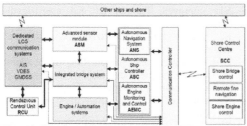

Figure 2: Overview of the high level modules [7]

Another factor in favour of the need for the automatic DSC control is the influence of the human factor. It is well known that non-automated data entry creates errors. Experience shows that manual entry of data into report forms, computer systems or AIS transceivers is a great source of errors. For AIS there are also issues related to the transmission of navigational data such as rate of turn and true heading that should be got from sensors external to the AIS. Many AIS transmitters are not connected to such sensors and send non-valid data or data derived internally from the position [7]. The same is true for DSC equipment under manual control.

According to Norwegian Maritime Authority (NMA) statistics from all ships Norwegian waters and all Norwegian-flagged ships worldwide, 65 per cent of all navigational accidents (collision and grounding) are caused by human error [1]. NMA statistic demonstrates that the majority of navigational accidents are caused by human errors (above 65%). The most significant causes for human errors are "Inadequate observation/inattention" (28%), "Poor judgment of ship movement" (17%) and "Fatigue/ work overload" (13%).

So the benefits mainly derive from the removal of the human element which may reduce associated errors.

5 LWE APPLICATION

Working group 6 (Digital Interfaces) of Technical Committee 80 "Maritime Navigation and Radiocommunication Equipment and Systems" of International Electrotechnical Commission (IEC) has designed a new computer network standard IEC 61162-450 [12] named as "Light-Weight Ethernet" (LWE) which is specially aimed to settle information provision on a ship's board [13]. This standard was developed coming from limitation on technology complexity and special needs from the maritime industry and up and down compatibility.

The family of IEC 61162 standards consists of the following parts:
- parts 1/2: Single talker and multiple listeners, low/high speed transmission (also known as NMEA 0183) [10];
- part 3: Serial data instrument network (also known as NMEA 2000);
- part 450: Multiple talkers and multiple listeners– Ethernet interconnection (also known as LWE);
- part 460: Multiple talkers and multiple listeners – Ethernet interconnection – Safety and security.

Presently existing interface NMEA 0183 doesn't directly support bidirectional regime and needs numerous links. An introduction of new SOLAS installations such as Voyage Data Recorder (VDR), AIS, ECDIS, Long Range Identification and Tracking (LRIT) system and other tools necessitated new data interchange methods between navigation and communication means on a bridge. The conception of Integrated Navigation System (INS) and e-navigation strategic plan obviously demand more advanced interfacing than NMEA 0183.

Standard NMEA 2000 is used to create a network of electronic devices, chiefly marine instruments, on small unconventional vessels. It provides data transfer rate up to 250 kbit/sec. Data rate significantly depends on the bus length.

Multilayer shipboard data network architecture [13] includes 4 on-board layers: Instrument, Process, Integrated Ship Control (ISC) and General Ship layers plus off ship layer. Instrument layer interconnects various sensors and actuators to the higher level components that are using them. Presently IEC 61162-1/2 belongs to this category. Process layer interconnects components of a vessel's specific control function, e.g. navigation, cargo and engine control, etc.

LWE standard is intended to provide interconnections on the lower layers (Instrument and Process) and may be used also on an ISC layer. It is chiefly intended for navigation and radiocommunication equipment.

Distinctive features of LWE standard are:
1 Applicability for embedded microcontrollers (AIS, DSC devices) as well as workstation (ECDIS, IBS);
2 Compatibility of existing interfaces (standard IEC 61162-1/2) and foreseeable equipment meeting IEC 61162-450 standard;
3 Support low and high speed as well as high volume data transfers;
4 Usability for ISC network to connect equipment on different parts of the ship on a distance of up to 1000 m.

Unlike commonly used Ethernet according IEEE 802 project for local area networks in the LWE using repeater hubs, i.e. repeaters of level 1 without message buffering is not foreseen. Routers and repeater hubs shall not be used to interconnect components of an IEC 61162-450 network. LWE is based on single Ethernet switch of level 2 [14] (see Figure 3). Elimination hubs make possible to avoid collisions and, as a result, the network capacity may be increased.

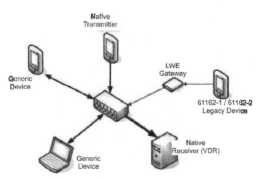

Figure 3. Connection of LWE devices to Ethernet switch [14]

In the Figure 3 a number of typical LWE devices connected to a single Ethernet switch are shown:
– legacy device according to IEC 61162-1/2 standard in a talker mode (transmit only device);
– native transmitter (transmit only device);
– native receiver (receive only device), e.g. VDR;
– generic device (both transmit and receive device).

The arrows indicate direction of data transfer and their thickness illustrates the network traffic intensity.

LWE eliminates numerous NMEA 0183 links between devices. In LWE only one junction is required for network device connection to send or receive information message to or from any network device. All functions that refer to buffering, multiplexing, converting and switching operations are executed by LWE switch.

LWE employs User Datagram Protocol (UDP). Datagram stands for a packet of information. A distinctive feature of UDP is the absence of error checking and requests for datagram retransmission in a case of error detection. All retransmissions slow down network information exchange.

In UDP information packets are just sent to the recipient. The sender won't wait to make sure the recipient received the packet – it will just continue sending the next packets. If the recipient misses some packets it can't ask for those packets again. There is no guarantee the recipient gets all the packets and there is no way to ask for a packet again, but losing this entire overhead means the network can communicate more quickly.

Another feature of LWE is multicast addressing. Multicasting identifies logical groups of recipients. A single message can then be sent to the group keeping the "one talker to many listeners" paradigm from IEC 61162-1 standard.

The use of multicast was chosen over broadcast for performance considerations. With broadcast solution every device would receive, read and process all messages to find those of interest to that particular device. This requires the processor to examine all packets even though the device is only interested in a fraction of the traffic.

A number of firms have already responded to the emergence of IEC 61162-450 standard and launched production of LWE switches. For example, the Magicplex 8 GIC LWE switch [15] produces buffer, multiplexing, converter operations for NMEA 0183 and IEC 61162-450 devices. Some characteristics of Magicplex 8 device are given below:
– NMEA 0183 connections can be set up either as input and output;
– multiplexer and buffer/converter united in one device;
– speed of all NMEA interfaces is adjustable from 1.200 Baud to 115.200 Baud;
– automatic bidirectional converting between IEC 61162 Ed.1-4 and the new Ethernet standard IEC 61162-450.

6 CONCLUSION

Reliable and addressed VHF communication using the mandatory DSC procedures requires the proper human machine interface (HMI) according to user needs. The HMI should be designed for clearly understood presentation of information and handling of communication tasks on the vessel's bridge.

Integration DSC-ECDIS-AIS allows creation of standardized, intuitional HMI interface for handling of VHF radio installation under strict observance of all operational DSC procedures. Integration implies not just the location nearby of radio communication monitor and ECDIS monitor in the IBS, but data integration. Joint utilization of data from several sources essentially simplifies an active address communication and transmitting vessel's identification in relation to current navigating conditions. Effect is achieved through eliminating manual operations on DSC replacing them by clear and understandable actions on ECDIS display according to intuitive computer operations.

The outlook for development and operation of unmanned ships makes requirements for automated control of radio installations, including DSC. Unmanned vessel must have a means for controlling the DSC equipment regardless of whether or not its data will be used by ECDIS. So it should be provided with connection of VHF equipment to a local vessel's networks.

LWE standard satisfies connection into onboard network for VHF transceiver on instrument layer. Ship-to-ship communication in general consists of various data coming in a voice (or DSC) over VHF radio. This will not be continuously needed and voice communication bitrate may be assumed as of 8 kbps per channel [8]. DSC sentence comprises about 30 bytes and shall be transmitted to/from DSC modem quite episodically when call transmission/receive occurs. Therefore no noticeable load on interface is foreseen.

REFERENCE

[1] Development of an e-Navigation Strategy Implementation Plan. Report of the Correspondence Group on e-Navigation to NAV 59, May 2013.

[2] Simplification of DSC equipment and procedures. Submitted by Finland / Sub-Committee on Radiocommunications and Search and Rescue 8/4/1, 27 November 2003.

[3] Miyusov M.V., Koshevoy V.M., Shishkin A.V. "Increasing Maritime Safety: Integration of the Digital Selective Calling VHF Marine Radiocommunication System and ECDIS", TransNav, the International Journal on Marine Navigation and Safety of Sea Transportation, Vol. 5, No. 2, pp. 159 – 161, 2011.

[4] Koshevoy V.M., Shishkin A.V. "Enhancement of VHF Radiotelephony in the Frame of Integrated VHF/DSC – ECDIS/AIS System". In: A. Weintrit (ed.), Navigational Problems, Marine Navigation and Safety of Sea Transportation. CRC Press/Balkena, London, UK, 2013.

[5] Koshevyy V., Shyshkin O. ECDIS Modernization for Enhancing Addressed VHF Communication. TransNav, the International Journal on Marine Navigation and Safety of Sea Transportation, Vol. 9, No. 3, pp. 327-331, 2015

[6] MUNIN. "D8.6: Final Report: Autonomous Bridge." 2015. http://www.unmanned-ship.org/munin/wp-content/uploads/2015/09/MUNIN-D8-6-Final-Report-Autonomous-Bridge-CML-final.pdf.

[7] Trudi Hogg & Samrat Ghosh () Autonomous merchant vessels: examination of factors that impact the effective implementation of unmanned ships, Australian Journal of Maritime & Ocean Affairs, Vol. 8, No 3, pp. 206-222, 2016. DOI: 10.1080/18366503.2016.1229244

[8] Rødseth Ø. J., Kvamstad B., Burmeister H.C., Porathe T. Communication architecture for an unmanned merchant ship / OCEANS-Bergen, MTS/IEEE, 1-9, 2013.

[9] http://www.nauticexpo.com/boat-manufacturer/radio-ais-receiver-26238.html

[10] NMEA 0183 Standard for Interfacing Marine Electronic Devices. Version 3.01. January 1, 88 p., 2002.

[11] Scoping exercise to establish the need for a review of the elements and procedures of the GMDSS. Shipboard User Needs. Submitted by the Nautical Institute / COMSAR 15/INF.3, 31 December 2010.

[12] IEC 61162-450 Maritime navigation and radiocommunication equipment and systems - Digital interfaces - Part 450: Multiple talkers and multiple listeners - Ethernet interconnection.

[13] Rødseth Ø.J., Christensen M.J.; Lee K. Design challenges and decisions for a new ship data network, ISIS 2011, Hamburg, 15th to 16th September 2011.

[14] Christensen M.J., Rødseth Ø.J. Lightweight Ethernet - a new standard for shipboard networks / Digital Ship, December 2010, pp. 31 - 32

[15] http://www.nomatronics.com/index.php?id=109&L=1

Performance Evaluation for Maritime Data Communication - LF band Radio Wave

S. Okuda, M. Toba & Y. Arai
Marine Technical College, Ashiya, Japan

ABSTRACT: The latest maritime data communication system consists of satellite-based and grand-based system. The former uses micro wave for off shore communication or broadcasting, and the latter uses VHF or LF band radio wave for coastal communication or broadcasting. According to authors' precedence studies, onboard data acquisition such as signal and noise levels, and/or reception rate were executed to discuss the performance of data communication and propose the stable and valid provision for LF and MF band radio wave. In this paper, it is concluded that high performance of LF and MF band radio waves' propagation should be essential for coastal navigation to keep safety and efficiency. It is meaningful for the future operation and the effective utilization about radio wave that the relationship with the configuration of stations and the characteristic of huge offshore structure will become to be clear. Furthermore it would be expected to improve the communication capability, and it should be contribute to safe and efficient navigation.

1 INTRODUCTION

The latest maritime data communication system consists of satellite-based and grand-based system. The former uses micro wave for off shore communication or broadcasting, and the latter uses VHF or LF band radio wave for coastal communication or broadcasting. Otherwise, in land communication the mobile data communication is applied with micro wave and VHF band and in near future 5G will be running for autonomous mobile navigation using IoT. As for the mobile phone and the data communication on land use, according to measuring on-site data including an effect of structure, the communication system is improved that it becomes to vanish a communication disturbance. For maritime navigation it is just difficult to cover distance 25 n.m. from shoreline entirely because of the cost of infrastructure. Therefore LF and/or MF band radio waves are used in general. The theoretical and practical characteristics of the radio propagation in the open sea are well known, but there is few data according to the effect of huge offshore structure such as a bridge and these are not known very well. So, in this paper, according to authors' precedence studies, onboard observation such as signal and noise levels, and/or reception rate were executed to discuss the performance of data communication and propose the

stable and valid provision for LF and/or MF band radio waves. According to analyzing the effect of the propagation characteristic near huge offshore structure by receiving and measuring radio signal data on board and on sailing, it should be grasped how much effect to be essential for coastal navigation to keep safety and efficiency.

DGPS detects pseudorange error between GPS satellite and the reference station whose position is known, converts the error into correction data, and broadcasts the correction data to user on board around the reference station. Each user on board receives the correction data using MF beacon receiver. Position accuracy is improved by fixed calculation using the correction data.

At present in Japan there are 27 DGPS stations which is the reference station, and the coverage is all coastal area except a few isolated islands. Table 1 shows DGPS specification operated by the Authority. They call user's attention concerning DGPS coverage (DGPS center).

1 Exception of some area at the Inland Sea about 200 km coverage.

2 Existence of difficult case to use by effect of terrain etc.

However authority makes no mention about an area or a phenomenon concretely. In addition the requirement of position accuracy is 2-5 m for the inland waterway phase described in FRP 2008.

Therefore the position accuracy is insufficiency. Consequently it needs to use DGPS (Difference GPS) for navigator sailing the Inland Sea.

Degradation of the reliability at fixed position using GPS means that the reliability of GPS signal information decreases. The reliability of GPS signal information depends on the transmitting signal of GPS satellite and receiving environment, and DGPS correction data signal is included for navigator sailing the Inland Sea. In our previous study, there is a possibility of anomalous propagation of DGPS correction data signal, and such case occurs in the Inland Sea. It leads to decrease of the reliability of GPS signal information.

Authors have evaluated the reliability of GPS signal information by measuring actual data and carrying out numeric simulation until now. In previous paper, it was analyzed the propagation characteristic of DGPS correction data signal by comparing with measuring its electric field intensity and carrying out numeric simulation. It was observed the interference by reflection signal on bridge at passing the big bridge which is an offshore structure, and it provided a similar result by simulation. In that case, we pointed out that a bit error might occur when signal and reflection signal is weakened mutually by interference. Furthermore we also analyzed it about the effect of land propagation.

In this paper, it is analyzed the propagation characteristic of DGPS correction data signal at passing through a big bridge. It is the propagation characteristic of signal from DGPS station which is very close to a big bridge. In the measurement of electric field intensity, it is observed a complicated variation near the bridge, but leaving a little from a bridge it is to become clear that there is the existence of standing wave according to the interference by combination with the direct signal and the reflection signal depending on an offshore structure.

Table 1. DGPS Specification in Japan

transmission rate	200 bps
transmitting power	75 W
coverage	200 km from DGPS station
transmission format	ITU-R M.823-1(RTCM SC-104)
message type	Type 3,7,9,16

2 INTERFERENCE AND PROPAGATION LOSS

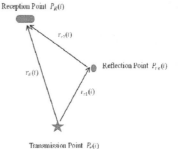

Figure 1. The Relationship with Transmitting, Reception and Reflection Points

The distance relationship among the transmission point, the reception point and the reflection point are expressed as follow.

Transmission Point : $P_T(i) = (x_T(i), y_T(i))$,

Reception Point : $P_R(i) = (x_R(i), y_R(i))$,

Reflection Point : $P_{re}(i) = (x_{re}(i), y_{re}(i))$.

The Range between Transmission and Reception Point, Transmission and Reflection Point, and Reflection and Reception Point are as followed;

$$R_{TR}(i) = |P_R(i) - P_T(i)| \qquad (1)$$

$$R_{Tre}(i) = |P_{re}(i) - P_T(i)| \qquad (2)$$

$$R_{reR}(i) = |P_R(i) - P_{re}(i)| \qquad (3)$$

Travel distance of Reflection Signal is

$$R_{re}(i) = R_{Tre}(i) + R_{reR}(i) \qquad (4)$$

Then, the interference between direct signal and reflection signal is expressed as follow;

Direct Signal is

$$e_{TR}(t) = E_D(i) \cdot \sin(\omega t + k \cdot R_{TR}(i)) \qquad (5)$$

Reflection Signal is

$$e_{re}(t)\varphi = E_{re}(i) \cdot \sin(\omega t + k \cdot R_{re}(i) + \varphi(i)) \qquad (6)$$

where,

　　Direct Signal Level 　　: $E_D(i)$

　　Reflection Signal Level 　: $E_{re}(i)$.

So, Reception Level is

$$e_R(t) = e_{TR}(i) + e_{re}(i) = \\ \sqrt{E_D(i)^2 + E_{re}(i)^2 + 2\cos(\varphi_{re}(i))} \cdot \sin(\omega t + \varphi'(i)) \qquad (7)$$

Amplitude at peak of Level is

$$E_{peak}(i) = E_D(i) + E_{re}(i) \tag{8}$$

where, $\varphi_{re}(i) = k\left(R_{re}(i) - R_{TR}(i)\right) + \varphi(i) = 2n\pi$.

Amplitude at dep of Level is

$$E_{dip}(i) = E_D(i) - E_{re}(i) \tag{9}$$

where,

$$\varphi_{re}(i) = k\left(R_{re}(i) - R_{TR}(i)\right) + \varphi(i) = \left(2n\pi + 1\right)\pi$$

Set Reflecting point, then it is possible to resolve $\varphi(i)$.

It is assumed that the antenna of transmission station is an ideal one. If it is able to estimate a propagation loss, it becomes to be able to estimate the variation of electric field intensity relatively.

From peak level and bottom level, it is easy to get direct signal level and reflection signal level, so this multi path levels will be confirmed.

$$E_D(i) = \sqrt{30P_o} / R_{TR}(i) \tag{10}$$

$$E_{re}(i) = S_{re}(i) \cdot \sqrt{30P_o} / R_{Tre}(i) \times (1 / R_{reR}(i)) \tag{11}$$

where,

Transmitting Power (Watts): P_o

Reflecting Factor   : $S_{re}(i)$

Distance (m)   : $R_*(i)$

Field intensity (electric field) (V/m) : $E_*(i)$

3 ONBOAED OBSERVATION

3.1 *Observation Method*

Onboard survey was executed using T/S Kaigi-Maru belonged in Marine Technical College for 27th – 29th July, 2016. Table 2 shows her principal and Figure 2 shows her photograph. We mounted set of measuring instruments shown Figure 3 and measured electric field intensity of DGPS correction data signal. It was able to collect signal data from Esaki DGPS station on outward voyage which is westbound on 27th and on homeward voyage which is eastbound on 29th.

Table 2. The Principal of T/S Kaigi-Maru

LOA	38.00 meters
B	6.80 meters
D	3.30 meters
GT	157 GT
Playing Limit	Coasting Area
H.P.	588.4 KW

Figure 2. T/S Kaigi-Maru.

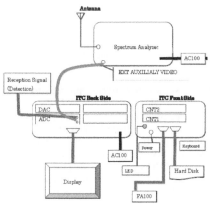

Figure 3. Set of Measuring Instruments

3.2 *Observation Results*

Figure 4 shows the result of electric field intensity at outward voyage (westbound) on expanded scale when passing through the Akashi-Kaikyo Traffic Route. W1, W2 and W3 which appear rapid variation are at just sailing under the bridge. So a periodic variation appears in a few minutes. As an overall trend, this neighborhood is the maximum of electric field intensity (except point W1, W2, W3) and is most close distance from the DGPS station.

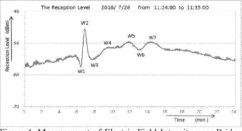

Figure 4. Measurement of Electric Field Intensity near Bridge (Westbound)

Figure 5 shows the result of electric field intensity at homeward voyage (eastbound), and it is similar to westbound (in Figure 4). Moreover what a rapid variation appears is similar to westbound at time of passing under the bridge. In the case of eastbound a periodic variation appears before passing the bridge. It is similar to west bound about the maximum of electric field intensity.

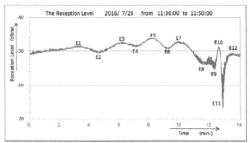

Figure 5. Measurement of Electric Field Intensity near Bridge (Eastbound)

Figure 6 shows her tracks plotted by AIS reception data. Each point (W1 – W7, E1 – E12) represents peak or bottom thought to be standing wave appeared, so the details will be described below. Antenna position of Esaki DGPS station is a hill overlooking Akashi Kaikyo Channel, and it is close to Osaka MATIS which is VTS (Vessel Traffic Service) operated by the Authority. Distance to 3P (Pier 3) which becomes the problem in this analysis is 2,471 meters.

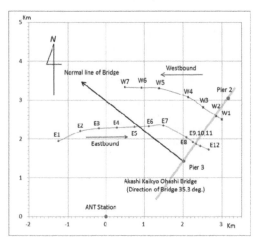

Figure 6. Kaigi-Maru Track at Westbound and Eastbound

4 DISCUSSION - *EXISTENCE OF STANDING WAVE* -

In this section, it is confirmed that there are some standing wave by the result of measurement and it is examined about a reflection characteristic from huge offshore structure.

If these points represent the standing wave by an interference with direct signal and reflected signal, it is related to the wavelength of carrier wave, the

propagation direction of reception signal wave and the direction of receiver which is her course. In Equation (12) and (13), the relationship with the direct signal wave and reflected signal wave, and the propagation direction of their composite signal wave and the wavelength of standing wave on observed course line, expresses as follow, where the angle of phase change at reflection is 180 degrees.

Ship's Course (COG) is $\theta_{COG}(t)$, Propagation Direction of Reception Signal Wave is $\theta_{RSW}(t)$ and Wave Length of Signal Wave $\lambda_S = 2\pi / f_S$, then Wave Length on her Course is follows.

$$\lambda_{COG}(t) = \lambda_S / \left(\cos(\theta_{RSW}(t) - \theta_{COG}(t)) \right) \quad (12)$$

So, we can find the propagation direction of Reception Signal Wave as Equation (10) and (11).

$$\theta_{RSW}(t) = \cos^{-1}\left(\lambda_S / \lambda_{COG}(t) \right) + \theta_{COG}(t) \quad (13)$$

Propagation Direction of Direct Signal Wave is $\theta_D(t)$ and Propagation Direction of Reflected Signal Wave is $\theta_{re}(t)$, then Electric Field Intensity Ratio of Direct and Reflected Wave as follows;

$$L_{D-re}(t) = |E_{re}| / |E_D| \quad (14)$$

In estimation, there is no effect of absorption loss, and the assumption that the phase change in reflection is π, then

$$\tan\left(\theta_{RSW}(t) \right) = \frac{\sin\theta_D - L_{D-re} \cdot \sin\theta_{re}}{\cos\theta_D - L_{D-re} \cdot \cos\theta_{re}} \quad (15)$$

We can find $\theta_{RSW}(t)$ by the result of observation data, so it is easy to the estimated ratio as follows;

$$L(t) = \frac{\tan\theta_{RSW} \cdot \cos\theta_D - \sin\theta_D}{\tan\theta_{RSW} \cdot \cos\theta_{re} - \sin\theta_{re}} \quad (16)$$

Figure 7 shows the Electric Field Intensity Ratio of Direct and Reflected Wave calculated by Equation (16).

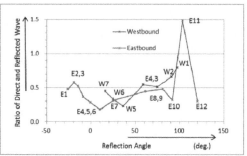

Figure 7. The Estimated Electric Field Intensity Ratio of Direct and Reflection Wave

The reflection angle "0 degrees" means the direction of normal line of Bridge as 305.3 degrees in true direction, and the angle of incidence (AOI)

from Antenna to Pier 3 is calculated as -60.5 degrees in reflection angle. (True direction from ANT to Pier 3 is 54.8 degrees)

It is classified these result as follow.

1 It is almost reasonable that the electric field intensity ratio is about 0.3 to 0.6 at westbound W3 – W7 and eastbound E7 – E1 is in the domain of back scattering at a distance from the bridge.

2 The wavelength of standing wave shortens than the wavelength of transmitting signal wave and the electric field intensity ratio is over 1.0 which is very large, and so it must be compared at 4 times, because it is considered that the reflection is complicated.

It is considered from mentioned above.

1 When a reception point is far from a bridge more than a half of the wavelength of carrier wave, it is confirmed that a standing wave occurs. From this, it is able to be evaluated that there is an effect of huge offshore structure on sailing course or not.

2 A phenomenon such as multiple reflections was caused, and the standing wave of higher harmonics might occur at neighborhood under the bridge by complexity of bridge pier's structure.

3 In case of Akashi Kaikyo Straight, it is seemed to be a cause that there is transmitting antenna close to the bridge and the direction of transmitting antenna to the bridge and the direction of the bridge's line are almost same.

4 The situation near under the bridge is considered enough in the case of Akashi Kaikyo closing to DGPS station. However it will be necessary to examine that there is an effect of the propagation under a bridge or not in case of long distance from transmitting antenna.

5 If it is to be utilized proposed method in this paper, there is a possibility to be able to evaluate an effect of huge offshore structure based on a collection data during sailing.

5 CONCLUSION

In this study we have investigated an effect of huge offshore structure. Consequently it is obvious that we can grasp the propagation situation on sailing course and we are quite sure of advantage of elucidation of mechanism. The investigation all sea areas should be essential. At first it is carrying out the evaluation on the regular traffic route at least. Based on this evaluation, it is considered that we can narrow down a sea area with poor propagation situation by a method to estimate a situation surrounding the sea area. We can confirm that the standing wave of LF and/or MF radio wave exists at on specific sea area. An effect of huge structure depends on the reflection characteristic of the primary reflector. To confirm the existence of the standing wave and to find its reflection characteristic by the modeling in this study are able to construct a simulation model of the status of the standing wave in target sea area if the effect of typical primary reflectors is found without carrying out all measurements in that sea area and it is able to proceed an evaluation of the reception situation and its improvement in that sea area. In addition, it is meaningful for the future operation and the effective utilization about radio wave that the relationship with the configuration of stations and the characteristic of huge offshore structure will become to be clear. Furthermore it would be expected to improve the communication capability and it should be contribute to safe and efficient navigation.

REFERENCES

Araki, T. 1977, Denji Bougai to Bousi Taisaku, Tokyo Denki University Press, 138-139

Department of Defense, Department of Homeland Security and Department of Transportation: 2008 Federal Radionavigation Plan, 2008

Japan Coast Guard DGPS center, http://www.kaiho.mlit.go.jp/syoukai/soshiki/toudai/dgps/menu.htm

Kalafus, R. M., Dierendonck, A. J. VAN & Pealer, N. A. 1986, Special Committee 104 Recommendations for Differential GPS Service, The Institute of Navigation, Global Positioning System Volume III, 101-116

Nishitani, Y. 1980, Hakuyou Denshi Kougaku Gairon, Seizann-Do, 174-175

Okuda, S. & Arai, Y. 2011. The Position Accuracy of DGPS Affected by the Propagation Characteristic on MF Beacon Wave, 2011 International Technical Meeting, 718-724 January 24-26

Okuda, S., Toba, M. & Arai, Y. 2012. The Propagation Characteristic of DGPS Correction Data Signal in Japan – Propagation Characteristic near Big Bridge -, 2012 International Technical Meeting, 1383-1389, January 29-31

Okuda, S., Toba, M. & Arai, Y. 2013. The Propagation Characteristic of DGPS Correction Data Signal at Inland Sea - Propagation Characteristic on LF/MF Band Radio Wave -, TransNav 2013, Marine Navigation and Safety of Sea Transportation: Navigational Problems, 279-285, June 19-21

Safar, J., Williams, P., Grant, A. & Vejrazka, F. 2016. Analysis, Modeling and Mitigation of Cross-Rate Interference in eLoran, Navigation, Journal of the Institute of Navigation, Vol. 63, No. 3, 295-319

Shinji, M. 1992, Musen Tsuushin no Denpa Denpan, The Institute of Electronics, Information and Communication Engineers, 25-49

Saito, Y. 1996, Digital Modulation Techniques for Wireless Communications, The Institute of Electronics, Information and Communication Engineers, 70-107

Yagitani, S., Nagano, I., Tabata, T., Yamagata, K., Iwasaki, T. & Isamu Matsumoto, I. 2004, Nighttime Ionospheric Propagation of MF Radio Waves Used for DGPS – Comparison between Observation and Theory -, IEICE Transactions on Communications, Vol. J87-B, No. 12, 2029-2037

Ships Manoeuvring - Theoretical Studies

Optimal Path Planning of an Unmanned Surface Vehicle in a Real- Time Marine Environment using a Dijkstra Algorithm

Y. Singh, S. Sharma, R. Sutton & D. Hatton
Plymouth University, Plymouth, United Kingdom

ABSTRACT: The growing need of ocean surveying and exploration for scientific and industrial application has led to the requirement of routing strategies for ocean vehicles which are optimal in nature. Most of the optimal path planning for marine vehicles had been conducted offline in a self-made environment. This paper takes into account a practical marine environment, i.e. Portsmouth Harbour, for finding an optimal path in terms of computational time between source and end points on a real time map for an USV. The current study makes use of a grid map generated from original and uses a Dijkstra algorithm to find the shortest path for a single USV.

1 INTRODUCTION

With the growing advances in navigation technologies, there is a greater need to explore oceans for resources as well as for the future needs. Autonomous unmanned vehicles have shown the potential towards various missions of scientific and military significance depending upon the requirement, environment and cost involved (Serreze et al., 2008 and Legrand et al., 2003). Unmanned vehicles can be classified into four categories namely, unmanned aerial vehicles (UAVs), unmanned underwater vehicles (UUVs), unmanned ground vehicles (UGVs) and unmanned surface vehicles (USVs). USVs are watercraft of small (<1 tonnes) or medium (100 tonnes) size in terms of water displacement.

The general architecture for an USV operation in a maritime environment has three basic systems namely, control and path planning, communication and monitoring and obstacle detection and avoidance (ODA), which are responsible for mission planning and execution as shown in figure 1. Path planning is one of the basic subsystems in the maritime operation of USVs to generate way-points for a safe navigation within a desired environment from start to end point. Research and development in areas of artificial intelligence has provided larger scope for development in this territory of marine navigation (Campbell et al., 2012). The abstraction of path planning for an USV is summarized in figure 2.

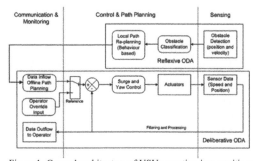

Figure 1. General architecture of USV operation in a maritime environment (Campbell et al., 2012)

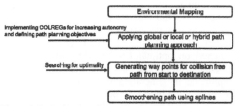

Figure 2. Path planning abstraction for USVs

Until now in path planning of an USV, global and local approaches have been adopted. In global approaches, the complete information of environment is well known while in the local approach only partial information about the environment is known. Under global approaches, grid map-based path planning techniques are the best known since they generate sub optimal trajectories with the fastest computation time.

1.1 Literature review

Dijkstra (1959) initiated the work in the area of grid map-based path planning algorithm by describing the shortest path between two nodes specified on a map. This was later improved by Hart et al. (1968) who introduced A*, which is an extended version of Dijkstra algorithm. In the last two decade, many variants of A* have been introduced by various researchers to improve the performance of robots working in various environments. Stentz (1995) introduced the first major improvement of A*, focused D* algorithm for real time path replanning which was later improved for partially unknown environment by induction of D* Lite (Koenig and Likhachev, 2002). Another improvement by fixing inficelicities of A* in a dynamic environment was introduced by Likhachev et al. (2005) through Anytime Dynamic A*. Since these algorithms do not consider the heading and dynamics of a robot in account, another major improvement was introduced in the form of Theta* (Nash et al., 2007). This algorithm accounts heading angle and yaw rate of a robot in the path planning, which is a necessity for USV path planning since it cannot follow an unrealistic path with sharp turns (Kruger et al., 2007; Prasanth Kumar et al., 2005; Yang et al., 2011). Advanced approaches like the ant colony algorithm (ACO) (Song, 2014) and particle swarm optimization (PSO) (Song et al., 2015) have been adopted for USV navigation but cannot generate trajectories in real time due to high computational load. Along with this, these algorithms do not give consideration to vehicle dynamics and turning radius.

1.2 Major contribution

Many studies in marine navigation have been conducted but most of them have been related to collision avoidance rather than the path planning problem (Tam et al., 2009). Even the studies conducted on optimal USV navigation have been struggling with the high computational load and are inapplicable in generating trajectory in real time. Until now in the literature, path planning approaches have been applied on a self-simulated Euclidean $SE(2)$ grid map with no consideration to real time environment. This study presents the use of the Dijkstra algorithm in a real time environment with minimum computational load to generate a trajectory within a real time operation. This approach is well suited for optimal USV navigation in a static environment with minimum computational requirement.

The paper has been organized in three sections. The section after the introductory material comprising of literature review and major contribution explains the methodology and the algorithm used for the study. The final section discusses the results and provides conclusions with recommendations towards future work.

2 METHODOLOGY

2.1 Dijkstra Algorithm

There are various variants of the Dijkstra algorithm. The variant used in this study fixes a source node which is the start point of the USV and finds the shortest paths from source node to all other nodes in the graph leading to shortest-path tree. In order to reduce the computational load in the original variant, a sparse graph i.e. graph with fewer edges approach has been adopted leading to more efficient storage of graph nodes. The algorithm is defined in Algorithm 1 (Ahuja, 1990).

Algorithm 1. Dijkstra(Graph, source)
1:   **function** Dijkstra(*Graph, source*):
2:     create vertex set Q
3:     **for each** vertex v in *Graph*:   // Initialization
4:       dist[v] ← INFINITY // Unknown distance from source to v
5:       prev[v] ← UNDEFINED // Previous node in optimal path from source
6:       add v to Q // All nodes initially in Q (unvisited nodes)
7:     dist[*source*] ← 0 // Distance from source to source
8:     **while** Q is not empty:
9:       u ← vertex in Q with min dist[u]   // Node with the least distance will be selected first
10:      remove u from Q
11:      **for each** neighbour v of u: // v is still in Q.
12:        alt ← dist[u] + length(u, v)
13:        **if** alt < dist[v]: // A shorter path to v has been found
14:          dist[v] ← alt
15:          prev[v] ← u
16:      **return** dist[], prev[]

2.2 Environmental mapping

Environmental mapping is the first step in the abstraction of path planning as shown in figure 2. In order to use a practical environment, Portsmouth harbor has been considered as shown in figure 3. The map is organised as a weighted occupancy map using a cell decomposition method (Latombe, 1991). This map represents obstacles as black and free space as white in a matrix of black and white as shown in figure 4. A 800×800 pixel map size has been used for the simulation with a resolution of 3.6 m/pixel.

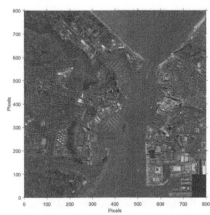

Figure 3. Aerial view of the simulation area (Source: Google Maps)

Figure 4. Grid map of the simulation area

3 SIMULATION

The proposed simulation was executed using MATLAB 2015a on Intel i5 2.80GHz quad core with a 16 GB RAM. In these simulations, computational time of the simulation for three different start nodes and a fixed goal node have been compared in order to determine the effectiveness of the algorithm in terms of computational time to find an optimal trajectory in a practical marine environment. The simulations are assumed to be used by Springer, a USV available with Plymouth University whose specifications have been given in Table 1. Figure 5 shows the Springer USV. Figure 6 shows the three cases of three different start nodes within the grid map having a fixed goal node. These starting nodes are chosen arbitrarily within grid map on different positions within the simulation area to show the effectiveness of the algorithm in finding different trajectories with least computational load.

Table 1. Specifications of Springer

Configuration	Values
Length (m)	4.2
Width (m)	2.3
Displacement (tonnes)	0.6
Maximum speed (m/s)	4

Figure 5. The *Springer* USV

Table 2. Performance analysis for three cases in terms of computational time

Cases	Computational Time (s)
Case 1 (Figure 6(a))	6.801
Case 2 (Figure 6(b))	5.579
Case 3 (Figure 6(c))	6.141

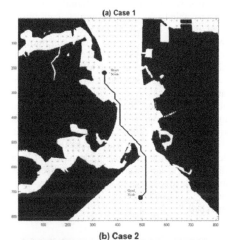

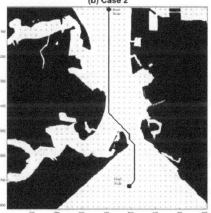

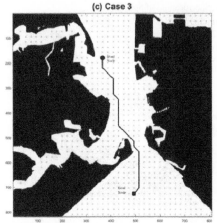

(c) Case 3

Figure 6. Simulation results for three cases

Table 2 shows the comparison of computational time for three cases as shown in figure 6. The results show that the trajectories generated by the Dijkstra algorithm within a huge grid map from any source nodes satisfy the computational efficiency. All cases are able to generate a complete path in less than 7 seconds which in turns lead to the generation of path in less than 1 second per metre length of the distance covered by USV. Henceforth, such an algorithm is applicable in a real time operation where faster optimal trajectories are needed to be generated.

Since the maximum speed of the USV for which the algorithm is designed is 4 m/s, henceforth, the proposed approach satisfies the dynamic constraints of the platform. Although various factors such as vehicle dynamics and heading angle have not been considered in the approach, the basic objective of the study towards generation of trajectory with minimum computational load has been accomplished.

4 CONCLUSION

In this paper, a computationally efficient Dijkstra algorithm to find a path between a source and goal node on a grid map is proposed. The performance was measured in terms of computational time for three different cases, where source points where chosen arbitrarily. The results show that the proposed approach satisfies the computational requirement of the path planning in a real time environment. In conclusion, this new approach is suitable for global path planning of an USV in a static environment. Towards future work, vehicle dynamics and environmental disturbances can be included in the grid map to better understand the applicability of this approach in a dynamic environment.

ACKNOWLEDGEMENTS

This research is supported by the doctoral grant of the Commonwealth Scholarship Commission, United Kingdom tenable for doctoral studies in the School of Engineering at Plymouth University.

REFERENCES

Ahuja, R.K.., Mehlhorn, K., Orlin, J.B., and Tarjan, R.E. 1990. Faster Algorithms for the Shortest Path Problem, *Journal of Association for Computing Machinery (ACM)*, 37 (2):213–223.

Campbell, S., Naeem, W. and Irwin, G.W. 2012. A review on improving the autonomy of unmanned surface vehicles through intelligent collision avoidance manoeuvres, *Annual Reviews in Control* 36, 267–83.

Dijkstra, E. 1959. A note on two problems in connexion with graphs, *Numer. Math.* 1, 269-271.

Hart, P.E., Nilsson, N.J., Rapheal, B. 1968. A formal basis for the heuristic determination of minimum cost paths, *IEEE Transactions on System and Scientific Cybernetics* 4, 100-107.

Koenig, S., and Likhachev, M. 2002. D* Lite, *AAAI/IAAI*, 476-483.

Kruger, D, Stolkin, R., Blum, A., and Briganti, J. 2007. Optimal AUV path planning for extended missions in complex, fast-flowing estuarine environments, *In Proceedings of the IEEE International Conference on Robotics and Automation (ICRA)*, 4265-4271.

Latombe, J. 1991. Robot Motion Planning, *Klumer Academic Publishers*, Norwell, USA.

Legrand, J., Alfonso, M., Bozzano, R., Goasguen, G., Lindh, H., Ribotti, A., Rodrigues, I., Tziavos, C. 2003. Monitoring the Marine environment: Operational Practices in Europe, *In Third International Conference on Euro GOOS*.

Likhachev, M., Ferguson, D.I., Gordon, G.J., Stentz, A. and Thrun, S. 2005. Anytime Dynamic A*: An Anytime, Replanning Algorithm, *ICAPS*, 262-271.

Nash, A., Daniel, K.., Koenig, S. And Felner, A. 2007. Theta*: any-angle path planning on grids, *In Proceedings of the National Conference on Artificial Intelligence*, 1177-1183.

Prasanth-Kumar, R., Dasgupta, A., Kumar, C. 2005. Real-time optimal motion planning for autonomous underwater vehicles, *Ocean Engineering* 32, 4265-4271.

Serreze, M.C. 2008. Arctic sea ice in 2008: standing on the threshold, *In MTS/IEEE OCEANS'08 Conference*.

Song, C.H. 2014. Global Path Planning Method for USV System Based on Improved Ant Colony Algorithm, *Applied Mechanics and Materials*, 568-570.

Song, L., Mao, Y., Xian, Z., Zhou, Y., and Du, K. 2015. A Study on Path Planning Algorithms Based upon Particle Swarm Optimization, *Journal of Information and Computational Science* 12, 673-680.

Stentz, A. 1995. The Focussed D* Algorithm for Real-Time Replanning, *IJCAI* 95, 1652-1659.

Tam, C., Bucknall, R., and Greig, A. 2009. Review of collision avoidance and path planning methods for ships in close range encounters, *Journal of Navigation* 59, 27-42.

Yang, Y., Wang, S., Wu, Z., and Wang, Y. 2011. Motion planning for multi-HUG formation in an environment with obstacles, *Ocean Engineering* 38, 2262-2269.

Performance of the Second-order Linear Nomoto Model in Terms of ZigZag Curve Parameters

J. Artyszuk

Maritime University of Szczecin, Szczecin, Poland

ABSTRACT: The second-order linear model of ship yaw motion, against the well-known first-order approximation, is hydrodynamically sound and therefore more important for many reasons. The zigzag (Kempf) test is one of standard benchmark manoeuvres to evaluate ship designs and provide informative comprehensive data for identifying motion mathematical models. This study produces and analyses universal charts of systematic interdependence between the second-order model parameters and the zigzag curve parameters. An essentially wide range of model parameters is concerned. This can enable a unique identification of the second-order linear model from the ship zigzag behaviour. For that purpose, some practical guidelines are also developed.

1 INTRODUCTION

Linear models of ship yaw motion (or ship steering), in particular of the second-order, can be analysed from the theoretical point of view, e.g. Tzeng & Chen (1999), focusing on some motion responses, as well as in practical aspects, e.g. Norrbin (1996), here being of ship design application. However, even restricting ourselves to directionally stable ships, just the few undertake the problem of what specific zigzag test performance can be purely attributed to the second-order linearity of motion and how close the second-order model stands to the reality. In the latter, however, the situation seems to be more complicated, since some ships reveal the nonlinear character of motion for rudder weak/mild manoeuvres, including directional stability or not, where essentially nonlinear models have to be used.

The present study, one of first studies in the history of ship manoeuvring hydrodynamics, is aimed to link the zigzag parameters to the four constants of the second-order yaw model.

Though tending to be systematic or comprehensive, because of its rather preliminary nature, the paper provides only qualitative guidelines that can be directly employed either by a mathematical model developer, while fixing some hydrodynamic coefficients based on recorded sea trials, or by a naval architect designing a ship to meet the IMO (2002a) manoeuvring standards. In the latter, an extensive use of the 10°/10°zigzag test

is made. The delivered piece of research gives a good starting point for more detailed further research and development, and improvement, of the zigzag relationship with the second-order linear model. This shall encourage and enhance the application of results.

2 METHODOLOGY

Under constant forward (surge) speed, the common coupled linear differential equations of sway and yaw motions for weak rudder steering of a ship can easily be transformed to the well-known uncoupled second-order linear equation of yaw motion, traditionally referred to as the second-order Nomoto model:

$$T_1 T_2 \frac{d^2 \omega'_z}{ds'^2} + (T_1 + T_2)\frac{d\omega'_z}{ds'} + \omega'_z = K\left(\delta + T_3 \frac{d\delta}{ds'}\right) \quad (1)$$

$$ds' = \frac{ds}{L} = dt\frac{v}{L} = \frac{dt}{t_L} \quad (2)$$

$$\frac{d\psi}{dt} = \omega_z, \quad \omega'_z = \omega_z \frac{L}{v} \quad \Rightarrow \quad \frac{d\psi}{ds'} = \omega'_z \quad (3)$$

where T_1, T_2, T_3 – dimensionless distance constants; K – dimensionless amplification (or gain) constant; s' = dimensionless distance; ω_z = dimensionless yaw velocity; δ = helm (rudder) angle [rad] as control variable; s = travelled absolute distance; t = elapsed absolute time; v = ship's speed; L = ship's length; t_L = time of travelling $1L$; ω_z = yaw velocity; ψ = heading [rad].

All the four introduced, in Equation 1, model constants, of somewhat direct kinematic interpretation, can effortlessly be linked to the hydrodynamic coefficients of the aforementioned background set of coupled equations, in both ways. The conversion of the kinematic formulation to the hydrodynamic one has been recently presented by this Author (Artyszuk 2016a).

Equation 1 is fully dimensionless and thus universal in that it describes the ship yaw motion independent of her size (represented by L) and the initial forward speed v. Approximately, for the same helm, the yaw dimensional velocity ω_z (resulting in heading ψ variation) increases proportionally to forward speed and inversely proportionally to ship's length. In the latter, the smaller ships exercise higher turning rates. However, in dimensionless terms, both small and large ships, belonging to the same type, behave similarly.

The zigzag test requires a special rudder steering, consisting of multiple counter-rudder orders as forced by actual heading, and of trapezoidal (finite constant rudder rate) nature due to onboard steering gears. Details of executing such a manoeuvre can be found in many references, e.g. in IMO (2002b). The plot of zigzag, in the so-called 10°/10° variant, that serves as the only reference in our present study, is illustrated in Figure 1. The four zigzag curve parameters (indices) of concern in the study: OS_1 – 1st overshoot angle; OS_2 – 2nd overshoot angle; s'_{10} – dimensionless distance travelled to reach the required heading variation, i.e. 10°, referred to as the initial turning ability in IMO (2002a) standards; s'_{full} – dimensionless distance covered to the 3rd counter-rudder, that is very close to the zigzag ' full period' are also depicted.

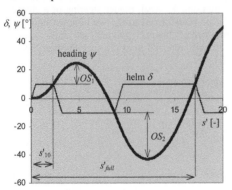

Figure 1. Definition of 10°/10° zigzag curve basic parameters.

In our considerations of a moving ship, the term (dimensionless) 'distance' has been so far and will be

still uniformly used, i.e. to denote the parameters of the model and the parameters of zigzag curve, although the wholly equivalent (dimensionless) 'time' can also be substituted for, that is, however, without any physical consequence. A direct 'time-referencing' is also the frequent case in ship studies, likewise in other transient or time response investigations of general control systems.

Since Equation 1 is linear, of well-established general analytical solution of some exponential functions, and we know the analytical form of rudder steering $\delta = \delta(s')$, we may attempt to derive the complete analytical thus exact solution for zigzag heading curve like in Figure 1. This essentially consists in a piecewise (segmental) combination of heading response to either linearly varied or constant rudder angle, one after another and so on. We may also find the final expressions for the four parameters of this curve that are of our primary interest in this report. However, the resulting expressions are too complex thus unclear to discover the wanted relationships with the model constants. Moreover, the obtained implicit algebraic equations require in many places a numerical solution that hinders computational automation and makes the usual ODE integration still a good choice. Hence, the numerical integration, the same as adopted in Artyszuk (2016b), that is presented below, is utilized in our study:

$$\frac{d\varepsilon'_z}{ds'} = \frac{1}{T_1 T_2}\left[-(T_1 + T_2)\frac{d\omega'_z}{ds'} - \omega'_z + K\left(\delta + T_3 \frac{d\delta}{ds'}\right)\right] \quad (4)$$

where ε'_z = dimensionless distance derivative of ω'_z,

$$\varepsilon'_z\big|_{i+1} = \varepsilon'_z\big|_i + \frac{d\varepsilon'_z}{ds'}\bigg|_i \cdot \Delta s', \; i = 0, ..., n \quad (5)$$

$$\omega'_z\big|_{i+1} = \omega'_z\big|_i + 0.5 \cdot \left(\varepsilon'_z\big|_i + \varepsilon'_z\big|_{i+1}\right)\cdot \Delta s' \quad (6)$$

$$\psi_{i+1} = \psi_i + 0.5 \cdot \left(\omega'_z\big|_i + \omega'_z\big|_{i+1}\right)\cdot \Delta s' \quad (7)$$

Equations 5 to 7 constitute Euler method for the primary/original variable – ε'_z – and trapezoid methods for the other secondary (or derived) variables. Of course, this numerical solution requires initial conditions for dimensionless yaw velocity and acceleration, and heading as well. The distance step $\Delta s'$ equal to 0.002 is used that is absolutely sufficient to keep the numerical accuracy within reasonable limit. Such a numerical solution was compared with the exact analytical solution, as discussed before, and exhibited visually no noticeable differences when the resulting charts were magnified to fit the area 15 cm x 15 cm, even if we investigate the more sensitive yaw velocity ω'_z or yaw acceleration ε'_z charts.

The distance range covered has been limited to ca. 20. The rudder rate, under the term $d\delta/ds'$, corresponds to 23 °/L that is the SOLAS value 2.3 °/s (one steering gear motor) for small ($L = 100$ m) and fast ($v = 10$ m/s) ship. The other extreme case of most rapid rudder (300 °/L) has not been investigated hereafter and the reader is asked to refer to Artyszuk (2016b), though it is fairly believed that the present findings will also be representative for this case.

The following discrete values of model (Eq. 1) constants are employed in the simulation, which are around the reference values investigated in Artyszuk (2016b), yet rounded to $\{K, T_1\} = \{5, 10\}$:

$$K = 2, 3, 4, \underline{5}, 6, 7, 8$$

$$T_1 = 4, 6, 8, \underline{10}, 12, 14, 16$$

They encompass fairly good range of possible change, though at many points they cause to exceed the ship design criteria set by IMO(2002a), especially concerning overshoot angles and initial turning ability in the 10°/10° zigzag test, as will be seen later on. The manoeuvring standards are 10÷20° for OS_1, 25÷40° for OS_2, and ≤2.5 for s'_{10}. The above criteria values for overshoot angles are made dependent on L/v ratio by the regulating body, but more operationally than hydrodynamically, since L/v has little effect on the zigzag indices, that is also obvious from the dimensionless treatment of the yaw equation (Eq. 1).

The values for the other two distance (time) constants – T_2 (much lower than T_1) and T_3 (close to T_2) are:

$$T_2/T_1 = 0.05, 0.10, \text{ and } 0.20$$

$$T_3/T_2 = 0.4 \text{ and } 1.6$$

Hence, due to positive distance constants T_1 to T_3, the present study concentrates on a directionally stable ship.

The intended research has been conducted in two stages. First, under the assumption of $T_2=T_3$, when our yaw model reduces in its behaviour to first-order, independent of T_2, see Artyszuk (2016b):

$$T_1 \frac{d\omega'_z}{ds'} + \omega'_z = K\delta \tag{8}$$

leading to 49 (= 7x7) simulation runs.

Next, the central and extreme values of K and T_1 have been combined with the listed above ratios for T_2/T_1 and T_3/T_2, thus meaning additional 54 runs (=3x3x3x2).

The formal (theoretical) proof for the convergence of Equation 1 into Equation 8 under $T_2=T_3$, apart from referring to the comparison of the resulting yaw responses, is still difficult and intriguing. One of difficulties likely lies in that T_2 adds to the left side (homogeneous part) of Equation 1 while T_3 affects the right side (as specific to the inhomogeneous linear differential equation).

3 NUMERICAL RESULTS AND DISCUSSION

Figures 2 and 3 present the overshoot angles and the distance indices in the zigzag heading curve under the theoretical condition of $T_2=T_3$.

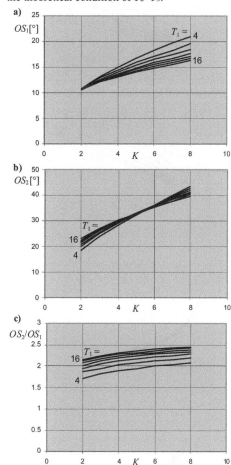

Figure 2. Overshoot angles in 1st-order reduction ($T_2=T_3$).

Both OS_1 and OS_2 are more dependent on K than on T_1, where they increase, almost linearly, with K. The much purer relationship is for OS_2. In the case of K as low as 2, the first overshoot OS_1 is not affected by T_1, displaying rather common value of ca. 10°. On the other hand, for higher K, a reduction in T_1 implies an increase of OS_1 from the initial value of ca. 16° by up to 5°. Figure 2c gives an impression of the very important ratio of the both overshoots in the zigzag test. Under some theoretical assumptions, essentially consisting of constant (step) rudder steering, that shall be of the order of 2. This ratio for real rudder steering varies slightly both with K and T_1.

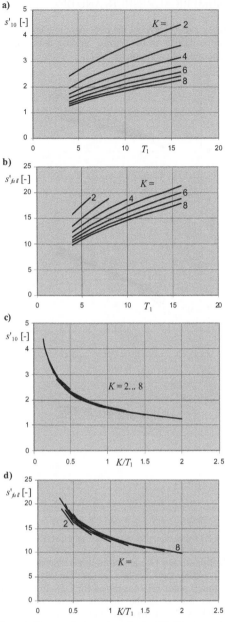

Figure 3. Distances in 1st-order reduction ($T_2 = T_3$).

Such simple relationships, as pointed out before for the overshoot angles, do not appear, when one moves his/her attention to the 'initial' distance s'_{10} and the 'full cycle' distance s'_{full} in Figure 3. Both significantly increase with T_1 and with inverse of K (the lower K, the higher distance indices). Whichever variable, T_1 or K, one takes as abscissa and the other for parameter, the visual patterns of such relationships in 2D plot view (for 2-variable functions) are similar to each other in their complexity. It seems very interesting to show the dependence of s'_{10} and s'_{full} against the K/T_1 ratio, that is practically insensitive to either K or T_1 at all. Such a unique relationship with K/T_1, however, can not be established for the previously analyzed overshoot angles.

The fully second-order linearity, for arbitrary values of the model constants in Equation 1, are gathered in the subsequent Figures 4 to 9. The selection of charts within these figures has been limited to the most interesting extreme values of either T_1 or K. All the presented responses for $T_3/T_2 \neq 1$ lie on the opposite sides of the corresponding earlier curves for the $T_3 = T_2$ condition, though not quite symmetrically. T_3/T_2 higher than 1 ($T_3 > T_2$) has rather low effect both on the overshoot angles, Figures 4 to 6, and the distance indices, Figures 7 to 9, slightly decreasing all of them. The previously formulated performance laws with K and T_1, see Figures 2 and 3, still apply. Nevertheless, the condition $T_3/T_2 < 1$, probably connected with $T_3 \to 0$ while $T_2 > 0$, substantially increases all the four zigzag curve parameters. In addition, the low value of T_3/T_2 amplifies the influence of T_2/T_1. The ratio of OS_2/OS_1, refer to Figure 6b, under 'unfavorable' very low T_3/T_2 can even increase up to more than 3.0.

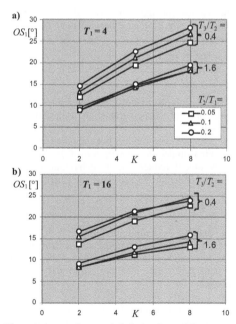

Figure 4. 1st overshoot angle in 2nd-order model.

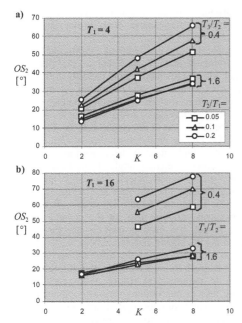

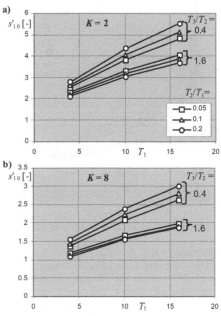

Figure 5. 2nd overshoot angle in 2nd-order model.

action, that is totally lost in the former case, also refer to Artyszuk (2016b).

Figure 7. Distance s'_{10} in 2nd-order model.

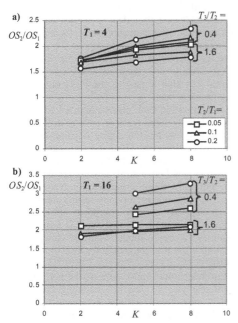

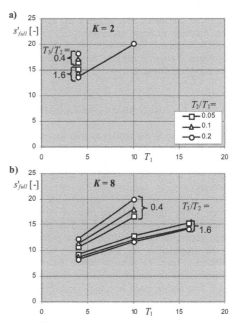

Figure 6. Ratio of overshoot angles in 2nd-order model.

Figure 8. Distance s'_{full} in 2nd-order model.

The hypothetical $T_3=0$ case is quite different from the case of $T_3\neq0$ with the very high rudder speed (very short rudder deflection time), that essentially denotes step rudder steering. In the latter case the system still inherits a significant 'impulse' from T_3

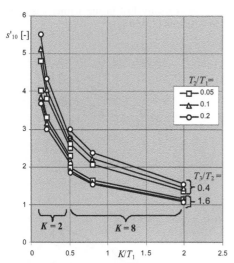

Figure 9. Distance s'_{10} vs. K/T_1 ratio in 2nd-order model.

The missing value points in Figure 5, 6 and 8 for T_3/T_2=0.4 occur due to the adopted integration limit of $s' \leq$ ca. 20, wherein the second overshoot angle OS_2 and the subsequent full cycle of heading variation (zigzag), contributing to s'_{full}, have not yet been developed. According to the considered range of model parameters, they just require more simulation distance to run.

Of some assistance still appears the relationship of the distance indices, especially s'_{10}, versus the ratio K/T_1 (compare Fig. 9 with Fig. 3c), even with the mentioned huge sensitivity for very low T_3, that is now clearly evidenced.

4 FINAL REMARKS AND CONCLUSIONS

The second-order linear model of ship yaw motion is much more sophisticated in zigzag response than the corresponding first-order model, though the former one partially draws on the latter (as set by the $T_2=T_3$ condition).

More or less significant independence exists between some zigzag indices, four of which have been carefully chosen and analyzed in this study, that are generated by any of these two models. This lack of correlation is crucial for the model identification that is nothing more than solving nonlinear algebraic equations with a few unknowns under a few inputs, each of which shall bring some new information.

In line with the performed research, such little connection exists with regard to any of the overshoot angles and any of the distance indices, since they depend on various model constants. Such an independence is also partially noticeable either between the both overshoot angles or between the two distance indices – s'_{10} and s'_{full} (let's say - short-term and long-term ones). They do not vary in the same degree.

Supplementing the Nomoto (1960) method for identifying K and T_1 constants of the first-order model, also commented on and partially criticized in Artyszuk (2016b), we may attempt to determine this model based on the provided results in much simpler way, using only one overshoot angle and one distance index. The obtained accuracy is of the same order. Depending on the magnitude of OS_1, the gain constant K can be directly determined using the first or the second overshoot angle. Then, based on any of the distance parameters, preferably the initial turning distance s'_{10}, the ratio K/T_1 (Fig. 3c,d), and thus T_1 is to be fixed. The role of K/T_1 as independent variable, also for the general second-order model, has been proved for the purpose of presenting the zigzag distance indices.

The identification of the arbitrary second-order model is much more demanding and difficult, but is closer to the physical reality. From each of the zigzag indices one obtains various combinations of the four model constants (yet of different sensitivity) that shall be next narrowed/resolved to satisfy other zigzag indices. Apart from the reported problems with so-called cancellation effect, this challenge with full-scale sea trial data is being left for the future research, also expecting some feedback from other scientists.

The next investigations shall also be directed to the theoretically interesting boundary case of T_3=0 that produces high impact of the T_2/T_1 ratio on the zigzag performance.

A final question shall also be here asked, how to detect the first- (less real) or second-order (more real) 'linearity' within the zigzag heading curve, or maybe just nonlinearity. Of course, this can partially be completed for example by means of checking the mutual relations between all indices for the zigzag curve, not only those discussed in detail in this study. The presented preliminary results are going to considerably serve this purpose. If such a discrimination stage is omitted, there is a risk to get a wrong (very rough) model from the viewpoint of simulating other manoeuvres, refer also to Artyszuk (2016b).

Of primary interest also appears the performance of nonlinear yaw models in simulating the zigzag test that may be of various nonlinearity and of various form (e.g. the so-called nonlinear Nomoto model). Just a few works are published on this subject, e.g. Sutulo & Guedes Soares (2005).

REFERENCES

Artyszuk, J. 2016a. Inherent Properties of Ship Manoeuvring Linear Models in View of the Full-mission Model

Adjustment. *TransNav, the International Journal on Marine Navigation and Safety of Sea Transportation*, vol. 10 (no. 4): 595-604. DOI: 10.12716/1001.10.04.08.

Artyszuk, J. 2016b. Peculiarities of zigzag behaviour in linear models of ship yaw motion. *Annual of Navigation*, vol. 26/2016: 23-38. DOI: 10.1515/aon-2016-0002.

IMO 2002a. *Standards for Ship Manoeuvrability*, Res. IMO MSC.137(76), doc. no. MSC 76/23/Add.1, Annex 6, Dec 4th. London: IMO.

IMO 2002b. *Explanatory Notes to the Standards for Ship Manoeuvrability*. MSC/Circ.1053, Dec 16th. London: IMO.

Nomoto, K. 1960. Analysis of Kempf's standard maneuver test and proposed steering quality indices. In *First Symposium on Ship Maneuverability*, May 24-25, DTMB Rep. 1461 (AD 442036). Washington: DTMB.

Norrbin, N.H. 1996. Further Notes on the Dynamic Stability Parameter and the Prediction of Manoeuvring Characteristics. In Chislett M.S. (ed.), *MARSIM '96 Proc., Marine Simulation and Ship Manoeuvrability;* Sep 9-13, DMI, Copenhagen. Rotterdam: A.A. Balkema.

Sutulo, S. & Guedes Soares, C. 2005. Numerical Study of Some Properties of Generic Mathematical Models of Directionally Unstable Ships. *Ocean Engineering*, vol. 32: 485-497.

Tzeng, C.Y. & Chen, J.F. 1999. Fundamental Properties of Linear Ship Steering Dynamic Models, *Journal of Marine Science and Technology* (NTOU), vol. 7 (no. 2):79-88.

Port and Routes Optimum Location

Models and Methods for Locating LNG Distributing Routes in the Baltic Sea Area

E. Chłopińska, A. Bąk & M. Gucma
Maritime University of Szczecin, Szczecin, Poland

ABSTRACT: The article presents modern logistic of distribution systems designed for the Baltic Sea area. Particular attention was paid to those models and methods which reflect the actual LNG demands. They take into consideration the particular bunkering station situated along the coast as well as the amount of LNG in each of them. The paper presents a proposal of an optimal LNG distribution network in the Baltic Sea area.

1 INTRODUCTION

According to the regulations of the Directive of the European Parliament and of the Council 2012/3/EU of 21 November 2012, which amends Council Directive 1999/32/EC concerning the level of sulfur in marine fuels into Sulphur Oxide-Emission Control Area the owners were faced with many challenges. One of the most serious problems which are facing shipowners providing services on the Baltic Sea, which is strictly controlled regarding sulfur emissions is the inability to bunker the LNG fuel (eg. The lack of access to the bunkering station).

The article proposes methods and models supporting in establishing the LNG bunker stations locations along the main shipping routes in the Baltic Sea.

2 THE LOCATION ROUTING PROBLEM

2.1 Models and method for LRP.

Fuel distribution models are built taken into account two interdependent factors, which can be e.g.: shipping routes and location of the distribution facilities. In this case, the problem of finding the optimal route is a combination of classic problem of path locations and the availability of the distribution objects. From a practical point of view, the problems concerning the location of the ship's route are part of distribution management, and from the mathematical point of view it concerns the modeling of object locations and ships' routes.

Starting from the production phase, by distribution managing, control of fuel flow, supply of raw materials to the customer as well as any unexpected events such as cancellation of service, route optimization systems must provide a suitable level of service warranting minimization of logistic costs. Thus, the structure of the transport systems and logistics network is essential, together with the locations of bunkering stations.

All parts of the system must interact with each other in order to function properly allowing to transport the LNG among the various nodes of the system with the highest performance and accuracy at the lowest possible cost.

In literature problems location of distribution centers and warehouses of raw materials are presented with its use of mathematical models and effective techniques and methods. Due to the complexity of the problem the LRP systems are mostly handled by heuristic methods [3]. It also results from the fact that solutions for the LRP systems are particularly difficult problems.

For example, the complexity of the problem is to determine the optimum solution e.g. to provide a fuel to storage using a cryogenic tank, which can be done in two ways:

− each store has assigned a tank with specified parameters with a given volume of fuel,
− tank supports more than one magazine on the route, which requires less fuel than the volume of the tank.

Of course, the real situation may be even more complex, because there are many possible solutions. In these cases, decisions concerning location centers or warehouses should be represented in the models in a realistic manner due to the fact that there are many possible limitations that should be taken into consideration.

LRP systems used in practice indicate diversity of application. Significant part of the literature points at the distribution of consumer goods (food products [2], the distribution of goods [10] [11], the distribution of press [13]), but there are also systems addressing health [6] [16] and army [8].

Considering various aspects of research it is possible to refer to a location defining eight aspects of the problem: [5]
— hierarchical structure,
— type of input data (deterministic or stochastic),
— planning period (single-period, multi-period),
— solution method (exact or heuristic),
— objective function,
— solution space (discrete, network or continuous),
— number of depots (single or multiple),
— number and types of units,
— route structure.

Problems of systems and routes positioning can also be classified as exceptional LPR cases:
— classical problem of location (a separate warehouse is directly assigned to each recipient of fuel);
— LRP system as a vehicle routing problem, VRP (permanent storage location).

The solution may be classified according to the methodology of the way to create a relation between the location of warehouses and the routes: [5]
1 In sequential methods the problem of distribution of the object is solved at first by minimizing the distance of radial spreading in relation to the route. A disadvantage in the use of these methods is the lack of feedback (route - location).
2 Clustering solution methods consist of the division of recipients and on the application of the method in sequence:
   — travelling salesman problem, TSP,
   — for each cluster location of warehouses using TSP or VRP is determined.
3 Iterative heuristics decompose the problem of location into two factors (sub-problems), where the solutions are transferred from one phase to another.
4 In hierarchical heuristics methods location is the main problem. A problem of routing maritime units is considered then. In order to use the LRP systems a multi-phased solution may be used with care to minimize its complexity based on procedures consisting of four algorithms combination: [9]
   — (1) location-allocation (2) route,
   — (1) route (2) location-allocation,
   — (1) saving (2) insertion,
   — (1) routes improvement (2) exchange.

The location planning process may provide better quality solutions than conventional models location.

LRP systems may reflect a realistic distribution of goods within the Baltic Sea area.

In conclusion it should be noted that the presented solutions of LPR systems for scheduling and location can result in obtaining the optimal solution for warehouse locations of fuel distribution for vessels along the shoreline of the Baltic Sea and may lead to significant savings in distribution costs.

3 LOW-SULFUR FUEL

3.1 *LNG*

Global energy consumption as the permanent growing trend which simultaneously contributes to the pollution of the environment is gaining more and more interest. The size of air pollution (Table 1) aroused great concern and restrictions aimed at its reduction were introduced.

Table 1. Total emission of main air pollutants.

SPECYFICATION	2005	2010	2013	2014
	in thousand tonnes			
Sulphur dioxide	1246	970	853	800
Nitrogen oxides[b]	851	874	774	723
Carbon dioxide	323373	334026	322440	310307
Carbon oxides	2738	3119	2868	2704
Volatile Non-methane organic compounds	879	949	898	888
Ammonia	274	274	270	265
Particulates	469	462	403	383

S o u r c e: data of the National Centre for Emissions Management — Institute of Environmental Protection — National Research Institute.

Permanent development of the maritime economy (Table 2) and increase in number of vessels providing services within the Baltic Sea area (contributing to increase in pollution) caused rapid pollution growth. Maritime transport is one of the major emitters of sulfur into the atmosphere.

Table 2. Maritime transport fleet as of 31 XII.

SPECIFICATION	2005	2010	2014	2015
Ships	**130**	**121**	**104**	**102**
cargo ships	121	107	9	8
ships for dry bulk transport	108	95	84	80
of which bulk cargo ships	76	69	60	5
tankers	13	12	5	5
maritime barges	—	—	4	3
ferries	7	11	7	7
passenger ships	2	3	4	7
DWT in thous	**2610**	**2942**	**2721**	**2515**
cargo ships	2582	2887	2684	2476
ships for dry bulk transport	2518	2816	2659	2453
of which bulk cargo ships	2101	2463	2337	2135
tankers	64	71	20	20
maritime barges	—	—	4	3
ferries	28	55	37	39
passenger ships	0,2	0,2	0,2	0,3

S o u r c e: [15]

In connection with the entry into force of the Directive of the European Parliament and of the Council 2012/33 / EU limiting fumes emission of the fuel driving vessels moving along SECAs (Sulphur Oxide Emission Control Area), LNG has become for shipowners providing their services over waters of the Baltic Sea, the coast of North America, the English Channel and the North Sea an alternative solution of fulfilling restrictions that have been imposed by the International Maritime Organisation (IMO) for emission control areas (Table 3).

Table 3. Fuel sulfur content limit.

Date set limit	Fuel sulfur content limit. [% m/m]	
	SO_x ECA	global
05.2005	1,5%	4,5%
07.2010	1,0 %	4,5%
01.2012	1,0 %	3,5%
01.2015	0,1%	3,5%
01.2020*	0,1%	0,5 %

* In 2018 standard of 0.5% sulfur in fuel will be evaluated by specially appointed by the IMO expert group, which will examine the possibility of its introduction due to future trends and availability of such fuels on the fuel-dim market. In the case of a negative assessment of the standard of 0.5% it will be effective only from 01.01.2025.
S o u r c e: [17].

Liquefied natural gas is the alternative for fuel with a low sulfur content. Large reserves of gas and its favorable environmental characteristics became an appropriate solution for the growing energy demand and to replacing the existing, more polluting fuels.

LNG due to its properties (cooling the gas to a temperature below 162° C) can be economically transported over long distances by specially designed LNG vessels, which are able to maintain the fuel in a liquid state with a constant low temperature. In addition, the delivery of natural gas to customers may be provided by using:
– using pipelines,
– fuel tankers,
– LNG bunker vessel with a capacity of 200 – 1 000 m³ or 1 000 – 10 000 m³,
– LNG terminal,
– LNG container vessel,
– gas carrier.

Transporting of natural gas in the form of LNG by vessels because the cost of transportation is more favorable than the constructing the gas pipe (gas supply) and related to this fact restrictions of area limits.

The supply chain is of primary importance for vessels using LNG. Good structure results in the smooth functioning of the entire system of fuel distribution and may contribute to significant savings (lower capital costs and operating costs).

Restrictions which owners of LNG units are exposed to (also related to investment costs) are as follows:
– location of LNG stations,
– availability of LNG stations,
– operating restrictions,

One of the most important infrastructure investments as energy in Poland is concerned is building LNG receiving terminal in Swinoujscie. This project aims to improve the Polish energy security and economic conditions for the import of gas. Opening the LNG terminal will allow for diversification of sources of gas supply to Poland and the use of maritime transport for the supply of gas will allow for agreements with suppliers from different places in the world.

The LNG terminal in Swinoujscie will enable reception of 5 billion m³ of gas per year, and in future its capacity will be increased to 7.5 billion m³. The main centre of the terminal consists of two vessels having a diameter of 80 m and a height of 52 m (each) to store natural gas with a total capacity of 320 thousand m³. The LNG terminal is equipped with the infrastructure which enables complete independent gas reception, re-gasification and transfer to the Polish gas transmission line.

3.2 The demand for LNG

The LNG supply chain efficiency for the terminal depends on the fleet of ships being able to transport LNG in an efficient and safe manner.

At present the costs of gas transportation by means of vessel are much higher in comparison with costs of transportation by means of land-based infrastructure. These expenses include not only the price of maritime transport but also the process of liquefying the gas, which is very expensive.

The years to come forecasts show that the cost of gas transportation by pipelines will grow due to the depletion of closely spaced gas deposits. For this reason a border competitiveness of the gas transportation by sea to gas transportation by pipeline shifts in favor of the use of LNG.

PGNiG has undersigned a contract with Qatar's Qatargas LNG supplier for the supply of 1 million tons of LNG per year to the LNG terminal for next 20 years. The first delivery of test gas was held in December 2015 by boat Q-flex, which provided 210 thousand m³ to the terminal.

Table 3. The amount of the supply of LNG to the terminal within one year.

The amount of the supply of LNG	The amount of the supply of LNG after	The amount of LNG after re-gasification	The amount of the supply of LNG to the terminal within one year. [%]
	[tys. m³]	[mln m³]	
1	210	120	2,4
42	210	5000	100

S o u r c e: Own elaboration based on [19]

415

Statement contained in Table 3 indicates that in case of LNG supply to the terminal by means of Q-flex vessels the annual demand for gas from the LNG terminal will be covered when the gas is delivered 42 times a year. This means that the gas terminal is able to receive a unit approximately every eight days. LNG transportation to Polish LNG terminal by Q-flex units is the most cost-effective solution because of the technical capabilities of these units.

The LNG terminal in Swinoujscie not only provides reception, re-gasification and gas transmission possibilities - it is also a natural gas warehouse. Two cryogenic tanks are able to store 320 thousand m^3 of gas in liquid form, what at the re-gasification process will provide approx. 5 billion m^3 of natural gas that can be introduced into the national gas network. In 2016 PGNiG SA disposed of seven warehouses to store high-methane gas in such places as Wierzchowice Kosakowo, Moglino, Strachocina, Brzeźnica, Husów and Swarzów. The total capacity of these reservoirs is 2.9 billion m^3 of gas. Poland's economy will be able to function for 58 days (Table 4) on the accumulated reserves of natural gas in 2016.

Table 4. The period of independence from supplies of gas.

Year	The period of consumption	reserves in Poland	the period of independence from supplies of gas
2016	18,03 [mld m^3]	2,9 [mld m^3]	58 [days]
		after the construction of the terminal:	
		7,9 [mld m^3]	159 [days]

S o u r c e: Own elaboration.

The data contained in Table 4 show the extent to which the construction of the LNG terminal will affect the energy security of Poland. The period of independence from the supply will increase from 58 to 159 days which represents an increase in the period of 101 days. The increase of storage space for natural gas in Poland, among other things, will result as follows:
– Poland's gas import independence will extend three times,
– profitable contracts for the purchase of gas due to the enormous storage capacity of raw materials will be achievable,
– exporting gas to neighboring countries will be possible,
– increasing the prestige of Poland as an important entity on the gas market in Central and Eastern Europe,
– increasing state sovereignty in international relations.
The above analysis directly shows the advantages of the LNG terminal in Swinoujscie. It turns out that these advantages not only have the physical aspect of increasing the storage area of natural gas and diversification of import sources - an important element of this project is to increase the prestige of Poland internationally as a country which is one of the undisputed leaders on the gas market in Central and Eastern Europe.

4 LNG DISTRIBUTION NETWORK OVER THE BALTIC SEA AREA

4.1 Supply chain of liquefied natural gas

Global use of liquefied natural gas is growing steadily due to the economic price of this fuel and its energy and environmentally friendly physicochemical properties. The LNG contains methane and the combustion process results in CO_2 and water and a slight amount of other adverse environmental pollution. The amounts of these compounds are significantly lower than at the time of combustion of oil or coal.

Supply chain of liquefied natural gas at first consists of researches. The next steps are the production of fuel, transportation of LNG to the liquefaction station, storage and transfer of purified fuel to the ship (loading at export terminals). Next there are unloading and re-gasification terminals (import) and the last element - the use of LNG. Operation consists of the fuel supply to the recipients over the short and long distances and applying it in different technologies, each of them having defined environmental impact and safety.

The LNG supply quality depends on many factors. Recipients of LNG from around the world pay special attention to product quality and delivery time. Effective management results in preserving the supply chain and increasing the number of new customers.

There is a high impact on the disruption of the supply from [14]:
– competitive advantage,
– modern technologies,
– globalization of the economy (global competition).

Ecological fuel in the form of LNG is a strategic product of all economies of the world. Due to its characteristics the fuel requires specific conditions of production, storage and transport. Limited resources of natural gas deposits led to the fact that the circulation of this specific fuel is limited to a few global supply chains. World LNG chains are exposed to far more threats than chains operating on a smaller scale (regional, national). For this reason supply chain management should include strategic, operational, and - most importantly - risk management [1].

Risk is the probability of loss for the operator, which results from the consequences of the decision [7].

The smooth functioning of the supply chain of fuel depends on the correct operation of all its cells / components (liquid risk management). The risk acceptable is of particular relevance here - in accordance with the principle of As Low As Reasonable Practicable. That is associated with the introduction of the most cost-effective measures to diminish the risk which should be defined.

Global use of LNG resulted in larger groups of supply chain risk for the following types of risk [4]:
– exchange,
– overcoming technical and technological barriers between consumers,
– differences in organizational cultures,
– asymmetrical threats.

The aspect of risk management is a very important issue in the fuel supply chain. In practice it refers to the maintenance of the highest possible level of safety (technical aspects) on human life, the environment and infrastructure (disaster prevention). Preventing fires, explosions and accidents of any type enables the efficient flow of LNG in the supply chain.

The most important issue is to preserve the external security of the LNG supply chain or prevent the terrorists' attacks.

It is also worth mentioning that the development and implementation of legal regulations applying to maritime transport security the International Maritime Organization has developed a Formal Safety Assessment, FSA, which is widely used in determining the costs of implementing changes arising in new technologies.

4.2 The proposed LRP for the basin of the Baltic Sea

This chapter presents possible optimization models of LNG warehouse location process along the coast of Baltic Sea with the use of LNG terminal in Swinoujscie.

Vessels can be powered by LNG via:
– fuel tankers,
– smoll LNG bunker vessel, capacity: 200 – 1 000 m³
– large LNG bunker vessel, capacity: 1 000 – 10 000 m³,
– LNG terminal,
– LNG container vessel,
– gas carrier.

The basic, common restrictions for the vessels LNG bunkering operations are:
– location of LNG stations,
– availability of LNG stations,
– operating restrictions,

Terminal LNG in Swinoujscie is structured at the point of long and constant coastline. For the purpose of the space contraction searches for LNG warehouse location and receiving "nearly" optimum results, the most profitable solution is heuristic methods usage.

In order to locate the warehouse of LNG in the first should be appointed:
– demand for LNG in the relevant area,
– distance to customers,
– capacity of the LNG tanks,
– cost of fuel.

This information is needed to create a combination of supply chain optimization of LNG on the relevant waters.

Raine Jokinen has presented a model of LNG distribution for on-land customers. There have been described LNG terminals construction location and sizes, of customers using the services of terminals and way of bunkering in respect of costs reduction (Figure 1):
– the cost of LNG transportation (in relation to the kind of transport),
– the cost of LNG [12].

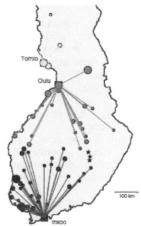

Figure 1. Optimal terminal locations and LNG supply.
S o u r c e: [12]

The optimization of the supply chain can therefore be applied for the vessels' supply along the coast of the Baltic Sea. The model is reflected in minimizing the costs associated with the delivery of fuel from the LNG terminal in Swinoujscie to the warehouses (to cover the LNG demand). It also provides the optimal configuration of the supply chain in relation to the terminal and warehouse location, naval units size and distribution to the customer.

L. Guerra, T. Murino and E. Romano [9] describe the issue of laying units routes, which is known in the literature as multisalesman problem. Issues concerning the evolutionary algorithms are algorithms involving the NP-hard problems, which can be easily formulated, but finding the optimal solution for a given problem is very difficult. Also characterized solution locations and routes using the Hamilton circuit and Clarke and Wright's algorithm (Figure 2).

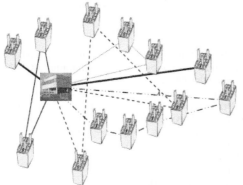

Figure 2. Clarke and Wright's algorithm final solution for the VRP.
S o u r c e: [9]

Compared to conventional methods of evolutionary algorithms analyze very efficiently in the area of solutions to the problem under consideration. The purpose of these algorithms is first of all to optimization tasks. Their particular utility is manifested in solving issues of a combinatorial.

The use of evolutionary algorithms to problems of location and route is appropriate if:
- use evolutionary algorithms to obtain results for a variety of variants of tasks to optimize the location and route, taking into account the many additional concepts, where is a possibility of economy practice,
- use of nontraditional construction methods of encoding tasks and their adaptation functions, which take into account additional assumptions,
- in order to obtain better results build new genetic operators and use selection methods.

5 CONCLUSIONS

Types of risks to the supply chain of fuel reducing sulfur emissions have been described. Defining the risk of limiting the correct functioning of the supply chain is essential in the process of planning the optimal location of the LNG storage facilities along the coast of the Baltic Sea. Defined kinds of LRP systems were directed at vessels using low-sulfuric fuel (LNG).

Kinds of the risk threatening the supply chain of fuel which are lowering emission of sulfur were also described. Defining the limiting risk and correct functioning of the supply chain is necessary in the process of planning the optimum localization of LNG magazines along the Baltic Sea coast.

REFERENCES

[1] A. Nerć-Pełka, „Obszary ryzyka w łańcuchach dostaw skroplonego gazu ziemnego.", Elektro-niczne czasopismo naukowe z dziedziny logistyki., t. 5(4), nr 2, ss. 1–8, 2009.
[2] C.D.T. Watson-Gandy i P. J Dohrn, „Depot location with van salesmen – a practical approach.", Omega 1, ss. 321–329, 1973.
[3] D. Tuzun i L. I. Burke, „A two-phase tabu search approach to the location routing problem.", European Journal of Operational Research, t. 166, ss. 87–99, mar. 1997.
[4] E. Gołembska, Podstawowe Problemy Logistyki Globalnej, Międzynarodowej, Eurologistyki., 1. wyd. Łódź: Wydawnictwo Naukowe Wyższej Szkoły Kupieckiej, 2007.
[5] G. Nagy i S. Salhi, „Location-routing: Issues, models and methods.", European Journal of Op-erational Research, t. 177, nr 649–672, 2006.
[6] I. Or i W.P. Pierskalla, „A transportation loca-tion-allocation model for regional blood bank-ing.", AIIE Transactions 11, ss. 86–94, 1979.
[7] J. Penc, Leksykon biznesu. Warszawa: Agencja Wydawnicza Placet., 1997.
[8] K.G. Murty i P.A. Djang, „The U.S. army na-tional guard's mobile training simulators location and routing problem.", Operations Research., t. 47, ss. 175–182, 1999.
[9] L. Guerra, T. Murino, i E. Romano, „A heuristic algorithm for the constrained location - routing problem.", International Journal of System Ap-plications, Engineering and Development, t. 4, nr 1, ss. 146–154, 2007.
[10] M.S. Daskin i J. Perl, „A unified warehouse location-routing methodology.", Journal of Busi-ness Logistics., t. 5, ss. 92–111, 1984.
[11] M. S. Daskin i J. Perl, „A warehouse location-routing problem.", Transportation Research, ss. 381–396, 1985.
[12] R. Jokinen, F. Pettersson, i H. Saxén, „An MILP model for optimization of a small-scale LNG supply chain along a coastline.", Applied Energy, nr 131, ss. 423–431, 2015.
[13] S.K. Jacobsen i O.B.G. Madsen, „A comparative study of heuristics for a two-level routing-location problem.", European Journal of Opera-tional Research, t. 5, ss. 378–387, 1980.
[14] S. Konecka, „Kryzysogenne kategorie ryzyka specyficzne dla procesów logistycznych i łańcu-chów dostaw.", Logistyka., t. 1, ss. 14–17, 2007.
[15] Statistical Offece in Szczecin, Statistical year-book of maritime economy. Szczecin: Central Statistical Offece in SzczecinGłówny Urząd Sta-tystyczny, 2016.
[16] Y. Chan, W.B. Carter, i M.D. Burnes, „A multi-ple-depot, multiple-vehicle, location-routing problem with stochastically processed de-mands.", Computers and Operations Research, t. 28, ss. 803–826, 2001.
[17] Data of PRS.
[18] Data of the National Centre for Emissions Man-agement - Institute of Environmental Protection - National Research Institute.
[19] www.polskielng.pl

Mathematical Approaches for Finding a Dry Port Optimum Location on the Level of Intermodal Transport Networks

D. Carp & V. Stîngă
Constanta Maritime University, Constanta, Romania

ABSTRACT: In an increasingly dynamic business environment, the goal of any supply chain is providing competitive services to customers, especially in terms of costs. This is why over the last years there has been a remarkable growth of interest regarding the establishment of the location of the intermodal transport networks hubs that carry out the goods transfer, consolidation and grouping. Throughout this study we will try to perform a theoretical transition from the usual intermediate storage platform to the concept of "dry port" (a key component of the intermodal transport chain) in order to determine the optimal location for establishing it. We will approach two distinct methods that can be used to determine the optimal location of a facility, a genetic algorithms and a linear programming, in order to establish which of these methods give us the more appropriate results in accordance with our economic set goal. Theoretical discussions will be transposed at different ports of Dobrudja, respectively seaports: Constanta, Mangalia, Midia and river ports: Murfatlar, Medgidia and Cernavoda. Although it seems a local approach, the optimal solution that was found has a particular significance due to the geographical position of these ports (the Rhine-Main-Danube Canal and the Danube-Black Sea Canal), but also due to their importance and their share in the transit of goods inwards Europe.

1 INTRODUCTION

The development of international trade of goods, the increase in the beneficiary requirements and the diversification of imports and exports have increased the importance of international transport and shipments. Transport costs have been meant to be reduced throughout the years, while transport represents a defining element in determining the final cost of a merchandise. Thus, it was tried to combine the advantages of transport in order to ensure significant cost reductions finally.

This is how the need for intermodal transport has emerged, which according to the common definition given by the European Commission, the European Conference of Ministers of Transport and Economic Commission for Europe of the United Nations (UN/ECE, 2001, p.17), is the *"transport of goods in one and the same loading unit or road vehicle, which uses successively two or more modes of transport without handling the goods when changing modes."*

Achieving intermodal transport, taking into account the globalization involves the existence of an appropriate infrastructure in order to achieve a "gate-to-gate" transport chain as productive as possible. It is supposed to allow for the passage of means of transport in terms of high performance, reduced parking in certain points of the transport networks and the use of modern facilities tailored for different categories of goods (Remes 2011).

As Floden (2007) states in his paper, one of the basic elements of the system of intermodal freight transport (along with the collection and distribution of goods and transport system flows of freight over long distances) is the intermodal transport terminal.

The intermodal transport terminal decisively contributes to the efficiency of the whole transport process, as it is an element that ensures effective transfer of load units (containers, swap bodies or semi-trailers) on a modal system of transport to another. Its role is more complex, as terminals represent an enclosed area, provided with special equipment and infrastructure, where various operations are carried out: transshipment, storage, and other logistics operations. For this reason, intermodal transport terminals are considered key elements in intermodal transport networks.

2 THE DRY PORT CONCEPT

In the course of time, the logistics system led to major changes in its elements being executed a gradual shift from platform usual intermediate storage, transport terminal, or more to the concept of "dry port".

As stated above, the transport terminal is a key element in intermodal transport networks, generally defined as nodes. The connections made between these hubs are basically the carried out transport, finally making up the intermodal transport network (Roso et al., 2008).

Globalization, the development of transport and the freight volume growth led to major improvements in transport terminals and to the enhancement of activities thereof, for the purpose of increasing efficiency of the entire transport chain. Thus, the development of terminals has become a crucial factor in ensuring competitive advantages of seaports.

In this regard the literature makes the transition from classical storage platform to more complex elements of the intermodal transport network designed to provide efficiency and profitability: intermodal terminal or dry port (Jaržemskis and Vasiliauskas, 2007).

Over time many definitions of the concept of "dry port" have been formulated, but a comprehensive definition was offered by Roso in 2008: *"a dry port is an inland intermodal terminal directly connected to a seaport by large means of transport (at least two different modes of transport), where customers can leave / pick up their standardized merchandise just like in a seaport"* (Roso, 2008).

The specialized literature shows several benefits of dry port sites, which are those that determine their competitive advantages (Cullinane, 2012). Indian Customs (2004) establishes the following advantages - thus, they:
– represent important points of concentration of goods over long distances;
– provide customs clearance (which is most often available near the centres of production and consumption);
– reduce the risks of theft from the actual amount of merchandise;
– ensure a low level of demurrage;
– influence the final cost of transportation;
– eliminate the need for customs inspections in seaports;
– contribute to increased trade flows;
– reduce he transit of empty cargo units (especially empty containers) etc.

Ciortescu (2016) identifies in his doctoral thesis the competitive advantages that a dry port area brings along in terms of its location. This causes a series of positive aspects both socially, economically, and in terms of environmental

protection. Thus, on a social level, it substantially reduces the number of road accidents by traffic decongestion, increases the level of safety and security of containers, and especially contributes to the increase in the number of jobs. Economically, it reduces the time required for handling operations/operating containers, especially reducing existing costs in container traffic. Regarding environmental protection, it creating a dry port contributes to the reduction of external costs of transport units loads (lowering the noise level in road infrastructure, it contributes to the road traffic decongestion at the gate of the seaport, it may contribute to the decrease of fuel consumption by using transport rail etc.).

The implementation of a dry port contributes to the massive reduction of the number of links to/from seaports (Beresford et.al., 2012). As can be seen in the figure below, a conventional transport is based on a large number of road or rail links at great distances from the seaports. Carriage to be achieved through one or more dry ports allows for short-distance transport, grouped and consolidated accordingly. They are supposed to increase productivity by increasing the capacity of intermodal terminals, along with the operation of very large capacity vessels (Roso and Lumsden, 2009; Roso, 2007).

The classification of dry port sites in terms of the distance between the sea ports that they serve and between themselves categorises them in: close, mid-range and distant dry port (Woxenius et al., 2004).

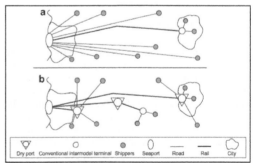

Figure 1. (a) Conventional hinterland transport and (b) implemented dry port concept (close, mid-range and distant dry port).
Source: Roso, V., 2007

3 MATHEMATICAL APPROACHES FOR FINDING A DRY PORT OPTIMUM LOCATION

In an ever-increasing dynamic business environment, the goal of any supply chain aims to provide competitive services to customers, especially in terms of costs. In this respect, an increasing importance is given to establishing the location of points in the intermodal transport networks that carry out the transfer, consolidation or group of commodities.

The location of a dry port requires a decision influenced by many factors (both qualitative and quantitative), which in the long run, affects the costs and revenues over which no later change can be made (Eroglu & Keskinturk, 2005). It must be located in such a manner as to ensure customer satisfaction by providing the shortest distance transport.

In this paper we address two different ways by which one can determine the optimal location, genetic algorithms and linear programming, trying to determine which of these methods give us results that correspond our proposed economic purpose.

The problem of determining the optimal location of a site of some kind is a matter of planning already addressed through operational research techniques, being first introduced in 1965 by Balinski (1965). This issue aims to minimize transportation costs, as well as fixed costs associated with the site, while also considering the restrictions and conditions data.

Establishing the optimal location for a dry port, can be done through several methods, such as:
- genetic algorithm;
- linear programming method.

In this article, we critically analyse the results of the two methods for the ports of Dobrudja.

4 CASE STUDY

Theoretical discussions presented above will be implemented in the Dobrudja ports, i.e. sea ports Constanta, Mangalia and Midia on the one hand and river ports Murfatlar, Medgidia and Cernavoda on the other hand. Although the problem is local, the optimal determined solution is important both due to the geographical site of these ports (Rhine-Main-Danube and Danube-Black Sea Canal) and to their significance and share of the transit of goods to inland Europe.

Taking into account these aspects, we will seek the best location of a dry port.

The analysed problem can solve both of the following situations simultaneously:
- **Situation 1**: For general goods arriving to Romanian seaports, the question of building a dry port arises from where the goods to proceed towards the three river ports (Murfatlar, Medgidia and Cernavoda) to serve their entire hinterland area.

Table 1. Data related to the first situation

City	Population	Distance to (km):		
		Constanta	Midia	Mangalia
Murfatlar	10,746	18	41	55
Medgidia	43,841	40	47	76
Cernavodă	20,105	62	86	102

- **Situation 2**: Various agricultural, industrial or mineral products obtained in different parts of Dobrudja, arrive to three river ports, where they are stored for a period of time and then are transported to the three sea ports (Constanta, Midia and Mangalia).

Table 2. Data related to the second situation

City	Population	Distance to (km):		
		Murfatlar	Medgidia	Cernavodă
Constanta	319,168	18	40	62
Midia	32,300*	41	47	86
Mangalia	33,434	55	76	102

* The data is for the population of Navodari, the closest city to Port of Midia.

The network made between the seaports/river ports and possible dry port locations is shown in the figure below.

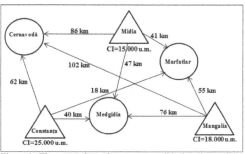

Figure 1. The network made among possible locations of the dry port and potential distributors

4.1 Method I: The genetic algorithm method

This search and optimization method was presented and analysed by Carp and Stinga (2016). The genetic algorithm method differs from other conventional search techniques due to its mechanism of natural selection and genetic inheritance.

The objective function used was:

$$min\left(\sum_{i\in I} x_i \cdot CI_i + \sum_{j=1}^{m} min_{i\in I_1}\left(d_{ij} \cdot \frac{P_j}{100} \right) \right)$$

where:
- $J = \{1, 2, ..., m\}$, is the set indexed by j, consists of potential providers;
- $I = \{1, 2, ..., n\}$, is the set indexed by i, made up of possible locations for intermediate storage platform;

- $I_1 = \{i \in I \,/\, x_i = 1\}$;
- $x_i = \begin{cases} 1, & \text{if there is a platform on } "i" \\ 0, & \text{if there is not a platform on } "i" \end{cases}$
- CI_i, represents the installation costs associated with each possible location of the platform;
- d_{ij}, is the distance between the storage platform and the distributors;
- P_j, is the existing population in every city where the storage platform can be placed.

The table below shows the results obtained using genetic algorithms using the MATLAB software.

Table 3. The results obtained by using genetic algorithms

The running time of the algorithm	Number of iterations	Value of the objective function	The structure of the solution chromosome
0.624989	100	43,316.16	(011000)
0.569121	100	29,243	(000100)
0.573037	100	23,910.3	(001000)
0.563188	100	36,181	(000010)
0.604099	100	26,934.28	(100000)
0.621351	100	19,405.86	(010000)
0.600968	100	36,181	(000010)
0.544985	100	26,934.28	(100000)

Comparing the results obtained by using the Matlab software, for the objective function values we concluded that the dry port will be located at Midia, as the objective function value is 19,405.86.

4.2 Method II: Linear programming method

The data allow us to take into consideration the following transportation problem, where the costs are direct connected to the distances between the ports:

	Murfatlar cost	Medgidia cost	Cernavoda cost	Demand Value $*10^3$
Constanta	36	80	124	3194
Midia	82	94	172	5833
Mangalia	110	152	204	5974
Offer Value$*10^3$	1225	7146	5026	

Since the total demand isn't equaled to the total offer, the problem is not balanced and we have to consider a virtual consumer as follows:

	Murfatlar cost	Medgidia cost	Cernavoda cost	Virtual consumer	Demand Value$*10^3$
Constanta	1225	1969	-	-	3194
	36	80	124		-
Midia	-	5177	656	-	5833
	82	94	172		-
Mangalia	-	-	4360	1604	5974
	110	152	204		-
Offer Value$*10^3$	1225	7146	5026	1604	

Initial solution is: $x_{11}=1225$, $x_{12}=1969$, $x_{22}=5177$, $x_{23}=656$, $x_{33}=4360$, $x_{34}=1604$. The significance of $x_{34}=1604$ is that the quantity 1604 offered at Mangalia will not be distributed.

The cost associated to the initial solution is $C=1225*36+1969*80+5177*94+656*172+4360*204$ or $C=1,690,530$.

Since the optimum criteria aren't satisfied, the initial solution was transformed as follows:

	Murfatlar cost	Medgidia cost	Cernavoda cost	Virtual consumer	Demand Value$*10^3$
Constanta	1225	1313	656	-	3194
	36	80	124		-
Midia	-	5833	-	-	5833
	82	94	172		-
Mangalia	-	-	4360	1604	5974
	110	152	204		-
Offer Value $*10^3$	1225	7146	5026	1604	

The cost associated to the second solution decreased, being $C=1,667,226$,

Since the optimum criteria aren't satisfied again, the solution was transformed as follows:

	Murfatlar cost	Medgidia cost	Cernavoda cost	Virtual consumer	Demand Value$*10^3$
Constanta	1225		1969	-	3194
	36	80	124		-
Midia	-	5833	-	-	5833
	82	94	172		-
Mangalia	-	1313	3047	1604	5974
	110	152	204		-
Offer Value $*10^3$	1225	7146	5026	1604	

The cost associated to the third solution decreased, being $C=1,657,722$.

The solution is not optim, so we transformed it in:

	Murfatlar cost	Medgidia cost	Cernavoda cost	Virtual consumer	Demand Value$*10^3$
Constanta	-		3194	-	3194
	36	80	124		-
Midia	-	5833	-	-	5833
	82	94	172		-
Mangalia	1225	1313	1822	1604	5974
	110	152	204		-
Offer Value $*10^3$	1225	7146	5026	1604	

The set of dual variables are solutions of the system:

$U_1+v_3=124$ $U_2+v_2=94$ $U_3+v_1=110$ $U_3+v_2=152$
$U_3+v_3=204$

A particular solution of it is the following: $U_1=0$ $U_2=22$ $U_3=80$ $v_1=30$ $v_2=72$ $v_3=124$ and now the criteria of optimum are completed, so this solution is the optimal one. The minimum cost is C=1,650,372.

Since the best connection and distribution is done from Mangalia, we can conclude that a dry port could locate there.

5 CONCLUSIONS

The two methods presented are not at all similar, or equivalent. Their conclusion are not identically because the objective functions are different: the genetic algorithm was related to the minimization of the total cost of installation and allocation of the consumption and the objective function of the transportation problem is the minimization of the transportation cost only. The best location could be established using the two complementary methods.

REFERENCES

[1] Balinski, M. L. (1965), *Integer Programming: Methods, uses, computations*, Management Science, 12, pp. 253-313

[2] Beresford, A., Pettit, S., Xu, Q. & Williams, S. (2012), *A study of dry port development in China*. Maritime Economics & Logistics, *14*(1), pp. 73-98

[3] Carp, D. & Stinga, V.G. (2016), *Algoritm genetic pentru stabilirea locaţiei optime a unei platforme de depozitare intermediară*, Conferinţa naţională TRANSLU' 16 Interacţiuni dintre transporturi şi dezvoltarea regional, Buletinul AGIR, Supliment 2/2016, pp. 76-81

[4] Ciortescu, C.G. (2016), *Cercetări privind implementarea dryport-ului în România*, PhD thesis, Iasi, Romania

[5] Cullinane, K., Bergqvist, R., & Wilmsmeier, G. (2012), *The dry port concept-Theory and practice*. Maritime Economics & Logistics, 14(1), 1

[6] Eroglu, E. & Keskinturk, T. (2005), *Warehouse location problem with genetic algorithm*, 35th International Conference on Computers and Industrial Engineering, 655-660 p.

[7] Flodén, J. (2007), *Modelling Intermodal Freight Transport-The Potential of Combined Transport in Sweden*, BAS Publising, Suedia, Doctoral thesis ISBN: 978-91-7246-252-6, 272 p.

[8] Indian Customs (2004), CBEC Manual, Setting up of ICDs/CFSs

[9] Jaržemskis, A. & Vasiliauskas, A. V. (2007), *Research on dry port concept as intermodal node*. Transport, *22*(3), pp. 207-213

[10] Remeş, C. (2011), *Globalizarea sistemului de transport*, Studia Universitatis "Vasile Goldis" Arad, Seria Ştiinţe Economice, Anul 21/2011 Partea a II – a, p. 541-545

[11] Roso, V., & Lumsden, K. (2009), *The dry port concept: moving seaport activities inland*. UNESCAP, Transport and Communications Bulletin for Asia and the Pacific, *5*(78), pp. 87-102

[12] Roso, V., Woxenius, J. & Lumsden, K. (2008), *The dry port concept- connecting seaports with their hinterland*. Journal of Transport Geography, 17(5), pp.338-345

[13] Roso, V. (2008) *Factors influencing implementation of a dry port*. International Journal of Physical Distribution and Logistics Management, 38(10), pp. 782-798

[14] Roso, V. (2007), *Evaluation of the dry port concept from an enviromental perspective: a note*. Transportation ResearchPart D, 12 (7), pp. 523-527

[15] Woxenius, J., Roso, V. & Lumsden, K. (2004), *The dry port concept – connecting seaports with their hinterland by rail*. In: ICLSP Conference Proceedings, Dalian

Magnetic Compasses

Contemporary Considerations of Change Regulations Regarding Use of Magnetic Compasses in the Aspect of the Technical Progress

E. Łusznikow & K. Pleskacz
Maritime University of Szczecin, Szczecin, Poland

ABSTRACT: Authors of papers to presents the urgent need to review the rules of daily operation of magnetic compasses, taking into account existing realities of modern navigation.
They also raise the question of educational policy in terms of changes to the regulations regarding the operation of magnetic compasses in the aspect of growing technical progress, justify the need of urgent review through presentation and analysis of the results of the carried out tests and justify its request through the presentation and analysis of research results.

1 INTRODUCTION

The main aims of navigation are determination of current ship's position and course. Until the mid–twentieth century, the navigation was based solely on magnetic compasses. The rapid scientific progress in the 20th century has allowed to develop and widely implemented to operate of the navigation gyro device, which has much higher accuracy than magnetic compasses. But at the same time: higher failure rate and dependence on an external power source.

For those reasons, according to the requirements of SOLAS (International Convention for the Safety of Life at Sea) - chapter V, for ship's navigation systems and equipment, all ships, irrespective of size, shall have:
- properly adjusted standard magnetic compass, or other means, independent of any power supply, to determine the ship's heading and display the reading at the main steering position;
- means of correcting heading and bearings to true at all times;
- all ships of 500 gross tonnage and upwards shall be fitted with a gyro-compass, or other means, to determine and display their heading by ship borne non-magnetic means and to transmit heading information for input to the radars, automatic identification systems or automatic tracking systems [1].

Resolution A.382(X), annex I, paragraph 3, requires, that each magnetic compass is properly compensated and its table of curve of residual deviations is available on board in the vicinity of the compass at all times [4].

While the convention STCW (Standards of Training, Certification and Watchkeeping) requires, that relieving officers shall personally satisfy themselves regarding the errors of gyro- and magnetic compasses [2].

All these rules and requirements have led to the development of recommendations on the use of magnetic compasses and gyrocompasses included in the collections of the implementing rules. A magnetic compass' error should be determined at least once a watch while the vessel is at sea and, when possible, after any major alteration of course. The observed error should be recorded in the ship's logbook [2][3][5].

2 THE MODERN ROLE OF THE SHIPS COMPASSES

The main objective of the current course control indicators is to detect the possible damage to one of the compasses and determine the level of confidence in these devices. Excessive differences of true courses obtained from both devices may indicate damage of one of them. With the advent of GPS (Global Positioning System) handling requirements for magnetic compasses have not changed. Although completely disappeared need to obtain the true course by using a magnetic compass in conditions of normal operation. The true course obtained with the help of this compass is less accurate. Calculations require more time due to the changing value of the declination and need to update the magnetic declination. Simplified actions used in the daily routine work in unlawfully way destroy the purpose of mutual control of the course indicators. The scientific and technical development at the end of the 20th century caused the solution to the problem of constant monitoring of the ship's position using global satellite system such as: GLONASS (Global Navigational Satellite System), GPS or Beidou (Big Dipper / COMPASS Navigation Satellite System). Therefore, the role of the compasses to determine the position of the vessel almost completely disappeared and has been limited to keep the ship on course only. Special limitation has been the role of magnetic compass. Magnetic compass is treated as a control–reserve indicator, giving you control of properly work of the gyrocompass and the ability steer safely the ship in the event of failure of the more unreliable gyro. Currently, with the use of satellite systems, in addition to continuous monitoring of the ship's position is possible also to determine course made good (CMG). Information about CMG is more useful for navigator than information about true course (TC), as included are drift and leeway.

Today can compare the courses obtained in three options:
– Gyro/GPS;
– Gyro/Magnetic compass;
– GPS/Magnetic compass.

It is obvious that the navigator will choose the option that requires the least afford and time, and at the same time, the required accuracy. This way is to compare the information's from gyro and GPS. Other combinations, which require the use of less accurate magnetic compass, are assumption rejected. Navigator selects the combination of GPS + gyrocompass almost intuitively, because almost does not require any calculations. Determination of the CMG by using the GPS system does not cause loss of energy and time. The same can be said about the true course obtained by gyro.

In contrast, for calculations in the configuration: the true course obtained by gyro and magnetic compass cannot pull such a statement. Determination in accordance with regulations the total error of magnetic compass, summing declination plus deviation and additionally: annual or more frequent compensation requires a lot of time. The time translates directly into cost of operation of the ship.

3 COMPLICATIONS IN THE USE OF MAGNETIC COMPASSES IN SHIPS EQUIPED WITH ECDIS

Chart Display and Information System (ECDIS) means a navigation information system which with adequate back-up arrangements can be accepted as complying with the up-to-date chart required by regulation V/20 of the 1974 SOLAS Convention, by displaying selected information from a system electronic navigational chart with positional information from navigation sensors to assist the mariner in route planning and route monitoring, and if required display additional navigation-related information [1].

According to the SOLAS, more and more units must be equipped with ECDIS systems, which can work independently or as a back-up arrangements have a set of paper charts.

And here comes the problem. Information about the value of the magnetic declination, which is necessary to calculate, the navigator draws from the chart. In the case of paper chart form of its presentation is clearly defined by international regulations.

However, in the case of vector charts, which are the basis of ECDIS, it is not so obvious.

International regulations clearly specify the form of its presentation. But do not specify the requirement to include information about the declination chart content (!).

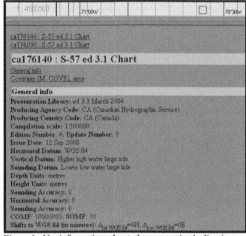

Figure 1. No information about the magnetic declination on chart of Canada and Lithuania
[*source: simulator TRANSAS ECDIS 3000-i*]

Some hydrographical offices do not present the declination in the content of the chart. They hiding behind the statement, that there is a wide range of other sources of information about the declination value independent of the ECDIS. Sometimes it can be dangerous. Not all vessels are equipped with continuous access to the Internet or appropriate GPS receiver Magnetic declination in Newfoundland area, the region presented in the first figure, is 20 degrees. But even when it is included in the chart legend, it still requires a navigator update. That is, data from your computer to be converted manually (!).

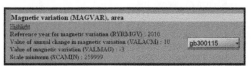

Figure 2. Presentation of the magnetic declination on the British vector's chart
[*source: simulator TRANSAS ECDIS 3000-i*]

4 ANALYSIS OF THE STATUS OF THE CONTROL INDICATORS ON SHIPS OF THE WORLD FLEET

Under ideal conditions, using hypothetically perfect devices actual courses obtained from both devices must be the same. Because there is no such situation in reality the difference between true courses obtained from gyro and a magnetic compass is usually not equal to zero. Modern statistical analyses of indications of gyro and magnetic compasses indicate that the value of the difference ($m_{\Delta TC}$) of true courses for modern compasses equals $m_{\Delta TC}$ = 2,4° [6].

This means, that with a probability of 68.3%, the difference of true courses, calculated on the basis of both compasses is within $-2,4° \leq \Delta TC \leq +2,4°$.

Generally accepted principles for the evaluation of navigation safety require a confidence level of 95%. On a constant course and speed and under normal weather conditions, the error of courses indicators should contain in the range $\pm 4,8° \approx \pm 5°$ [7]. In such limits would expect incompatibility between true courses set out with gyro and magnetic compass.

To know the actual state of the course control indicators on the ships of the world fleet, records of 37 logbooks from 17 different countries of the world were analyzed [9]. Extracts from the logs were obtained from the captains of the ships calling at Szczecin.

All the vessels from which information has been obtained were merchant ships under different flags and manned by crews from different countries. Randomly selected entries were chosen from five consecutive days when the ship was in operation on the open sea or on the approach to the port. The term „single entry" means records relating to a single, specific hour of observation, which is a single line entry in the log book.

Statistical processing of data from the logbooks led to surprising results. It was found that 100% of true courses obtained from a gyrocompass and a magnetic compass written down in the examined logbooks had exactly the same values (Table 1).

Table 1. Comparison of true courses obtained from observations of a magnetic compass (TC_M) and gyrocompass (TC_G)

No.	Ship's name	Number of observations	$\Delta TC = |TCM - TCG|$ [°]
1	Fast Sam	60	0
2	Celine	44	0
3	Kapitan Zhikharev	56	0
4	OW Scandinavia	59	0
5	Transwing	57	0
6	Zillertal	51	0
7	Steinau	50	0
8	Lady Elena	58	0
9	Gas Evoluzione	120	0
10	Walka Młodych	56	0
11	Flottbek	51	0
12	Transmar	60	0
13	Hans Lehmann	96	0
14	Karina G	55	0
15	Clare Christine	53	0
16	Pitztal	72	0
17	Crystal Topaz	104	0
18	Taganrogskiy Zaliv	120	0
19	Ametyst	75	0
20	Finland	60	0
21	Fjordstraum	111	0
22	Flinterhaven	60	0
23	SV.Knyaz Vladimir	92	0
24	Fast Sam	49	0
25	Ostanhav	60	0
26	Gas Arctic	54	0
27	Frisian Ocean	51	0
28	Leonid Leonov	113	0
29	RMS Ratingen	59	0
30	Yigt Bay 1	102	0
31	Elizabeth	85	0

source: [8]

Such a coincidence is contrary with scientific worldview view.

Systematic errors can be compensated for, but not completely excluded. Random errors, by definition, cannot be compensated for by definition.

A mean error ($m_{\Delta TC}$) of true course difference obtained from a magnetic compass and gyrocompass can be calculated from the formula [8]:

$$m_{\Delta TC} = \sqrt{m_M^2 + m_V^2 + m_D^2 + m_G^2 + m_O^2 + m_{GE}^2} \qquad (1)$$

where:

m_M – error of the magnetic compass;
m_V – error of the declination;
m_D – error of the deviation;
m_G – error of the gyrocompass;
m_{GE} – error of the gyrocompass correction
m_O – error of observer.

Unthinkable that all the navigators of the ship on 31 of 17 countries in the world did not understand, that what they write in logbooks is false information. And can even find harder information – dangerous for navigation.

In such a situation, inevitably raises the question about the reasons for this state of affairs?

To understand and respond to as formulated question should make a detailed analysis of the practice of filling logbooks compared with the state of the real requirements of officers on the watch, the content of the training, that they pass and stereotypes negatively changing good sea practice.

The collected statistical data shows that the convergence of the actual true course obtained by gyro and magnetic compass within the meaning of officers is quite normal fact.

An analysis of the difference in true courses for one ship steering a specific course is considered by navigators an artificial problem, and they know how to avoid it. All they do is to „slightly adjust" the readout from the less accurate magnetic compass to align with the more accurate true course produced by the gyrocompass. Thus, mandatory routine calculations of magnetic compass corrections are neglected. Naturally, such „adjustment" of magnetic compass correction does not have much to do with reality. Magnetic compass correction cannot be carried out accurately using a randomly estimated value; it is a scientific concept, a sum of declinations and variation. If ships' compass corrections were determined in accordance with procedures taught at training institutions, then true courses obtained from a gyrocompass and a magnetic compass would not be equal, and the value of the calculated difference between the true courses would allow estimation of how reliable the indicators are.

Everything points to the fact that modern a navigator ignores his knowledge in the daily routine work and what is very important, good sea practice, which he learnt in the educational process.

In order to determine what values reach realistic corrections of gyro and magnetic compasses and thus the actual true courses obtained by both compasses, measurements were made on various ships during entry and exit from the port of Szczecin. The measurements were carried out while the ship was in the line of leading and helmsman steered on leading line too. Then the axis of symmetry of the ship was covered with the leading line.

All measurements took place for each time a permission of the Master.

The measurements were carried out by the pilot Cpt. Piotr Szelepajło in a short period of time, about four months, which ensured a minimum change of declination. He has been previously instructed on the rules required when making measurements and details to pay attention. To maximize eliminate the influence of drift, were selected days of good hydro meteorological conditions.

Table 2. Comparing true couses obtained from various course indicators

No.	Ship's name	True course obtained by gyro compass TC_G [°]	True course obtained by magnetic compass TC_M [°]	ΔTC = |TCM–TCG| [°]
1	Athos	176	177	1,0
2	Alva	173	174	1,0
3	Kapitan Zykharev	168	173,5	5,5
4	Sardius	355	357	2,0
5	Fast Sam	353,5	354	0,5
6	Celine	142	147	5,0
7	Viscaria	139	141,5	2,5
8	Vita Theresa	143	138,5	4,5
9	Flinterhaven	141	142	1,0
10	Ostanhav	143	142	1,0
11	Gas Arctic	142	145,5	3,5
12	Fast Julia	141	139,5	1,5
13	Kniyaz Vladimir	140	133	7,0
14	Magda D	141	145,5	4,5
15	Brovig Wind	140	114	4,0
16	Ovi Victoria	141,5	136,5	5,0
17	Grand	140	143	3,0
18	RMS Ratingen	139	140	1,0
19	Aasheim	141	144	3,0
20	Wilson Bar	322	325	3,0
21	Valentin Pikul	321,5	316,5	5,0
22	Wilson Ghent	322	328,5	6,5
23	Frisian Ocean	320	323	3,0
24	Leonid Leonov	322	316	6,0
25	Karl-Erik	320,5	318,5	2,0
26	Solvegk	321,5	327	5,5
27	Iris Bolten	322	322,5	0,5
28	Karina C	323	324	1,0
29	Volgo-Balt	244	323 321	2,0
30	Osterems	323	323,5	0,5
31	Elizabeth	321	323	2,0
32	Frisian Summer	321	317	4,0
33	Yigt Bey 1	321	322	1,0
34	Wilson Brugge	320	321	1,0

source: [8]

The analysis of the obtained results it is clear that differences ΔTC = 0°, commonly written on ships of the world fleet – do not occur. In 16 cases, the difference were 5 ° or more, 9 ships were in the range of 3° to 5° and also 9 difference were less than 3°.

Received data allow document improper conduct in relation to the course control indicators by the navigators.

Until the end of the last century, with unreliable equipment device – gyrocompass, magnetic compass ensures the possibility of emergency navigation to the port. Each practitioner, who spent several years at sea, the gyrocompass broke down at least once. Such system was a major advantage of using layout the gyro + magnetic compass. One ensured, in normal conditions of high accuracy, but the second ensured opportunity for safe navigation in the event of a failure the first.

At the end of the last century, in connection with the dissemination of the GPS system, there is a new

430

configuration: GPS + gyrocompass + magnetic compass. In the event of failure of the gyrocompass, it is possible to control according to the magnetic compass, after determining out the course and at the same time – CMG, based on GPS. In this case, the indication of magnetic compass can be loaded with any mistake.

And if so, raises the question of what to fight for the accuracy of the magnetic compass, even when the variation of the order is 30 ° in those circumstances, it does not matter.

Rigid frame precision magnetic compass and procedures for their maintenance are unnecessary, especially since all the parameters will be retained only for the first good storm, during which will change the magnetic field of the ship's hull.

The same applies to the obligation of the annual compensation of magnetic compass deviation. This causes unnecessary costs, efforts and waste of time. If additional costs do not have the chance to return, may not be considered appropriate.

This is confirmed by the results of expert methods of research conducted by the authors among 212 managers deck department: chief officers and captains. Only 56% of respondents said they regularly check the magnetic compass, and 76% – gyro. Striking is also the percentage of 14% of the crew members, who have never witnessed the conduct compensation magnetic compass and 7% of those, with 16 years experience, who had never met deviator while working on the ship.

In summary, one can clearly say that after more than 20 years since the entry of satellite positioning systems to universal service should review and modify provisions relating to the operation of the magnetic compass.

These requirements can be greatly reduced and will entail negative consequences in terms of safety of navigation.

5 CONCLUSION

On the question of why still held are high requirements for magnetic compasses, which carrying a high cost? There is only one answer. Maritime community has not reached so far, all the opportunities that brought widespread introduction of the GPS system. The delay in the use of scientific and technological progress is a natural phenomenon. However, in this case it is more than twenty years.

Personal experiences and interviews with Captains can see that sense intuitively unnecessary annual compensation of the deviation. All time they are under time pressure. Therefore, more and more often, resorted to measures not fully in accordance with the provisions of the regulations. About improper practices shows that 9% of respondents declared, that table deviation magnetic compass in force on the last ship has not been developed either by deviator or by crewmember (!). The signature was illegible.

Should be changed outdated regulations, forcing captains around the world, to unacceptable and illegal attempts circumvention provisions.

Taking into account the existing realities in the navigation is required urgent review of the rules for daily use a magnetic compass.

It should also raise the issue of educational policy in this regard. If this knowledge will not be distributed in the schools, and we will continue to teach outdated rules, this vulnerability will grow. Currently, these negative rules and regulations are in force in all educational institutions throughout the world. This is due to the provisions of the STCW.

And here we must make radical changes.

REFERENCES

[1] The International Convention for the Safety of Life at Sea (SOLAS). Consolidated Edition (2004), IMO, London;

[2] International convention on Standard of Training, Certification and Watch keeping for Seafarers. (1978), IMO, London;

[3] IMO MSC/Circ/1061, Guidance for the operational use of integrated bridge system (IBS), (2003), IMO, London;

[4] IMO Resolution A.382(x), Magnetic compasses carriage and performance, (2003), IMO, London;

[5] IMO MSC/Circ.982, Guidelines on ergonomic criteria for bridge equipment and layout, (2003), IMO, London;

[6] Jagniszczak I., Łusznikow E., Bezpieczeństwo nawigacji, Fundacja Promocji Przemysłu Okrętowego i Gospodarki Morskiej, (2010), Gdańsk;

[7] Łushnikov E., Pleskacz K., Analysis's of problems related to the use of ship's course indicators, Scientific Journals, (2012), Maritime University of Szczecin, 29(101), p. 122–126;

[8] Łusznikow E., Pleskacz K., Analiza stanu kontroli dokładności i niezawodność wskaźników kursu, jako czynnik bezpieczeństwa nawigacji, p. 764– 769;

[9] Łusznikow E., Pleskacz K., Prezentacja wyników badań rzeczywistej dokładności żyrokompasów i kompasów magnetycznych, (2014), Logistyka No. 4, p. 770–776.

AUTHOR INDEX

Printed and bound by CPI Group (UK) Ltd, Croydon, CR0 4YY

24/10/2024

01778293-0009